AF809332

Random Walk Intersections

Large Deviations and Related Topics

Mathematical
Surveys
and
Monographs

Volume 157

Random Walk Intersections

Large Deviations and Related Topics

Xia Chen

American Mathematical Society
Providence, Rhode Island

EDITORIAL COMMITTEE

Jerry L. Bona
Michael G. Eastwood
Ralph L. Cohen, Chair
J. T. Stafford
Benjamin Sudakov

2000 *Mathematics Subject Classification.* Primary 60F05, 60F10, 60F15, 60F25, 60G17, 60G50, 60J65, 81T17, 82B41, 82C41.

This work was supported in part by NSF Grant DMS-0704024

For additional information and updates on this book, visit
www.ams.org/bookpages/surv-157

Library of Congress Cataloging-in-Publication Data

Chen, Xia, 1956–
Random walk intersections : large deviations and related topics / Xia Chen.
p. cm.— (Mathematical surveys and monographs ; v. 157)
Includes bibliographical references and index.
ISBN 978-0-8218-4820-3 (alk. paper)
1. Random walks (Mathematics) 2. Large deviations. I. Title.

QA274.73.C44 2009
519.2′82–dc22 2009026903

Copying and reprinting. Individual readers of this publication, and nonprofit libraries acting for them, are permitted to make fair use of the material, such as to copy a chapter for use in teaching or research. Permission is granted to quote brief passages from this publication in reviews, provided the customary acknowledgment of the source is given.

Republication, systematic copying, or multiple reproduction of any material in this publication is permitted only under license from the American Mathematical Society. Requests for such permission should be addressed to the Acquisitions Department, American Mathematical Society, 201 Charles Street, Providence, Rhode Island 02904-2294 USA. Requests can also be made by e-mail to reprint-permission@ams.org.

© 2010 by the American Mathematical Society. All rights reserved.
The American Mathematical Society retains all rights
except those granted to the United States Government.
Printed in the United States of America.

∞ The paper used in this book is acid-free and falls within the guidelines
established to ensure permanence and durability.
Visit the AMS home page at http://www.ams.org/

10 9 8 7 6 5 4 3 2 1 15 14 13 12 11 10

To the memory of my great grandmother Ding, Louyi

Contents

Preface

This book aims to provide a systematic account for some recent progress on the large deviations arising from the area of sample path intersections, including calculation of the tail probabilities of the intersection local times, the ranges and the intersections of the ranges of random walks and Brownian motions. The phrase "related topics" appearing in the title of the book mainly refers to the weak law and the law of the iterated logarithm for these models. The former is the reason for certain forms of large deviations known as moderate deviations; while the latter appears as an application of the moderate deviations.

Quantities measuring the amount of self-intersection of a random walk, or of mutual intersection of several independent random walks have been studied intensively for more than twenty years; see *e.g.* [**57**], [**59**], [**124**], [**125**], [**116**], [**22**], [**131**], [**86**], [**135**][**136**], [**17**], [**90**], [**11**], [**10**], [**114**]. This research is often motivated by the role that these quantities play in renormalization group methods for quantum field theory (see *e.g.* [**78**], [**51**], [**52**], [**64**]); our understanding of polymer models (see *e.g.* [**134**], [**19**],[**96**], [**98**] [**162**], [**165**], [**166**],[**167**],[**63**], [**106**], [**21**], [**94**], [**93**]); or the analysis of stochastic processes in random environments (see *e.g.* [**107**], [**111**],[**43**], [**44**] [**82**], [**95**], [**4**], [**42**] [**79**], [**83**]).

Sample path intersection is also an important subject within the probability field. It has been known ([**48**], [**138**], [**50**]) that sample path intersections have a deep link to the problems of cover times and thick points through tree-encoding techniques. In addition, it is impossible to write a book on sample path intersection without mentioning the influential work led by Lawler, Schramm and Werner ([**118**], [**119**], [**120**], [**117**]) on the famous intersection exponent problem and on other Brownian sample path properties in connection to the Stochastic Loewner Evolution, which counts as one of the most exciting developments made in the fields of probability in recent years.

Contrary to the behavior patterns investigated by Lawler, Schramm and Werner, where the sample paths avoid each other and are loop-free, most of this book is concerned with the probability that the random walks and Brownian motions intersect each other or themselves with extreme intensity. When these probabilities decay at exponential rates, the problem falls into the category of large deviations. In recent years, there has been some substantial input about the new tools and new ideas for this subject. The list includes the method of high moment asymptotics, sub-additivity created by moment inequality, and the probability in Banach space combined with the Feynman-Kac formula. Correspondent to the progress in methodology, established theorems have been accumulated into a rather complete picture of this field. These developments make it desirable to write a monograph on this subject which has not been adequately exposed in a systematic way.

This book was developed from the lecture notes of a year-long graduate course at the University of Tennessee. Making it accessible to non-experts with only basic knowledge of stochastic processes and functional analysis has been one of my guidelines in writing it. To make it reasonably self-contained, I added Chapter 1 for the general theory of large deviations. Most of the theorems listed in this chapter are not always easy to find in literature. In addition, a few exercises are included in the "Notes and comments" section in each chapter, an effort to promote active reading. Some of them appear as extensions of, or alternative solutions to the main theorems addressed in the chapter. Others are not very closely related to the main results on the topic, such as the exercises concerning small ball probabilities, but are linked to our context by sharing similar ideas and treatments. The challenging exercises are marked with the word "hard". The mainspring of the book does not logically depend on the results claimed in the exercises. Consequently, skipping any exercise does not compromise understanding the book.

The topics and results included in the book do reflect my taste and my involvement on the subject. The "Notes and comments" section at the end of each chapter is part of the effort to counterbalance the resulted partiality. Some relevant works not included in the other sections may appear here. In spite of that, I would like to apologize in advance for any possible inaccuracy of historic perspective appearing in the book.

In the process of investigating the subject and writing the book, I benefitted from the help of several people. It is my great pleasure to acknowledge the contributions, which appear throughout the whole book, made by my collaborators R. Bass, W. Li, P. Mörters and J. Rosen in the course of several year's collaboration. I would like to express my special thanks to D. Khoshnevisan, from whom I learned for the first time the story about intersection local times. I thank A. Dembo, J. Denzler, A. Dorogovtsev, B. Duplantier, X. B. Feng, S. Kwapien, J. Rosinski, A. Freire, J-F. Le Gall, D. S. Wu, M. Yor for discussion, information, and encouragement. I appreciate the comments from the students who attended a course based on a preliminary version of this book, Z. Li, J. Grieves and F. Xing in particular, whose comments and suggestions resulted in a considerable reduction of errors. I am grateful to M. Saum for his support in resolving the difficulties I encountered in using latex.

I would like to thank the National Science Foundation for the support I received over the years and also the Department of Mathematics and Department of Statistics of Standford University for their hospitality during my sabbatical leave in Fall, 2007. A substantial part of the manuscript was written during my visit at Stanford. Last and most importantly, I wish to express my gratitude to my family, Lin, Amy and Roger, for their unconditional support.

CHAPTER 1

Basics on large deviations

In this chapter we introduce some general theorems on large deviations which will be frequently used in this book. In most of the cases, the state space we deal with is the real line. Indeed, a substantial portion of the discussion is limited to the random variables taking non-negative values. Sometimes, the underlying stochastic processes are sub-additive (see Section 1.3). Unlike most textbooks on this subject, we put more attention on the tail probability

$$\mathbb{P}\{Y_n \ge \lambda\}$$

than the probability of the form

$$\mathbb{P}\{Y_n \in A\}.$$

The unique structure of the models we deal with in this book requires some non-conventional treatments. The topics we chose in this chapter reflect this demand. As a consequence, most theorems introduced in Section 1.2 and in Section 1.3 are non-standard and are not usually seen in the textbooks on large deviations.

1.1. Gärtner-Ellis theorem

In the area of large deviations, we are concerned about asymptotic computation of small probabilities on an exponential scale. The general form of large deviation can be roughly described as

$$\mathbb{P}\{Y_n \in A\} \approx \exp\{-b_n I(A)\} \qquad (n \to \infty)$$

for a random sequence $\{Y_n\}$, a positive sequence $\{b_n\}$ with $b_n \to \infty$, and a coefficient $I(A) \ge 0$. In the application, we are often concerned with the probability that the random variable(s) takes large values. Since the remarkable works by Donsker and Varadhan (and others) in the 1970s and 1980s, this area has developed into a relatively complete system. There have been several standard approaches in dealing with large deviation problems. Perhaps the most useful tool is the Gärtner-Ellis theorem.

We have no intention to state the large deviation theory in its full generality. Let $\{Y_n\}$ be a sequence of real random variables and let $\{b_n\}$ be a positive sequence such that $b_n \longrightarrow \infty$.

ASSUMPTION 1.1.1. *For each $\theta \in \mathbb{R}$, the logarithmic moment generating function $\Lambda(\theta)$, defined as the limit*

$$\Lambda(\theta) = \lim_{n\to\infty} \frac{1}{b_n} \log \mathbb{E} \exp\Big\{\theta b_n Y_n\Big\} \qquad \theta \in \mathbb{R} \tag{1.1.1}$$

exists as an extended real number. Further, the origin belongs to the interior $\mathcal{D}_\Lambda^o$ of the domain $\mathcal{D}_\Lambda = \{\theta \in \mathbb{R};\ \ \Lambda(\theta) < \infty\}$ of the function $\Lambda(\theta)$.

By Hölder inequality, $\Lambda(\theta)$ is a convex function. Consequently, $\mathcal{D}_\Lambda^o = (a, b)$ for some $-\infty \le a < 0 < b \le \infty$ and $\Lambda(\theta)$ is continuous in its domain $\mathcal{D}_\Lambda$. Define the *Fenchel-Legendre transform* $\Lambda^*(\lambda)$ of $\Lambda(\theta)$ as

$$\Lambda^*(\lambda) = \sup_{\theta\in\mathbb{R}} \Big\{\theta\lambda - \Lambda(\theta)\Big\} \qquad \lambda \in \mathbb{R}. \tag{1.1.2}$$

To discuss the role played by the function $\Lambda^*(\cdot)$, we introduce the following definition.

DEFINITION 1.1.2. *A function I: $\mathbb{R} \longrightarrow [0, \infty]$ is called a rate function if it is lower semi-continuous: For any $l > 0$, the level set*

$$I_l = \{\lambda \in \mathbb{R};\ \ I(\lambda) \le l\}$$

is a closed set. Further, a rate function is said to be good if every level set is compact in $\mathbb{R}$.

We point out an equivalent statement of lower semi-continuity: For any $\lambda_n, \lambda \in \mathbb{R}$ with $\lambda_n \to \lambda$,

$$\liminf_{n\to\infty} I(\lambda_n) \ge I(\lambda). \tag{1.1.3}$$

We now claim that under Assumption 1.1.1, the function $\Lambda^*(\cdot)$ is a good rate function. Indeed, by definition

$$\Lambda^*(\lambda_n) \ge \theta\lambda_n - \Lambda(\theta) \qquad \theta \in \mathbb{R}\ \ n = 1, 2, \cdots,$$

which leads to

$$\liminf_{n\to\infty} \Lambda^*(\lambda_n) \ge \theta\lambda - \Lambda(\theta) \qquad \theta \in \mathbb{R}.$$

Taking the supremum over $\theta \in \mathbb{R}$ on the right-hand side proves (1.1.3).

Let $l > 0$ be fixed. By Assumption 1.1.1 and by the continuity of $\Lambda(\theta)$ in its domain there is $\delta > 0$ such that

$$c \equiv \sup_{|\theta|\le\delta} \Lambda(\theta) < \infty.$$

Consequently, for any λ with $\Lambda^*(\lambda) \le l$,

$$l \ge \sup_{|\theta|\le\delta} \Big\{\theta\lambda - \Lambda(\theta)\Big\} \ge \delta|\lambda| - c.$$

Therefore, the level set

$$\Lambda_a^* = \{\lambda \in \mathbb{R},\ \ \Lambda^*(\lambda) \le l\}$$

is compact.

In addition, the fact that $\Lambda^*(\lambda)$ is the conjugate of the convex function $\Lambda(\theta)$ makes $\Lambda^*(\lambda)$ a convex function.

DEFINITION 1.1.3. *A convex function Λ: $\mathbb{R} \longrightarrow (-\infty.\infty]$ is essentially smooth if:*

(1) *$\mathcal{D}_\Lambda^o = (a, b)$ is non-empty.*

(2) $\Lambda(\theta)$ *is differentiable in* $\mathcal{D}_\Lambda^o$.
(3) $\Lambda(\cdot)$ *is steep:*

$$\lim_{\theta\to a^+}\Lambda'(\theta)=\lim_{\theta\to b^-}\Lambda'(\theta)=\infty.$$

THEOREM 1.1.4. *Let Assumption 1.1.1 hold. For any closed set* $F\subset\mathbb{R}$,

$$\limsup_{n\to\infty}\frac{1}{b_n}\log\mathbb{P}\{Y_n\in F\}\le-\inf_{\lambda\in F}\Lambda^*(\lambda). \tag{1.1.4}$$

If we further assume that the logarithmic moment function $\Lambda(\theta)$ *is essentially smooth, then for any open set* $G\subset\mathbb{R}$,

$$\liminf_{n\to\infty}\frac{1}{b_n}\log\mathbb{P}\{Y_n\in G\}\ge-\inf_{\lambda\in G}\Lambda^*(\lambda). \tag{1.1.5}$$

Proof. We first prove the upper bound (1.1.4). Without loss generality we may assume that $\inf\limits_{\lambda\in F}\Lambda^*(\lambda)>0$. Let $0<l<\inf\limits_{\lambda\in F}\Lambda^*(\lambda)$ be fixed but arbitrary. By the convexity, lower semi-continuity and goodness of $\Lambda^*(\cdot)$ there are real numbers $\alpha<\beta$ such that $\{\lambda\in\mathbb{R};\ \Lambda^*(\lambda)\le l\}=[\alpha,\beta]$. Clearly, $F\cap[\alpha,\beta]=\emptyset$. Therefore, there are $\alpha'<\alpha$ and $\beta'>\beta$ such that $F\cap(\alpha',\beta')=\emptyset$. Obviously, $\Lambda^*(\alpha'),\Lambda^*(\beta')>l$. We claim that

$$\Lambda^*(\alpha')=\sup_{\theta<0}\Big\{\theta\alpha'-\Lambda(\theta)\Big\},$$

$$\Lambda^*(\beta')=\sup_{\theta>0}\Big\{\theta\beta'-\Lambda(\theta)\Big\}.$$

In fact, the relation

$$\Lambda^*(\alpha')=\sup_{\theta\ge0}\Big\{\theta\alpha'-\Lambda(\theta)\Big\}$$

would lead to $\Lambda^*(\alpha')\le\Lambda^*(\lambda)$ for any $\lambda>\alpha'$. This is impossible when $\lambda\in[\alpha,\beta]$.

In addition,

$$\mathbb{P}\{Y_n\in F\}\le\mathbb{P}\{Y_n\notin(\alpha',\beta')\}=\mathbb{P}\{Y_n\le\alpha'\}+\mathbb{P}\{Y_n\ge\beta'\}.$$

Consequently,

$$\limsup_{n\to\infty}\frac{1}{b_n}\log\mathbb{P}\{Y_n\in F\}$$
$$\le\max\Big\{\limsup_{n\to\infty}\frac{1}{b_n}\log\mathbb{P}\{Y_n\le\alpha'\},\ \ \limsup_{n\to\infty}\frac{1}{b_n}\log\mathbb{P}\{Y_n\ge\beta'\}\Big\}.$$

For any $\theta>0$,

$$\mathbb{P}\{Y_n\ge\beta'\}\le e^{-\theta\beta' b_n}\mathbb{E}\exp\big\{\theta b_nY_n\big\}.$$

Hence,

$$\limsup_{n\to\infty}\frac{1}{b_n}\log\mathbb{P}\{Y_n\ge\beta'\}\le-\theta\beta'+\Lambda(\theta).$$

Taking infimum over $\theta>0$ on the right hand side gives that

$$\limsup_{n\to\infty}\frac{1}{b_n}\log\mathbb{P}\{Y_n\ge\beta'\}\le-\Lambda^*(\beta')\le-l.$$

Similarly, one can show that

$$\limsup_{n\to\infty} \frac{1}{b_n} \log \mathbb{P}\{Y_n \le \alpha'\} \le -l.$$

Summarizing our argument, we have

$$\limsup_{n\to\infty} \frac{1}{b_n} \log \mathbb{P}\{Y_n \in F\} \le -l.$$

Letting $l \to \inf_{\lambda\in F} \Lambda^*(\lambda)$ on the right hand side leads to (1.1.4).

The proof of lower bound (1.1.5) is harder. To do this we claim that for any closed (bounded or unbounded) interval $I \subset \mathbb{R}$ and $\theta \in \mathcal{D}_\Lambda^o$,

$$\limsup_{n\to\infty} \frac{1}{b_n} \log \mathbb{E}\Big(\exp\big\{\theta b_n Y_n\big\} 1_{\{Y_n\in I\}}\Big) \le \sup_{\lambda\in I}\Big\{\theta\lambda - \Lambda^*(\lambda)\Big\}. \tag{1.1.6}$$

Due to similarity, we only deal with the case $\theta \ge 0$. We first consider the case when $I = [a, b]$ is bounded. For $\epsilon > 0$, let $a = x_0 < x_1 < \cdots < x_m = b$ be a partition of I such that $x_j - x_{j-1} < \epsilon$. Thus,

$$\mathbb{E}\Big(\exp\big\{\theta b_n Y_n\big\} 1_{\{Y_n\in I\}}\Big) \le \sum_{j=1}^m \exp\big\{\theta b_n x_j\big\} \mathbb{P}\{Y_n \in [x_{j-1}, x_j]\}.$$

Consequently, by (1.1.4) we get

$$\begin{aligned}
&\limsup_{n\to\infty} \frac{1}{b_n} \log \mathbb{E}\Big(\exp\big\{\theta b_n Y_n\big\} 1_{\{Y_n\in I\}}\Big) \\
&\le \max_{1\le j\le m}\Big\{\theta x_j - \inf_{x_{j-1}\le\lambda\le x_j} \Lambda^*(\lambda)\Big\} \\
&\le \theta\epsilon + \sup_{\lambda\in I}\Big\{\theta\lambda - \Lambda^*(\lambda)\Big\}.
\end{aligned}$$

Letting $\epsilon \to 0^+$ leads to (1.1.6).

We now prove (1.1.6) in the case when I is unbounded. By duality,

$$\Lambda(\theta) = \sup_{\lambda\in\mathbb{R}}\Big\{\lambda\theta - \Lambda^*(\lambda)\Big\} \qquad \theta \in \mathbb{R}.$$

Consequently, (1.1.6) is equivalent to Assumption 1.1.1 when $I = (-\infty, \infty)$. The remaining case is when I is a half line. For similarity we let I be unbounded from above. Given a large $M > 0$, write $I_M = I \cap (-\infty, M]$. We have

$$\begin{aligned}
&\mathbb{E}\Big(\exp\big\{\theta b_n Y_n\big\} 1_{\{Y_n\in I\}}\Big) \\
&\le \mathbb{E}\Big(\exp\big\{\theta b_n Y_n\big\} 1_{\{Y_n\in I_M\}}\Big) + \mathbb{E}\Big(\exp\big\{\theta b_n Y_n\big\} 1_{\{Y_n\ge M\}}\Big).
\end{aligned}$$

By what has been established,

$$\begin{aligned}
&\limsup_{n\to\infty} \frac{1}{b_n} \log \mathbb{E}\Big(\exp\big\{\theta b_n Y_n\big\} 1_{\{Y_n\in I\}}\Big) \\
&\le \max\bigg\{ \limsup_{n\to\infty} \frac{1}{b_n} \log \mathbb{E}\Big(\exp\big\{\theta b_n Y_n\big\} 1_{\{Y_n\ge M\}}\Big),\ \sup_{\lambda\in I}\Big\{\theta\lambda - \Lambda^*(\lambda)\Big\}\bigg\}.
\end{aligned}$$

According to the assumption there is $\theta' > \theta$ such that $\Lambda(\theta') < \infty$. Using Chebyshev inequality and (1.1.4) one can show that

$$\lim_{M\to\infty}\limsup_{n\to\infty}\frac{1}{b_n}\log\mathbb{E}\Big(\exp\big\{\theta b_n Y_n\big\}1_{\{Y_n\ge M\}}\Big) = -\infty.$$

Letting $b \to \infty$ on the right hand side of the previous display leads to (1.1.6).

We now come to the proof of (1.1.5). It suffices to show that for any $\lambda_0 \in G$,

$$\liminf_{n\to\infty}\frac{1}{b_n}\log\mathbb{P}\{Y_n\in G\}\ge -\Lambda^*(\lambda_0). \tag{1.1.7}$$

We may assume that $\Lambda^*(\lambda_0) < \infty$. By essential smoothness of $\Lambda(\cdot)$ and by the definition of $\Lambda^*(\cdot)$ there is a $\theta_0 \in \mathcal{D}_\Lambda^o$ such that $\Lambda^*(\theta) = \lambda_0\theta_0 - \Lambda(\theta_0)$ and $\lambda_0 = \Lambda'(\theta_0)$.

Let $\delta > 0$ be so small that $(x_0 - \delta,\ x_0 + \delta) \subset G$.

$$\begin{aligned}\mathbb{P}\{Y_n\in G\} &\ge \mathbb{P}\{Y_n\in(x_0-\delta,\ x_0+\delta)\}\\ &\ge e^{-(\theta_0\lambda_0+|\theta_0|\delta)b_n}\mathbb{E}\Big(\exp\big\{\theta_0 b_n Y_n\big\}1_{\{Y_n\in(x_0-\delta,\ x_0+\delta)\}}\Big).\end{aligned}$$

Given $\lambda \neq \lambda_0$, $\Lambda^*(\lambda) > \lambda\theta_0 - \Lambda(\theta_0)$, for otherwise $\lambda = \Lambda'(\theta_0) = \lambda_0$. The goodness and the lower semi-continuity of $\Lambda^*(\cdot)$ imply that

$$\Lambda(\theta_0) > \sup_{\lambda\notin(x_0-\delta,\ x_0+\delta)}\Big\{\lambda\theta_0-\Lambda^*(\lambda)\Big\}.$$

Write

$$\begin{aligned}\mathbb{E}\exp\big\{\theta_0 b_n Y_n\big\} &= \mathbb{E}\Big(\exp\big\{\theta_0 b_n Y_n\big\}1_{\{Y_n\in(x_0-\delta,\ x_0+\delta)\}}\Big)\\ &\quad+\mathbb{E}\Big(\exp\big\{\theta_0 b_n Y_n\big\}1_{\{Y_n\notin(x_0-\delta,\ x_0+\delta)\}}\Big).\end{aligned}$$

By Assumption 1.1.1 and by (1.1.6), we get

$$\begin{aligned}\Lambda(\theta_0)\le\max\bigg\{&\liminf_{n\to\infty}\frac{1}{b_n}\log\mathbb{E}\Big(\exp\big\{\theta_0 b_n Y_n\big\}1_{\{Y_n\in(x_0-\delta,\ x_0+\delta)\}}\Big),\\ &\sup_{\lambda\notin(x_0-\delta,x_0+\delta)}\Big\{\lambda\theta_0-\Lambda^*(\lambda)\Big\}\bigg\}.\end{aligned}$$

Consequently,

$$\liminf_{n\to\infty}\frac{1}{b_n}\log\mathbb{E}\Big(\exp\big\{\theta_0 b_n Y_n\big\}1_{\{Y_n\in(x_0-\delta,\ x_0+\delta)\}}\Big)\ge\Lambda(\theta_0).$$

Thus,

$$\liminf_{n\to\infty}\frac{1}{b_n}\log\mathbb{P}\{Y_n\in G\}\ge-\theta_0\lambda_0-|\theta_0|\delta+\Lambda(\theta_0) = -\Lambda^*(\theta_0)-|\theta_0|\delta.$$

Finally, letting $\delta \to 0^+$ on the right hand side proves (1.1.7). □

The limit form described in (1.1.4) and (1.1.5) is called *large deviation principle* (LDP) in the literature. Theorem 1.1.4 is known as the *Gärtner-Ellis theorem* on large deviations. In application, the main focus is on the event that the random variables deviate away from their equilibrium state, or on the event that the random variables take large values. In the following we consider a historically important example. Let $\{X_k\}_{k\geq 1}$ be a real independent and identically distributed (i.i.d.) sequence such that there is a $c>0$ such that

$$\mathbb{E}\exp\big\{c|X_1|\big\}<\infty. \tag{1.1.8}$$

If we take Y_n to be the sample average

$$\overline{X}_n=\frac{1}{n}\{X_1+\cdots+X_n\}\qquad n=1,2\cdots \tag{1.1.9}$$

$b_n=n$, then Assumption 1.1.1 holds with

$$\Lambda(\theta)=\log\mathbb{E}\exp\big\{\theta X_1\big\}.$$

Unfortunately, $\Lambda(\theta)$ is not essentially smooth in general. Indeed, it is straightforward to check that $\Lambda(\theta)$ fails to be steep as X_1 is bounded. Nevertheless, the large deviation principle known as the *Cramér's large deviation principle* (Theorem 2.3.6, p. 44, [**47**]) claims that (1.1.4) and (1.1.5) (with $b_n=n$) hold under (1.1.8).

Further,

$$\log\mathbb{E}\exp\big\{\theta X_1\big\}\geq\theta\mathbb{E}X_1$$

using Jensen inequality. Hence, $\Lambda^*(\mathbb{E}X_1)=0$. On the other hand, assume that $\lambda\in\mathbb{R}$ satisfies $\Lambda^*(\lambda)=0$. We have that

$$\lambda\theta\leq\log\mathbb{E}\exp\big\{\theta X_1\big\}\qquad\theta\in\mathbb{R}.$$

In view of the fact that

$$\lim_{\theta\to 0}\theta^{-1}\log\mathbb{E}\exp\big\{\theta X_1\big\}=\mathbb{E}X_1,$$

letting $\theta\to 0^+$ and letting $\theta\to 0^-$ give, respectively, $\lambda\leq\mathbb{E}X_1$ and $\lambda\geq\mathbb{E}X_1$.

Summarizing our argument, $\Lambda^*(\lambda)=0$ if and only if $\lambda=\mathbb{E}X_1$. By the goodness of the rate function $\Lambda^*(\cdot)$, therefore, for given $\epsilon>0$,

$$\inf_{|\lambda-\mathbb{E}X_1|\geq\epsilon}\Lambda^*(\lambda)>0.$$

This observation shows that the probability that the sample average deviates away from the sample average $\mathbb{E}X_1$ has a genuine exponential decay. We summarize our discussion in the following theorem (Cramér's large deviation principle).

THEOREM 1.1.5. *Under the assumptions (1.1.8),*

$$\limsup_{n\to\infty}\frac{1}{n}\log\mathbb{P}\{\overline{X}_n\in F\}\leq-\inf_{x\in F}\Lambda^*(x),$$

$$\liminf_{n\to\infty}\frac{1}{n}\log\mathbb{P}\{\overline{X}_n\in G\}\geq-\inf_{x\in G}\Lambda^*(x)$$

for any closed set $F\subset\mathbb{R}$ and any open set $G\subset\mathbb{R}$.

In particular, for any $\epsilon>0$,

$$\limsup_{n\to\infty}\frac{1}{n}\log\mathbb{P}\Big\{|\overline{X}_n-\mathbb{E}X_1|\geq\epsilon\Big\}<0.$$

For the inverse of the Gärtner-Ellis theorem, we state the following *Varadhan's integral lemma.*

THEOREM 1.1.6. *Let $\Lambda^*(\cdot)$ be a good rate function and let $Q(\lambda)$ be a continuous function on $\mathbb{R}$.*

(1) *Assume that (1.1.5) holds for every open set G. Then*

$$\liminf_{n\to\infty}\frac{1}{b_n}\log\mathbb{E}\exp\big\{b_nQ(Y_n)\big\}\ge\sup_{\lambda\in\mathbb{R}}\Big\{Q(\lambda)-\Lambda^*(\lambda)\Big\}.$$

(2) *Assume that (1.1.4) holds for every closed set F and that*

$$\limsup_{n\to\infty}\frac{1}{b_n}\log\mathbb{E}\exp\big\{(1+\epsilon)b_nQ(Y_n)\big\}<\infty$$

for some $\epsilon>0$. Then for every $\theta\in\mathbb{R}$,

$$\limsup_{n\to\infty}\frac{1}{b_n}\log\mathbb{E}\exp\big\{b_nQ(Y_n)\big\}\le\sup_{\lambda\in\mathbb{R}}\Big\{Q(\lambda)-\Lambda^*(\lambda)\Big\}.$$

Proof. We omit the proof of part (2) since it is similar to the proof of (1.1.6).

To prove part (1), let $\lambda_0\in\mathbb{R}$ be fixed but arbitrary. By continuity, for given $\epsilon>0$ there is a $\delta>0$ such that $|Q(\lambda)-Q(\lambda_0)|<\epsilon$ if $|\lambda-\lambda_0|<\delta$. Hence,

$$\mathbb{E}\exp\big\{b_nQ(Y_n)\big\}\ge\exp\big\{b_n(Q(\lambda_0)-\epsilon)\big\}\mathbb{P}\{|Y_n-\lambda_0|<\delta\}.$$

Consequently,

$$\begin{aligned}\liminf_{n\to\infty}\frac{1}{b_n}\log\mathbb{E}\exp\big\{b_nQ(Y_n)\big\}&\ge Q(\lambda_0)-\epsilon-\inf_{|\lambda-\lambda_0|<\delta}\Lambda^*(\lambda)\\&\ge Q(\lambda_0)-\epsilon-\Lambda^*(\lambda_0).\end{aligned}$$

Letting $\epsilon\to0^+$, and then taking supremum over $\lambda_0\in\mathbb{R}$ on the right hand side, we have established the desired lower bound. □

The general theory of large deviations has been extended to the random variables taking values in abstract topological spaces. Let Y_n be a sequence of random variables taking values in a separable Banach space B. To extend the Gärtner-Ellis theorem, we assume the existence of the limit

$$\Lambda(f)\equiv\lim_{n\to\infty}\frac{1}{b_n}\log\mathbb{E}\exp\Big\{b_nf(Y_n)\Big\}\qquad f\in B^*$$

instead of (1.1.1), where B^* is the topological dual of B. In addition to some smoothness assumptions on $\Lambda(f)$, it is required that $\{Y_n\}$ be *exponentially tight* . That is, for any $l>0$ there is a compact set $K\subset B$ such that

$$\limsup_{n\to\infty}\frac{1}{b_n}\log\mathbb{P}\{Y_n\notin K\}\le -l.$$

In the finite dimensional setting, the exponential tightness automatically holds under the existence of $\Lambda(f)$.

In dealing with the large deviations in infinite dimensional spaces, the main challenge often lies in the issue of exponential tightness. In the rest of this section we introduce a result of de Acosta [**1**].

Recall that a set $K \subset B$ is said to be *positively balanced*, if $\lambda x \in K$ whenever $x \in K$ and $0 \le \lambda \le 1$. *The Minkowski functional* of a convex and positively balanced set K is defined by

$$q_K(x) = \inf\{\lambda > 0; \quad x \in \lambda K\}$$

with the customary convention that $\inf \varphi = \infty$. The Minkowski functional $q_K(\cdot)$ is sub-additive and positively homogeneous:

$$q_K(x+y) \le q_K(x) + q_K(y) \quad \text{and} \quad q_K(\lambda x) = \lambda q_K(x) \tag{1.1.10}$$

where $x, y \in B$ and $\lambda \ge 0$.

A family $\{\mu_\alpha; \quad \alpha \in \Theta\}$ of probability measures on B is said to be *uniformly tight*, if for any $\epsilon > 0$, there is a compact set $K \subset B$ such that

$$\mu_\alpha(K) \ge 1 - \epsilon \qquad \alpha \in \Theta.$$

The following result is given in Theorem 3.1, [**1**]. We state it without proof.

THEOREM 1.1.7. *Let $\{\mu_\alpha, \quad \alpha \in \Theta\}$ be a family of probability measures on the separable Banach space B and assume that $\{\mu_\alpha; \quad \alpha \in \Theta\}$ is uniformly tight and that*

$$\sup_{\alpha\in\Theta} \int_B \exp\left\{\lambda ||x||\right\} \mu_\alpha(dx) < \infty \qquad \forall \lambda > 0.$$

There is a convex, positively balanced and compact set $K \subset B$ such that

$$\sup_{\alpha\in\Theta} \int_B \exp\left\{q_K(x)\right\} \mu_\alpha(dx) < \infty.$$

1.2. LDP for non-negative random variables

In this section we assume that $\{Y_n\}$ take non-negative values. Recall that the full large deviation principle is stated as: For every closed set $F \subset \mathbb{R}^+$,

$$\limsup_{n\to\infty} \frac{1}{b_n} \log \mathbb{P}\{Y_n \in F\} \le -\inf_{\lambda\in F} I(\lambda) \tag{1.2.1}$$

and, for every open set $G \subset \mathbb{R}^+$,

$$\liminf_{n\to\infty} \frac{1}{b_n} \log \mathbb{P}\{Y_n \in G\} \ge -\inf_{\lambda\in G} I(\lambda). \tag{1.2.2}$$

In application, (1.2.1) and (1.2.2) are often replaced by our concern of the tail probability of the form

$$\mathbb{P}\{Y_n \ge \lambda\} \qquad \lambda > 0.$$

Under some mild conditions on the rate function $I(\lambda)$, the following theorem shows that large deviation principle is determined by the asymptotic behavior of tail probabilities.

THEOREM 1.2.1. *Assume that the rate function $I(\lambda)$ is strictly increasing and continuous on $\mathbb{R}^+$. The following two statements are equivalent:*

(1) *The large deviation principle stated by (1.2.1) and (1.2.2) holds, respectively, for every closed set $F \subset \mathbb{R}^+$ and for every open set $G \subset \mathbb{R}^+$.*

(2) *For every* $\lambda > 0$,

$$\lim_{n\to\infty} \frac{1}{b_n} \log \mathbb{P}\{Y_n \geq \lambda\} = -I(\lambda). \tag{1.2.3}$$

Proof. Assume that (1) holds. Taking $F = [\lambda, \infty)$ in (1.2.1) gives that

$$\limsup_{n\to\infty} \frac{1}{b_n} \log \mathbb{P}\{Y_n \geq \lambda\} \leq -\inf_{x\geq\lambda} I(x) = -I(\lambda).$$

Similarly, by (1.2.2) with $G = (\lambda, \infty)$ we get

$$\liminf_{n\to\infty} \frac{1}{b_n} \log \mathbb{P}\{Y_n \geq \lambda\} \geq -\inf_{x>\lambda} I(x) = -I(\lambda)$$

where the last step follows from the continuity.

Assume that (2) holds. For a closed set $F \subset \mathbb{R}^+$, let $\lambda_0 = \inf F$. We have that

$$\mathbb{P}\{Y_n \in F\} \leq \mathbb{P}\{Y_n \geq \lambda_0\}.$$

By (1.2.3) we have

$$\limsup_{n\to\infty} \frac{1}{b_n} \log \mathbb{P}\{Y_n \in F\} \leq -I(\lambda_0) = -\inf_{\lambda\in F} I(\lambda)$$

where the last step follows from the monotonicity of $I(\cdot)$.

To establish (1.2.2), let $\lambda_0 \in G$. There is a $\delta > 0$ such that $[\lambda_0, \lambda_0 + \delta) \subset G$. By the fact that

$$\mathbb{P}\{Y_n \geq \lambda_0\} \leq \mathbb{P}\{Y_n \geq \lambda_0 + \delta\} + \mathbb{P}\{Y_n \in [\lambda_0, \lambda_0 + \delta)\}$$

and by (1.2.3) we have

$$-I(\lambda_0) \leq \max\left\{ -I(\lambda_0 + \delta),\ \liminf_{n\to\infty} \frac{1}{b_n} \log \mathbb{P}\{Y_n \in [\lambda_0, \lambda_0 + \delta)\}\right\}.$$

By assumption we have that $I(\lambda_0 + \delta) > I(\lambda_0)$. Consequently,

$$\liminf_{n\to\infty} \frac{1}{b_n} \log \mathbb{P}\{Y_n \in G\} \geq -I(\lambda_0).$$

Taking supremum over $\lambda_0 \in G$ on the right hand side gives (1.2.2). □

The following theorem shows that under certain conditions, the tail probability of the fixed sum of independent non-negative random variables is dominated by the tail probabilities of individual terms.

THEOREM 1.2.2. *Let $Z_1(n), \cdots, Z_l(n)$ be independent non-negative random variables with $l \geq 2$ being fixed.*

(*a*) *If there are constant $C_1 > 0$ and $0 < a \leq 1$ such that*

$$\limsup_{n\to\infty} \frac{1}{b_n} \log \mathbb{P}\{Z_j(n) \geq \lambda\} \leq -C_1\lambda^a \qquad \forall\lambda > 0$$

for $j = 1, \cdots, l$, then

$$\limsup_{n\to\infty} \frac{1}{b_n} \log \mathbb{P}\{Z_1(n) + \cdots + Z_l(n) \geq \lambda\} \leq -C_1\lambda^a \qquad \forall\lambda > 0.$$

(b) *If there are constants $C_2 > 0$ and $b \geq 2$ such that*

$$\limsup_{n\to\infty} \frac{1}{b_n} \log \mathbb{E} \exp\Big\{\theta b_n Z_j(n)\Big\} \leq C_2 \theta^b \qquad \forall \theta > 0$$

for $j = 1, \cdots, l$, then

$$\limsup_{n\to\infty} \frac{1}{b_n} \log \mathbb{E} \exp\Big\{\theta b_n \sqrt{Z_1^2(n) + \cdots + Z_l^2(n)}\Big\} \leq C_2 \theta^b \qquad \forall \theta > 0.$$

Proof. Clearly, part (a) needs only to be proved in the case $l = 2$. Given $0 < \delta < \lambda$, let $0 = x_0 < x_1 < \cdots < x_N = \lambda$ be a partition of $[0, \lambda]$ such that $x_k - x_{k-1} < \delta$. Then

$$\mathbb{P}\big\{Z_1(n) + Z_2(n) \geq \lambda\big\} \leq \sum_{k=1}^N \mathbb{P}\{Z_1(n) \geq x_{k-1}\}\mathbb{P}\{Z_2(n) \geq \lambda - x_k\}.$$

Consequently,

$$\limsup_{n\to\infty} \frac{1}{b_n} \log \mathbb{P}\big\{Z_1(n) + Z_2(n) \geq \lambda\big\} \leq -C_1 \min_{1\leq k\leq N} \Big\{x_{k-1}^a + (\lambda - x_k)^a\Big\}.$$

By the fact $0 < a \leq 1$, $x_{k-1}^a + (\lambda - x_k)^a \geq (\lambda - x_k + x_{k-1})^a \geq (\lambda - \delta)^a$. So we have

$$\limsup_{n\to\infty} \frac{1}{b_n} \log \mathbb{P}\big\{Z_1(n) + Z_2(n) \geq \lambda\big\} \leq -C_1(\lambda - \delta)^a.$$

Letting $\delta \to 0^+$ on the right hand side proves part (a).

We now come to part (b). For any $\epsilon > 0$, there are finitely many vectors

$$\mathbf{t}_j = (t_{j,1}, \cdots, t_{j,l}) \in (\mathbb{R}^+)^l \qquad j = 1, \cdots, N$$

such that

$$\sqrt{t_{j,1}^2 + \cdots + t_{j,l}^2} = 1 \qquad j = 1, \cdots, N,$$

$$\sqrt{x_1^2 + \cdots + x_l^2} \leq (1+\epsilon) \max_{1\leq j\leq N} \sum_{k=1}^l t_{j,k} x_k \qquad \forall x_1, \cdots, x_l \in (\mathbb{R}^+)^l.$$

Hence,

$$\begin{aligned}
&\mathbb{E} \exp\Big\{\theta b_n \sqrt{Z_1^2(n) + \cdots + Z_l^2(n)}\Big\} \\
&\leq \sum_{j=1}^N \mathbb{E} \exp\Big\{(1+\epsilon)\theta b_n \sum_{k=1}^l t_{j,k} Z_k(n)\Big\} \\
&= \sum_{j=1}^N \prod_{k=1}^l \mathbb{E} \exp\Big\{(1+\epsilon)\theta b_n t_{j,k} Z_k(n)\Big\}.
\end{aligned}$$

Consequently, by the fact that $b \geq 2$,

$$\begin{aligned}
&\limsup_{n\to\infty} \frac{1}{b_n} \log \mathbb{E} \exp\Big\{\theta b_n \sqrt{Z_1^2(n) + \cdots + Z_l^2(n)}\Big\} \\
&\leq C_2(1+\epsilon)^b \theta^b \max_{1\leq j\leq N} \sum_{k=1}^l t_{j,k}^b \leq C_2(1+\epsilon)^b\theta^b.
\end{aligned}$$

Letting $\epsilon \to 0^+$ on the right hand side proves part (b). □

A basic assumption of the Gärtner-Ellis theorem (Theorem 1.2.4 below) for non-negative random variables is the existence of the logarithmic moment generating function on $\mathbb{R}^+$. Correspondent to Definition 1.1.3, we adopt the following notion of essential smoothness for the functions defined on $\mathbb{R}^+$.

DEFINITION 1.2.3. *A convex function* $\Lambda(\theta)$: $\mathbb{R}^+ \longrightarrow [0,\infty]$ *is said to be essentially smooth on* $\mathbb{R}^+$, *if*

(1) *There is a* $\theta_0 > 0$ *such that* $\Lambda(\theta) < \infty$ *for every* $0 \le \theta \le \theta_0$.
(2) $\Lambda(\theta)$ *is differentiable in the interior* $D_\Lambda^o = (0,a)$ $(0 < a \le \infty)$ *of the domain* $D_\Lambda = \{\theta \in \mathbb{R}^+;\ \Lambda(\theta) < \infty\}$.
(3) *The function* $\Lambda(\theta)$ *is steep at the right end (of the domain) and is flat at the left end, i.e.,*

$$\lim_{\theta\to a^-} \Lambda'(\theta) = \infty \quad \text{and} \quad \Lambda'(0^+) \equiv \lim_{\theta\to 0^+} \frac{\Lambda(\theta)-\Lambda(0)}{\theta} = 0.$$

The following theorem appears as a version of Gaïtner-Ellis large deviation.

THEOREM 1.2.4. *Assume that for all* $\theta \ge 0$, *the limit*

$$\Lambda(\theta) = \lim_{n\to\infty} \frac{1}{b_n} \log \mathbb{E}\exp\Big\{\theta b_n Y_n\Big\} \tag{1.2.4}$$

exists as an extended real number, and that the function $\Lambda(\theta)$ *is essentially smooth on* $\mathbb{R}^+$.

(1) *The function*

$$I(\lambda) = \sup_{\theta>0}\Big\{\theta\lambda - \Lambda(\theta)\Big\} \qquad \lambda \ge 0 \tag{1.2.5}$$

is strictly increasing and continuous on $\mathbb{R}^+$. *Consequently, the LDP given defined by (1.2.1) and (1.2.2) and the LDP defined by (1.2.3) are equivalent.*

(2) *The equivalent forms (1.2.1), (1.2.2) and (1.2.3) hold.*

Proof. Let ξ be an independent random variable with distribution

$$\mathbb{P}\{\xi = -1\} = \mathbb{P}\{\xi = 1\} = \frac{1}{2}.$$

We have

$$\mathbb{E}\exp\Big\{\theta b_n \xi Y_n\Big\} = \frac{1}{2}\Big\{\mathbb{E}\exp\Big\{-\theta b_n Y_n\Big\} + \mathbb{E}\exp\Big\{\theta b_n Y_n\Big\}\Big\} \qquad \theta \in \mathbb{R}.$$

Consequently,

$$\lim_{n\to\infty}\frac{1}{b_n}\log\mathbb{E}\exp\Big\{\theta b_n\xi Y_n\Big\} = \Lambda(|\theta|) \qquad \theta\in\mathbb{R}.$$

Notice that $\Lambda(|\theta|)$ is essentially smooth on $\mathbb{R}$ according to Definition 1.1.3. Applying Theorem 1.1.4 to the sequence $\{\xi Y_n\}$ we have the large deviation principle given in (1.2.1) and (1.2.2) with the rate function given by

$$I(\lambda) = \sup_{\theta\in\mathbb{R}}\Big\{\theta\lambda - \Lambda(|\theta|)\Big\} = \sup_{\theta>0}\Big\{\theta\lambda - \Lambda(\theta)\Big\} \qquad \forall\lambda \ge 0. \tag{1.2.6}$$

By Theorem 1.2.1 it remains to show that $I(\lambda)$ is strictly increasing and continuous on $\mathbb{R}^+$. From (1.2.5) and by the steepness of $\Lambda(\theta)$, $I(\lambda)$ is well defined on $\mathbb{R}^+$, non-negative and non-decreasing. Let $\lambda > 0$ be fixed. By essential smoothness of $\Lambda(\theta)$ and by (1.2.6), the function

$$h(\theta) = \lambda\theta - \Lambda(\theta) \qquad \theta \geq 0$$

is bounded from above and attains its supremum $I(\lambda)$ at some $\theta \geq 0$. Further, $\theta > 0$, for otherwise (observe that $\Lambda(0) = 0$) $\Lambda(\widetilde{\theta}) \geq \lambda\widetilde{\theta}$ for all $\widetilde{\theta} > 0$, which contradicts the assumption that $\Lambda'(0^+) = 0$ (again, we use the fact $\Lambda(0) = 0$ here).

In summary, for any $\lambda > 0$ there is a $\theta > 0$, such that $\Lambda'(\theta) = \lambda$ and that $I(\lambda) = \lambda\theta - \Lambda(\theta)$.

Let $0 < \lambda_1 < \lambda_2 < a$ and find $\theta_1 > 0$ such that $I(\lambda_1) = \lambda_1\theta_1 - \Lambda(\theta_1)$. We have

$$I(\lambda_2) \geq \theta_1\lambda_2 - \Lambda(\theta_1) > \theta_1\lambda_1 - \Lambda(\theta_1) = I(\lambda_1).$$

In addition, the above argument also leads to the fact that $I(\lambda) > 0 = I(0)$ for any $\lambda > 0$. Hence, $I(\cdot)$ is strictly increasing on $\mathbb{R}^+$.

Being increasing and lower semi-continuous, $I(\cdot)$ is left continuous. To establish continuity for $I(\lambda)$ in $\mathbb{R}^+$, therefore, all we need to show is that for any $0 \leq \lambda_0 < \lambda_n$ with $\lambda_n \to \lambda_0^+$, $I(\lambda_n) \longrightarrow I(\lambda_0)$. Indeed, find $\theta_n > 0$ such that

$$I(\lambda_n) = \theta_n\lambda_n - \Lambda(\theta_n) \quad \text{and} \quad \Lambda'(\theta_n) = \lambda_n \qquad n = 1, 2, \cdots.$$

In particular, the sequence $\{\Lambda'(\theta_n)\}$ is bounded. By essential smoothness of $\Lambda(\cdot)$, $\{\theta_n\}$ is bounded.

Hence

$$0 \leq I(\lambda_n) - I(\lambda_0) = \theta_n\lambda_n - \Lambda(\theta_n) - I(\lambda_0) \leq \theta_n(\lambda_n - \lambda_0) \longrightarrow 0$$

as $n \to \infty$. □

The major step for establishing a large deviation principle by the Gärtner-Ellis theorem is to compute the exponential moment generating function

$$\Lambda(\theta) = \lim_{n\to\infty} \frac{1}{b_n} \log \mathbb{E} \exp\Big\{\theta b_n Y_n\Big\}.$$

When the exponential moment generating function is too difficult to deal with, or when the exponential generating function blows up, we may look for some other moment functions instead. For example, we may consider the large deviations under the existence of the limit

$$\Lambda_p(\theta) \equiv \lim_{n\to\infty} \frac{1}{b_n} \log \mathbb{E} \exp\Big\{\theta b_n Y_n^{1/p}\Big\} \qquad \theta \geq 0 \tag{1.2.7}$$

where $p > 0$ is fixed.

COROLLARY 1.2.5. *Assume that for all $\theta > 0$, the limit $\Lambda_p(\theta)$ given in (1.2.7) exists as an extended real number, and that $\Lambda_p(\theta)$ is essentially smooth on $\mathbb{R}^+$ according to Definition 1.2.3. Then*

$$\lim_{n\to\infty} \frac{1}{b_n} \log \mathbb{P}\{Y_n \geq \lambda\} = -I_p(\lambda) \qquad \lambda > 0 \tag{1.2.8}$$

where

$$I_p(\lambda) = \sup_{\theta>0}\Big\{\theta\lambda^{1/p} - \Lambda_p(\theta)\Big\} \qquad \lambda > 0. \tag{1.2.9}$$

Proof. Replacing Y_n by $Y_n^{1/p}$ in Theorem 1.2.4 completes the proof. □

In view of the Taylor expansion,

$$\mathbb{E}\exp\Big\{\theta b_n Y_n^{1/p}\Big\} = \sum_{m=0}^{\infty}\frac{\theta^m}{m!}b_n^m\mathbb{E}Y_n^{m/p}, \tag{1.2.10}$$

one may attempt to estimate $\mathbb{E}Y_n^{m/p}$ when establishing (1.2.7) by "standard" approach becomes technically difficult. When $1/p$ is not integer, however, it is not very pleasant to deal with the (possibly) fractional power m/p. To resolve this problem, we introduce the following lemma.

LEMMA 1.2.6. *Let $p > 0$ be fixed and let $\Psi\colon [0,\infty) \longrightarrow [0,\infty]$ be a non-decreasing lower semi-continuous function. Assume that the domain of Ψ has the form*

$$\mathcal{D}_\Psi \equiv \{\theta;\ \Psi(\theta) < \infty\} = [0, a)$$

where $0 < a \le \infty$, and that $\Psi(\theta)$ is continuous on $\mathcal{D}_\Psi$.

(1) *The lower bound*

$$\liminf_{n\to\infty}\frac{1}{b_n}\log\left\{\sum_{m=0}^{\infty}\frac{\theta^m}{m!}b_n^m\Big(\mathbb{E}Y_n^m\Big)^{1/p}\right\} \ge \Psi(\theta) \qquad \theta > 0 \tag{1.2.11}$$

holds if and only if

$$\liminf_{n\to\infty}\frac{1}{b_n}\log\mathbb{E}\exp\Big\{\theta b_n Y_n^{1/p}\Big\} \ge p\Psi\Big(\frac{\theta}{p}\Big) \qquad \theta > 0. \tag{1.2.12}$$

(2) *The upper bound*

$$\limsup_{n\to\infty}\frac{1}{b_n}\log\left\{\sum_{m=0}^{\infty}\frac{\theta^m}{m!}b_n^m\Big(\mathbb{E}Y_n^m\Big)^{1/p}\right\} \le \Psi(\theta) \qquad \theta > 0 \tag{1.2.13}$$

holds if and only if

$$\limsup_{n\to\infty}\frac{1}{b_n}\log\mathbb{E}\exp\Big\{\theta b_n Y_n^{1/p}\Big\} \le p\Psi\Big(\frac{\theta}{p}\Big) \qquad \theta > 0. \tag{1.2.14}$$

Proof. We first prove "(1.2.11) $\Longrightarrow$ (1.2.12)". We may assume that in (1.2.12), the right hand side is positive. By the expansion (1.2.10) we have

$$\mathbb{E}\exp\Big\{\theta b_n Y_n^{1/p}\Big\} \ge \frac{\theta^{[pm]+1}}{([pm]+1)!}b_n^{[pm]+1}\mathbb{E}Y_n^{\frac{[pm]+1}{p}} \qquad m = 0, 1, \cdots.$$

By Jensen inequality

$$b_n^{[pm]+1}\mathbb{E}Y_n^{\frac{[pm]+1}{p}} \ge \Big\{b_n^{pm}\mathbb{E}Y_n^m\Big\}^{\frac{[pm]+1}{pm}}.$$

Therefore, as $b_n^{pm}\mathbb{E}Y_n^m \ge 1$ we have

$$b_n^{[pm]+1}\mathbb{E}Y_n^{\frac{[pm]+1}{p}} \ge b_n^{pm}\mathbb{E}Y_n^m.$$

For any $0 < \delta < \epsilon$, by Stirling formula there is $c > 0$ such that

$$\frac{\theta^{[pm]+1}}{([pm]+1)!} \ge c^p \frac{\theta^{pm}}{(m!)^p}\Big(\frac{1+\delta}{p(1+\epsilon)}\Big)^{pm}.$$

Thus,

$$(1+\delta)^{-m}\Big(\mathbb{E}\exp\Big\{\theta b_n Y_n^{1/p}\Big\}\Big)^{1/p} \ge c\frac{1}{m!}\Big(\frac{\theta}{p(1+\epsilon)}\Big)^m b_n^m\Big(\mathbb{E}Y_n^m\Big)^{1/p}$$

whenever

$$m \in Q \equiv \Big\{m;\ \ b_n^m\Big(\mathbb{E}Y_n^m\Big)^{1/p} \ge 1\Big\}.$$

Summing up over m,

$$\frac{1+\delta}{\delta}\Big(\mathbb{E}\exp\Big\{\theta b_n Y_n^{1/p}\Big\}\Big)^{1/p} \ge c\sum_{m\in Q}\frac{1}{m!}\Big(\frac{\theta}{(1+\epsilon)p}\Big)^m b_n^m\Big(\mathbb{E}Y_n^m\Big)^{1/p}.$$

Consequently,

$$\begin{aligned}(1.2.15)\qquad &\frac{1+\delta}{\delta}\Big(\mathbb{E}\exp\Big\{\theta b_n Y_n^{1/p}\Big\}\Big)^{1/p} + c\exp\Big\{\frac{\theta}{(1+\epsilon)p}\Big\}\\ &\ge c\sum_{m=1}^{\infty}\frac{1}{m!}\Big(\frac{\theta}{(1+\epsilon)p}\Big)^m b_n^m\Big(\mathbb{E}Y_n^m\Big)^{1/p}.\end{aligned}$$

Thus,

$$\max\Big\{0,\ \ \frac{1}{p}\liminf_{n\to\infty}\frac{1}{b_n}\log\mathbb{E}\exp\Big\{\theta b_n Y_n^{1/p}\Big\}\Big\} \ge \Psi\Big(\frac{\theta}{(1+\epsilon)p}\Big).$$

Letting $\epsilon \to 0^+$ on the right hand side, by the lower semi-continuity of $\Psi(\cdot)$, we get

$$\max\Big\{0,\ \ \frac{1}{p}\liminf_{n\to\infty}\frac{1}{b_n}\log\mathbb{E}\exp\Big\{\theta b_n Y_n^{1/p}\Big\}\Big\} \ge \Psi\Big(\frac{\theta}{p}\Big).$$

Since the right hand side is positive, we have proved (1.2.12).

The same estimate can be used to establish "(1.2.14) $\Longrightarrow$ (1.2.13)". To proceed we may assume that the left hand side of (1.2.13) is positive.

Replacing θ by $(1+\epsilon)p\theta$ in (1.2.15) gives

$$(1.2.16)\qquad \frac{1+\delta}{\delta}\Big(\mathbb{E}\exp\Big\{(1+\epsilon)p\theta b_n Y_n^{1/p}\Big\}\Big)^{1/p} + ce^{\theta} \ge c\sum_{m=1}^{\infty}\frac{\theta^m}{m!}b_n^m\Big(\mathbb{E}Y_n^m\Big)^{1/p}.$$

By (1.2.14) (with θ being replaced by $(1+\epsilon)p\theta$), we get

$$\max\Big\{0,\ \Psi\big((1+\epsilon)\theta\big)\Big\} \ge \limsup_{n\to\infty}\frac{1}{b_n}\log\Big\{\sum_{m=0}^{\infty}\frac{\theta^m}{m!}b_n^m\Big(\mathbb{E}Y_n^m\Big)^{1/p}\Big\}.$$

Since the right hand side is positive, we must have

$$\limsup_{n\to\infty}\frac{1}{b_n}\log\Big\{\sum_{m=0}^{\infty}\frac{\theta^m}{m!}b_n^m\Big(\mathbb{E}Y_n^m\Big)^{1/p}\Big\} \le \Psi\big((1+\epsilon)\theta\big).$$

To continue, we may assume that $\theta \in \mathcal{D}_\Psi$ (for otherwise (1.2.13) is trivial). By continuity of $\Psi(\cdot)$, letting $\epsilon \to 0^+$ on the right hand side gives (1.2.13).

We now prove "(1.2.12) $\Longrightarrow$ (1.2.11)". We may assume that the right hand side of (1.2.11) is positive. For any $k \ge 0$,

$$\frac{\theta^{[p^{-1}k]+1}}{([p^{-1}k]+1)!} b_n^{[p^{-1}k]+1} \Big[\mathbb{E} Y_n^{[p^{-1}k]+1}\Big]^{1/p} \le \sum_{m=0}^{\infty} \frac{\theta^m}{m!} b_n^m \Big(\mathbb{E} Y_n^m\Big)^{1/p}.$$

By Jensen inequality, Stirling formula and an argument similar to the one used for (1.2.15), we can prove that for any $0 < \delta < \epsilon$ there is $C > 0$ independent of n such that

$$\begin{aligned}&\mathbb{E}\exp\Big\{\frac{p\theta}{1+\epsilon} b_n Y_n^{1/p}\Big\} \\ &\le \exp\Big\{\frac{p\theta}{1+\epsilon}\Big\} + C\frac{1+\delta}{\delta}\bigg(\sum_{m=0}^{\infty} \frac{\theta^m}{m!} b_n^m \Big(\mathbb{E} Y_n^m\Big)^{1/p}\bigg)^p.\end{aligned} \tag{1.2.17}$$

By (1.2.12) (with θ being replaced by $\dfrac{p\theta}{1+\epsilon}$), we get

$$p\Psi\Big(\frac{\theta}{1+\epsilon}\Big) \le \max\bigg\{0,\ \ p\liminf_{n\to\infty} \frac{1}{b_n} \log\bigg\{\sum_{m=0}^{\infty} \frac{\theta^m}{m!} b_n^m \Big(\mathbb{E} Y_n^m\Big)^{1/p}\bigg\}\bigg\}.$$

Since the left hand side is positive, we have

$$\liminf_{n\to\infty} \frac{1}{b_n} \log\bigg\{\sum_{m=0}^{\infty} \frac{\theta^m}{m!} b_n^m \Big(\mathbb{E} Y_n^m\Big)^{1/p}\bigg\} \ge \Psi\Big(\frac{\theta}{1+\epsilon}\Big).$$

By the lower semi-continuity of $\Psi(\cdot)$, letting $\epsilon \to 0^+$ on the right hand side gives (1.2.11).

Finally, "(1.2.13) $\Longrightarrow$ (1.2.14)" also follows from the estimate given in (1.2.17). $\square$

An immediate application of Lemma 1.2.6 is the following Gärtener-Ellis-type theorem.

THEOREM 1.2.7. *Assume that for each $\theta \ge 0$, the limit*

$$\Psi(\theta) \equiv \lim_{n\to\infty} \frac{1}{b_n} \log\bigg\{\sum_{m=0}^{\infty} \frac{\theta^m}{m!} b_n^m \Big(\mathbb{E} Y_n^m\Big)^{1/p}\bigg\} \tag{1.2.18}$$

exists as an extended real number. Assume that the function $\Psi(\theta)$ is essentially smooth on $\mathbb{R}^+$ (Definition 1.2.3). For each $\lambda > 0$,

$$\lim_{n\to\infty} \frac{1}{b_n} \log \mathbb{P}\{Y_n \ge \lambda\} = -I_\Psi(\lambda) \tag{1.2.19}$$

where the rate function $I_\Psi(\cdot)$ is defined by

$$I_\Psi(\lambda) = p\sup_{\theta>0}\big\{\theta\lambda^{1/p} - \Psi(\theta)\big\} \quad \lambda > 0. \tag{1.2.20}$$

Proof. By Lemma 1.2.6, the condition posed in Corollary 1.2.5 is satisfied by $\Lambda_p(\theta) = p\Psi(\theta/p)$. $\square$

We now consider the case of a single random variable.

THEOREM 1.2.8. *Let $Y \geq 0$ be a random variable such that*

$$\lim_{m\to\infty} \frac{1}{m} \log\left(\frac{1}{(m!)^\gamma} \mathbb{E}Y^m\right) = \kappa \tag{1.2.21}$$

for some $\gamma > 0$ and $\kappa \in \mathbb{R}$. Then

$$\lim_{t\to\infty} t^{-1/\gamma} \log \mathbb{P}\{Y \geq t\} = -\gamma e^{-\kappa/\gamma}. \tag{1.2.22}$$

Proof. We check the condition posed in Theorem 1.2.6 with $Y_t = Y/t$, $b_t = t^{1/\gamma}$ and $p = 2\gamma$. Indeed, for any $\theta > 0$,

$$\log\left\{\sum_{m=0}^\infty \frac{\theta^m}{m!} t^{m/(2\gamma)} \Big(\mathbb{E}Y^m\Big)^{\frac{1}{2\gamma}}\right\} \sim \log\left\{\sum_{m=0}^\infty \frac{\theta^m}{m!} t^{m/(2\gamma)} \Big((m!)^\gamma e^{\kappa m}\Big)^{\frac{1}{2\gamma}}\right\}$$
$$= \log\left\{\sum_{m=0}^\infty \frac{1}{\sqrt{m!}} \Big(\theta t^{\frac{1}{2\gamma}} e^{\frac{\kappa}{2\gamma}}\Big)^m\right\}.$$

Consider the decomposition

$$\sum_{m=0}^\infty \frac{1}{\sqrt{m!}} \Big(\theta t^{\frac{1}{2\gamma}} e^{\frac{\kappa}{2\gamma}}\Big)^m$$
$$= \sum_{m=0}^\infty \frac{1}{\sqrt{(2m)!}} \Big(\theta t^{\frac{1}{2\gamma}} e^{\frac{\kappa}{2\gamma}}\Big)^{2m} + \sum_{m=0}^\infty \frac{1}{\sqrt{(2m+1)!}} \Big(\theta t^{\frac{1}{2\gamma}} e^{\frac{\kappa}{2\gamma}}\Big)^{2m+1}.$$

By the Stirling formula, we get

$$(2m)! = \big(1+o(1)\big)^m (2^m m!)^2 \quad \text{and} \quad (2m+1)! = \big(1+o(1)\big)^m (2^m m!)^2$$

as $m \to \infty$.

Thus,

$$\log\left\{\sum_{m=0}^\infty \frac{1}{\sqrt{m!}} \Big(\theta t^{\frac{1}{2\gamma}} e^{\frac{\kappa}{2\gamma}}\Big)^m\right\}$$
$$\sim \log\left\{\sum_{m=0}^\infty \frac{1}{2^m m!} \Big(\theta t^{\frac{1}{2\gamma}} e^{\frac{\kappa}{2\gamma}}\Big)^{2m}\right\} = \frac{1}{2}\theta^2 t^{1/\gamma} e^{\kappa/\gamma}.$$

Summarizing our discussion,

$$\lim_{t\to\infty} t^{-1/\gamma} \log\left\{\sum_{m=0}^\infty \frac{\theta^m}{m!} t^{m/(2\gamma)} \Big(\mathbb{E}Y^m\Big)^{\frac{1}{2\gamma}}\right\} = \frac{1}{2}\theta^2 e^{\kappa/\gamma}.$$

Therefore, (1.2.22) follows from Theorem 1.2.6 and the fact that

$$I_\Psi(\lambda) = 2\gamma \sup_{\theta>0}\left\{\theta\lambda^{\frac{1}{2\gamma}} - \frac{1}{2}\theta^2 e^{\kappa/\gamma}\right\} = \lambda^{1/\gamma}\gamma e^{-\kappa/\gamma} \qquad (\lambda > 0).$$

(so $I_\Psi(1) = \gamma e^{\kappa/\gamma}$ appearing on the right hand side of (1.2.22)). □

The following theorem appears as an inverse to Theorem 1.2.7.

THEOREM 1.2.9. *Let $I(\lambda)$ be a non-decreasing rate function $I(\lambda)$ on $\mathbb{R}^+$ with $I(0) = 0$.*

(1) *Assume that*

$$\liminf_{n\to\infty}\frac{1}{b_n}\log\mathbb{P}\{Y_n\ge\lambda\}\ge -I(\lambda)\qquad(\lambda>0). \tag{1.2.23}$$

Then for every $\theta>0$

$$\liminf_{n\to\infty}\frac{1}{b_n}\log\left\{\sum_{m=0}^\infty\frac{\theta^m}{m!}b_n^m\Big(\mathbb{E}Y_n^m\Big)^{1/p}\right\}\ge\sup_{\lambda>0}\Big\{\theta\lambda^{1/p}-p^{-1}I(\lambda)\Big\}. \tag{1.2.24}$$

(2) *Assume that*

$$\limsup_{n\to\infty}\frac{1}{b_n}\log\mathbb{P}\{Y_n\ge\lambda\}\le -I(\lambda)\qquad(\lambda>0). \tag{1.2.25}$$

Then for every $\theta>0$ *satisfying*

$$\lim_{l\to\infty}\limsup_{n\to\infty}\frac{1}{b_n}\log\left\{\sum_{m=0}^\infty\frac{\theta^m}{m!}b_n^m\Big(\mathbb{E}\big(Y_n^m1_{\{Y_n\ge l\}}\big)\Big)^{1/p}\right\}=0 \tag{1.2.26}$$

we have

$$\limsup_{n\to\infty}\frac{1}{b_n}\log\left\{\sum_{m=0}^\infty\frac{\theta^m}{m!}b_n^m\Big(\mathbb{E}Y_n^m\Big)^{1/p}\right\}\le\sup_{\lambda>0}\Big\{\theta\lambda^{1/p}-p^{-1}I(\lambda)\Big\}. \tag{1.2.27}$$

Proof. For any $\lambda>0$,

$$\mathbb{E}Y_n^m\ge\lambda^m\mathbb{P}\{Y_n\ge\lambda\}.$$

Consequently,

$$\sum_{m=0}^\infty\frac{\theta^m}{m!}b_n^m\Big(\mathbb{E}Y_n^m\Big)^{1/p}\ge\exp\big\{b_n\theta\lambda^{1/p}\big\}\Big(\mathbb{P}\{Y_n\ge\lambda\}\Big)^{1/p}.$$

By (1.2.23)

$$\liminf_{n\to\infty}\frac{1}{b_n}\log\left\{\sum_{m=0}^\infty\frac{\theta^m}{m!}b_n^m\Big(\mathbb{E}Y_n^m\Big)^{1/p}\right\}\ge\theta\lambda^{1/p}-p^{-1}I(\lambda).$$

Taking supremum over $\lambda>0$ on the right hand side gives (1.2.24).

To prove (1.2.27), write

$$\begin{aligned}\sum_{m=0}^\infty\frac{\theta^m}{m!}b_n^m\Big(\mathbb{E}Y_n^m\Big)^{1/p}\\ &=\sum_{m-0}^\infty\frac{\theta^m}{m!}b_n^m\Big(\mathbb{E}\big(Y_n^m1_{\{Y_n<l\}}\big)\Big)^{1/p}\\ &+\sum_{m=0}^\infty\frac{\theta^m}{m!}b_n^m\Big(\mathbb{E}\big(Y_n^m1_{\{Y_n\ge l\}}\big)\Big)^{1/p}.\end{aligned} \tag{1.2.28}$$

Given $\epsilon>0$, partition the interval $[0,l]$ into $0=\lambda_0<\cdots<\lambda_N=l$ such that for each $1\le i\le l$ the length of the sub-interval $A_i=[\lambda_{i-1},\lambda_i]$ is less than ϵ. So we have

$$\begin{aligned}&\sum_{m=0}^{\infty}\frac{\theta^m}{m!}b_n^m\Big(\mathbb{E}\big(Y_n^m 1_{\{Y_n<l\}}\big)\Big)^{1/p}\\ &\le\sum_{i=1}^{N}\sum_{m=0}^{\infty}\frac{\theta^m}{m!}b_n^m\lambda_i^{m/p}\Big(\mathbb{P}\{Y_n\in A_i\}\Big)^{1/p}\\ &\le\sum_{i=1}^{N}\exp\Big\{\theta b_n(\lambda_{i-1}+\epsilon)\Big\}\Big(\mathbb{P}\{Y_n\ge\lambda_{i-1}\}\Big)^{1/p}.\end{aligned}$$

Consequently, by (1.2.25), we get

$$\begin{aligned}&\limsup_{n\to\infty}\frac{1}{b_n}\log\Big\{\sum_{m=0}^{\infty}\frac{\theta^m}{m!}b_n^m\Big(\mathbb{E}\big(Y_n^m 1_{\{Y_n<l\}}\big)\Big)^{1/p}\Big\}\\ &\le\theta\epsilon+\max_{1\le i\le N}\Big\{\theta\lambda_{i-1}^{1/p}-p^{-1}I(\lambda_{i-1})\Big\}\\ &\le\theta\epsilon+\sup_{\lambda>0}\Big\{\theta\lambda^{1/p}-p^{-1}I(\lambda)\Big\}.\end{aligned}$$

Letting $\epsilon\to 0^+$ on the right hand side, we get

$$\limsup_{n\to\infty}\frac{1}{b_n}\log\Big\{\sum_{m=0}^{\infty}\frac{\theta^m}{m!}b_n^m\Big(\mathbb{E}\big(Y_n^m 1_{\{Y_n<l\}}\big)\Big)^{1/p}\Big\}\le\sup_{\lambda>0}\Big\{\theta\lambda^{1/p}-p^{-1}I(\lambda)\Big\}.$$

By the decomposition (1.2.28), we get

$$\begin{aligned}&\limsup_{n\to\infty}\frac{1}{b_n}\log\Big\{\sum_{m=0}^{\infty}\frac{\theta^m}{m!}b_n^m\Big(\mathbb{E}Y_n^m\Big)^{1/p}\Big\}\\ &\le\max\Big\{\sup_{\lambda>0}\Big\{\theta\lambda^{1/p}-p^{-1}I(\lambda)\Big\},\\ &\limsup_{n\to\infty}\frac{1}{b_n}\log\Big\{\sum_{m=0}^{\infty}\frac{\theta^m}{m!}b_n^m\Big(\mathbb{E}\big(Y_n^m 1_{\{Y_n\ge l\}}\big)\Big)^{1/p}\Big\}\Big\}.\end{aligned}$$

In view of (1.2.25), letting $l\to\infty$ on the right hand side leads to (1.2.26). □

The condition (1.2.26) is called *uniform exponential integrability*. It can be examined through the following lemma.

LEMMA 1.2.10. *Given $\theta>0$, the assumption (1.2.26) holds if there is $\epsilon>0$ such that either one of the following happens:*

$$\limsup_{n\to\infty}\frac{1}{b_n}\log\mathbb{E}\exp\Big\{p(1+\epsilon)\theta b_n Y_n^{1/p}\Big\}<\infty,\tag{1.2.29}$$

$$\limsup_{n\to\infty}\frac{1}{b_n}\log\Big\{\sum_{m=0}^{\infty}\frac{(1+\epsilon)^m\theta^m}{m!}b_n^m\Big(\mathbb{E}Y_n^m\Big)^{1/p}\Big\}<\infty.\tag{1.2.30}$$

Proof. The conclusion follows from the second part of Lemma 1.2.6 and a standard application of Chebyshev inequality. □

1.3. LDP by sub-additivity

A random sequence $\{Y_n\}_{n\geq 1}$ is said to be *sub-additive*, if for any $n, m \geq 1$, there is a random variable Y_m' such that

$$Y_{n+m} \leq Y_n + Y_m',$$

$Y_m' \overset{d}{=} Y_m$ and Y_m' is independent of $\{Y_1, \cdots, Y_n\}$.

In particular, a *deterministic sub-additive sequence* $\{a(n)\}$ is defined by the inequality $a(n+m) \leq a(n) + a(m)$.

LEMMA 1.3.1. *For any deterministic sub-additive sequence* $\{a(n)\}$, *the equality*

$$\lim_{n\to\infty} n^{-1}a(n) = \inf_{m\geq 1} m^{-1}a(m)$$

holds in the extended real line $[-\infty, \infty)$.

Proof. All we need is to show that

$$\limsup_{n\to\infty} n^{-1}a(n) \leq \inf_{m\geq 1} m^{-1}a(m). \tag{1.3.1}$$

Let $m \geq 1$ be fixed but arbitrary. For any big n, write $n = km + r$, where $k \geq 1$ and $0 \leq r < m$ are integers. By sub-additivity,

$$a(n) \leq ka(m) + a(r).$$

Consequently,

$$\limsup_{n\to\infty} n^{-1}a(n) \leq m^{-1}a(m).$$

Taking infimum over m on the right hand side leads to (1.3.1). □

Given a sub-additive random sequence $\{Y_n\}_{n\geq 1}$ and $\theta > 0$, we have that

$$\mathbb{E}\exp\{\theta Y_{n+m}\} \leq \mathbb{E}\exp\{\theta Y_n\}\mathbb{E}\exp\{\theta Y_m\}.$$

Consequently,

$$a(n) \equiv \log\mathbb{E}\exp\{\theta Y_n\} \qquad n = 1, 2, \cdots$$

is a deterministic sub-additive sequence. Consequently, we have the following corollary.

COROLLARY 1.3.2. *Let* $\{Y_n\}_{n\geq 1}$ *be sub-additive. For any* $\theta > 0$, *the equality*

$$\lim_{n\to\infty} \frac{1}{n}\log\mathbb{E}\exp\{\theta Y_n\} = \inf_{m\geq 1} \frac{1}{m}\log\mathbb{E}\exp\{\theta Y_m\}$$

holds in the extended real line $[-\infty, \infty]$.

In the following theorem, we establish exponential integrability for Y_n.

THEOREM 1.3.3. *Let* $\{Y_n\}$ *be a sub-additive random sequence such that* $Y_1 \leq C$ *a.s. for some constant* $C > 0$. *Let* $c_n \geq 1$ *be a deterministic sequence such that the normalized sequence*

$$\max_{k\leq n} Y_k/c_n \qquad n = 1, 2, \cdots$$

is stochastically bounded. Then there is a $\theta > 0$ *such that*

$$\sup_n \mathbb{E}\exp\Big\{\theta Y_n/c_n\Big\} < \infty. \tag{1.3.2}$$

Furthermore, if

$$\liminf_{n\to\infty} c_{mn}/c_n > 1 \tag{1.3.3}$$

for some $m > 1$*, then (1.3.2) holds for all* $\theta > 0$.

Proof. We first show that for any $s, t > 0$ and $n \ge 1$,

$$\max_{1\le k\le n} \mathbb{P}\Big\{Y_k \ge s + t + C\Big\} \le \mathbb{P}\Big\{\max_{1\le k\le n} Y_k \ge s\Big\}\max_{1\le k\le n}\mathbb{P}\Big\{Y_k \ge t\Big\}. \tag{1.3.4}$$

Set

$$\mathcal{T}_s = \inf\Big\{j \ge 1;\quad Y_j \ge s\Big\}.$$

Notice that for any $j \ge 1$,

$$Y_j - Y_{j-1} \le Y_1' \stackrel{d}{=} Y_1 \le C \qquad a.s.$$

For any $1 \le k \le n$,

$$\mathbb{P}\Big\{Y_k \ge s + t + C\Big\} = \mathbb{P}\Big\{\mathcal{T}_s \le k-1;\ Y_k \ge s+t+C\Big\}$$
$$\le \sum_{j=1}^{k-1}\mathbb{P}\Big\{\mathcal{T}_s = j,\ Y_k - Y_{j-1} \ge t + C\Big\} \le \sum_{j=1}^{k-1}\mathbb{P}\Big\{\mathcal{T}_s = j,\ Y_k - Y_j \ge t\Big\}$$

where the second step follows from the fact that $Y_{\mathcal{T}_s - 1} \le s$. By sub-additivity, $Y_k - Y_j \le Y_{k-j}' \stackrel{d}{=} Y_{k-j}$ and Y_{k-j}' is independent of $\{\mathcal{T}_s = j\}$. Therefore,

$$\mathbb{P}\Big\{Y_k \ge s+t+C\Big\} \le \sum_{j=1}^{k-1}\mathbb{P}\{\mathcal{T}_s = j\}\mathbb{P}\{Y_{k-j}\ge t\}$$
$$\le \mathbb{P}\{\mathcal{T}_s \le n\}\max_{1\le k\le n}\mathbb{P}\Big\{Y_k \ge t\Big\} = \mathbb{P}\Big\{\max_{1\le k\le n} Y_k \ge s\Big\}\max_{1\le k\le n}\mathbb{P}\Big\{Y_k \ge t\Big\}.$$

Taking maximum over k on the left hand side leads to (1.3.4).

For any $N > 0$ and integers $m, n \ge 1$, by (1.3.4),

$$\max_{1\le k\le n}\mathbb{P}\Big\{Y_k \ge m(Nc_n + C)\Big\} \le \bigg(\mathbb{P}\Big\{\max_{1\le k\le n} Y_k \ge Nc_n\Big\}\bigg)^m.$$

In particular,

$$\mathbb{P}\Big\{Y_n \ge m(Nc_n + C)\Big\} \le \bigg(\mathbb{P}\Big\{\max_{1\le k\le n} Y_k \ge Nc_n\Big\}\bigg)^m.$$

By assumption, one can make $N > 0$ sufficiently large, so that

$$\mathbb{P}\Big\{\max_{1\le k\le n} Y_k \ge Nc_n\Big\} \le e^{-2} \qquad n = 1, 2, \cdots.$$

Therefore,

$$\sup_{n\ge 1}\mathbb{P}\Big\{Y_n \ge m(Nc_n + C)\Big\} \le e^{-2m} \qquad m = 1, 2, \cdots.$$

Consequently, (1.3.2) holds for some $\theta > 0$.

Assuming (1.3.3), we now show that (1.3.2) holds for every $\theta > 0$. Let $\theta_0 > 0$ satisfy (1.3.2). By (1.3.3) there is $m > 1$ such that

$$c_{[n/m]}/c_n \le \theta_0/\theta$$

for all $n \ge 1$. By sub-additivity we have that

$$\mathbb{E}\exp\left\{\theta c_n^{-1}Y_n\right\} \le \left(\mathbb{E}\exp\left\{\theta c_n^{-1}Y_{[n/m]}\right\}\right)^m.$$

By the choice of m and by Hölder inequality, the right hand side is bounded uniformly over $n \ge 1$. □

The notion of sub-additivity can be extended to the stochastic processes with continuous time. A stochastic process Z_t $(t \ge 0)$ is said to be *sub-additive*, if for any $s, t \ge 0$, $Z_{s+t} \le Z_s + \widetilde{Z}_t$ for a random variable $\widetilde{Z}_t$ independent of $\{Z_u;\ \ 0 \le u \le s\}$ with $\widetilde{Z}_t \overset{d}{=} Z_t$. With a completely parallel argument we have

LEMMA 1.3.4. *For any deterministic sub-additive function* $\{a(t)\}$ $(t \in \mathbb{R}^+)$, *the equality*

$$\lim_{t\to\infty} t^{-1}a(t) = \inf_{s>0} s^{-1}a(s)$$

holds in the extended real line $[-\infty, \infty)$.

We may restate Theorem 1.3.3 in the setting of continuous time. Instead, we give the following slightly different version.

THEOREM 1.3.5. *For any non-decreasing sub-additive process Z_t with continuous path and with $Z_0 = 0$,*

$$\mathbb{E}\exp\{\theta Z_t\} < \infty \qquad \forall \theta, t > 0. \tag{1.3.5}$$

Consequently,

$$\lim_{t\to\infty}\frac{1}{t}\log\mathbb{E}\exp\left\{\theta Z_t\right\} = \Psi(\theta) \tag{1.3.6}$$

exists with $0 \le \Psi(\theta) < \infty$ for every $\theta > 0$.

Proof. Clearly, we need only to establish (1.3.5). By sample path continuity and by monotonicity, an argument used in achieving (1.3.4) leads to

$$\mathbb{P}\{Z_t \ge a + b\} \le \mathbb{P}\{Z_t \ge a\}\mathbb{P}\{Z_t \ge b\} \tag{1.3.7}$$

holds for any $a, b > 0$.

Consequently,

$$\mathbb{P}\{Z_t \ge ma\} \le \left(\mathbb{P}\{Z_t \ge a\}\right)^m \qquad m = 1, 2, \cdots.$$

For any $a > 0$, by the fact that $Z_0 = 0$ and by sample path continuity, one can have $\mathbb{P}\{Z_t \ge a\} \le e^{-2}$ by making $t > 0$ sufficiently small. Consequently, for any $\theta > 0$,

$$\mathbb{E}\exp\{\theta Z_t\} < \infty$$

for small t. This can be extended to arbitary $t > 0$ by sub-additivity. □

1.4. Notes and comments

Section 1.1.

The earliest recorded work in large deviation theory is due to Cramér ([**41**]) and was published in 1938. The literature on large deviations is massive and it is impossible to list even a small portion of it. We point to the fundamental roles played by Donsker and Varadhan, and Freidlin and Wentzell in the birth of the modern theory of large deviations. The idea that the limit of the logarithmic moment generating function decides the large deviation goes back to Cramér ([**41**]). It has been formulated by Gärtner ([**84**]) and Ellis ([**73**]) into a general theorem later known as the Gärtner-Ellis theorem (Theorem 1.1.4). There are many excellent book accounts available in the theory of large deviations. We mention here the books by Varadhan [**159**], Freidlin and Wentzell [**80**], Ellis [**74**], Stroock [**156**], Deuschel and Stroock [**53**], Bucklew [**20**], Dembo and Zeitouni [**47**], den Hollander [**97**], Feng and Kurtz [**77**]. Finally, we refer an interested reader to the recent survey by Varadhan [**161**] for an overview on the latest development in the area of large deviations.

Most of the material in this section comes from the book by Dembo and Zeitouni ([**47**]).

For the large deviations in infinite dimensional space, a challenging part is to establish the exponential tightness. Theorem 1.1.7 provides a practical way of examining the exponential tightness. This useful result is due to de Acosta ([**1**]).

EXERCISE 1.4.1. *Let $\{Y_n\}$ and $\{Z_n\}$ are two sequences of real random variables such that*

$$\lim_{n\to\infty}\frac{1}{b_n}\log\mathbb{P}\{|Y_n - Z_n| \ge \epsilon\} = -\infty.$$

Show that if Y_n obeys the large deviation principle with the scale b_n and the good rate function $I(\cdot)$, then the same large deviation principle holds for Z_n.

EXERCISE 1.4.2. *Recall that a Poisson process N_t is a stochastic process taking non-integer values such that*

(1) $N_0 = 0$,
(2) *For any $t > 0$,*

$$\mathbb{P}\Big\{N_t = m\Big\} = e^{-t}\frac{t^m}{m!} \qquad m = 0, 1, \cdots,$$

(3) *For any $s, t \in \mathbb{R}^+$ with $s < t$, $N_t - N_s$ is independent of $\{N_u;\ \ 0 \le u \le s\}$ and has same distribution as N_{t-s}.*

Prove the following LDP: For any closed set $F \subset \mathbb{R}$ and open set $G \in \mathbb{R}$,

$$\limsup_{t\to\infty}\frac{1}{t}\log\mathbb{P}\Big\{N_t/t \in F\Big\} \le -\inf_{\lambda\in F} I(\lambda),$$

$$\liminf_{t\to\infty}\frac{1}{t}\log\mathbb{P}\Big\{N_t/t \in G\Big\} \ge -\inf_{\lambda\in G} I(\lambda)$$

where

$$I(\lambda)=\begin{cases}\lambda\log\lambda-\lambda+1 & \lambda\ge 0,\\ \\ +\infty & \lambda<0.\end{cases}$$

In particular, compute the limit

$$\lim_{t\to\infty}\frac{1}{t}\log\mathbb{P}\Big\{N_t/t\ge\lambda\Big\}$$

for every $\lambda>0$.

Comment. Exercise 1.4.2 serves as a counterexample at an attempt to weaken the conditions assumed in Theorem 1.2.1 and Theorem 1.2.4. In this example, the LDP given by (1.2.1)-(1.2.2) and the LDP defined by (1.2.3) have different rate functions, and none of them are strictly increasing on $\mathbb{R}^+$.

EXERCISE 1.4.3. *Prove the upper bound of Varadhan's integral lemma given in part (2) of Theorem 1.1.6.*

Section 1.2.

Much material in this section exists in some recent research papers instead of standard textbooks. Theorem 1.2.2 was essentially obtained in the paper by Bass, Chen and Rosen ([**8**]), Theorem 1.2.7 appeared in Chen ([**27**], [**28**]). Theorem 1.2.8 is due to König and Mörters ([**114**]). A weaker version of Theorem 1.2.8 is also established in [**114**]). We put it into the following exercise.

EXERCISE 1.4.4. *Let* $Y\ge 0$ *be a random variable such that*

$$\limsup_{m\to\infty}\frac{1}{m}\log\Big(\frac{1}{(m!)^\gamma}\mathbb{E}Y^m\Big)\le\kappa$$

for some $\gamma>0$ *and* $\kappa\in\mathbb{R}$. *Prove directly that*

$$\limsup_{t\to\infty}t^{-1/\gamma}\log\mathbb{P}\{Y\ge t\}\le-\gamma e^{-\kappa/\gamma}.$$

Hint: You may use Chebyshev inequality and Stirling formula.

We now consider the asymptotic behavior of the probability

$$\mathbb{P}\{Y\le\epsilon\}\qquad(\epsilon\to 0^+).$$

The study in this direction is often referred to as the problem on *small ball probability,* since very often $Y=||X||$ for some variable X taking values in a Banach space. We refer to the survey paper by Li and Shao ([**133**]) for an overview in this area. Generally speaking, large deviation principle and small ball probability deal with drastically different situations. One is for the probability that the random variables take large values and another is for the probability that the random variables take small values. In the following exercise, we seek a small ball version of the Gärtner-Ellis theorem.

EXERCISE 1.4.5 (hard). *Let $Y \geq 0$ be a random variable such that*

$$\lim_{t\to\infty} \frac{1}{t} \log \mathbb{E} \exp\Big\{ - t^{\gamma} Y \Big\} = -\kappa$$

for some $\gamma > 1$ and $\kappa > 0$. Prove that

$$\lim_{\epsilon\to 0^+} \epsilon^{\frac{1}{\gamma-1}} \log \mathbb{P}\{Y \leq \epsilon\} = -(\gamma - 1)\gamma^{-\frac{\gamma}{\gamma-1}} \kappa^{\frac{\gamma}{\gamma-1}}. \tag{1.4.1}$$

This result is known as exponential Tauberian theorem (see Theorem 3.5, [**133**] *for a more general form).*

Hint: Use Chebyshev inequality to get the upper bound.

The lower bound can be proved in a way similar to some argument used in the proof of Theorem 1.1.4. In particular, you may need to prove that for any interval $I \subset \mathbb{R}^+$ and $\theta > 0$,

$$\limsup_{\epsilon\to 0^+} \epsilon^{\frac{1}{\gamma-1}} \log \mathbb{E}\Big(\exp\Big\{ - \theta \epsilon^{-\frac{\gamma}{\gamma-1}} Y \Big\} 1_{\{Y\in \epsilon I\}} \Big)$$
$$\leq - \inf_{\lambda \in I} \Big\{ \theta\lambda + (\gamma - 1)\gamma^{-\frac{\gamma}{\gamma-1}} \kappa^{\frac{\gamma}{\gamma-1}} \lambda^{-\frac{1}{\gamma-1}} \Big\}.$$

Section 1.3.

Argument by sub-additivity has become a sophisticated tool in the general frame work of large deviations. Very often in literature, it is the deterministic sub-additivity formulated in Lemma 1.3.1 that is used to prove the existence of the logarithmic moment generating function.

EXERCISE 1.4.6. *Prove (1.3.7).*

CHAPTER 2

Brownian intersection local times

2.1. Introduction

Recall that a d-dimensional *Brownian motion* $W(t)$ is defined as a stochastic process in $\mathbb{R}^d$ with the following properties:

(1) For any $t > 0$, $W(t)$ is a normal random variable with mean $\mathbf{0}$ and covariance matrix $t\mathbf{I}_d$, where $\mathbf{I}_d$ is the $d \times d$ identity matrix.

(2) For any $s < t$, the increment $W(t) - W(s)$ is independent of the family

$$\Big\{W(u); \quad u \le s\Big\}$$

and has a distribution the same as $W(t-s)$.

(3) The random path $t \mapsto W(t)$ is continuous on $t \in \mathbb{R}^+$.

We follow the convention that $W(0) = 0$ most of the time, with the exception when $W(t)$ is viewed as a Markov process. Brownian motion is also known as Wiener process. Due to its importance, Brownian motion has been studied extensively. It is virtually impossible, in the scope of this book, to list even the major results in literature about Brownian motion. We refer the reader to the books of Ito and McKean [**100**], Revuz and Yor [**141**], Rogers and Williams [**144**], [**145**] .

The notion of *intersection local times* has been introduced to measure the intensity of the Brownian path intersections. For a d-dimensional Brownian motion $W(t)$, an integer $p \ge 2$ and a p-multiple time set

$$A \subset \Big\{(t_1, \cdots, t_p) \in (\mathbb{R}^+)^p; \quad t_1 < \cdots < t_p\Big\}$$

the random quantity $\beta(A)$ symbolically written as

$$\beta(A) = \int_A \prod_{j=2}^p \delta_0\big(W(s_j) - W(s_{j-1})\big) ds_1 \cdots ds_p \tag{2.1.1}$$

measures the amount of p-multiple self-intersections over the time set A. Here and everywhere in the book, $\delta_0(\cdot)$ represents *Dirac function*[1] defined as a (generalized) probability density with the property that

$$\int_{\mathbb{R}^d} f(x)\delta_0(x)dx = f(0)$$

for every bounded and continuous function $f(\cdot)$ on $\mathbb{R}^d$. For any $x \in \mathbb{R}^d$, $\delta_x(y) = \delta_0(y - x)$.

[1]Throughout the book, the Dirac function is used only for symbolic expression.

The random measure $\beta(A)$ is called the (p-multiple) self-intersection local time of the Brownian motion $W(t)$.

Let $W_1(t), \cdots, W_p(t)$ be independent d-dimensional Brownian motions. For any $A \subset (\mathbb{R}^+)^p$, the random quantity heuristically written as

$$\alpha(A) = \int_A \prod_{j=2}^{p} \delta_0\big(W_{j-1}(s_{j-1}) - W_j(s_j)\big)ds_1 \cdots ds_p \tag{2.1.2}$$

measures the amount of mutual intersection over the time set A. In literature, $\alpha\big([0,t_1] \times \cdots \times [0,t_p]\big)$ is called (p-multiple) mutual intersection local time of the Brownian motions $W_1(t), \cdots, W_p(t)$.

In addition to its mathematical importance, the study of intersection local times is also motivated by the needs from physics. Physicists are concerned about the geometric shape of the polymer which is often described by a suitable random path (such as a Brownian curve). The geometry of a polymer is decided by the intensity that the random path intersects itself. Write $[0,t]^2_< = \{(r,s);\ \ 0 \le r \le s \le t\}$. Then the quantity

$$\beta\big([0,t]^2_<\big) = \iint_{\{0 \le r < s \le t\}} \delta_0\big(W(r) - W(s)\big)drds$$

measures (double) self-intersection up to time t. The case when $\beta\big([0,t]^2_<\big)$ is large corresponds to a "contracting" polymer; while the case when $\beta\big([0,t]^2_<\big)$ is small corresponds to a "spread-out" polymer. The geometric shape of a polymer is often influenced by the environment (media). If the environment encourages attraction among the molecules, then the polymer is contracting. In this case, the polymer is called *self-attracting polymer.* In the opposite case, the polymer is spread-out and is called *self-repelling polymer.*

For example, the probability measure $\widehat{\mathbb{P}}_\lambda$ on $C\big\{[0,t], \mathbb{R}^2\big\}$ defined as

$$\widehat{\mathbb{P}}_\lambda(B) = \widehat{C}_\lambda^{-1}\mathbb{E}\bigg(\exp\Big\{\lambda\beta\big([0,t]^2_<\big)\Big\}1_{\{W(\cdot)\in B\}}\bigg), \tag{2.1.3}$$

where $B \in C\big\{[0,t], \mathbb{R}^2\big\}$ is Borel measurable, is regarded as the distribution of the curve representing a two-dimensional self-attracting planar polymer, where $\lambda > 0$ represents certain media parameters (such as temperature) and $\widehat{C}_\lambda$ is the normalizing constant making $\widehat{\mathbb{P}}_\lambda$ a probability measure. Indeed, the curves with higher self-intersection are given more distributional weight and the degree of this favoritism is decided by the value of λ.

Similarly, a two-dimensional self-repelling planar polymer is modeled by the distribution

$$\widetilde{\mathbb{P}}_\lambda(B) = \widetilde{C}_\lambda^{-1}\mathbb{E}\bigg(\exp\Big\{ - \lambda\beta\big([0,t]^2_<\big)\Big\}1_{\{W(\cdot)\in B\}}\bigg). \tag{2.1.4}$$

An interested reader is referred to an excellent survey paper by van der Hofstad and König ([**94**]) for a systematic account on a link between polymers and sample path intersections.

Intersection local times of Brownian motions need to be properly defined. For example, we shall soon find out that (Proposition 2.3.6) the quantity $\beta\big([0,t]^2_<\big)$

appearing in (2.1.3) and (2.1.4) blows up and we shall fix this problem in Section 2.4. In addition, some basic but important properties of intersection local times, such as integrability and Le Gall's moment formula, need to be installed. We devote this chapter to carry out these goals.

2.2. Mutual intersection local time

Prior to the construction of intersection local time $\alpha(A)$ (formally given in (2.1.2)), we need to see when the independent d-dimensional Brownian motions $W_1(t), \cdots, W_p(t)$ intersect. This problem was completely solved in 1950s by Dvoretzsky, Erdös and Kakutani ([**66**] and [**67**]). We introduce their result without proof.

THEOREM 2.2.1. *The set of intersection*

$$\bigcap_{j=1}^{p} \{x \in \mathbb{R}^d;\quad x = W_j(t) \text{ for some } \ t \geq 0\}$$

contains points different from 0 if and only if $p(d-2) < d$.

In the following discussion we restrict to the case $p(d-2) < d$, which contains three sub-cases:

1. $d = 1$, $p = 2, 3, \cdots$.
2. $d = 2$, $p = 2, 3, \cdots$.
3. $d = 3$, $p = 2$.

A function playing an important role in deciding whether Brownian paths intersect is *Green's function* (killed at exponential time) $G(x)$ defined as

$$G(x) = \int_0^\infty e^{-t} p_t(x) dt \qquad x \in \mathbb{R}^d \tag{2.2.1}$$

where

$$p_t(x) = \frac{1}{(2\pi t)^{d/2}} e^{-|x|^2/2t} \qquad x \in \mathbb{R}^d \tag{2.2.2}$$

is the density function of the random variable $W(t)$. A crucial fact (Theorem A.1, Appendix) is that

$$\int_{\mathbb{R}^d} G^p(x) dx < \infty \tag{2.2.3}$$

under $p(d-2) < d$.

Given $\epsilon > 0$, define a random measure $\alpha_\epsilon(\cdot)$ on $(\mathbb{R}^+)^p$ as follows:

$$\alpha_\epsilon(A) = \int_{\mathbb{R}^d} \left[\int_A ds_1 \cdots ds_p \prod_{j=1}^p p_\epsilon\big(W_j(s_j) - x\big) \right] dx \qquad A \subset (\mathbb{R}^+)^p \tag{2.2.4}$$

where $A \subset (\mathbb{R}^+)^p$.

Roughly speaking, we shall show that as $\epsilon \to 0^+$, $\alpha_\epsilon(\cdot)$ converges on all bounded p-dimensional boxes and we then show that the limiting family can be extended into a random measure which is later called the mutual intersection local time generated by the Brownian motions $W_1(t), \cdots, W_p(t)$.

For two functions f and g on $\mathbb{R}^d$, the *convolution* $f * g$ is defined as

$$(f * g)(x) = \int_{\mathbb{R}^d} f(y)g(x-y)dy = \int_{\mathbb{R}^d} f(x-y)g(y)dy \qquad x \in \mathbb{R}^d$$

whenever the involved integrals are well defined. For any probability density $h(x)$ on $\mathbb{R}^d$ and any $\epsilon > 0$, it is easy to see that the function

$$h_\epsilon(x) = \epsilon^{-d} h(\epsilon^{-1}x)$$

is a probability density on $\mathbb{R}^d$. The probability density function $h_\epsilon(x) \equiv \epsilon^{-d} h(\epsilon^{-1}x)$ converges to the Dirac function $\delta_0(x)$ in the sense that

$$\lim_{\epsilon \to 0^+} \int_{\mathbb{R}^d} f(x) h_\epsilon(x) dx = f(0)$$

for every bounded and continuous function $f(\cdot)$ on $\mathbb{R}^d$.

LEMMA 2.2.2. *For any function $f \in \mathcal{L}^p(\mathbb{R}^d)$ with $p \geq 1$, and for any probability density $h(x)$ on $\mathbb{R}^d$,*

$$\lim_{\epsilon \to 0^+} \int_{\mathbb{R}^d} |f(x) - f * h_\epsilon(x)|^p dx = 0.$$

Proof. First we claim that for $\delta > 0$ there exists $N > 0$ such that

$$\limsup_{\epsilon \to 0^+} \int_{\{|x| \geq N\}} |f * h_\epsilon(x)|^p dx < \delta. \tag{2.2.5}$$

Indeed, by Jensen inequality,

$$\begin{aligned}
&\int_{\{|x|\geq N\}} |f * h_\epsilon(x)|^p dx \\
&\leq \int_{\{|x|\geq N\}} \int_{\mathbb{R}^d} h_\epsilon(y) |f(x-y)|^p dy dx \\
&= \int_{\mathbb{R}^d} h_\epsilon(y) \left[\int_{\{|x|\geq N\}} |f(x-y)|^p dx \right] dy \\
&= \int_{\{|y|\geq 1\}} h_\epsilon(y) \left[\int_{\{|x|\geq N\}} |f(x-y)|^p dx \right] dy \\
&+ \int_{\{|y|<1\}} h_\epsilon(y) \left[\int_{\{|x|\geq N\}} |f(x-y)|^p dx \right] dy.
\end{aligned}$$

For the first term on the right hand side, it is bounded by the quantity

$$\begin{aligned}
&\left\{ \int_{\mathbb{R}^d} |f(x)|^p dx \right\} \left\{ \int_{\{|y|\geq 1\}} h_\epsilon(y) dy \right\} \\
&= \left\{ \int_{\mathbb{R}^d} |f(x)|^p dx \right\} \left\{ \int_{\{|x|\geq \epsilon^{-1}\}} h(x) dx \right\},
\end{aligned}$$

which tends to zero as $\epsilon \to 0^+$.

As for the second term, it is dominated by

$$\int_{\{|x|\geq N-1\}} |f(x)|^p dx,$$

which can be sufficiently small if N is sufficiently large.

In view of (2.2.5), therefore, it remains to show that for any fixed $N > 0$, we get

$$\lim_{\epsilon\to 0+}\int_{\{|x|\le N\}}|f(x) - f * h_\epsilon(x)|^p dx = 0. \tag{2.2.6}$$

The validity of this claim would be immediate if f is uniformly continuous. When this is not the case, for any given $\delta > 0$ one can pick up a uniformly continuous function $\bar{f}$ on $\mathbb{R}^d$ such that

$$\int_{\mathbb{R}^d}|f(x) - \bar{f}(x)|^p dx < \delta.$$

By Jensen inequality, one can show that

$$\int_{\mathbb{R}^d}|(f * h_\epsilon)(x) - (\bar{f} * h_\epsilon)(x)|^p dx \le \int_{\mathbb{R}^d}|f(x) - \bar{f}(x)|^p dx < \delta.$$

Thus, the desired (2.2.6) follows from the fact that it is satisfied by $\bar{f}$. □

For any $A \subset \mathbb{R}^+$ and any integer $m \ge 2$, set

$$A_<^m = \big\{(s_1, \cdots, s_m) \in A^m;\quad s_1 < \cdots < s_m\big\}.$$

THEOREM 2.2.3. *For any bounded Borel sets $A_1, \cdots, A_p \subset \mathbb{R}^+$, the limit*

$$\lim_{\epsilon\to 0+}\alpha_\epsilon(A_1\times\cdots\times A_p) = \alpha(A_1\times\cdots\times A_p) \tag{2.2.7}$$

exists in $\mathcal{L}^m(\Omega, \mathcal{A}, \mathbb{P})$ for all $m > 0$. Further, for any integer $m \ge 1$,

$$\begin{aligned} &\mathbb{E}\Big[\alpha\big(A_1\times\cdots\times A_p\big)^m\Big] \\ &= \int_{(\mathbb{R}^d)^m} dx_1\cdots dx_m \prod_{j=1}^p \sum_{\sigma\in\Sigma_m}\int_{(A_j)_<^m} ds_1\cdots ds_m \\ &\qquad\times\prod_{k=1}^m p_{s_k - s_{k-1}}(x_{\sigma(k)} - x_{\sigma(k-1)}) \end{aligned} \tag{2.2.8}$$

where Σ_m is the permutation group over $\{1, \cdots, m\}$ and we adopt the convention that $s_0 = 0$ and $x_{\sigma(0)} = 0$.

Proof. The proof consists of three steps. First we show that the right hand side of (2.2.8) is finite. Set, for a given $t \in \mathbb{R}^+$,

$$V_t = \int_{(\mathbb{R}^d)^m} dx_1\cdots dx_m\bigg[\sum_{\sigma\in\Sigma_m}\int_{[0,t]_<^m} ds_1\cdots ds_m \prod_{k=1}^m p_{s_k - s_{k-1}}(x_{\sigma(k)} - x_{\sigma(k-1)})\bigg]^p.$$

By Hölder inequality and by the fact that A_j is bounded, we need only to prove that $V_t < \infty$ for any $t > 0$.

Indeed, for any $\sigma \in \Sigma_m$,

$$\begin{aligned}
&\int_{[0,t]^m_<} ds_1 \cdots ds_m \prod_{k=1}^m p_{s_k - s_{k-1}}(x_{\sigma(k)} - x_{\sigma(k-1)})\\
&\le \prod_{k=1}^m \int_0^t p_s(x_{\sigma(k)} - x_{\sigma(k-1)})ds\\
&= \prod_{k=1}^m g_t(x_{\sigma(k)} - x_{\sigma(k-1)}) \qquad \text{(say)}.
\end{aligned}$$

By Jensen inequality, therefore, we have

$$\begin{aligned}
(2.2.9) \qquad V_t &\le (m!)^{p-1} \sum_{\sigma \in \Sigma_m} \int_{(\mathbb{R}^d)^m} dx_1 \cdots dx_m \left[\prod_{k=1}^m g_t(x_{\sigma(k)} - x_{\sigma(k-1)}) \right]^p\\
&= (m!)^p \left(\int_{\mathbb{R}^d} g_t^p(x) dx \right)^m = (m!)^p \left(\int_{\mathbb{R}^d} \left[\int_0^t p_s(x) ds \right]^p dx \right)^m.
\end{aligned}$$

To proceed, we need to show that the integral on the right hand side is finite. Indeed, by Gaussian integration

$$\begin{aligned}
&\int_{\mathbb{R}^d} \left[\int_0^t p_s(x) ds \right]^p dx\\
&= \int_{[0,t]^p} ds_1 \cdots ds_p \int_{\mathbb{R}^d} \prod_{j=1}^p p_{s_j}(x) dx\\
&= (2\pi)^{-\frac{d(p-1)}{2}} \int_{[0,t]^p} ds_1 \cdots ds_p \left(\sum_{j=1}^p \prod_{\substack{1 \le l \le p\\ l \ne j}} s_l \right)^{-d/2}.
\end{aligned}$$

According to arithmetic-geometric mean inequality,

$$\frac{1}{p} \sum_{j=1}^p \prod_{\substack{1 \le l \le p\\ l \ne j}} s_l \ge \sqrt[p]{\prod_{j=1}^p \prod_{\substack{1 \le l \le p\\ l \ne j}} s_l} = \prod_{j=1}^p s_j^{\frac{p-1}{p}}.$$

In view of the fact that $\dfrac{d(p-1)}{2p} < 1$, we have

$$(2.2.10) \qquad \int_{\mathbb{R}^d} \left[\int_0^t p_s(x) ds \right]^p dx \le (2\pi)^{-\frac{d(p-1)}{2}} \left\{ \int_0^t s^{-\frac{d(p-1)}{2p}} ds \right\}^p < \infty.$$

As the second step, we establish the identity

$$\begin{aligned}
(2.2.11) \qquad &\mathbb{E}\left[\alpha_\epsilon (A_1 \times \cdots \times A_p)^m \right]\\
&= \int_{(\mathbb{R}^d)^m} dx_1 \cdots dx_m \prod_{j=1}^p \int_{(\mathbb{R}^d)^m} dz_1 \cdots dz_m \left(\prod_{k=1}^m p_\epsilon(z_k - x_k) \right)\\
&\times \sum_{\sigma \in \Sigma_m} \int_{(A_j)^m_<} ds_1 \cdots ds_m \prod_{k=1}^m p_{s_k - s_{k-1}}(z_{\sigma(k)} - z_{\sigma(k-1)}).
\end{aligned}$$

By Fubini theorem, we get

$$\mathbb{E}\Big[\alpha_\epsilon\big(A_1\times\cdots\times A_p\big)^m\Big]$$
$$=\int_{(\mathbb{R}^d)^m}dx_1\cdots dx_m\prod_{j=1}^p\int_{A_j^m}ds_1\cdots ds_m\mathbb{E}\prod_{k=1}^m p_\epsilon\big(W(s_k)-x_k\big).$$

For each $1\le j\le p$, write

$$A_j^m=\bigcup_{\sigma\in\Sigma_m}(A_j)_<^m\circ\sigma+\mathcal{N}$$

where $(A_j)_<^m\circ\sigma=\big\{(s_1,\cdots,s_m)\in A_j^m;\ s_{\sigma(1)}<\cdots<s_{\sigma(m)}\big\}$ and $\mathcal{N}\subset A_j^m$ is a null set (with respect to Lebesgue measure on $(\mathbb{R}^+)^m$).

Observe that for any $\sigma\in\Sigma_m$, we get

$$\prod_{k=1}^m p_\epsilon\big(W(s_k)-x_k\big)=\prod_{k=1}^m p_\epsilon\big(W(s_{\sigma(k)})-x_{\sigma(k)}\big).$$

Thus,

$$\int_{A_j^m}ds_1\cdots ds_m\mathbb{E}\prod_{k=1}^m p_\epsilon\big(W(s_k)-x_k\big)$$
$$=\sum_{\sigma\in\Sigma_m}\int_{(A_j)_<^m\circ\sigma}ds_1\cdots ds_m\mathbb{E}\prod_{k=1}^m p_\epsilon\big(W(s_{\sigma(k)})-x_{\sigma(k)}\big)$$
$$=\sum_{\sigma\in\Sigma_m}\int_{(A_j)_<^m}ds_1\cdots ds_m\mathbb{E}\prod_{k=1}^m p_\epsilon\big(W(s_k)-x_{\sigma(k)}\big).$$

where the last equality follows from re-indexing the time variables.

For any $(s_1,\cdots,s_m)\in(A_j)_<^m$, the probability density of the random vector $\big(W(s_1),\cdots,W(s_m)\big)$ is

$$\prod_{k=1}^m p_{s_k-s_{k-1}}(z_k-z_{k-1})\qquad z_1,\cdots,z_m\in\mathbb{R}^d$$

where we adopt the convention that $s_0=0$ and $z_0=0$. Thus, (2.2.11) follows from the following computation:

$$\mathbb{E}\prod_{k=1}^m p_\epsilon\big(W(s_k)-x_{\sigma(k)}\big)$$
$$=\int_{(\mathbb{R}^d)^m}dz_1\cdots dz_m\Big(\prod_{k=1}^m p_\epsilon(z_k-x_{\sigma(k)})\Big)\Big(\prod_{k=1}^m p_{s_k-s_{k-1}}(z_k-z_{k-1})\Big)$$
$$=\int_{(\mathbb{R}^d)^m}dz_1\cdots dz_m\Big(\prod_{k=1}^m p_\epsilon(z_{\sigma(k)}-x_{\sigma(k)})\Big)\Big(\prod_{k=1}^m p_{s_k-s_{k-1}}(z_{\sigma(k)}-z_{\sigma(k-1)})\Big)$$
$$=\int_{(\mathbb{R}^d)^m}dz_1\cdots dz_m\Big(\prod_{k=1}^m p_\epsilon(z_k-x_k)\Big)\Big(\prod_{k=1}^m p_{s_k-s_{k-1}}(z_{\sigma(k)}-z_{\sigma(k-1)})\Big)$$

where the second equality follows from index permutation and the third from the fact that

$$\prod_{k=1}^{m} p_\epsilon\big(z_{\sigma(k)} - x_{\sigma(k)}\big) = \prod_{k=1}^{m} p_\epsilon\big(z_k - x_k\big).$$

Before starting the third step, we reduce our task to a suitable point. For $1 \le j \le p$, set

$$f_j(x_1, \cdots, x_m) = \sum_{\sigma\in\Sigma_m} \int_{(A_j)^m_<} ds_1 \cdots ds_m \prod_{k=1}^{m} p_{s_k - s_{k-1}}\big(x_{\sigma(k)} - x_{\sigma(k-1)}\big)$$

and

$$f_{\epsilon,j}(x_1, \cdots, x_m) = \int_{(\mathbb{R}^d)^m} dz_1 \cdots dz_m \Big(\prod_{k=1}^{m} p_\epsilon(z_k - x_k)\Big) f_j(z_1, \cdots, z_m).$$

By (2.2.9) $f_j \in \mathcal{L}^p(\mathbb{R}^{dm})$. By Lemma 2.2.2, $f_{\epsilon,j}$ converges to f_j in $\mathcal{L}^p(\mathbb{R}^{dm})$ as $\epsilon \to 0^+$. Observe that the multilinear form

$$L(g_1, \cdots, g_p) = \int_{(\mathbb{R}^d)^m} dx_1 \cdots dx_m \prod_{j=1}^{p} g_j(x_1, \cdots x_m) \qquad g_j \in \mathcal{L}^p(\mathbb{R}^{dm})$$

is continuous on the product space $\bigotimes_{j=1}^p \mathcal{L}^p(\mathbb{R}^{dm})$. Thus, the right hand side of (2.2.11) converges to the right hand side of (2.2.8). Therefore, it remains to establish the convergence of $\alpha_\epsilon(A_1 \times \cdots \times A_p)$ (as $\epsilon \to 0^+$) in $\mathcal{L}^m(\Omega, \mathcal{A}, \mathbb{P})$ for all integers $m \ge 1$. By (2.2.11) and the convergence of the right hand side, we have

$$\sup_{\epsilon>0} \mathbb{E}\Big[\alpha_\epsilon(A_1 \times \cdots \times A_p)^m\Big] < \infty \qquad m = 1, 2, \cdots.$$

Therefore, all we need to prove is the convergence (2.2.7) in $\mathcal{L}^2(\Omega, \mathcal{A}, \mathbb{P})$. To this end, all we need is to prove that the limit

$$\lim_{\epsilon,\epsilon'\to 0^+} \mathbb{E}\Big\{\alpha_\epsilon(A_1 \times \cdots \times A_p) \cdot \alpha_{\epsilon'}(A_1 \times \cdots \times A_p)\Big\}$$

exists. That is the objective of the third step.

A slight modification of the computation for (2.2.11) (with $m = 2$) gives that

$$\mathbb{E}\Big\{\alpha_\epsilon(A_1 \times \cdots \times A_p) \cdot \alpha_{\epsilon'}(A_1 \times \cdots \times A_p)\Big\}$$
$$= \iint_{\mathbb{R}^d\times\mathbb{R}^d} dxdy \prod_{j=1}^{p} \iint_{\mathbb{R}^d\times\mathbb{R}^d} dz_1 dz_2 p_\epsilon(z_1 - x) p_{\epsilon'}(z_2 - y) f_j(z_1, z_2)$$

where the function

$$f_j(x, y) = \int_{(A_j)^2_<} \Big\{p_s(x)p_{t-s}(y - x) + p_s(y)p_{t-s}(x - y)\Big\} dsdt$$

is in $\mathcal{L}^p(\mathbb{R}^d \times \mathbb{R}^d)$.

A slight modification of Lemma 2.2.2 gives that for each $1 \le j \le p$,

$$\begin{aligned}&\iint_{\mathbb{R}^d\times\mathbb{R}^d} dxdy\bigg| f_j(x,y)\\ &-\iint_{\mathbb{R}^d\times\mathbb{R}^d} dz_1dz_2p_\epsilon(z_1-x)p_{\epsilon'}(z_2-y)f_j(z_1,z_2)\bigg|^p\\ &\longrightarrow 0 \quad (\epsilon,\epsilon'\to 0^+).\end{aligned}$$

This leads to the requested conclusion. □

The equality (2.2.8), together with a variety of its versions developed later in this section, is called *Le Gall's moment identity* and will play a crucial role in this book. By Hölder inequality, an immediate consequence of Le Gall's moment identity is the following inequality:

$$\mathbb{E}\Big[\alpha(A_1\times\cdots\times A_p)^m\Big] \le \prod_{j=1}^p\Big\{\mathbb{E}\Big[\alpha(A_j^p)^m\Big]\Big\}^{1/p} \qquad m=1,2,\cdots. \tag{2.2.12}$$

The next thing we intend to do is to extend the limiting family

$$\Big\{\alpha(A_1\times\cdots\times A_p);\quad \text{bounded } A_1,\cdots,A_p\subset\mathbb{R}^+\Big\}$$

into a random measure on $(\mathbb{R}^+)^p$. Strictly speaking, for each $(A_1,\cdots,A_p)$ the notation $\alpha(A_1\times\cdots\times A_p)$ represents an equivalent class of random variables which are equal to each other with probability 1. Here is the challenge we face: we lose a null exceptional event every time we check the measure properties for a set of subsets of $(\mathbb{R}^+)^p$. The uncountable union of these exceptional events is not necessarily a null event again.

Our approach is to show the random measure can be generated by a continuous distribution function. Our major tool is Le Gall's moment identity. Here we introduce the notation

$$\mathbf{t}=(t_1,\cdots,t_p) \qquad t_1,\cdots,t_p\in\mathbb{R}^+.$$

For two multiple time parameters $\mathbf{s},\mathbf{t}\in(\mathbb{R}^+)^p$, write

$$||\mathbf{t}-\mathbf{s}||=\sqrt{(t_1-s_1)^2+\cdots+(t_p-s_p)^2}$$

and $[\mathbf{s},\mathbf{t}]=[s_1,t_1]\times\cdots\times[s_p,t_p]$ if $\mathbf{s}\le\mathbf{t}$ (which means $s_j\le t_j$ for $j=1,\cdots,p$).

LEMMA 2.2.4. *Let $T>0$ be fixed. There is a constant $C=C(T)>0$ such that for any $\mathbf{s},\mathbf{t}\in[0,T]^p$ and for any $m=1,2,\cdots$,*

$$\mathbb{E}\Big|\alpha([\mathbf{0},\mathbf{t}])-\alpha([\mathbf{0},\mathbf{s}])\Big|^m \le (m!)^pC^m||\mathbf{t}-\mathbf{s}||^{\frac{2p-d(p-1)}{2p}m}.$$

Proof. We fist consider the case $\mathbf{s}\le\mathbf{t}$. Write $[\mathbf{0},\mathbf{t}]\setminus[\mathbf{0},\mathbf{s}]$ into the union of p disjoint boxes. For each $1\le j\le p$, the jth box is of the form $A_1\times\cdots\times A_p$, where

$A_l = [0, s_l]$ or $[0, t_l]$ for $l \neq j$, and $A_j = [s_j, t_j]$. By (2.2.8) in Theorem 2.2.3, for $l = 1, \cdots, p$

$$\mathbb{E}\Big[\alpha\big(A_l^p\big)^m\Big] = \int_{(\mathbb{R}^d)^m} dx_1 \cdots dx_m \bigg[\sum_{\sigma\in\Sigma_m} \int_{(A_l)^m_<} du_1 \cdots du_m \times \prod_{k=1}^m p_{u_k - u_{k-1}}(x_{\sigma(k)} - x_{\sigma(k-1)})\bigg]^p.$$

For $l \neq j$, we use the estimate given in (2.2.9) and (2.2.10), which gives that

$$\mathbb{E}\Big[\alpha\big(A_l^p\big)^m\Big] \le (m!)^p C^m \qquad m = 1, 2, \cdots \tag{2.2.13}$$

where $C = C(T) > 0$ represents a constant independent of $\mathbf{s}$, $\mathbf{t}$ and m (to simplify our notation, we allow C to be different from place to place).

For $l = j$, observe that

$$\begin{aligned}&\int_{(A_j)^m_<} du_1 \cdots du_m \prod_{k=1}^m p_{u_k - u_{k-1}}(x_{\sigma(k)} - x_{\sigma(k-1)})\\ &\le \Big(\int_{s_j}^{t_j} p_u(x_{\sigma(1)})du\Big) \prod_{k=2}^m \int_{[0, t_j - s_j]} p_u(x_{\sigma(k)} - x_{\sigma(k-1)})du.\end{aligned}$$

By an estimate modified from the one used for (2.2.9), we get

$$\begin{aligned}&\mathbb{E}\Big[\alpha\big(A_j^p\big)^m\Big]\\ &\le (m!)^p \bigg\{\int_{\mathbb{R}^d} \Big[\int_{s_j}^{t_j} p_u(x)du\Big]^p dx\bigg\}\bigg\{\int_{\mathbb{R}^d} dx \Big[\int_0^{t_j - s_j} p_u(x)du\Big]^p dx\bigg\}^{m-1}.\end{aligned}$$

Modifying the estimate used for (2.2.10), we obtain

$$\begin{aligned}&\int_{\mathbb{R}^d} \Big[\int_0^{t_j - s_j} p_u(x)du\Big]^p dx\\ &\le (2\pi)^{-\frac{d(p-1)}{2}} p^{-d/2} \Big(\frac{2p}{2p - d(p-1)}\Big)^p (t_j - s_j)^{\frac{2p - d(p-1)}{2}}\end{aligned}$$

and

$$\begin{aligned}&\int_{\mathbb{R}^d} \Big[\int_{s_j}^{t_j} p_u(x)du\Big]^p dx\\ &\le (2\pi)^{-\frac{d(p-1)}{2}} p^{-d/2} \int_{s_j}^{t_j} s^{-\frac{d(p-1)}{2p}} ds\\ &= (2\pi)^{-\frac{d(p-1)}{2}} p^{-d/2} \Big(\frac{2p}{2p - d(p-1)}\Big)^p \Big(t_j^{\frac{2p - d(p-1)}{2p}} - s_j^{\frac{2p - d(p-1)}{2p}}\Big)^p\\ &\le (2\pi)^{-\frac{d(p-1)}{2}} p^{-d/2} \Big(\frac{2p}{2p - d(p-1)}\Big)^p (t_j - s_j)^{\frac{2p - d(p-1)}{2}}.\end{aligned}$$

Summarizing what we have accomplished, we have

$$\mathbb{E}\Big[\alpha\big(A_j^p\big)^m\Big] \le (m!)^p C^m (t_j - s_j)^{\frac{2p - d(p-1)}{2} m}. \tag{2.2.14}$$

Thus, by (2.2.12), (2.2.13) and (2.2.14) we have

$$\mathbb{E}\Big[\alpha(A_1 \times \cdots \times A_p)^m\Big] \le (m!)^p C^m (t_j - s_j)^{\frac{2p-d(p-1)}{2p}m}.$$

This leads to

$$\mathbb{E}\Big[\alpha([\mathbf{0},\mathbf{t}]) - \alpha([\mathbf{0},\mathbf{s}])\Big]^m \le (m!)^p C^m ||\mathbf{t}-\mathbf{s}||^{\frac{2p-d(p-1)}{2p}m}.$$

Observe that when $\mathbf{s} \le \mathbf{t}$,

$$\alpha([\mathbf{0},\mathbf{t}]) - \alpha([\mathbf{0},\mathbf{s}]) \ge 0.$$

Hence, we have proved the desired conclusion in the special case when $\mathbf{s} \le \mathbf{t}$.

For the general $\mathbf{s} = (s_1, \cdots, s_p)$ and $\mathbf{t} = (t_1, \cdots, t_p)$, write

$$\mathbf{s}' = (s_1 \wedge t_1, \cdots, s_p \wedge t_p), \qquad \mathbf{t}' = (s_1 \vee t_1, \cdots, s_p \vee t_p).$$

By monotonicity of the functional $\alpha([\mathbf{0},\mathbf{t}])$ in $\mathbf{t}$,

$$\Big|\alpha([\mathbf{0},\mathbf{t}]) - \alpha([\mathbf{0},\mathbf{s}])\Big| \le \alpha([\mathbf{0},\mathbf{t}']) - \alpha([\mathbf{0},\mathbf{s}']).$$

Based on what we have proved

$$\begin{aligned}\mathbb{E}\Big|\alpha([\mathbf{0},\mathbf{t}]) - \alpha([\mathbf{0},\mathbf{s}])\Big|^m &\le (m!)^p C^m ||\mathbf{t}'-\mathbf{s}'||^{\frac{2p-d(p-1)}{2p}m} \\ &= (m!)^p C^m ||\mathbf{t}-\mathbf{s}||^{\frac{2p-d(p-1)}{2p}m}.\end{aligned}$$

□

Two processes $\{X_t\}_{t\in\Theta}$ and $\{Y_t\}_{t\in\Theta}$ are said to be a *modification* to each other, if $\mathbb{P}\{X_t = Y_t\} = 1$ for every $t \in \Theta$. By *Kolmogorov's continuity theorem* (Theorem D.7, Appendix) and by Lemma 2.2.4, the process

$$\Big\{\alpha([\mathbf{0},\mathbf{t}]);\quad \mathbf{t} \in [0,T]^p\Big\}$$

has a continuous modification

$$\alpha^T = \Big\{\alpha^T([\mathbf{0},\mathbf{t}])\quad \mathbf{t} \in [0,T]^p\Big\}.$$

For any $T < T'$, we have that $\mathbb{P}\Big\{\alpha^T([\mathbf{0},\mathbf{t}]) = \alpha^{T'}([\mathbf{0},\mathbf{t}])\Big\} = 1$ for all $\mathbf{t} \in [0,T]^p$. By continuity,

$$\mathbb{P}\Big\{\alpha^T([\mathbf{0},\mathbf{t}]) = \alpha^{T'}([\mathbf{0},\mathbf{t}])\ \text{ for all } \mathbf{t} \in [0,T]^p\Big\} = 1.$$

Let $0 < T_1 < T_2 < \cdots < T_k < \cdots$ be an increasing sequence such that $T_k \to \infty$ as $k \to \infty$. Then outside a null event, the limit

$$\tilde{\alpha}([\mathbf{0},\mathbf{t}]) \equiv \lim_{k\to\infty} \alpha^{T_k}([\mathbf{0},\mathbf{t}])$$

exists for every $\mathbf{t} \in (\mathbb{R}^+)^p$ and, as a function of $\mathbf{t}$, $\tilde{\alpha}([\mathbf{0},\mathbf{t}])$ is continuous on $(\mathbb{R}^+)^p$. In addition, the process $\Big\{\tilde{\alpha}([\mathbf{0},\mathbf{t}]);\quad \mathbf{t} \in (\mathbb{R}^+)^p\Big\}$ is a modification of $\Big\{\alpha([\mathbf{0},\mathbf{t}]);\quad \mathbf{t} \in (\mathbb{R}^+)^p\Big\}$.

By the measure extension theorem, the continuous distribution function $\tilde{\alpha}([\mathbf{0},\mathbf{t}])$ can be extended uniquely into a (random) measure $\tilde{\alpha}(\cdot)$ on $(\mathbb{R}^+)^p$. We call $\tilde{\alpha}(\cdot)$ the *mutual intersection local time* generated by the independent Brownian motions $W_1(t), \cdots, W_p(t)$.

It is straightforward to see that for any bounded sets $A_1, \cdots, A_p \subset \mathbb{R}^+$,

$$\mathbb{P}\Big\{\alpha(A_1 \times \cdots \times A_p) = \tilde{\alpha}(A_1 \times \cdots \times A_p)\Big\} = 1.$$

To simplify our notation, from now on we use $\alpha(A)$, rather than $\tilde{\alpha}(A)$, to denote this continuous version. Our construction of the intersection local time $\alpha(A)$ constitutes the notation

$$\alpha\big([0, t_1] \times \cdots \times [0, t_p]\big) = \int_{\mathbb{R}^d} \left[\prod_{j=1}^p \int_0^{t_j} \delta_x\big(W_j(s)\big) ds\right] dx. \tag{2.2.15}$$

We now briefly describe an alternative way to construct intersection local time without proof. We start with a general setting in which $Z_{\mathbf{t}}$ ($\mathbf{t} \in (\mathbb{R}^+)^p$) is a multi-parameter stochastic process (also called a random field in literature) taking values in $\mathbb{R}^N$. For any $\mathbf{t} \in (\mathbb{R}^+)^p$, the random measure

$$\mu_{\mathbf{t}}(B) = \int_{[\mathbf{0}, \mathbf{t}]} 1_{\{Z_{\mathbf{s}} \in B\}} d\mathbf{s} \qquad B \subset \mathbb{R}^N$$

defined on $\mathbb{R}^N$ is known as the *occupation measure* of the random field $Z_{\mathbf{t}}$. An important problem is the absolute continuity of $\mu_{\mathbf{t}}(\cdot)$ with respect to the Lebesgue measure on $\mathbb{R}^N$. Assume that with probability 1, $\mu_{\mathbf{t}}(\cdot)$ is absolutely continuous for every $\mathbf{t} \subset (\mathbb{R}^+)^p$ (see some exercises at the end of the chapter on how the absolute continuity is examined). Then the associated density $L(\mathbf{t}, x)$ ($x \in \mathbb{R}^N$) is called *local time* of the random field $Z_{\mathbf{t}}$.

In summary, a crucial property which defines the local time is that

$$\int_{\mathbb{R}^N} f(x) L(\mathbf{t}, x) dx = \int_{[\mathbf{0}, \mathbf{t}]} f(Z_{\mathbf{s}}) d\mathbf{s} \qquad \mathbf{t} \in (\mathbb{R}^+)^p$$

for every bounded function on $\mathbb{R}^N$.

Here we encounter an obstacle: Notice that for each $\mathbf{t} \in (\mathbb{R}^+)^p$, the notation $L(\mathbf{t}, \cdot)$ represents an equivalent class (under the Lebesgue measure on $\mathbb{R}^N$) of the densities. For this reason, it is not quite meaningful at this point to discuss, for example, $L(\mathbf{t}, 0)$. On the other hand, if there is a member $\widetilde{L}(\mathbf{t}, x)$ ($(\mathbf{t}, x) \in (\mathbb{R}^+)^p \times \mathbb{R}^N$) among the the equivalent class of the functions $L(\mathbf{t}, x)$, which is continuous in x, then for each $\mathbf{t}$ the function $\widetilde{L}(\mathbf{t}, \cdot)$ is well defined. For the purpose of simplification, the words "local time" and the notation $L(\mathbf{t}, x)$ are identified with this continuous version $\widetilde{L}(\mathbf{t}, x)$ if it exists.

The local time $L(\mathbf{t}, x)$ is then heuristically written as

$$L(\mathbf{t}, x) = \int_{[\mathbf{0}, \mathbf{t}]} \delta_x(Z_{\mathbf{t}}) d\mathbf{t} \qquad x \in \mathbb{R}^N, \quad \mathbf{t} \in (\mathbb{R}^+)^p$$

to reflect the fact that it measures the "time" that $Z_{\mathbf{t}}$ spends near the site x during the "period" $[\mathbf{0}, \mathbf{t}]$.

As we have seen, an effective way to install continuity for $L(\mathbf{t}, x)$ is to use the Kolmogorov continuity theorem. For instance, if we can show that for any compact set $Q \subset (\mathbb{R}^+)^p \times \mathbb{R}^N$ there are $m \geq 1$ and $n > N + p$ and $M > 0$ such that

$$\mathbb{E}\big|L(\mathbf{s}, x) - L(\mathbf{t}, y)\big|^m \leq M\big\{|x-y|^2 + ||\mathbf{s}-\mathbf{t}||^2\big\}^{n/2} \qquad (\mathbf{s}, x), (\mathbf{t}, x) \in Q,$$

then by the Kolmogorov continuity theorem (Theorem D.7, Appendix) and by the extension argument we just made not long time ago, $L(\mathbf{t}, x)$ ($(\mathbf{t}, x) \in (\mathbb{R}^+)^p \times \mathbb{R}^N$) admits a modification jointly continuous in $\mathbf{t} \in (\mathbb{R}^+)^p$ and $x \in \mathbb{R}^N$.

In this way, the *local time*

$$L(t, x) = \int_0^t \delta_x\big(W(s)\big)ds \qquad x \in \mathbb{R} \tag{2.2.16}$$

of the 1-dimensional Brownian motion $W(t)$ is constructed as a stochastic process jointly continuous in $t \in \mathbb{R}^+$ and $x \in \mathbb{R}$.

In the next example, take $N = (p-1)d$ and let $Z_\mathbf{t}$ be the random field defined as

$$Z_\mathbf{t} = \Big(W_1(t_1) - W_2(t_2), \cdots, W_{p-1}(t_{p-1}) - W_p(t_p)\Big)$$

where $W_1(t), \cdots, W_p(t)$ are independent d-dimensional Brownian motions. Under $p(d-2) < d$, the local time

$$L(\mathbf{t}, x) = \int_{[\mathbf{0},\mathbf{t}]} \delta_x\Big(W_1(s_1) - W_2(s_2), \cdots, W_{p-1}(s_{p-1}) - W_p(s_p)\Big)ds_1 \cdots ds_p$$

is defined with the continuity claimed in the general discussion. We define the intersection local time as $\alpha\big([\mathbf{0}, \mathbf{t}]\big) = L(\mathbf{t}, 0)$. In other words, the intersection local time can be viewed as a local time (or the local time at $0 \in \mathbb{R}^{d(p-1)}$, more precisely) of some multi-parameter process. This definition justifies the notation introduced in (2.1.2).

In the following we show that the two definitions of $\alpha\big([\mathbf{0}, \mathbf{t}]\big)$ are equivalent. Momentarily, we denote the intersection local time defined as local time by $\bar{\alpha}\big([\mathbf{0}, \mathbf{t}]\big)$. Write

$$h(x_1, \cdots, x_{p-1}) = \int_{\mathbb{R}^d} p_1(-x) \prod_{j=1}^{p-1} p_1\Big(\sum_{k=j}^{p-1} x_k - x\Big)dx.$$

It is easy to verify that $h(\cdot)$ is a probability density on $\mathbb{R}^{d(p-1)}$. Define, for $\epsilon > 0$,

$$h_\epsilon(x_1, \cdots, x_{p-1}) = \epsilon^{-d(p-1)} h(\epsilon^{-1}x_1, \cdots, \epsilon^{-1}x_{p-1}).$$

For any bounded set $\mathbf{t} \in (\mathbb{R}^+)^p$,

$$\begin{aligned}
&\int_{[\mathbf{0},\mathbf{t}]} h_\epsilon\big(W_1(s_1) - W_2(s_2), \cdots, W_{p-1}(s_{p-1}) - W(s_p)\big)ds_1 \cdots ds_p \\
&= \int_{[\mathbf{0},\mathbf{t}]} ds_1 \cdots ds_p \int_{\mathbb{R}^d} p_\epsilon(-x) \prod_{j=1}^{p-1} p_\epsilon\Big(\sum_{k=j}^{p-1}\big(W_k(s_k) - W_{k+1}(s_{k+1})\big) - x\Big)dx \\
&= \int_{[\mathbf{0},\mathbf{t}]} ds_1 \cdots ds_p \int_{\mathbb{R}^d} p_\epsilon(-x) \prod_{j=1}^{p-1} p_\epsilon\Big(\big(W_j(s_j) - W_p(s_p)\big) - x\Big)dx \\
&= \int_{[\mathbf{0},\mathbf{t}]} ds_1 \cdots ds_p \int_{\mathbb{R}^d} \prod_{j=1}^{p} p_\epsilon\big(W_j(s_j) - x\big)dx;
\end{aligned}$$

or

$$\int_{\mathbb{R}^{d(p-1)}} h_\epsilon(x)L(\mathbf{t},x)dx = \alpha_\epsilon\big([\mathbf{0},\mathbf{t}]\big). \tag{2.2.17}$$

Let $\epsilon \to 0^+$ on the both sides. By Theorem 2.2.3 and by the continuity of $L(A,x)$ as the function of $x \in \mathbb{R}^{d(p-1)}$, we have $\bar{\alpha}\big([\mathbf{0},\mathbf{t}]\big) = \alpha\big([\mathbf{0},\mathbf{t}]\big)$.

Once again, the continuous distribution function $\alpha\big([\mathbf{0},\mathbf{t}]\big)$ generates a unique measure $\alpha(A)$ on $(\mathbb{R}^+)^p$.

We now turn our attention to establishing properties for the intersection local time $\alpha(A)$.

PROPOSITION 2.2.5. *The measure $\alpha(\cdot)$ is supported on the set*

$$\{(t_1,\cdots,t_p) \in (\mathbb{R}^+)^p;\quad W_1(t_1) = \cdots = W_p(t_p)\}.$$

Proof. By continuity of Brownian trajectories, the set

$$\Theta = \{(t_1,\cdots,t_p) \in (\mathbb{R}^+)^p;\quad W_1(t_1) = \cdots = W_p(t_p)\}$$

is a closed set. Let $A = [s_1,t_1] \times \cdots \times [s_p,t_p] \subset (\mathbb{R}^+)^p$ be a deterministic box with rational coordinates $s_1,\cdots,s_p;t_1,\cdots,t_p$ and write

$$B_A = \Big\{\omega;\quad A\cap\Theta = \emptyset\Big\}.$$

On the event B_A, $\lim_{\epsilon\to 0^+}\alpha_\epsilon(A) = 0$, which implies that $\alpha(A) = 0$. Observe that the fact that Θ is a closed set leads to

$$\Theta^c = \bigcup\{A;\quad A\cap\Theta = \emptyset\}$$

where the right hand side is a countable union. Hence, we have that $\alpha(\Theta^c) = 0$. □

PROPOSITION 2.2.6. *For any $a,t > 0$,*

$$\alpha\big([0,at]^p\big) \overset{d}{=} a^{\frac{2p-d(p-1)}{2}}\alpha\big([0,t]^p\big).$$

Proof. By definition and integration variable substitution,

$$\begin{aligned}
\alpha_\epsilon\big([0,at]^p\big) &= a^p\int_{\mathbb{R}^d} dx\prod_{j=1}^p\int_0^t p_\epsilon\big(W_j(as)-x\big)ds\\
&\overset{d}{=} a^p\int_{\mathbb{R}^d} dx\prod_{j=1}^p\int_0^t p_\epsilon\big(\sqrt{a}W_j(s)-x\big)ds\\
&= a^{\frac{2p-dp}{2}}\int_{\mathbb{R}^d} dx\prod_{j=1}^p\int_0^t p_{\epsilon/\sqrt{a}}\big(W_j(s)-(x/\sqrt{a})\big)ds\\
&= a^{\frac{2p-d(p-1)}{2}}\int_{\mathbb{R}^d} dx\prod_{j=1}^p\int_0^t p_{\epsilon/\sqrt{a}}\big(W_j(s)-x)\big)ds\\
&= a^{\frac{2p-d(p-1)}{2}}\alpha_{\epsilon/\sqrt{a}}\big([0,t]^p\big)
\end{aligned}$$

where the equality in law follows from the Brownian scaling. By Theorem 2.2.3 and letting $\epsilon \to 0^+$ on both sides leads to the desired conclusion. □

Let $L_1(t,x),\cdots,L_p(t,x)$ be the local times of the independent 1-dimensional Brownian motions $W_1(t),\cdots,W_p(t)$, respectively. We then have that for any $t_1,\cdots,t_p>0$,

$$\alpha_\epsilon\big([0,t_1]\times\cdots\times[0,t_p]\big)=\int_{-\infty}^{\infty}\bigg[\prod_{j=1}^p\int_{-\infty}^{\infty}p_\epsilon(y-x)L_j(t_j,y)dy\bigg]dx.$$

A deterministic argument by Lemma 2.2.2 establishes the fact that when $d=1$,

$$\alpha\big([0,t_1]\times\cdots\times[0,t_p]\big)=\int_{-\infty}^{\infty}\bigg[\prod_{j=1}^p L_j(t_j,x)\bigg]dx. \tag{2.2.18}$$

We have seen the fundamental role played by Le Gall's moment identity in constructing the intersection local time $\alpha(A)$. In Chapter 3, we shall see an even greater role by this powerful tool in establishing the large deviations for $\alpha([0,t]^p)$. In the following we demonstrate that Le Gall's moment identity takes a simpler form if the Brownian motions run up to exponential times. To this end, we first introduce the following analytic lemma.

LEMMA 2.2.7. *For any continuous functions $\varphi_1(t),\cdots,\varphi_m(t)$ satisfying*

$$\int_0^\infty e^{-t}|\varphi_k(t)|dt<\infty\qquad k=1,\cdots,m$$

and for any $\lambda>0$, we get

$$\int_0^\infty dte^{-\lambda t}\int_{[0,t]^m_<}ds_1\cdots ds_m\prod_{k=1}^m\varphi_k(s_k-s_{k-1})=\prod_{k=1}^m\int_0^\infty e^{-\lambda t}\varphi_k(t)dt$$

where we follow the convention $s_0=0$.

Proof. By the substitution

$$t_k=s_k-s_{k-1},\quad (k=1,\cdots,m)\ \ t_{m+1}=t-s_m,$$

we have

$$\begin{aligned}&\int_0^\infty dte^{-t}\int_{[0,t]^m_<}ds_1\cdots ds_m\prod_{k=1}^m\varphi_k(s_k-s_{k-1})\\&=\int_0^\infty\cdots\int_0^\infty dt_1\cdots dt_{m+1}e^{-t_{m+1}}\prod_{k=1}^m e^{-t_k}\varphi(t_k)\\&=\prod_{k=1}^m\int_0^\infty e^{-t}\varphi_k(t)dt.\end{aligned}$$

□

Let $\tau_1,\cdots\tau_p$ be independent exponential times with parameter 1. Assume also the independence between $\{\tau_1,\cdots\tau_p\}$ and the Brownian motions $\{W_1,\cdots,W_p\}$.

THEOREM 2.2.8. *For any any integer $m \geq 1$, we have*

$$\mathbb{E}\Big[\alpha\big([0,\tau_1]\times\cdots\times[0,\tau_p]\big)^m\Big] = \int_{(\mathbb{R}^d)^m} dx_1\cdots dx_m\bigg[\sum_{\sigma\in\Sigma_m}\prod_{k=1}^m G(x_{\sigma(k)}-x_{\sigma(k-1)})\bigg]^p, \tag{2.2.19}$$

where

$$G(x)=\int_0^\infty e^{-t}p_t(x)dt \qquad x\in\mathbb{R}^d$$

and $x_{\sigma(0)}=0$.

Proof. First, it is easy to see from (2.2.3) that the right hand side of the equation is finite. By Fubini's theorem and by (2.2.8), we get

$$\begin{aligned}
&\mathbb{E}\Big[\alpha\big([0,\tau_1]\times\cdots\times[0,\tau_p]\big)^m\Big]\\
&=\int_0^\infty\cdots\int_0^\infty dt_1\cdots dt_p e^{-(t_1+\cdots+t_p)}\int_{(\mathbb{R}^d)^m}dx_1\cdots dx_m\\
&\times\prod_{j=1}^p\sum_{\sigma\in\Sigma_m}\int_{[0,t_j]^m_<}ds_1\cdots ds_m\prod_{k=1}^m p_{s_k-s_{k-1}}(x_{\sigma(k)}-x_{\sigma(k-1)})\\
&=\int_{(\mathbb{R}^d)^m}dx_1\cdots dx_m\prod_{j=1}^p\sum_{\sigma\in\Sigma_m}\int_0^\infty dte^{-t}\\
&\times\int_{[0,t]^m_<}ds_1\cdots ds_m\prod_{k=1}^m p_{s_k-s_{k-1}}(x_{\sigma(k)}-x_{\sigma(k-1)}).
\end{aligned}$$

Finally, by Lemma 2.2.7,

$$\begin{aligned}
&\int_0^\infty dte^{-t}\int_{[0,t]^m_<}ds_1\cdots ds_m\prod_{k=1}^m p_{s_k-s_{k-1}}(x_{\sigma(k)}-x_{\sigma(k-1)})\\
&=\prod_{k=1}^m\int_0^\infty e^{-t}p_t(x_{\sigma(k)}-x_{\sigma(k-1)})dt=\prod_{k=1}^m G(x_{\sigma(k)}-x_{\sigma(k-1)}).
\end{aligned}$$

□

As an application of Le Gall's moment identity, we establish the following exponential integrability for $\alpha([0,1]^p)$ (See Remark 3.3.5 in the next chapter for a more precise version).

THEOREM 2.2.9. *There is a constant $\lambda>0$ such that*

$$\mathbb{E}\exp\Big\{\lambda\alpha\big([0,1]^p\big)^{\frac{2}{d(p-1)}}\Big\}<\infty.$$

Proof. Notice that $\tau_{\min}=\min\{\tau_1,\cdots,\tau_p\}$ is exponential with parameter p. By the scaling property given in Proposition 2.2.6 and independence between Brownian motions and exponential times, we get

$$
\begin{aligned}
(2.2.20)\qquad &\mathbb{E}\Big[\alpha\big([0,\tau_1]\times\cdots\times[0,\tau_p]\big)^m\Big]\\
&\geq \mathbb{E}\Big[\alpha\big([0,\tau_{\min}]^p\big)^m\Big] = \mathbb{E}\tau_{\min}^{\frac{2p-d(p-1)}{2}m}\mathbb{E}\Big[\alpha\big([0,1]^p\big)^m\Big]\\
&= p^{-\frac{2p-d(p-1)}{2}m-1}\Gamma\Big(1+\frac{2p-d(p-1)}{2}m\Big)\mathbb{E}\alpha\big([0,1]^p\big)^m.
\end{aligned}
$$

On the other hand, by (2.2.19) and by Jensen inequality, we get

$$
\begin{aligned}
(2.2.21)\qquad &\mathbb{E}\Big[\alpha\big([0,\tau_1]\times\cdots\times[0,\tau_p]\big)^m\Big]\\
&\leq (m!)^{p-1}\sum_{\sigma\in\Sigma_m}\int_{(\mathbb{R}^d)^m}dx_1\cdots dx_m\prod_{k=1}^m G^p(x_{\sigma(k)}-x_{\sigma(k-1)})\\
&= (m!)^p\int_{(\mathbb{R}^d)^m}dx_1\cdots dx_m\prod_{k=1}^m G^p(x_k-x_{k-1})\\
&= (m!)^p\Big(\int_{\mathbb{R}^d}G^p(x)dx\Big)^m.
\end{aligned}
$$

In view of (2.2.3), an estimate by Stirling formula based on (2.2.20) and (2.2.21) leads to the bound

$$
\mathbb{E}\Big[\alpha\big([0,1]^p\big)^m\Big]\leq (m!)^{\frac{d(p-1)}{2}}C^m \qquad m=1,2,\cdots.
$$

The rest follows from a standard application of Taylor expansion. □

We now allow the Brownian motions $W_1(t),\cdots,W_p(t)$ to start somewhere other than at the origin. When $W_1(0)=y_1,\cdots,W_p(0)=y_p$, we adopt the notations $\mathbb{E}_{\bar{y}}$ and $\mathbb{P}_{\bar{y}}$ for, respectively, the expectation and the probability associated to the random path of $\big(W_1(t),\cdots,W_p(t)\big)$ starting at $\bar{y}=(y_1,\cdots,y_p)$. In the important special case $y_1=\cdots=y_p=0$, we still use "$\mathbb{E}$" and "$\mathbb{P}$" instead. Under the law $\mathbb{P}_{\bar{y}}$, by the same procedure one can construct the intersection local time $\alpha(A)$. We list some similar properties here.

Similar to the identity in law given in Proposition 2.2.6, for any $a,t>0$, we get

$$
(2.2.22)\qquad \mathcal{L}_{\bar{y}}\Big\{\alpha\big([0,at]^p\big)\Big\} = \mathcal{L}_{\bar{y}/\sqrt{a}}\Big\{a^{\frac{2p-d(p-1)}{2}}\alpha\big([0,t]^p\big)\Big\}.
$$

Under $\mathbb{P}_{\bar{y}}$, Le Gall's moment identity takes the form

$$
\begin{aligned}
(2.2.23)\qquad &\mathbb{E}_{\bar{y}}\Big[\alpha\big(A_1\times\cdots\times A_p\big)^m\Big]\\
&= \int_{(\mathbb{R}^d)^m}dx_1\cdots dx_m\prod_{j=1}^p\sum_{\sigma_j\in\Sigma_m}\int_{(A_j)^m_<}ds_1\cdots ds_m\\
&\qquad\times\prod_{k=1}^m p_{s_k-s_{k-1}}(x_{\sigma_j(k)}-x_{\sigma_j(k-1)})
\end{aligned}
$$

where $A_1,\cdots,A_p\subset(\mathbb{R}^+)^p$ are bounded, $s_0=0$ and $\sigma_j(0)=y_j$ for $j=1,\cdots,p$.

Taking $A_1 = \cdots = A_p = A$ and applying Hölder inequality to (2.2.23), we get

$$\mathbb{E}_{\bar{y}}\Big[\alpha\big(A^p\big)^m\Big] \le \prod_{j=1}^p \Big\{ \int_{(\mathbb{R}^d)^m} dx_1 \cdots dx_m \Big[\sum_{\sigma_j \in \Sigma_m} \int_{A_<^m} ds_1 \cdots ds_m \times \prod_{k=1}^m p_{s_k - s_{k-1}}\big(x_{\sigma_j(k)} - x_{\sigma_j(k-1)}\big)\Big]^p\Big\}^{1/p}.$$

Notice that for each $1 \le j \le p$, the jth factor on the right hand side does not depend on the value of $x_{\sigma_j(0)}$. Consequently, the right hand side is equal to $\mathbb{E}\Big[\alpha\big(A^p\big)^m\Big]$. Thus,

$$\mathbb{E}_{\bar{y}}\Big[\alpha\big(A^p\big)^m\Big] \le \mathbb{E}\Big[\alpha\big(A^p\big)^m\Big] \quad \forall \bar{y} \in (\mathbb{R}^d)^p. \tag{2.2.24}$$

This inequality shows that for the Brownian motions $W_1(t), \cdots, W_p(t)$, the best strategy to get maximal intersection within the fixed period is to start at the same point.

2.3. Self-intersection local time

Let $W(t)$ be a d-dimensional Brownian motion. The main goal in this section is to construct the random measure $\beta(A)$ on $(\mathbb{R}^+)^p_<$ symbolically given by (2.1.1) or, equivalently, by

$$\beta(A) = \int_{\mathbb{R}^d} dx \int_A ds_1 \cdots ds_p \prod_{j=1}^p \delta_x\big(W(s_j)\big) \quad A \subset (\mathbb{R}^+)^p_<. \tag{2.3.1}$$

Let the numbers $0 < a_1 < b_1 < a_2 < b_2 < \cdots < a_{p-1} < b_p$ be fixed and write

$$W^{(j)}(t) = W(a_j + t) \quad 0 \le t \le b_j - a_j \quad j = 1, \cdots, p.$$

We assign each of the independent Brownian motions $W_1(t), \cdots, W_p(t)$ an initial distribution μ with the density

$$h(x) = C_d e^{-|x|} \quad x \in \mathbb{R}^d$$

where $C_d > 0$ normalizes $h(x)$ into a density function.

Write $T = [0, b_1 - a_1] \times \cdots \times [0, b_p - a_p]$. We view the random fields

$$\Big(W^{(1)}(t_1), \cdots, W^{(p)}(t_p)\Big) \quad \text{and} \quad \Big(W_1(t_1), \cdots, W_p(t_p)\Big) \quad (t_1, \cdots, t_p) \in T$$

as two random variables with values in the space $C\{T, (\mathbb{R}^d)^p\}$ and with the distributions, respectively, $\gamma_1(dw_1, \cdots, dw_p)$ and $\gamma_2(dw_1, \cdots, dw_p)$.

LEMMA 2.3.1. *$\gamma_1(\cdot)$ is absolutely continuous with respect to $\gamma_2(\cdot)$ with the associated Random-Nikodym derivative*

$$\Theta(w_1, \cdots, w_p) = C_d^{-p} \exp\Big\{\sum_{j=1}^p |w_j(0)|\Big\} p_{a_1}\big(w_1(0)\big) \times \prod_{j=2}^p p_{a_j - b_{j-1}}\big(w_j(0) - w_{j-1}(b_{j-1} - a_{j-1})\big). \tag{2.3.2}$$

In particular,

$$\int_{C\{T,(\mathbb{R}^d)^p\}} \Theta^2(w_1,\cdots,w_p)\gamma_2(dw_1,\cdots,dw_p) < \infty. \tag{2.3.3}$$

Proof. For each $1 \le j \le p$, let $0 = s_0^{(j)} < s_1^{(j)} < \cdots < s_{n_j}^{(j)} = b_j - a_j$ be a partition of $[0, b_j - a_j]$. It is straightforward to verify that the random vector

$$\Big(W^{(1)}(s_0^{(1)}),\cdots,W^{(1)}(s_{n_1}^{(1)});\cdots;W^{(p)}(s_0^{(p)}),\cdots,W^{(p)}(s_{n_p}^{(p)})\Big)$$

has the probability density

$$\begin{aligned}
&f(x_0^{(1)},\cdots,x_{n_1}^{(1)};\cdots;x_0^{(p)},\cdots,x_{n_p}^{(p)})\\
&=\prod_{j=1}^p p_{a_j-b_{j-1}}(x_0^{(j)}-x_{n_{j-1}}^{(j-1)})\prod_{k=1}^{n_j} p_{s_k^{(j)}-s_{k-1}^{(j)}}(x_k^{(j)}-x_{k-1}^{(j)})
\end{aligned}$$

where $b_0 = 0$ and $x_{n_0}^{(0)} = 0$; and that the random vector (with the given initial distribution)

$$\Big(W_1(s_0^{(1)}),\cdots,W_1(s_{n_1}^{(1)});\cdots;W_p(s_0^{(p)}),\cdots,W_p(s_{n_p}^{(p)})\Big)$$

has the probability density

$$\begin{aligned}
&g(x_0^{(1)},\cdots,x_{n_1}^{(1)};\cdots;x_0^{(p)},\cdots,x_{n_p}^{(p)})\\
&=\prod_{j=1}^p C_d e^{-|x_0^{(j)}|}\prod_{k=1}^{n_j} p_{s_k^{(j)}-s_{k-1}^{(j)}}(x_k^{(j)}-x_{k-1}^{(j)}).
\end{aligned}$$

Hence,

$$\begin{aligned}
&\frac{f(x_0^{(1)},\cdots,x_{n_1}^{(1)};\cdots;x_0^{(p)},\cdots,x_{n_p}^{(p)})}{g(x_0^{(1)},\cdots,x_{n_1}^{(1)};\cdots;x_0^{(p)},\cdots,x_{n_p}^{(p)})}\\
&=C_d^{-p}\exp\Big\{\sum_{j=1}^p |x_0^{(j)}|\Big\}p_{a_1}(x_0^{(1)})\prod_{j=2}^p p_{a_j-b_{j-1}}(x_0^{(j)}-x_{n_{j-1}}^{(j-1)}).
\end{aligned}$$

This leads to the identity

$$\gamma_1(A) = \int_A \Theta(w_1,\cdots,w_p)\gamma_2(dw_1,\cdots,dw_p)$$

for any $A \subset C\{T,(\mathbb{R}^d)^p\}$ of the form

$$A = \Big\{(w_1,\cdots,w_p);\ \big(w_1(s_0^{(1)}),\cdots,w_1(s_{n_1}^{(1)});\cdots;w_p(s_0^{(p)}),\cdots,w_p(s_{n_p}^{(p)})\big)\in B\Big\}.$$

Notice that $\Theta(w_1,\cdots,w_p)$ does not depend on the time set we chose. By standard measure extension, the identity holds for all $A \subset C\{T,(\mathbb{R}^d)^p\}$.

Finally, a straightforward calculation gives

$$\int_{C\{T,(\mathbb{R}^d)^p\}} \Theta^2(w_1,\cdots,w_p)\gamma_2(dw_1,\cdots,dw_p) = \mathbb{E}\Theta^2(W_1,\cdots,W_p) < \infty.$$

□

For each $\epsilon > 0$, define the random measure $\beta_\epsilon(\cdot)$ on $(\mathbb{R}^+)^p_<$:

$$\beta_\epsilon(A) = \int_{\mathbb{R}^d}\left[\int_A ds_1\cdots ds_p \prod_{j=1}^p p_\epsilon\big(W(s_j)-x\big)\right]dx \qquad A \subset (\mathbb{R}^+)^p_<.$$

A measure on a complete metric space is called *Radon measure* if the measure value of every compact set is finite.

THEOREM 2.3.2. *Assume $p(d-2) < d$. With probability 1 there exists a Radon measure $\beta(\cdot)$ on $(\mathbb{R}^+)^p_<$ such that:*

(1) *For any compact subset of $(\mathbb{R}^+)^p_<$ of the form $A_1 \times \cdots \times A_p$, where $A_1, \cdots, A_p$ are finite closed intervals,*

$$\lim_{\epsilon\to 0^+} \mathbb{E}\big|\beta_\epsilon(A_1\times\cdots\times A_p) - \beta(A_1\times\cdots\times A_p)\big|^m = 0 \tag{2.3.4}$$

for all $m > 0$.

(2) *The measure is supported on*

$$\big\{(s_1,\cdots,s_p)\in(\mathbb{R}^+)^p_<;\quad W(s_1)=\cdots=W(s_p)\big\}.$$

In addition, $\beta(\{s_j = t\}) = 0$ for any $1 \le j \le p$ and any $t \ge 0$.

Proof. Let the numbers $0 \le a_1 < b_1 < \cdots < a_p < b_p$ be given and write

$$I = [\mathbf{a},\mathbf{b}] = [a_1,b_1]\times\cdots\times[a_p,b_p]. \tag{2.3.5}$$

We first construct a random measure $\beta^I(\cdot)$ on I which will appear as the limitation of $\beta(\cdot)$ on I. Notice that for any $t > 0$, we have

$$\Big\{\beta_\epsilon(t+A);\quad A\subset(\mathbb{R}^+)^p_<\Big\} \stackrel{d}{=} \Big\{\beta_\epsilon(A)\quad A\subset(\mathbb{R}^+)^p_<\Big\}$$

where $t + A = \big\{(t+s_1,\cdots,t+s_p);\quad (s_1,\cdots,s_p)\in A\big\}$. We may assume $a_1 > 0$, for otherwise we may consider $[1+a_1,1+b_1]\times\cdots\times[1+a_p,1+b_p]$ instead.

We adopt the notations used in Lemma 2.3.1. Let $\mathbf{t}\in I$ and define the function ξ_ϵ on $C\{I,(\mathbb{R}^d)^p\}$ as

$$\xi_\epsilon(w_1,\cdots,w_p) = \int_{\mathbb{R}^d}\left[\prod_{j=1}^p\int_0^{t_j-a_j} p_\epsilon\big(w_j(s)-x\big)ds\right]dx.$$

By time shifting, one can see that

$$\beta_\epsilon([\mathbf{a},\mathbf{t}]) = \xi_\epsilon\big(W^{(1)},\cdots,W^{(p)}\big).$$

According to Lemma 2.3.1, for any $\epsilon,\epsilon' > 0$, we have

$$\begin{aligned}
&\mathbb{E}\Big|\beta_\epsilon\big([\mathbf{a},\mathbf{t}]\big) - \beta_{\epsilon'}\big([\mathbf{a},\mathbf{t}]\big)\Big|^m\\
&= \int_{C\{T,(\mathbb{R}^d)^p\}} \big|\xi_\epsilon(w_1,\cdots,w_p) - \xi_{\epsilon'}(w_1,\cdots,w_p)\big|^m\\
&\qquad\qquad \times\Theta(w_1,\cdots,w_p)\gamma_2(dw_1,\cdots,dw_p)\\
&\le \left\{\int_{C\{T,(\mathbb{R}^d)^p\}} \Theta^2(w_1,\cdots,w_p)\gamma_2(dw_1,\cdots,dw_p)\right\}^{1/2}\\
&\quad\times\Big\{\mathbb{E}\big|\alpha_\epsilon([\mathbf{0},\mathbf{t}-\mathbf{a}]) - \alpha_{\epsilon'}([\mathbf{0},\mathbf{t}-\mathbf{a}])\big|^{2m}\Big\}^{1/2}.
\end{aligned}$$

In view of Theorem 2.2.3, there exists a non-negative random variable $\beta^I([\mathbf{a},\mathbf{t}])$, such that

$$\lim_{\epsilon\to 0^+}\beta_\epsilon([\mathbf{a},\mathbf{t}])=\beta^I([\mathbf{a},\mathbf{t}]) \tag{2.3.6}$$

in all positive moments.

For any $\mathbf{s},\mathbf{t}\in[\mathbf{a},\mathbf{b}]$, by Lemma 2.3.1 and a procedure the same as above, we get

$$\begin{aligned}&\mathbb{E}\Big|\beta_\epsilon([\mathbf{a},\mathbf{t}])-\beta_\epsilon([\mathbf{a},\mathbf{s}])\Big|^m\\ &\le\bigg\{\int_{C\{T,(\mathbb{R}^d)^p\}}\Theta^2(w_1,\cdots,w_p)\gamma_2(dw_1,\cdots,dw_p)\bigg\}^{1/2}\\ &\times\Big\{\mathbb{E}\big|L_\epsilon(\mathbf{t}-\mathbf{a},0)-L_\epsilon(\mathbf{s}-\mathbf{a},0)\big|^{2m}\Big\}^{1/2}.\end{aligned}$$

By Lemma 2.2.4, for any $M>0$ there is a $C_I>0$ such that for any $\mathbf{s},\mathbf{t}\in[0,M]^p$,

$$\mathbb{E}\big|\beta^I([\mathbf{a},\mathbf{s}])-\beta^I([\mathbf{a},\mathbf{t}])\big|^m\le(m!)^pC_I^m||\mathbf{s}-\mathbf{t}||^{\frac{2p-d(p-1)}{2}m}\qquad m=1,2,\cdots.$$

By the Kolmogorov continuity theorem (Theorem D.7, Appendix), the process

$$\Big\{\beta^I([\mathbf{a},\mathbf{t}]);\ \ \mathbf{t}\in I\Big\}$$

has a continuous modification (which is still denoted by $\beta^I([\mathbf{a},\mathbf{t}])$). The continuous distribution function $\beta^I([\mathbf{a},\mathbf{t}])$ then generates a Borel measure $\beta^I(A)$ on I, which is extended to a measure on $(\mathbb{R}^+)^p$ but supported by I. Clearly, the continuity of $\beta^I([\mathbf{a},\mathbf{t}])$ leads to $\beta^I(\{s_j=t\})=0$ for any $t\ge 0$ and $1\le j\le p$. In addition, an argument used in the proof of Proposition 2.2.5 shows that the measure $\beta^I(\cdot)$ is supported on

$$\big\{(s_1,\cdots,s_p)\in(\mathbb{R}^+)^p_<;\ \ W(s_1)=\cdots=W(s_p)\big\}.$$

Take the countable collection $\{I_k\}_{k\ge 1}$ of compact rectangles of the form given in (2.3.5) such that

(i) $(\mathbb{R}^+)^p_<=\bigcup_{k=1}^\infty I_k$;

(ii) if $j\neq k$, $I_j\cap I_k$ is contained in a finite union of "hyperplanes" $\{s_j=t\}$;

(iii) any compact set of $(\mathbb{R}^+)^p$ intersects only a finite number of the rectangles I_k.

By our discussion, for each k, there is a finite measure $\beta^{I_k}(\cdot)$ supported on I_k and on the time set of p-multiple points such that $\beta^{I_k}(\cdot)$ takes zero measure values at the hyperplanes of the form $\{(s_1,\cdots,s_p);\ s_j=t\}$, and that as $\epsilon\to 0^+$, $\beta_\epsilon(I')\longrightarrow\beta^{I_k}(I')$ in all moments for any sub-rectangle $I'\subset I_k$. Define the measure β on $(\mathbb{R}^+)^p_<$ by

$$\beta(A)=\sum_{k=1}^\infty\beta^{I_k}(A)\qquad A\subset(\mathbb{R}^+)^p_<.$$

One can easily see how (2.3.4) follows from (i) and (ii). Finally, the property (iii) ensures that $\beta(\cdot)$ is a Radon measure. □

The random measure $\beta(A)$ is called *p-multiple self-intersection local time* of the Brownian motion $W(t)$.

Similarly to (2.2.17), we have

$$\beta_\epsilon(A) = \int_A h_\epsilon\big(W(s_1) - W(s_2), \cdots, W(s_{p-1}) - W(s_p)\big) ds_1 \cdots ds_p, \tag{2.3.7}$$

which justifies the notation given in (2.1.1).

From the proof of Theorem 2.3.2, we have seen some connections between the mutual intersection local time $\alpha(\cdot)$ and the self-intersection local time $\beta(\cdot)$. Compared with Proposition 2.2.6, the following proposition shows that $\beta(\cdot)$ has a scaling rate same as $\alpha(\cdot)$.

PROPOSITION 2.3.3. *Assume $p(d-2) < d$. For any $t > 0$, and $A \in (\mathbb{R}^+)^p_<$,*

$$\beta(tA) \overset{d}{=} t^{\frac{2p-d(p-1)}{2}} \beta(A), \tag{2.3.8}$$

$$\beta(t + A) \overset{d}{=} \beta(A) \tag{2.3.9}$$

where $t + A = \{(t + s_1, \cdots, t + s_p);\quad (s_1, \cdots, s_p) \in A\}$.

Proof. By a standard argument of monotonic measure extension, we need only to consider the case when A is of the form given in (2.3.5). Indeed, by variable substitution, we get

$$\begin{aligned}
\beta_\epsilon(tA) &= t^p \int_{\mathbb{R}^d} dx \prod_{j=1}^p \int_{a_j}^{b_j} p_\epsilon\big(W(ts) - x\big) ds \\
&\overset{d}{=} t^p \int_{\mathbb{R}^d} dx \prod_{j=1}^p \int_{a_j}^{b_j} p_\epsilon\big(\sqrt{t} W(s) - x\big) ds \\
&= t^{p - \frac{dp}{2}} \int_{\mathbb{R}^d} dx \prod_{j=1}^p \int_{a_j}^{b_j} p_{\epsilon/\sqrt{t}}\big(W(s) - (x/\sqrt{t})\big) ds \\
&= t^{\frac{2p-d(p-1)}{2}} \int_{\mathbb{R}^d} dx \prod_{j=1}^p \int_{a_j}^{b_j} p_{\epsilon/\sqrt{t}}\big(W(s) - x)\big) ds \\
&= t^{\frac{2p-d(p-1)}{2}} \beta_{\epsilon/\sqrt{t}}(A)
\end{aligned}$$

where the equality in law is implied by the Brownian scaling. Thus, (2.3.8) follows from part (1) of Theorem 2.3.2. The proof of (2.3.9) is similar, where the key ingredient is the increment stationary of Brownian motion and translation invariance of the Lebesgue integral on $\mathbb{R}^d$. □

The next proposition shows that the connection is even more direct when it comes to the double intersections.

PROPOSITION 2.3.4. *Assume that $p = 2$ and $d \le 3$. For any $0 \le a < b < c$, we have*

$$\beta\big([a, b] \times [b, c]\big) \overset{d}{=} \alpha\big([0, b - a] \times [0, c - b]\big).$$

Proof. Let $b < b' < c$ and notice that as $p = 2$,

$$\beta_\epsilon\big([a,b]\times[b',c]\big) = \int_a^b\int_{b'}^c p_{2\epsilon}\Big(\big(W(r)-W(b)\big)-\big(W(s)-W(b)\big)\Big)dsdr$$

$$= \int_{b'-b}^{b'-a}\int_0^{c-b'} p_{2\epsilon}\Big(\big(W(b'-r)-W(b')\big)-\big(W(b'+s)-W(b')\big)\Big)dsdr$$

$$= \int_{b'-b}^{b'-a}\int_0^{c-b'} p_{2\epsilon}\big(W_1(r)-W_2(s)\big)dsdr$$

where $W_1(r) = -\Big(W(b')-W(b'-r)\Big)$ $(0\le r\le b'-a)$ and $W_2(s) = W(b'+s)-W(b)$ $(0\le s\le c-b')$ are independent Brownian motions. Letting $\epsilon\to 0^+$ gives

$$\beta\big([a,b]\times[b',c]\big) \overset{d}{=} \alpha([b'-b,b'-a]\times[0,c-b']).$$

Letting $b'\to b^+$ leads to the desired conclusion. □

In another special case $d = 1$, the self-intersection local time can be written in terms of the integral of the local time $L(t,x)$ of the 1-dimensional Brownian motion $W(t)$. By an analysis (omitted here) similar to the one used for (2.2.18) we have

PROPOSITION 2.3.5. *As $d = 1$ and $p \ge 2$,*

$$\beta([0,t]^p_<) = \frac{1}{p!}\int_{-\infty}^{\infty} L^p(t,x)dx \qquad t>0.$$

Here comes the bad news. Sharply contrary to the 1-dimensional case, the self-intersection local time of a multi-dimensional Brownian motion can blow up even in a finite time period.

PROPOSITION 2.3.6. *As $d\ge 2$, for any $0\le a<b$, we get*

$$\beta\big([a,b]^p_<\big) = \infty \qquad a.s.$$

Proof. By Proposition 2.3.3, we may only consider the case $a = 0$ and $b = 1$. Here we only consider the case $d = 2$. A different treatment for the case $d\ge 3$ is needed and we include it in Exercise 2.5.9. Set

$$I = I_0^0 = \Big[0,\frac{1}{2p}\Big]\times\Big[\frac{2}{2p},\frac{3}{2p}\Big]\times\cdots\times\Big[\frac{2(p-1)}{2p},\frac{2p-1}{2p}\Big],$$

and more generally for any $k\ge 0$, $l = 0,1,\cdots,2^k-1$, set

$$I_l^k = l2^{-k} + 2^{-k}I_0^0$$
$$= \Big[l2^{-k},\ l2^{-k}+\frac{1}{2p}2^{-k}\Big]\times\cdots\times\Big[l2^{-k}+\frac{2(p-1)}{2p}2^{-k},\ l2^{-k}+\frac{2p-1}{2p}2^{-k}\Big].$$

Then for each k, the random variables $\beta(I_l^k)$ $(l = 0,1,\cdots,2^k-1)$ are independent and identically distributed. By Proposition 2.3.3, we get

$$\beta(I_l^k) \overset{d}{=} 2^{-k}\beta(I).$$

It follows that

$$\mathbb{E}\Big[\sum_{l=0}^{2^k-1}\beta(I_l^k)\Big] = \mathbb{E}\big(\beta(I)\big) \quad\text{and}\quad \mathrm{Var}\Big[\sum_{l=0}^{2^k-1}\beta(I_l^k)\Big] = 2^{-k}\mathrm{Var}\,\big(\beta(I)\big).$$

Hence,

$$\sum_{k=0}^{\infty}\mathbb{E}\Big|\sum_{l=0}^{2^k-1}\beta(I_l^k)-\mathbb{E}\Big[\sum_{l=0}^{2^k-1}\beta(I_l^k)\Big]\Big|\le\sum_{k=1}^{\infty}\Big\{\mathrm{Var}\Big[\sum_{l=0}^{2^k-1}\beta(I_l^k)\Big]\Big\}^{1/2}<\infty.$$

Consequently,

$$\sum_{k=0}^{\infty}\sum_{l=0}^{2^k-1}\beta(I_l^k)=\infty\qquad a.s.$$

as $d=2$. Finally, the conclusion follows from the fact that $I_l^k\subset[0,1]_<^p$ and that the intersection of any two members of the family $\{I_l^k\}$ is contained in a finite union of "hyperplanes" $\{s_j=t\}$. □

The reason behind it is that when $d\ge 2$, the short range intersection is too strong compared with long range intersection. In the case $d=2$, a way to fix it is the renormalization method, which will be discussed in the next section.

2.4. Renormalization

In this section, we consider the case $d=p=2$. Because of the problem pointed out in Proposition 2.3.6, the polymer models suggested in (2.1.3) and (2.1.4) are not mathematically defined. To fix this problem, we construct the so-called *renormalized self-intersection local time* $\gamma([0,t]_<)$ heuristically written as

$$\gamma([0,t]_<^2)=\iint_{\{0\le r<s\le t\}}\delta_0\big(W(r)-W(s)\big)drds-\mathbb{E}\iint_{\{0\le r<s\le t\}}\delta_0\big(W(r)-W(s)\big)drds. \tag{2.4.1}$$

The idea is that if we replace $\beta([0,t]_<^2)$ by $\gamma([0,t]_<^2)$ in (2.1.3) and (2.1.4) the models stay intact in physics. Of course, the normalizing constants $\widehat{C}$ and $\widetilde{C}$ in (2.1.3) and (2.1.4) have to be redefined properly so that $\widehat{P}(\cdot)$ and $\widetilde{P}(\cdot)$ remain to be probability measures. This procedure is called *renormalization.*

Let $t>0$ be fixed and set

$$A_l^k=\Big[\frac{2l}{2^{k+1}}t,\frac{2l+1}{2^{k+1}}t\Big)\times\Big[\frac{2l+1}{2^{k+1}}t,\frac{2l+2}{2^{k+1}}t\Big)\qquad l=0,1,\cdots,2^k-1\quad k=0,1,\cdots \tag{2.4.2}$$

(See Figure 1).

THEOREM 2.4.1. *The series*

$$\sum_{k=0}^{\infty}\Big\{\sum_{l=0}^{2^k-1}\Big(\beta(A_l^k)-\mathbb{E}\beta(A_l^k)\Big)\Big\}$$

convergences with probability 1 and in $\mathcal{L}^2(\Omega,\mathcal{A},\mathbb{P})$.

Proof. By Proposition 2.3.3 and Proposition 2.3.4 we have that

$$\beta(A_l^k)\overset{d}{=}2^{-(k+1)}\beta\big([0,t]\times[t,2t]\big)\overset{d}{=}2^{-(k+1)}\alpha\big([0,t]^2\big). \tag{2.4.3}$$

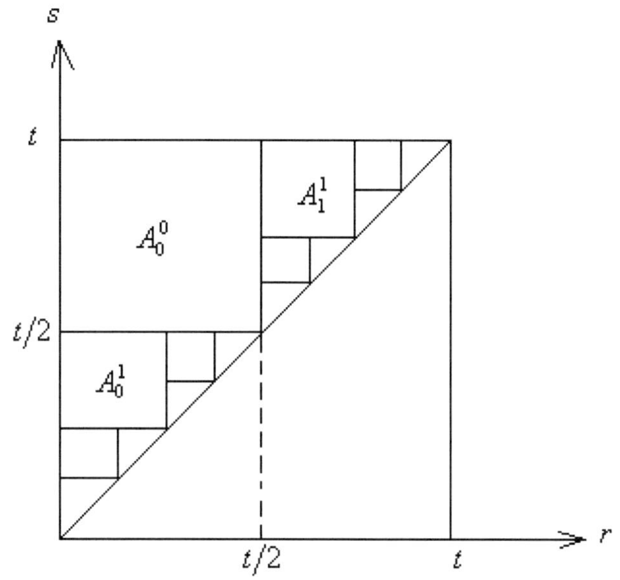

FIGURE 1. triangular approximation

In addition, for each $k \geq 0$, the finite sequence

$$\beta(A_l^k), \qquad l = 0, 1, \cdots, 2^k - 1$$

is independent. Consequently,

$$\mathrm{Var}\left(\sum_{l=0}^{2^k-1} \beta(A_l^k)\right) = 2^{-k-2}\mathrm{Var}\left(\alpha([0,t]^2)\right) = C2^{-k}.$$

Finally, the conclusion follows from the estimate

$$\left\{\mathbb{E}\left[\sum_{k=0}^{\infty}\left|\sum_{l=0}^{2^k-1}\left(\beta(A_l^k) - \mathbb{E}\beta(A_l^k)\right)\right|\right]^2\right\}^{1/2}$$
$$\leq \sum_{k=0}^{\infty}\left\{\mathrm{Var}\left(\sum_{l=0}^{2^k-1}\beta(A_l^k)\right)\right\}^{1/2} < \infty.$$

□

The approach used in the proof is called *triangular approximation* , due to the geometric intuition from Figure 1.

Write

$$\gamma([0,t]_<^2) = \sum_{k=0}^{\infty}\left\{\sum_{l=0}^{2^k-1}\left(\beta(A_l^k) - \mathbb{E}\beta(A_l^k)\right)\right\} \qquad t \geq 0. \tag{2.4.4}$$

Clearly, $\mathbb{E}\gamma([0,t]_<^2) = 0$. By Proposition 2.3.3, we have

$$\gamma([0,at]_<^2) \stackrel{d}{=} a\gamma([0,t]_<^2) \qquad a > 0. \tag{2.4.5}$$

We introduce the notation $\gamma(A) = \beta(A) - \mathbb{E}\beta(A)$ and some related properties without further notification whenever it makes obvious sense. Given $t' < t$, for example, we can define $\gamma\big([t',t]^2_<\big)$ by performing the same approximation on the triangle $[t',t]^2_<$. It is not hard to establish

$$\gamma\big([t',t]^2_<\big) \overset{d}{=} \gamma\big([0,t-t']^2_<\big) \tag{2.4.6}$$

using Proposition 2.3.3.

In the rest of the section, we solve the following two problems: First, we establish the exponential integrability for $\gamma\big([0,t]^2_<\big)$, in order to give mathematical justification to the polymer models in (2.1.3) and (2.1.4) (with $\beta\big([0,t]^2_<\big)$ being replaced by $\gamma\big([0,t]^2_<\big)$). Second, we prove that the family

$$\Big\{\gamma\big([0,t]^2_<\big);\quad t \geq 0\Big\}$$

yields a continuous modification.

With regard to our first task, we may take $t = 1$ in the following theorem, due to the scaling property given in (2.4.5).

THEOREM 2.4.2. *Let $\gamma([0,t]^2_<)$ be defined by (2.4.4). There is $\lambda > 0$ such that*

$$\mathbb{E}\exp\Big\{\lambda\big|\gamma\big([0,1]^2_<\big)\big|\Big\} < \infty. \tag{2.4.7}$$

Proof. Again, our approach is the triangular approximation based on Figure 1 with $t = 1$.

Let A^k_l be defined by (2.4.2) with $t = 1$. By (2.4.3) one can see that

$$\gamma(A^k_l) \overset{d}{=} 2^{-(k+1)}\Big\{\alpha\big([0,1]^2 - \mathbb{E}\alpha\big([0,1]^2\Big\}.$$

By Theorem 2.2.9 (with $d = p = 2$) there is a $\lambda_1 > 0$ such that

$$\mathbb{E}\exp\Big\{\lambda_1\alpha\big([0,1]^2\big)\Big\} < \infty.$$

By Taylor expansion, there is a $C > 0$ such that for sufficiently small $\lambda > 0$, we have

$$\mathbb{E}\exp\Big\{\lambda\Big(\alpha\big([0,1]^2\big) - \mathbb{E}\alpha\big([0,1]^2\big)\Big)\Big\} \leq e^{C\lambda^2}.$$

Fix $a \in (0,1)$ and a small $\lambda > 0$. For each $N \geq 1$ set

$$b_1 = 2\lambda, \quad b_N = 2\lambda\prod_{j=2}^{N}\Big(1 - 2^{-a(j-1)}\Big) \qquad N = 2,3,\cdots.$$

By Hölder inequality, we have

$$\begin{aligned}
&\mathbb{E}\exp\left\{b_N\sum_{k=0}^{N}\sum_{l=0}^{2^k-1}\Big(\beta(A_l^k)-\mathbb{E}\beta(A_l^k)\Big)\right\}\\
&\le\left[\mathbb{E}\exp\left\{b_{N-1}\sum_{k=0}^{N-1}\sum_{l=0}^{2^k-1}\Big(\beta(A_l^k)-\mathbb{E}\beta(A_l^k)\Big)\right\}\right]^{1-2^{-a(N-1)}}\\
&\times\left[\mathbb{E}\exp\left\{2^{a(N-1)}b_N\sum_{l=0}^{2^N-1}\Big(\beta(A_l^N)-\mathbb{E}\beta(A_l^N)\Big)\right\}\right]^{2^{-a(N-1)}}\\
&\le\mathbb{E}\exp\left\{b_{N-1}\sum_{k=0}^{N-1}\sum_{l=0}^{2^k-1}\Big(\beta(A_l^k)-\mathbb{E}\beta(A_l^k)\Big)\right\}\\
&\times\left[\mathbb{E}\exp\left\{2^{a(N-1)}b_N\Big(\beta(A_0^N)-\mathbb{E}\beta(A_0^N)\Big)\right\}\right]^{2\cdot 2^{(1-a)(N-1)}}.
\end{aligned}$$

Notice that $b_N\le 2\lambda$. We have

$$\begin{aligned}
&\mathbb{E}\exp\left\{2^{a(N-1)}b_N\Big(\beta(A_0^N)-\mathbb{E}\beta(A_0^N)\Big)\right\}\\
&=\mathbb{E}\exp\left\{2^{-N-1+a(N-1)}b_N\Big(\alpha([0,1]^2)-\mathbb{E}\alpha([0,1]^2)\Big)\right\}\\
&\le\exp\left\{Cb_N^2 2^{-2N+2a(N-1)}\right\}.
\end{aligned}$$

Summarizing what we have gives us

$$\begin{aligned}
&\mathbb{E}\exp\left\{b_N\sum_{k=0}^{N}\sum_{l=0}^{2^k-1}\Big(\beta(A_l^k)-\mathbb{E}\beta(A_l^k)\Big)\right\}\\
&\le\mathbb{E}\exp\left\{b_{N-1}\sum_{k=0}^{N-1}\sum_{l=0}^{2^k-1}\Big(\beta(A_l^k)-\mathbb{E}\beta(A_l^k)\Big)\right\}\exp\left\{C'2^{(a-1)N}\right\}.
\end{aligned}$$

Repeating the above procedure gives us

$$\begin{aligned}
&\mathbb{E}\exp\left\{b_N\sum_{k=0}^{N}\sum_{l=0}^{2^k-1}\Big(\beta(A_l^k)-\mathbb{E}\beta(A_l^k)\Big)\right\}\\
&\le\exp\left\{C'\sum_{k=0}^{N}2^{(a-1)k}\right\}\le\exp\left\{C'\Big(1-2^{a-1}\Big)^{-1}\right\}<\infty.
\end{aligned}$$

Observe that

$$b_\infty=2\lambda\prod_{j=2}^{\infty}\Big(1-2^{-a(j-1)}\Big)>0.$$

By Fatou's lemma, letting $N\to\infty$ we have

$$\mathbb{E}\exp\left\{b_\infty\gamma([0,1]_<^2)\right\}\le\exp\left\{C'\Big(1-2^{a-1}\Big)^{-1}\right\}<\infty.$$

In a similar way, we can show that

$$\mathbb{E}\exp\Big\{-b_\infty\gamma\big([0,1]^2_<\big)\Big\}<\infty.$$

We have completed the proof. □

For our second task, we first prove

LEMMA 2.4.3. *For any $T>0$ there is a $c=c(T)>0$ such that*

$$\sup_{\substack{s,t\in[0,T]\\ s\neq t}}\mathbb{E}\exp\left\{c\frac{\big|\gamma\big([0,t]^2_<\big)-\gamma\big([0,s]^2_<\big)\big|}{|t-s|^{1/2}}\right\}<\infty.$$

Proof. For $0\le s<t\le T$, we have

$$\gamma\big([0,t]^2_<\big)-\gamma\big([0,s]^2_<\big)=\gamma\big([s,t]^2_<\big)+\gamma\big([0,s]\times[s,t]\big).$$

Notice that

$$\gamma\big([s,t]^2_<\big)\overset{d}{=}(t-s)\gamma\big([0,1]^2_<\big).$$

By Theorem 2.4.2, there is a $c_1>0$ such that

$$\sup_{\substack{s,t\in[0,T]\\ s<t}}\mathbb{E}\exp\left\{c_1\frac{\big|\gamma\big([s,t]^2_<\big)\big|}{|t-s|}\right\}<\infty.$$

In addition, by Proposition 2.3.4, we get

$$\gamma\big([0,s]\times[s,t]\big)\overset{d}{=}\alpha\big([0,s]\times[0,t-s]\big)-\mathbb{E}\alpha\big([0,s]\times[0,t-s]\big).$$

Checking (2.2.12), we see that

$$\begin{aligned}\mathbb{E}\Big[\alpha\big([0,s]\times[0,t-s]\big)^m\Big]&\le\Big\{\mathbb{E}\Big[\alpha\big([0,s]^2\big)^m\Big]\Big\}^{1/2}\Big\{\mathbb{E}\Big[\alpha\big([0,t-s]^2\big)^m\Big]\Big\}^{1/2}\\&=s^{m/2}(t-s)^{m/2}\mathbb{E}\Big[\alpha\big([0,1]^2\big)^m\Big]\le T^{m/2}(t-s)^{m/2}\mathbb{E}\Big[\alpha\big([0,1]^2\big)^m\Big]\end{aligned}$$

for $m=1,2,\cdots$. By Theorem 2.2.9, there is a $c_2>0$ such that

$$\sup_{\substack{s,t\in[0,T]\\ s<t}}\mathbb{E}\exp\left\{c_2\frac{\big|\gamma\big([0,s]\times[s,t]\big)\big|}{|t-s|^{1/2}}\right\}<\infty.$$

The proof is completed. □

By theorem D.6 in the Appendix and by the extension argument used for $\alpha(A)$, there is a continuous modification $\big\{\widetilde{\gamma}\big([0,t]^2_<\big)\big\}$ of the random family $\big\{\gamma\big([0,t]^2_<\big)\big\}$, such that for any $T>0$ there is a $c=c(T)>0$ such that

$$\mathbb{E}\exp\Big\{c\sup_{0\le s,t\le T}\big|\widetilde{\gamma}\big([0,t]^2_<\big)-\widetilde{\gamma}\big([0,s]^2_<\big)\big|\Big\}<\infty. \tag{2.4.8}$$

Notice that $\gamma\big([0,s]^2_<\big)=0$ as $s=0$. In particular,

$$\mathbb{E}\exp\Big\{c\sup_{0\le t\le T}\big|\widetilde{\gamma}\big([0,t]^2_<\big)\big|\Big\}<\infty. \tag{2.4.9}$$

For the sake of simplicity, from now on we use $\gamma([0,t]^2_<)$, rather than $\widetilde{\gamma}([0,t]^2_<)$, to denote this continuous process, and we call it *renormalized self-intersection local time* generated by the planar Brownian motion $W(t)$.

2.5. Notes and comments

Section 2.1.

For the mathematical description of random polymers, we point out the monographs by Vanderzande ([**162**]) and, more recently, by den Hollander ([**98**]). In addition, we mention the excellent exposition on this subject by Westwater ([**165**], [**166**], [**167**]).

Section 2.2.

We refer the interested reader to the book by Lawler ([**116**]) and the monograph notes by Le Gall ([**129**]) for the general information on the intersections and self-intersections of Brownian motions and random walks. In addition to the historic works by Dvoretzsky, Erdös and Kakutani ([**66**], [**67**]), we mention the survey paper by Khoshnevisan ([**110**]) for an elementary proof of Theorem 2.2.1.

The first work on the intersection local times of independent Brownian motions is the paper by Wolpert ([**168**]). Dynkin ([**68**]) developed a way to construct additive functionals of several independent Markov processes, which includes Brownian intersection local times as a special case. In [**69**], Dynkin treated the special case of Brownian motions. Our approach for constructing $\alpha(A)$ is modified from Le Gall's work [**129**].

The alternative treatment which views intersection local time as local time can be found in Geman and Horowitz [**85**] and in Geman, Horowitz and Rosen [**86**]. The following two exercises below describe some of the ideas appearing in these works.

EXERCISE 2.5.1. *For $\epsilon > 0$, let the random measure $\alpha_\epsilon(A)$ be defined by (2.2.4) and let $\tau_1, \cdots, \tau_p$ be i.i.d. exponential times with parameter 1 such that $\{\tau_1, \cdots, \tau_p\}$ and $\{W_1(t), \cdots, W_p(t)\}$ are independent.*

(1) *Establish the following version of Le Gall's moment identity:*

$$\begin{aligned}(2.5.1)\qquad &\mathbb{E}\Big[\alpha_\epsilon\big([0,\tau_1]\times\cdots\times[0,\tau_p]\big)^m\Big]\\ &= \int_{(\mathbb{R}^d)^m} dx_1\cdots dx_m\bigg[\int_{(\mathbb{R}^d)^m} dz_1\cdots dz_m\Big(\prod_{k=1}^m p_\epsilon(z_k-x_k)\Big)\\ &\qquad\times\sum_{\sigma\in\Sigma_m}\prod_{k=1}^m G(z_{\sigma(k)}-z_{\sigma(k-1)})\bigg]^p.\end{aligned}$$

Hint: Read the computation for (2.2.11)

(2) *Prove that for any $\epsilon > 0$ and $m = 1, 2, \cdots$,*

$$(2.5.2)\qquad \mathbb{E}\Big[\alpha_\epsilon\big([0,\tau_1]\times\cdots\times[0,\tau_p]\big)^m\Big] \le \mathbb{E}\Big[\alpha\big([0,\tau_1]\times\cdots\times[0,\tau_p]\big)^m\Big].$$

(3) *Prove that*

$$\lim_{\epsilon\to 0^+} \alpha_\epsilon\big([0,\tau_1]\times\cdots\times[0,\tau_p]\big) = \alpha\big([0,\tau_1]\times\cdots\times[0,\tau_p]\big) \tag{2.5.3}$$

in $\mathcal{L}^m(\Omega,\mathcal{A},\mathbb{P})$ *for all* $m>0$.

EXERCISE 2.5.2. *Establish the absolute continuity of the occupation time*

$$\mu_t(B) = \int_0^t 1_B\big(W(s)\big)ds \qquad B\subset\mathbb{R},\quad t\geq 0$$

of the 1-dimensional Brownian motion $W(t)$ *by showing*

$$\mathbb{P}\Big\{\text{For all } t\geq 0,\ \mu_t(\cdot) \text{ is absolutely continuous with respect to the Lebesgue measure on } \mathbb{R}\Big\} = 1.$$

Hint: To avoid losing an exceptional event for each $t\in\mathbb{R}^+$*, you may introduce the bounded measure*

$$\mu(B) = \int_0^\infty e^{-t}1_B\big(W(t)\big)dt \qquad B\subset\mathbb{R}.$$

Since for any $t\geq 0$, $\mu_t(\cdot)\leq e^t\mu(\cdot)$. *Hence, all you need is to show that*

$$\mathbb{P}\Big\{\mu(\cdot) \text{ is absolutely continuous with respect to the Lebesgue measure on } \mathbb{R}\Big\} = 1. \tag{2.5.4}$$

The tool for (2.5.4) is Fourier analysis. Use the Plancherel-Parseval theorem (Theorem B.3, Appendix) to reduce the problem to examining

$$\int_{\mathbb{R}} \mathbb{E}|\widehat{\mu}(\lambda)|^2 d\lambda < \infty. \tag{2.5.5}$$

Does (2.5.5) hold when $d\geq 2$*? Why?*

EXERCISE 2.5.3 (hard). *Let* $p(d-2)<d$ *and let* $W_1(t),\cdots,W_p(t)$ *be independent d-dimensional Brownian motions. For each* $\mathbf{t}=(t_1,\cdots,t_p)\in(\mathbb{R}^+)^p$*, define the random measure*

$$\mu_{\mathbf{t}}(B) = \int_{[\mathbf{0},\mathbf{t}]} 1_B\big(W_1(s_1)-W_2(s_2),\cdots,W_{p-1}(s_{p-1})-W_p(s_p)\big)ds_1\cdots ds_p$$

where $B\subset\mathbb{R}^{d(p-1)}$.

Prove that

$$\mathbb{P}\Big\{\mu_A(\cdot) \text{ is absolutely continuous with respect to the Lebesgue measure on } \mathbb{R}^{d(p-1)} \text{ for every measurable and bounded } A\subset(\mathbb{R}^+)^p\Big\} = 1.$$

Hint: Define the random measure $\mu(\cdot)$ *on* $\mathbb{R}^{d(p-1)}$ *as*

$$\mu(B) = \int_0^\infty \cdots \int_0^\infty ds_1 \cdots ds_p e^{-(s_1+\cdots+s_p)}$$
$$\times 1_B\big(W_1(s_1) - W_2(s_2), \cdots, W_{p-1}(s_{p-1}) - W_p(s_p)\big) \qquad B \subset \mathbb{R}^{d(p-1)}$$

and then establish the identity

$$\int_{(\mathbb{R}^d)^{p-1}} \mathbb{E}|\widehat{\mu}(\lambda_1, \cdots, \lambda_{p-1})|^2 d\lambda_1 \cdots d\lambda_{p-1} = (2\pi)^{d(p-1)} \int_{\mathbb{R}^d} G^p(x)dx.$$

EXERCISE 2.5.4. *Let* $L(t,x)$ *be the local time of the 1-dimensional Brownian motion* $W(t)$. *Prove the identity in law that for any* $t > 0$, *we get*

$$\Big\{L(t,x);\quad x \in \mathbb{R}\Big\} \overset{d}{=} \Big\{\sqrt{t}L(1, x/\sqrt{t});\quad x \in \mathbb{R}\Big\}. \tag{2.5.6}$$

Hint: Compare two occupation measures

$$\int_0^t 1_B\big(W(s)\big)ds \quad \textit{and} \quad \int_0^1 1_B\big(\sqrt{t}W(s)\big)ds \qquad (B \subset \mathbb{R}).$$

EXERCISE 2.5.5. *Let* $L(t,x)$ *be the local time of the 1-dimentional Brownian motion* $W(t)$ *and define, for every* $\epsilon > 0$,

$$L(t,x,\epsilon) = \int_0^t p_\epsilon\big(W(s) - x\big)ds \qquad x \in \mathbb{R}, \quad t \geq 0.$$

(1) *Prove that for each integer* $m \geq 1$,

$$\mathbb{E}L^m(t,x,\epsilon) = m! \int_{[0,t]^m_<} ds_1 \cdots ds_m \int_{\mathbb{R}^m} dy_1 \cdots dy_m$$
$$\times \prod_{k=1}^m p_\epsilon(y_k - x)p_{s_k - s_{k-1}}(y_k - y_{k-1}).$$

(2) *Prove that for any* $x \in \mathbb{R}$ *and* $t > 0$, $L(t,x,\epsilon)$ *converges to* $L(t,x)$ *in* $\mathcal{L}^m(\Omega, \mathcal{A}, \mathbb{P})$ *for all integers* $m \geq 1$, *and that*

$$\mathbb{E}L^m(t,x) = m! \int_{[0,t]^m_<} ds_1 \cdots ds_m p_{s_1}(x) \prod_{k=2}^m p_{s_k - s_{k-1}}(0). \tag{2.5.7}$$

EXERCISE 2.5.6. *We point out the elementary fact that two random variables* X *and* Y *have the same distribution if* $\mathbb{E}e^{\lambda|X|} < \infty$ *for every* $\lambda > 0$ *and if* $\mathbb{E}X^m = \mathbb{E}Y^m$ *for* $m = 1, 2, \cdots$. *Let* $L(t,x)$ *be the local time of a 1-dimensional Brownian motion* $W(t)$. *Establish the identity in law*

$$L(1,0) \overset{d}{=} |W(1)| \tag{2.5.8}$$

by showing that

$$\mathbb{E}L^m(1,0) = \mathbb{E}|W(1)|^m \qquad m = 1, 2, \cdots.$$

Hint: Let τ be an exponential time with parameter 1 and assume independence between τ and $W(t)$. First compute $\mathbb{E}L^m(\tau,0)$. In the end, you may need the identity

$$m! = \frac{2^m}{\sqrt{\pi}}\Gamma\Big(\frac{m+2}{2}\Big)\Gamma\Big(\frac{m+1}{2}\Big).$$

EXERCISE 2.5.7. *Le Gall's moment identity can be extended to the general Markov processes. Recall that a stochastic process $\{X(t);\ t\in\mathbb{R}^+\}$ with the state space E being called a time-homogeneous Markov process if for any $s,t\in\mathbb{R}^+$, we have*

$$\mathbb{P}\big\{X(t)\in A\big|X(u);\ 0\le u\le s\big\} = \mathbb{P}\big\{X(t-s)\in A\big|X(0)\big\}$$

for every measurable subset $A\subset E$.

To avoid technical difficulty, the state space E is countable in the following discussion. The function

$$p_t(x,y) = \mathbb{P}\{X(t)=y|X(0)=x\} \qquad x,y\in E \qquad t\ge 0$$

is called the transition probability.

Let $X_1(t),\cdots,X_p(t)$ be independent copies of $X(t)$ and assume that $X_1(0)=\cdots=X_p(0)=x_0$ for some deterministic point $x_0\in E$. Define the mutual intersection local time (with respect to $\pi(\cdot)$)

$$I\big([0,t_1]\times\cdots\times[0,t_p]\big) = \int_{[0,t_1]\times\cdots\times[0,t_p]} 1_{\{X_1(s_1)=\cdots=X_p(s_p)\}}ds_1\cdots ds_m. \tag{2.5.9}$$

(1) *Given an integer $m\ge 1$, derive Le Gall's moment identities*

$$\mathbb{E}\Big[I\big([0,t_1]\times\cdots\times[0,t_p]\big)^m\Big] = \sum_{x_1,\cdots,x_m\in E}\prod_{j=1}^p\sum_{\sigma\in\Sigma_m}\int_{[0,t_j]^m_<}\prod_{k=1}^m p_{s_k-s_{k-1}}(x_{\sigma(k-1)},x_{\sigma(k)})ds_1\cdots ds_m \tag{2.5.10}$$

$$\mathbb{E}\Big[I\big([0,\tau_1]\times\cdots\times[0,\tau_p]\big)^m\Big] = \lambda^p\sum_{x_1,\cdots,x_m\in E}\Big[\sum_{\sigma\in\Sigma_m}\prod_{k=1}^m G_\lambda(x_{\sigma(k-1)},x_{\sigma(k)})\Big]^p \tag{2.5.11}$$

where $\sigma(0)=0$, $s_0=0$, $\tau_1,\cdots,\tau_p$ are independent i.i.d. exponential times with the parameter $\lambda>0$ (i.e., $\mathbb{E}\tau_1=\lambda^{-1}$), and finally,

$$G_\lambda(x,y) = \int_0^\infty e^{-\lambda t}p_t(x,y)dt \qquad x,y\in E.$$

(2) *Prove that for any $t_1,\cdots,t_p>0$,*

$$\mathbb{E}\Big[I\big([0,t_1]\times\cdots\times[0,t_p]\big)^m\Big] \le \prod_{j=1}^p\Big\{\mathbb{E}\Big[I\big([0,t_j]^p\big)^m\Big]\Big\}^{1/p}. \tag{2.5.12}$$

Section 2.3.

The existence and continuity of self-intersection local time of Brownian motion has been investigated by Rosen ([**146**], [**147**]), by Yor ([**172**], [**173**]) and by Le Gall ([**129**]). The study has been extended to the setting of diffusion processes by Rosen ([**150**]) and to the setting of Lévy processes by Le Gall ([**126**]). The method used here is due to Le Gall ([**129**]). By a different method, Rosen ((2.15), [**147**]) estimated the moment of $\beta(I)$ for I in a class wider than the one defined in (2.3.5), in the setting of fractional Brownian motions.

EXERCISE 2.5.8. *We demonstrate here that when $p = 2$, the construction of $\beta(A)$ can be simplified and the identity in law given in Proposition 2.3.4 can be slightly strengthened.*

(1) *Let $h(x)$ be a bounded function on $\mathbb{R}^d$ and let $W(t)$ and $\widetilde{W}(t)$ be two independent d-dimensional Brownian motions. Prove that for any two non-negative numbers $a < b$, we get*

$$\Big\{\iint_{[a,s]\times[b,t]} h\big(W(v)-W(u)\big)dudv;\quad a\le s\le b \text{ and } t\ge b\Big\}$$
$$\stackrel{d}{=}\Big\{\iint_{[b-s,\,b-a]\times[0,\,t-b]} h\big(\widetilde{W}(v)-W(u)\big)dudv;\quad a\le s\le b \text{ and } t\ge b\Big\}.$$

(2) *In the special case $p = 2$ and $d \le 3$, simplify the construction of the random measure $\beta(A)$ in the special case $p = 2$ and $d \le 3$ and prove that for any $0 \le a < b$, we get*

$$\Big\{\beta\big([a,s]\times[b,t]\big);\quad a\le s\le b,\ t\ge b\Big\}$$
$$\stackrel{d}{=}\Big\{\alpha\big([b-s,\,b-a]\times[0,\,t-b]\big);\quad a\le s\le b,\ b\ge 0\Big\}.$$

EXERCISE 2.5.9. *Prove Proposition 2.3.6 in the case $d \ge 3$.*

Hint: You may need the following reverse Markov inequality: For any $0<\theta<1$ and any random variable $X \in \mathcal{L}^2(\Omega, \mathcal{A}, \mathbb{P})$,

$$\mathbb{P}\{|X|\ge\theta||X||_1\}\ge(1-\theta)^2\frac{||X||_1}{||X||_2}$$

where for any $\gamma>0$, $||X||_\gamma = \big(\mathbb{E}|X|^\gamma\big)^{1/\gamma}$.

Section 2.4.

The idea of renormalization goes back to the paper by Varadhan ([**160**]), where for the first time the method of triangular approximation was introduced for some related but different models. This brilliant idea was later adopted by Le Gall ([**122**], [**129**], [**130**]) to the setting of self-intersection local time ([**129**]). The method used in this section comes from [**129**] and [**130**].

An alternative way of renormalization is by the Tanaka formula. We refer to Rosen and Yor ([**148**] , [**170**], [**153**]) for this approach. See the paper [**70**] by Dynkin for a different treatment.

The renormalization of the p-multiple self-intersection local times is far more complicated as $p \geq 3$. Given a planar Brownian motion $W(t)$, the stochastic process $\gamma([0,t]^p_<)$ with the symbolic form

$$\gamma([0,t]^p_<) = \int_{[0,t]^p_<} \overline{\prod_{j=2}^{p} \delta_0\big(W(s_{j-1}) - W(s_j)\big) ds_1 \cdots ds_p} \qquad \forall t \geq 0 \tag{2.5.13}$$

can be properly defined and is known as the *renormalized p-multiple self-intersection local time*. Here $\overline{X} \equiv X - \mathbb{E}X$ for any random variable X. We refer the interested reader to the works by Rosen ([**149**]), Rosen and Yor ([**153**]), Le Gall ([**129**]) for a mathematical construction of $\gamma([0,t]^p_<)$.

In the case $d \geq 3$, the self-intersection local time of a d-dimensional Brownian motion cannot be renormalized in the same way as when $d = 2$. Indeed, Yor ([**171**]), Calais and Yor ([**22**]) have proved that the random variables

$$\begin{cases} \big(\log(1/\epsilon)\big)^{-1/2}\Big\{\beta_\epsilon([0,1]^2_<) - \mathbb{E}\beta_\epsilon([0,1]^2_<)\Big\} & \text{if } d = 3, \\ \epsilon^{\frac{d-3}{2}}\Big\{\beta_\epsilon([0,1]^2_<) - \mathbb{E}\beta_\epsilon([0,1]^2_<)\Big\} & \text{if } d \geq 4 \end{cases}$$

weakly converge to symmetric normal distributions as $\epsilon \to 0^+$. In particular, the blow-up of the renormalized quantity

$$\beta_\epsilon([0,1]^2_<) - \mathbb{E}\beta_\epsilon([0,1]^2_<)$$

(as $\epsilon \to 0^+$) shows that the renormalization fails in the case $d \geq 3$. In addition, Calais and Yor's observation links to the central limit theorem stated later in Theorem 5.5.3 in a natural way. See also the paper [**99**] by Hu and Nualart for an extension of Calais and Yor's work to the setting of fractional Brownian motions.

CHAPTER 3

Mutual intersection: large deviations

3.1. High moment asymptotics

One of the major goals of this chapter is to provide a precise estimate to the tail probability

$$\mathbb{P}\Big\{\alpha\big([0,1]^p\big) \ge t\Big\}$$

as $t \to \infty$. By Theorem 1.2.8, the problem is in connection to the study of the *high moment asymptotics* posted as

$$\mathbb{E}\Big[\alpha([0,1]^p)^m\Big] \qquad (m \to \infty).$$

By comparing (2.2.8) and (2.2.19), it seems that the high moment asymptotics for

$$\mathbb{E}\Big[\alpha\big([0,\tau_1] \times \cdots \times [0,\tau_p]\big)^m\Big]$$

more tractable. The question is: What can we say about the quantity

$$\int_{(\mathbb{R}^d)^m} dx_1 \cdots dx_m \Big[\sum_{\sigma \in \Sigma_m} \prod_{k=1}^m G(x_{\sigma(k)} - x_{\sigma(k-1)})\Big]^p$$

when $m \to \infty$?

We generalize this problem to the study of

$$\int_{E^m} \pi(dx_1) \cdots \pi(dx_m) \Big[\sum_{\sigma \in \Sigma_m} \prod_{k=1}^m K(x_{\sigma(k-1)}, x_{\sigma(k)})\Big]^p$$

where $(E, \mathcal{E}, \pi)$ is a measure space, $K(x,y)$ is a non-negative measurable function on $E \times E$, and $x_{\sigma(0)} = x_0$ is an arbitrary but fixed point in E.

THEOREM 3.1.1. *Let $p > 1$ be fixed. Assume that the kernel satisfies:*

(1) *Irreducibility: $K(x,y) > 0$ for every $(x,y) \in E \times E$.*
(2) *Symmetry: $K(x,y) = K(y,x)$ for every $(x,y) \in E \times E$.*
(3) *Integrability:*

$$\varrho \equiv \sup_f \iint_{E\times E} K(x,y) f(x) f(y) \pi(dx)\pi(dy) < \infty$$

where the supremum is taken over all f satisfying

$$\int_E |f(x)|^{\frac{2p}{2p-1}} \pi(dx) = 1.$$

Then

$$\liminf_{m\to\infty}\frac{1}{m}\log\frac{1}{(m!)^p}\int_{E^m}\pi(dx_1)\cdots\pi(dx_m)\Big[\sum_{\sigma\in\Sigma_m}\prod_{k=1}^m K(x_{\sigma(k-1)},x_{\sigma(k)})\Big]^p$$

$$\geq p\log\varrho. \tag{3.1.1}$$

Proof. Let $f\in\mathcal{L}^{\frac{2p}{2p-1}}(E,\mathcal{E},\pi)$ be a bounded, non-negative function such that

$$\delta\equiv\inf_{x;\ f(x)>0}K(x_0,x)>0 \tag{3.1.2}$$

and that

$$\int_E f^{\frac{2p}{2p-1}}(x)\pi(dx)=1.$$

Let $g(x)=f^{\frac{2(p-1)}{2p-1}}(x)$ and $h(x)=f^{\frac{p}{2p-1}}(x)$. Then $f(x)=\sqrt{g(x)}h(x)$ and

$$\int_E g^{\frac{p}{p-1}}(x)\pi(dx)=\int_E h^2(x)\pi(dx)=1.$$

By boundedness of $f(x)$, there is $\epsilon>0$ such that $g(x)\geq\epsilon h^2(x)$. By Hölder inequality, we get

$$\Big\{\int_{E^m}\pi(dx_1)\cdots\pi(dx_m)\Big[\sum_{\sigma\in\Sigma_m}\prod_{k=1}^m K(x_{\sigma(k-1)},x_{\sigma(k)})\Big]^p\Big\}^{1/p} \tag{3.1.3}$$

$$\begin{aligned}
&\geq\int_{E^m}\pi(dx_1)\cdots\pi(dx_m)\Big(\prod_{k=1}^m g(x_k)\Big)\sum_{\sigma\in\Sigma_m}\prod_{k=1}^m K(x_{\sigma(k-1)},x_{\sigma(k)})\\
&=m!\int_{E^m}\pi(dx_1)\cdots\pi(dx_m)\Big(\prod_{k=1}^m g(x_k)\Big)\prod_{k=1}^m K(x_{k-1},x_k)\\
&=m!\int_{E^m}\pi(dx_1)\cdots\pi(dx_m)K(x_0,x_1)\sqrt{g(x_1)}\\
&\quad\times\Big[\prod_{k=2}^m\sqrt{g(x_{k-1})}K(x_{k-1},x_k)\sqrt{g(x_k)}\Big]\sqrt{g(x_m)}\\
&\geq\delta\epsilon m!\int_{E^m}\pi(dx_1)\cdots\pi(dx_m)h(x_1)\\
&\quad\times\Big[\prod_{k=2}^m\sqrt{g(x_{k-1})}K(x_{k-1},x_k)\sqrt{g(x_k)}\Big]h(x_m)\\
&=\delta\epsilon m!\langle h,T^{m-1}h\rangle
\end{aligned}$$

where the linear operator T: $\mathcal{L}^2(E,\mathcal{E},\pi)\longrightarrow\mathcal{L}^2(E,\mathcal{E},\pi)$ is defined as

$$(T\tilde{h})(x)=\sqrt{g(x)}\int_E K(x,y)\sqrt{g(y)}\tilde{h}(y)\pi(dy)\qquad\tilde{h}\in\mathcal{L}^2(E,\mathcal{E},\pi).$$

For any $h_1,h_2\in\mathcal{L}^2(E,\mathcal{E},\pi)$ with $||h_1||=||h_2||=1$, we get

$$\begin{aligned}
\langle h_1,Th_2\rangle&=\iint_{E\times E}K(x,y)\sqrt{g(x)}h_1(x)\sqrt{g(y)}h_2(y)\pi(dx)\pi(dy)\\
&=\iint_{E\times E}K(y,x)\sqrt{g(x)}h_1(x)\sqrt{g(y)}h_2(y)\pi(dx)\pi(dy)=\langle Th_1,h_2\rangle.
\end{aligned}$$

That is, T is symmetric. In addition,

$$\langle h_1, Th_2\rangle = \frac{1}{4}\Big\{\langle h_1+h_2, T(h_1+h_2)\rangle - \langle h_1-h_2, T(h_1-h_2)\rangle\Big\}.$$

Notice that

$$\begin{aligned}&\Big|\langle h_1 \pm h_2, T(h_1 \pm h_2)\rangle\Big|\\ &\le \int\int_{E\times E} K(x,y)\sqrt{g(x)}\big|h_1(x)\pm h_2(x)\big|\sqrt{g(y)}\big|h_1(y)\pm h_2(y)\big|\pi(dx)\pi(dy).\end{aligned}$$

By Hölder inequality, we get

$$\begin{aligned}&\int_E \big|\sqrt{g(x)}\big(h_1(x)\pm h_2(x)\big)\big|^{\frac{2p}{2p-1}}\pi(dx)\\ &\le \Big\{\int_E g^{\frac{p}{p-1}}(x)\pi(dx)\Big\}^{\frac{p-1}{2p-1}}\Big\{\int_E |h_1(x)\pm h_2(x)|^2\pi(dx)\Big\}^{\frac{p}{2p-1}}\\ &= \Big\{\int_E f^{\frac{2p}{2p-1}}(x)\pi(dx)\Big\}^{\frac{p-1}{2p-1}}\Big\{\int_E |h_1(x)\pm h_2(x)|^2\pi(dx)\Big\}^{\frac{p}{2p-1}}\\ &= \Big\{\int_E |h_1(x)\pm h_2(x)|^2\pi(dx)\Big\}^{\frac{p}{2p-1}}.\end{aligned}$$

Consequently,

$$\Big|\langle h_1 \pm h_2, T(h_1 \pm h_2)\rangle\Big| \le \varrho\int_E |h_1(x)\pm h_2(x)|^2\pi(dx).$$

Hence,

$$\langle h_1, Th_2\rangle \le \frac{1}{4}\varrho\{||h_1+h_2||^2 + ||h_1-h_2||^2\} = \frac{\varrho}{2}\big(||h_1||^2 + ||h_2||^2\big) = \varrho.$$

Therefore, T is a bounded linear operator. This, together with symmetry, implies that T is self-adjoint.

According to Theorem E.2 in the Appendix, the self-adjoint operator T admits the spectral integral representation

$$T = \int_{-\infty}^{\infty}\lambda E(d\lambda). \tag{3.1.4}$$

Using Corollary E.5 in the Appendix, we have

$$T^{m-1} = \int_{-\infty}^{\infty}\lambda^{m-1}E(d\lambda). \tag{3.1.5}$$

By (E.15) in the Appendix, the above representations lead to

$$\langle h, Th\rangle = \int_{-\infty}^{\infty}\lambda\mu_h(d\lambda), \tag{3.1.6}$$

$$\langle h, T^{m-1}h\rangle = \int_{-\infty}^{\infty}\lambda^{m-1}\mu_h(d\lambda) \tag{3.1.7}$$

where μ_h is a measure on $\mathbb{R}$ with

$$\mu_h(\mathbb{R}) = \int_E h^2(x)\pi(dx) = 1$$

(i.e., μ_h is a probability measure).

When m is odd (so $m-1$ is even), by Jensen inequality, we have

$$\langle h, T^{m-1}h\rangle \geq \Big(\int_{-\infty}^{\infty} \lambda\mu_h(d\lambda)\Big)^{m-1} = \Big(\langle h, Th\rangle\Big)^{m-1}$$
$$= \Big(\iint_{E\times E} K(x,y)\sqrt{g(x)}h(x)\sqrt{g(y)}h(y)\pi(dx)\pi(dy)\Big)^{m-1}$$
$$= \Big(\iint_{E\times E} K(x,y)f(x)f(y)\pi(dx)\pi(dy)\Big)^{m-1}.$$

Summarizing our argument, when m is odd, we have

$$\int_{E^m} \pi(dx_1)\cdots\pi(dx_m)\Big[\sum_{\sigma\in\Sigma_m}\prod_{k=1}^m K(x_{\sigma(k-1)}, x_{\sigma(k)})\Big]^p$$
$$\geq (\delta\epsilon)^p(m!)^p\Big(\iint_{E\times E} K(x,y)f(x)f(y)\pi(dx)\pi(dy)\Big)^{p(m-1)}.$$

When m is even, a slight modification of (3.1.3) gives us

$$\Big\{\int_{E^m} \pi(dx_1)\cdots\pi(dx_m)\Big[\sum_{\sigma\in\Sigma_m}\prod_{k=1}^m K(x_{\sigma(k-1)}, x_{\sigma(k)})\Big]^p\Big\}^{1/p}$$
$$\geq c(\delta,\epsilon)m!\langle h, T^{m-2}h\rangle$$

for some $c(\delta,\epsilon) > 0$. So the argument based on spectral representation leads to the lower bound

$$\int_{E^m} \pi(dx_1)\cdots\pi(dx_m)\Big[\sum_{\sigma\in\Sigma_m}\prod_{k=1}^m K(x_{\sigma(k-1)}, x_{\sigma(k)})\Big]^p$$
$$\geq c^p(\delta,\epsilon)(m!)^p\Big(\iint_{E\times E} K(x,y)f(x)f(y)\pi(dx)\pi(dy)\Big)^{p(m-2)}.$$

Thus, we conclude that

$$\liminf_{m\to\infty}\frac{1}{m}\log\frac{1}{(m!)^p}\int_{E^m} \pi(dx_1)\cdots\pi(dx_m)\Big[\sum_{\sigma\in\Sigma_m}\prod_{k=1}^m K(x_{\sigma(k-1)}, x_{\sigma(k)})\Big]^p$$
$$\geq p\log\iint_{E\times E} K(x,y)f(x)f(y)\pi(dx)\pi(dy).$$

Finally, the desired conclusion follows from the following two facts:

First, the supremum in the definition of ϱ can be taken over the non-negative f.

Second, by irreducibility assumption the set of functions f satisfying (3.1.2) is dense in non-negative functions in $\mathcal{L}^{\frac{2p}{2p-1}}(E,\mathcal{E},\pi)$. □

The upper bound is much more difficult. Indeed, we are able to establish the upper bound only in the case when E is a finite set.

THEOREM 3.1.2. *Fix $p > 1$, let E be a finite set and let $K\colon E \times E \longrightarrow \mathbb{R}^+$ be non-negative function such that $K(x,y) = K(y,x)$ for any $x, y \in E$. Let π be a non-negative function on E. Then*

$$\begin{aligned}
&\limsup_{m\to\infty} \frac{1}{m} \log \frac{1}{(m!)^p} \sum_{x_1,\cdots,x_m\in E} \pi(x_1)\cdots\pi(x_m) \\
&\times \Big[\sum_{\sigma\in\Sigma_m} \prod_{k=1}^m K(x_{\sigma(k-1)}, x_{\sigma(k)})\Big]^p \qquad (3.1.8)\\
&\le p \log \sup_f \sum_{x,y\in E} K(x,y) f(x) f(y) \pi(x)\pi(y)
\end{aligned}$$

where the supremum is taken over all functions f on Ω satisfying

$$\sum_{x\in E} |f(x)|^{\frac{2p}{2p-1}} \pi(x) = 1.$$

Proof. We may assume that $\pi(x) > 0$ for every $x \in E$, for otherwise we can remove all zero points of π from E. Let

$$\mu = L_m^{\mathbf{x}} = \frac{1}{m} \sum_{k=1}^m \delta_{x_k}$$

be the empirical measure generated by $\mathbf{x} = (x_1, \cdots, x_m)$. Notice that for each $\sigma \in \Sigma_m$, we have

$$\sum_{y_1,\cdots,y_m\in E} 1_{\{L_m^{\mathbf{y}}=\mu\}} 1_{\{\mathbf{x}\circ\sigma=\mathbf{y}\}} = 1.$$

We have

$$\begin{aligned}
&\sum_{\sigma\in\Sigma_m} \prod_{k=1}^m K(x_{\sigma(k-1)}, x_{\sigma(k)}) \\
&= \sum_{y_1,\cdots,y_m\in E} 1_{\{L_m^{\mathbf{y}}=\mu\}} \sum_{\sigma\in\Sigma_m} 1_{\{\mathbf{x}\circ\sigma=\mathbf{y}\}} \prod_{k=1}^m K(x_{\sigma(k-1)}, x_{\sigma(k)}) \\
&= \sum_{y_1,\cdots,y_m\in E} 1_{\{L_m^{\mathbf{y}}=\mu\}} \prod_{k=1}^m K(y_{k-1}, y_k) \sum_{\sigma\in\Sigma_m} 1_{\{\mathbf{x}\circ\sigma=\mathbf{y}\}}.
\end{aligned}$$

We claim that

$$\sum_{\sigma\in\Sigma_m} 1_{\{\mathbf{x}\circ\sigma=\mathbf{y}\}} = \prod_{x\in E} (m\mu(x))!.$$

Indeed, for each $x \in E$ there are, respectively, exactly $m\mu(x)$ of $x_1, \cdots, x_m$ and exactly $m\mu(x)$ of $y_1, \cdots, y_m$ which are equal to x. Therefore, there are $(m\mu(x))!$ ways to match each x-valued component of $\mathbf{y}$ with each x-valued component of $\mathbf{x}$. Thus, the claim follows from multiplication principle.

Consequently,

$$\sum_{\sigma\in\Sigma_m}\prod_{k=1}^m K(x_{\sigma(k-1)},x_{\sigma(k)})$$
$$=\prod_{x\in E}(m\mu(x))!\sum_{y_1,\cdots,y_m\in E}1_{\{L_m^{\mathbf{y}}=\mu\}}\prod_{k=1}^m K(y_{k-1},y_k).$$

Let $q>1$ be the conjugate number of p and define $\varphi_\mu(x)=\mu(x)^{1/q}\pi(x)^{1/p}$. Then

$$\sum_{y_1,\cdots,y_m\in E}\varphi_\mu(y_1)\cdots\varphi_\mu(y_m)\prod_{k=1}^m K(y_{k-1},y_k)$$
$$\geq \sum_{y_1,\cdots,y_m\in E}\varphi_\mu(y_1)\cdots\varphi_\mu(y_m)1_{\{L_m^{\mathbf{y}}=\mu\}}\prod_{k=1}^m K(y_{k-1},y_k)$$
$$=\Big(\prod_{x\in E}\varphi_\mu(x)^{m\mu(x)}\Big)\sum_{y_1,\cdots,y_m\in E}1_{\{L_m^{\mathbf{y}}=\mu\}}\prod_{k=1}^m K(y_{k-1},y_k)$$

where the last step follows from the fact that when $L_m^{\mathbf{y}}=\mu$, there are $m\mu(x)$ factors in the product $\varphi_\mu(x_1)\cdots\varphi_\mu(x_m)$ which are equal to $\varphi_\mu(x)$ for any $x\in E$.

Summarizing the above steps, we have

$$\frac{1}{(m!)^p}\sum_{x_1,\cdots,x_m\in E}\pi(x_1)\cdots\pi(x_m)\Big[\sum_{\sigma\in\Sigma_m}\prod_{k=1}^m K(x_{\sigma(k-1)},x_{\sigma(k)})\Big]^p$$
$$\leq \sum_{x_1,\cdots,x_m\in E}\pi(x_1)\cdots\pi(x_m)\Big[\frac{1}{m!}\Big(\prod_{x\in E}(m\mu(x))!\Big)\Big(\prod_{x\in E}\varphi_\mu(x)^{m\mu(x)}\Big)^{-1}\Big]^p$$
$$\times\Big[\sum_{y_1,\cdots,y_m\in E}\varphi_\mu(y_1)\cdots\varphi_\mu(y_m)\prod_{k=1}^m K(y_{k-1},y_k)\Big]^p.$$

Define $g_\mu(x)=\mu(x)^{1/q}\pi(x)^{-1/q}$. Then $\varphi_\mu(x)=g_\mu(x)\pi(x)$ for every $x\in E$ and

$$\sum_{x\in E}g_\mu^q(x)\pi(x)=\sum_{x\in E}\mu(x)=1,\qquad \sup_{x\in E}g_\mu(x)\leq\sup_{x\in E}\pi(x)^{-1/q}.$$

Consequently,

$$\frac{1}{(m!)^p}\sum_{x_1,\cdots,x_m\in E}\pi(x_1)\cdots\pi(x_m)\Big[\sum_{\sigma\in\Sigma_m}\prod_{k=1}^m K(x_{\sigma(k-1)},x_{\sigma(k)})\Big]^p \tag{3.1.9}$$
$$\leq \sum_{x_1,\cdots,x_m\in E}\pi(x_1)\cdots\pi(x_m)\Big[\frac{1}{m!}\Big(\prod_{x\in E}(m\mu(x))!\Big)\Big(\prod_{x\in E}\varphi_\mu(x)^{m\mu(x)}\Big)^{-1}\Big]^p$$
$$\times\Big[\sup_g\sum_{y_1,\cdots,y_m\in E}\pi(y_1)\cdots\pi(y_m)\Big(\prod_{k=1}^m g(y_k)\Big)\prod_{k=1}^m K(y_{k-1},y_k)\Big]^p$$

where the supremum on the right hand side is taken for all non-negative functions g on E satisfying

$$\sum_{x\in E}g^q(x)\pi(x)=1,\qquad \sup_{x\in E}g(x)\leq c$$

and where $c = \sup_{x\in E} \pi(x)^{-1/q}$.

For each g,

$$\begin{aligned}
&\sum_{y_1,\cdots,y_m\in E} \pi(y_1)\cdots\pi(y_m)\Big(\prod_{k=1}^m g(y_k)\Big)\prod_{k=1}^m K(y_{k-1},y_k)\\
&= \sum_{y_1,\cdots,y_m\in E} \pi(y_1)\cdots\pi(y_m)K(x_0,y_1)\sqrt{g(y_1)}\\
&\times\left[\sqrt{g(y_{k-1})}\prod_{k=2}^m K(y_{k-1},y_k)\sqrt{g(x_k)}\right]\sqrt{g(x_m)}\\
&\le c\big(\sup_{x,y\in E} K(x,y)\big)\Big(\sum_{x\in E}\pi(x)\Big)\sum_{y_1,\cdots,y_m\in E}\pi(y_1)\cdots\pi(y_m)h_0(y_1)\\
&\times\left[\sqrt{g(y_{k-1})}\prod_{k=2}^m K(y_{k-1},y_k)\sqrt{g(x_k)}\right]h_0(x_m)\\
&= c\big(\sup_{x,y\in E} K(x,y)\big)\Big(\sum_{x\in E}\pi(x)\Big)\langle h_0, T^{m-1}h_0\rangle
\end{aligned}$$

where $h_0(y) \equiv \left(\sum_{x\in E}\pi(x)\right)^{-1/2}$ on E and the bounded self-adjoint linear operator T: $\mathcal{L}^2(E,\mathcal{E},\pi) \longrightarrow \mathcal{L}^2(E,\mathcal{E},\pi)$ is defined by

$$(Th)(x) = \sqrt{g(x)}\sum_{y\in E} K(x,y)\sqrt{g(y)}h(y)\pi(y) \qquad h \in \mathcal{L}^2(E,\mathcal{E},\pi).$$

Similarly to (3.1.7), we have

$$\langle h_0, T^{m-1}h_0\rangle = \int_{-\infty}^{\infty}\lambda^{m-1}\mu_{h_0}(d\lambda).$$

By the fact that $||h_0|| = 1$, μ_{h_0} is a probability measure on $\mathbb{R}$. By Theorem E.3 in the Appendix, μ_{h_0} is supported on the interval

$$\Big[\inf_{||h||=1}\langle h, Th\rangle, \quad \sup_{||h||=1}\langle h, Th\rangle\Big].$$

Thus,

$$\begin{aligned}
\langle h_0, T^{m-1}h_0\rangle &\le \left(\max\left\{\Big|\inf_{||h||=1}\langle h,Th\rangle\Big|, \ \Big|\sup_{||h||=1}\langle h,Th\rangle\Big|\right\}\right)^{m-1}\\
&= \sup_{||h||=1}\left|\sum_{x,y\in E} K(x,y)\sqrt{g(x)}h(x)\sqrt{g(y)}h(y)\pi(x)\pi(y)\right|^{m-1}.
\end{aligned}$$

Write $f(x) = \sqrt{g(x)}h(x)$. Then $f(x) \ge 0$ and

$$\sum_{x\in E} f^{\frac{2p}{2p-1}}(x)\pi(x) \le \Big\{\sum_{x\in E} g^q(x)\pi(x)\Big\}^{\frac{p-1}{2p-1}}\Big\{\sum_{x\in E} h^2(x)\pi(x)\Big\}^{\frac{p}{2p-1}} = 1.$$

Consequently,

$$\langle h_0, T^{m-1}h_0\rangle \le \varrho^{m-1}$$

where

$$\varrho = \sup\Big\{ \sum_{x,y\in E} K(x,y)f(x)f(y)\pi(x)\pi(y);\ \ \sum_{x\in E}|f(x)|^{\frac{2p}{2p-1}}\pi(x) = 1\Big\}.$$

Summarizing our argument, we have

$$\begin{aligned}\limsup_{m\to\infty}\frac{1}{m}\log\sup_g \sum_{y_1,\cdots,y_m\in E}&\pi(y_1)\cdots\pi(y_m)\\ \times\Big(\prod_{k=1}^m g(y_k)\Big)&\prod_{k=1}^m K(y_{k-1},y_k)\le\log\varrho.\end{aligned}\tag{3.1.10}$$

In view of (3.1.9), it remains to show that

$$\begin{aligned}\limsup_{m\to\infty}\frac{1}{m}\log&\sum_{x_1,\cdots,x_m\in E}\pi(x_1)\cdots\pi(x_m)\\ &\times\Big[\frac{1}{m!}\Big(\prod_{x\in E}(m\mu(x))!\Big)\Big(\prod_{x\in E}\varphi_\mu(x)^{m\mu(x)}\Big)^{-1}\Big]^p\le 0.\end{aligned}\tag{3.1.11}$$

Let $\mathcal{P}_m(E)$ be the set of the probability measures ν on E such that for each $x\in E$, $\nu(x)=k/n$ for some integer $0\le k\le m$.

$$\begin{aligned}&\sum_{x_1,\cdots,x_m\in E}\pi(x_1)\cdots\pi(x_m)\Big[\frac{1}{m!}\Big(\prod_{x\in E}(m\mu(x))!\Big)\Big(\prod_{x\in E}\varphi_\mu(x)^{m\mu(x)}\Big)^{-1}\Big]^p\\ &=\sum_{\nu\in\mathcal{P}_m(E)}1_{\{\mu=\nu\}}\sum_{x_1,\cdots,x_m\in E}\pi(x_1)\cdots\pi(x_m)\\ &\times\Big[\frac{1}{m!}\Big(\prod_{x\in E}(m\nu(x))!\Big)\Big(\prod_{x\in E}\varphi_\nu(x)^{m\nu(x)}\Big)^{-1}\Big]^p\\ &\le\sum_{\nu\in\mathcal{P}_m(E)}\sum_{x_1,\cdots,x_m\in E}\pi(x_1)\cdots\pi(x_m)\\ &\times\Big[\frac{1}{m!}\Big(\prod_{x\in E}(m\nu(x))!\Big)\Big(\prod_{x\in E}\varphi_\nu(x)^{m\nu(x)}\Big)^{-1}\Big]^p.\end{aligned}$$

By Stirling formula, we get

$$m!\sim\sqrt{2\pi m}m^m e^{-m}\quad\text{and}\quad(m\nu(x))!\le C\sqrt{m\nu(x)}\big(m\nu(x)\big)^{m\nu(x)}e^{-m\nu(x)}$$

where $x\in E$. Hence,

$$\frac{1}{m!}\Big(\prod_{x\in E}(m\nu(x))!\Big)\le Cm^{\#(E)/2}\prod_{x\in E}\nu(x)^{m\nu(x)}.$$

Recall that $\varphi_\nu(x)=\nu(x)^{1/q}\pi(x)^{1/p}$. Therefore,

$$\begin{aligned}&\sum_{x_1,\cdots,x_m\in E}\pi(x_1)\cdots\pi(x_m)\Big[\frac{1}{m!}\Big(\prod_{x\in E}(m\nu(x))!\Big)\Big(\prod_{x\in E}\varphi_\nu(x)^{m\nu(x)}\Big)^{-1}\Big]^p\\ &\le C^pm^{p\#(E)/2}\sum_{x_1,\cdots,x_m\in E}\pi(x_1)\cdots\pi(x_m)\prod_{x\in E}\Big(\frac{\nu(x)}{\pi(x)}\Big)^{m\nu(x)}\\ &=C^pm^{p\#(E)/2}\sum_{x_1,\cdots,x_m\in E}\nu(x_1)\cdots\nu(x_m)=C^pm^{p\#(E)/2}.\end{aligned}$$

Consequently,

$$\begin{aligned}&\sum_{x_1,\cdots,x_m\in E}\pi(x_1)\cdots\pi(x_m)\Big[\frac{1}{m!}\Big(\prod_{x\in E}(m\mu(x))!\Big)\Big(\prod_{x\in E}\varphi_\mu(x)^{m\mu(x)}\Big)^{-1}\Big]^p\\ &\le C^pm^{p\#(E)/2}\#\{\mathcal{P}_m(E)\}.\end{aligned}$$

Finally, (3.1.11) follows from the fact that $\#\{\mathcal{P}_m(E)\}$ is equal to the number of the non-negative lattice solutions $(k(x);\ \ x\in E)$ of the equation

$$\sum_{x\in E}k(x)=m$$

and that the latter is equal to

$$\binom{m+\#(E)-1}{\#(E)-1}.$$

□

3.2. High moment of $\alpha([0,\tau_1]\times\cdots\times[0,\tau_p])$

We now return to the intersection local time $\alpha(A)$ of independent d-dimensional Brownian motions $W_1(t),\cdots,W_p(t)$. Throughout we assume that $p(d-2)<d$. Define

$$\rho=\sup\Big\{\iint_{\mathbb{R}^d\times\mathbb{R}^d}G(x-y)f(x)f(y)dxdy;\ \ \int_{\mathbb{R}^d}|f(x)|^{\frac{2p}{2p-1}}dx=1\Big\}. \tag{3.2.1}$$

LEMMA 3.2.1. *Under $p(d-2)<d$, we get*

$$0<\rho\le\Big(\int_{\mathbb{R}^d}G^p(x)dx\Big)^{1/p}<\infty.$$

Proof. The lower bound $\rho>0$ is obvious. We now prove the upper bound. Let the non-negative f on $\mathbb{R}^d$ satisfy

$$\int_{\mathbb{R}^d}f^{\frac{2p}{2p-1}}(x)dx=1.$$

By Hölder inequality, we get

$$\begin{aligned}
&\iint_{\mathbb{R}^d\times\mathbb{R}^d} G(x-y)f(x)f(y)dxdy \\
&= \iint_{\mathbb{R}^d\times\mathbb{R}^d} G(x-y)\big(f(x)f(y)\big)^{\frac{1}{2p-1}}\big(f(x)f(y)\big)^{\frac{2(p-1)}{2p-1}}dxdy \\
&\le \Big(\iint_{\mathbb{R}^d\times\mathbb{R}^d} G^p(x-y)\big(f(x)f(y)\big)^{\frac{p}{2p-1}}dxdy\Big)^{1/p} \\
&\times \Big(\iint_{\mathbb{R}^d\times\mathbb{R}^d} \big(f(x)f(y)\big)^{\frac{2p}{2p-1}}dxdy\Big)^{1/q} \\
&= \Big(\iint_{\mathbb{R}^d\times\mathbb{R}^d} G^p(x-y)\big(f(x)f(y)\big)^{\frac{p}{2p-1}}dxdy\Big)^{1/p}
\end{aligned}$$

where $q = p(p-1)^{-1}$ is the conjugate number of p.

By translation invariance and the Fubini theorem, we get

$$\begin{aligned}
&\iint_{\mathbb{R}^d\times\mathbb{R}^d} G^p(x-y)\big(f(x)f(y)\big)^{\frac{p}{2p-1}}dxdy \\
&= \int_{\mathbb{R}^d} G^p(x)\Big(\int_{\mathbb{R}^d}\big(f(x+y)f(y)\big)^{\frac{p}{2p-1}}dy\Big)dx.
\end{aligned}$$

By Cauchy-Schwarz inequality, we get

$$\begin{aligned}
&\int_{\mathbb{R}^d}\big(f(x+y)f(y)\big)^{\frac{p}{2p-1}}dy \\
&\le \Big(\int_{\mathbb{R}^d} f(x+y)^{\frac{2p}{2p-1}}dy\Big)^{1/2}\Big(\int_{\mathbb{R}^d} f(y)^{\frac{2p}{2p-1}}dy\Big)^{1/2} \\
&= \int_{\mathbb{R}^d} f(y)^{\frac{2p}{2p-1}}dy = 1
\end{aligned}$$

where the second step partially follows from shifting invariance. Thus,

$$\iint_{\mathbb{R}^d\times\mathbb{R}^d} G^p(x-y)\big(f(x)f(y)\big)^{\frac{p}{2p-1}}dxdy \le \int_{\mathbb{R}^d} G^p(x)dx.$$

□

Recall that $\tau_1,\cdots,\tau_p$ are i.i.d. exponential times of parameter 1 and that $\tau_1,\cdots,\tau_p$ are independent of $W_1(t),\cdots,W_p(t)$. The following is the main result of this section.

THEOREM 3.2.2. *Under $p(d-2)<d$, we have*

$$\lim_{m\to\infty}\frac{1}{m}\log\frac{1}{(m!)^p}\mathbb{E}\Big[\alpha([0,\tau_1]\times\cdots\times[0,\tau_p])^m\Big] = p\log\rho.$$

Proof. By (2.2.19), the lower bound of (3.2.2) is a direct consequence of Theorem 3.1.1, and the upper bound is equivalent to

$$\limsup_{m\to\infty}\frac{1}{m}\log\frac{1}{(m!)^p}\int_{(\mathbb{R}^d)^m}dx_1\cdots dx_m \times\Big[\sum_{\sigma\in\Sigma_m}\prod_{k=1}^m G(x_{\sigma(k)}-x_{\sigma(k-1)})\Big]^p\le p\log\rho. \tag{3.2.2}$$

Based on Theorem 3.1.2, our approach relies on compactification and discretization. The fact (Theorem A.1, Appendix) that the function $G(x)$ is discontinuous at 0 when $d\ge 2$ makes the matter delicate.

The compactification procedure depends on the following folding method. Let $N>0$ be large but fixed, then

$$\begin{aligned}&\int_{(\mathbb{R}^d)^m}dx_1\cdots dx_m\Big[\sum_{\sigma\in\Sigma_m}\prod_{k=1}^m G(x_{\sigma(k)}-x_{\sigma(k-1)})\Big]^p\\
&=\sum_{z_1,\cdots,z_m\in\mathbb{Z}^d}\int_{([-N,N]^d)^m}dy_1\cdots dy_m\\
&\times\Big[\sum_{\sigma\in\Sigma_m}\prod_{k=1}^m G\big((2Nz_{\sigma(k)}+y_{\sigma(k)})-(2Nz_{\sigma(k-1)}+y_{\sigma(k-1)})\big)\Big]^p\\
&=\int_{([-N,N]^d)^m}dy_1\cdots dy_m\sum_{z_1,\cdots,z_m\in\mathbb{Z}^d}\\
&\times\Big[\sum_{\sigma\in\Sigma_m}\prod_{k=1}^m G\big(2N(z_{\sigma(k)}-z_{\sigma(k-1)})+(y_{\sigma(k)}-y_{\sigma(k-1)})\big)\Big]^p.\end{aligned}$$

Fix $y_1,\cdots,y_m\in[-N,N]^d$. By triangular inequality,

$$\begin{aligned}&\Big\{\sum_{z_1,\cdots,z_m\in\mathbb{Z}^d}\Big[\sum_{\sigma\in\Sigma_m}\prod_{k=1}^m G\big(2N(z_{\sigma(k)}-z_{\sigma(k-1)})+(y_{\sigma(k)}-y_{\sigma(k-1)})\big)\Big]^p\Big\}^{1/p}\\
&\le\sum_{\sigma\in\Sigma_m}\Big\{\sum_{z_1,\cdots,z_m\in\mathbb{Z}^d}\prod_{k=1}^m G^p\big(2N(z_{\sigma(k)}-z_{\sigma(k-1)})+(y_{\sigma(k)}-y_{\sigma(k-1)})\big)\Big\}^{1/p}\\
&=\sum_{\sigma\in\Sigma_m}\prod_{k=1}^m G_N(y_{\sigma(k)}-y_{\sigma(k-1)})\end{aligned}$$

where

$$G_N(x)=\Big\{\sum_{z\in\mathbb{Z}^d}G^p(2Nz+x)\Big\}^{1/p}.$$

Clearly, $G_N(x)$ is symmetric and periodic, where the periodicity is defined by the property

$$G_N(x+2Ny)=G_N(x)\qquad x\in\mathbb{R}^d\quad y\in\mathbb{Z}^d.$$

Notice that

$$\int_{[-N,N]^d}G_N^p(x)dx=\sum_{z\in\mathbb{Z}^d}\int_{[-N,N]^d}G^p(2Nz+x)dx=\int_{\mathbb{R}^d}G^p(x)dx<\infty.$$

Summarizing what we have, we see that

$$\begin{aligned}(3.2.3)\qquad &\int_{(\mathbb{R}^d)^m} dx_1\cdots dx_m\Big[\sum_{\sigma\in\Sigma_m}\prod_{k=1}^m G(x_{\sigma(k)}-x_{\sigma(k-1)})\Big]^p\\ &\le \int_{([-N,N]^d)^m} dy_1\cdots dy_m\Big[\sum_{\sigma\in\Sigma_m}\prod_{k=1}^m G_N(y_{\sigma(k)}-y_{\sigma(k-1)})\Big]^p.\end{aligned}$$

For an integer $n\ge 1$, partition $[-N,N]^d$ uniformly into 2nd small boxes with the side length $2\epsilon = N/2^n$. Let E be the set of the geometric centers of these small boxes. Write

$$Q_\epsilon = [-\epsilon,\epsilon]^d.$$

Then

$$\begin{aligned}&\int_{([-N,N]^d)^m} dx_1\cdots dx_m\Big[\sum_{\sigma\in\Sigma_m}\prod_{k=1}^m G_N(x_{\sigma(k)}-x_{\sigma(k-1)})\Big]^p\\ &=\sum_{y_1,\cdots,y_m\in E}\int_{Q_\epsilon^m} dz_1\cdots dz_m\\ &\quad\times\Big[\sum_{\sigma\in\Sigma_m}\prod_{k=1}^m G_N\big((z_{\sigma(k)}-z_{\sigma(k-1)})+(y_{\sigma(k)}-y_{\sigma(k-1)})\big)\Big]^p.\end{aligned}$$

By triangular inequality,

$$\begin{aligned}&\int_{Q_\epsilon^m} dz_1\cdots dz_m\Big[\sum_{\sigma\in\Sigma_m}\prod_{k=1}^m G_N\big((z_{\sigma(k)}-z_{\sigma(k-1)})+(y_{\sigma(k)}-y_{\sigma(k-1)})\big)\Big]^p\\ &\le\Big[\sum_{\sigma\in\Sigma_m}\Big\{\int_{Q_\epsilon^m} dz_1\cdots dz_m\prod_{k=1}^m G_N^p\big((z_k-z_{k-1})+(y_{\sigma(k)}-y_{\sigma(k-1)})\big)\Big\}^{1/p}\Big]^p\\ &\le(2\epsilon)^{md}\Big[\sum_{\sigma\in\Sigma_m}\prod_{k=1}^m G_{N,\epsilon}(y_{\sigma(k)}-y_{\sigma(k-1)})\Big]^p\end{aligned}$$

where

$$G_{N,\epsilon}(x)=\Big\{\frac{1}{(2\epsilon)^d}\sup_{y\in Q_\epsilon}\int_{Q_\epsilon} G_N^p(x+y+z)dz\Big\}^{1/p}.$$

Clearly, $G_{N,\epsilon}(x)$ is symmetric and enjoys the periodicity described by

$$G_{N,\epsilon}(x+2Ny)=G_{N,\epsilon}(x)\qquad x\in\mathbb{R}^d\ \ y\in\mathbb{Z}^d.$$

Summarizing our discussion since (3.2.3), we have

$$\begin{aligned}&\int_{(\mathbb{R}^d)^m} dx_1\cdots dx_m\Big[\sum_{\sigma\in\Sigma_m}\prod_{k=1}^m G(x_{\sigma(k)}-x_{\sigma(k-1)})\Big]^p\\ &\le(2\epsilon)^{md}\sum_{y_1,\cdots,y_m\in E}\Big[\sum_{\sigma\in\Sigma_m}\prod_{k=1}^m G_{N,\epsilon}(y_{\sigma(k)}-y_{\sigma(k-1)})\Big]^p.\end{aligned}$$

Applying Theorem 3.1.2 (with $\pi(x)\equiv(2\epsilon)^d$) to the right hand side gives us

$$\limsup_{m\to\infty}\frac{1}{m}\log\frac{1}{(m!)^p}\int_{(\mathbb{R}^d)^m}dx_1\cdots dx_m\Big[\sum_{\sigma\in\Sigma_m}\prod_{k=1}^m G(x_{\sigma(k)}-x_{\sigma(k-1)})\Big]^p$$
$$\le p\log\rho_{N,\epsilon}$$

where

$$\rho_{N,\epsilon}=\sup\Big\{(2\epsilon)^{2d}\sum_{x,y\in E}G_{N,\epsilon}(x-y)f(x)f(y);\ \sum_{x\in E}|f(x)|^{\frac{2p}{2p-1}}(2\epsilon)^d=1\Big\}. \tag{3.2.4}$$

Finally, (3.2.2) follows from Lemma 3.2.3 and Lemma 3.2.4 below. □

LEMMA 3.2.3. *Let $\rho_{N,\epsilon}$ be defined in (3.2.4), then*

$$\limsup_{\epsilon\to0^+}\rho_{N,\epsilon}\le\rho_N$$

where

$$\rho_N=\sup\Big\{\iint_{\{[-N,N]^d\times[-N,N]^d\}}G_N(u-v)f(u)f(v)dudv;\ \int_{[-N,N]^d}|f(u)|^{\frac{2p}{2p-1}}du=1\Big\}. \tag{3.2.5}$$

Proof. For fixed n_0 write $2\delta=N/2^{n_0}$. Set

$$G_N^{(1)}(x)=G_N(x)1_{\{x\in Q_\delta\}},\qquad G_N^{(2)}(x)=G_N(x)1_{\{x\notin Q_\delta\}}.$$

Then $G_N^{(1)}$ and $G_N^{(2)}$ are symmetric and periodic. By triangular inequality, we get

$$G_{N,\epsilon}(x)\le G_{N,\epsilon}^{(1)}(x)+G_{N,\epsilon}^{(2)}(x)$$

where

$$G_{N,\epsilon}^{(i)}(x)=\Big\{\frac{1}{(2\epsilon)^d}\sup_{y\in Q_\epsilon}\int_{Q_\epsilon}\big[G_N^{(i)}(x+y+z)\big]^pdz\Big\}^{1/p}\qquad i=1,2.$$

Consequently,

$$\rho_{N,\epsilon}\le\rho_{N,\epsilon}^{(1)}+\rho_{N,\epsilon}^{(2)} \tag{3.2.6}$$

where

$$\rho_{N,\epsilon}^{(i)}=\Big\{(2\epsilon)^{2d}\sum_{x,y\in E}G_{N,\epsilon}^{(i)}(x-y)f(x)f(y);\ \sum_{x\in E}|f(x)|^{\frac{2p}{2p-1}}(2\epsilon)^d=1\Big\}\qquad i=1,2.$$

Modifying the argument in the proof of Lemma 3.1.1, one can prove

$$\rho_{N,\epsilon}^{(1)}\le(2\epsilon)^d\Big\{\sum_{x\in E}\big[G_{N,\epsilon}^{(1)}(x)\big]^p\Big\}^{1/p}.$$

Indeed, the periodicity of the function $G_{N,\epsilon}^{(1)}(x)$ allows the argument by translation invariance to be carried out in the proof of Lemma 3.1.1.

By the definition of $G_{N,\epsilon}^{(1)}(x)$, we get

$$\rho_{N,\epsilon}^{(1)} \le \left\{ \sum_{x\in E} \sup_{y\in Q_\epsilon} \int_{Q_\epsilon} \left[G_N^{(1)}(x+y+z)\right]^p dz \right\}^{1/p}.$$

Notice that for any $y \in Q_\epsilon$,

$$\int_{Q_\epsilon} \left[G_N^{(1)}(x+y+z)\right]^p dz = \int_{y+Q_\epsilon} \left[G_N^{(1)}(x+z)\right]^p dz$$
$$\le \int_{Q_{2\epsilon}} \left[G_N^{(1)}(x+z)\right]^p dz = \int_{x+Q_{2\epsilon}} \left[G_N^{(1)}(z)\right]^p dz.$$

Observe that the family $\{x+Q_{2\epsilon},\ \ x\in E\}$ is contained in $[-N-\epsilon, N+\epsilon]^d$ and that every point in $[-N-\epsilon, N+\epsilon]^d$ is covered by this family at most 2^d times. Thus,

$$\begin{aligned} \rho_{N,\epsilon}^{(1)} &\le \left\{ \sum_{x\in E} \int_{x+Q_{2\epsilon}} \left[G_N^{(1)}(z)\right]^p dz \right\}^{1/p} \\ &\le \left\{ 2^d \int_{[-N-\epsilon,N+\epsilon]^d} \left[G_N^{(1)}(z)\right]^p dz \right\}^{1/p} \\ &= 2^{d/p} \left\{ \int_{Q_\delta} G_N^p(x)dx \right\}^{1/p}. \end{aligned} \tag{3.2.7}$$

By the fact that $G_N(x)$ is continuous and bounded outside Q_δ, there is a constant $K>0$ such that $G_N^{(2)}(x) \le K$ for any $x\in\mathbb{R}^d$. Write $G^*(x) = \min\{G_N(x), K\}$. Then $G^*(x)$ is a continuous, symmetric and periodic function such that $G_N^{(2)} \le G^* \le G_N$. In particular,

$$\rho_{N,\epsilon}^{(2)} \le \rho_\epsilon^* \tag{3.2.8}$$

where

$$\rho_\epsilon^* = \sup\left\{ (2\epsilon)^{2d} \sum_{x,y\in E} G_\epsilon^*(x-y)f(x)f(y);\ \ \sum_{x\in E} |f(x)|^{\frac{2p}{2p-1}}(2\epsilon)^d = 1 \right\}$$

and

$$G_\epsilon^*(x) = \left\{ \frac{1}{(2\epsilon)^d} \sup_{y\in Q_\epsilon} \int_{Q_\epsilon} \left[G^*(x+y+z)\right]^p dz \right\}^{1/p} \qquad x\in\mathbb{R}^d.$$

By the continuity of G^* one can easily establish

$$\lim_{\epsilon\to 0^+} \int_{[-N,N]^d} \left|G_\epsilon^*(u) - G^*(u)\right|^p du = 0.$$

By triangular inequality, and by the continuity of G^* again, we get

$$\Delta_\epsilon = \sup_{|u-v|\le 2\epsilon} \left|G_\epsilon^*(u) - G_\epsilon^*(v)\right| \longrightarrow 0 \qquad (\epsilon\to 0^+).$$

Fix the non-negative function $f(x)$ on E with

$$\sum_{x\in E} f^{\frac{2p}{2p-1}}(x)(2\epsilon)^d = 1.$$

We extend f into a function on $[-N,N]^d$ by letting $f(u)=f(x)$ for u in the small box centered by x. We have

$$\int_{[-N,N]^d} f^{\frac{2p}{2p-1}}(u)du = 1$$

and

$$\begin{aligned}
&(2\epsilon)^{2d}\sum_{x,y\in E} G^*_\epsilon(x-y)f(x)f(y)\\
&\le \iint_{\{[-N,N]^d\times[-N,N]^d\}} G^*(u-v)f(u)f(v)dudv\\
&+\iint_{\{[-N,N]^d\times[-N,N]^d\}} |G^*_\epsilon(u-v)-G^*(u-v)|f(u)f(v)dudv\\
&+\Delta_\epsilon\iint_{\{[-N,N]^d\times[-N,N]^d\}} f(u)f(v)dudv.
\end{aligned}$$

For the first term, by the relation $G^*\le G_N$,

$$\iint_{\{[-N,N]^d\times[-N,N]^d\}} G^*(u-v)f(u)f(v)dudv \le \rho_N.$$

For the second term, a procedure similar to the proof of Lemma 3.1.1 leads to the conclusion that

$$\begin{aligned}
&\iint_{\{[-N,N]^d\times[-N,N]^d\}} |G^*_\epsilon(u-v)-G^*(u-v)|f(u)f(v)dudv\\
&\le\left\{\int_{[-N,N]^d}\left|G^*_\epsilon(u)-G^*(u)\right|^p du\right\}^{1/p}.
\end{aligned}$$

As for the third term,

$$\begin{aligned}
&\iint_{\{[-N,N]^d\times[-N,N]^d\}} f(u)f(v)dudv = \left(\int_{[-N,N]^d} f(u)du\right)^2\\
&\le (2N)^{d/p}\left(\int_{[-N,N]^d} f^{\frac{2p}{2p-1}}(u)du\right)^{\frac{2p-1}{p}} = (2N)^{d/p}.
\end{aligned}$$

Summarizing our argument, we have

$$\rho^*_\epsilon \le \rho_N + \left\{\int_{[-N,N]^d}\left|G^*_\epsilon(u)-G^*(u)\right|^p du\right\}^{1/p} + \Delta_\epsilon(2N)^{d/p}.$$

By (3.2.8),

$$\limsup_{\epsilon\to0^+}\rho^{(2)}_{N,\epsilon}\le\rho_N.$$

In view of (3.2.6) and (3.2.7),

$$\limsup_{\epsilon\to0^+}\rho_{N,\epsilon}\le\rho_N + 2^{d/p}\left\{\int_{Q_\delta} G^p_N(x)dx\right\}^{1/p}.$$

Letting $\delta\to0^+$ leads to the desired conclusion. □

LEMMA 3.2.4. *Let ρ and ρ_N be defined as in (3.1.1) and (3.2.5), respectively.*

$$\limsup_{N\to\infty} \rho_N \le \rho. \tag{3.2.9}$$

Proof. Let $M > 0$ be fixed and let $N > M$. Set

$$G^{(1)}(x) = G(x)1_{[-M,M]^d}(x), \qquad G^{(2)}(x) = G(x) - G^{(1)}(x).$$

By triangular inequality, we get

$$G_N(x) \le K_N^{(1)}(x) + K_N^{(2)}(x) \qquad x \in \mathbb{R}^d$$

where

$$K_N^{(i)}(x) = \Big\{ \sum_{z\in\mathbb{Z}^d} \big[G^{(i)}(2Nz + x)\big]^p \Big\}^{1/p} \qquad i = 1, 2.$$

Consequently,

$$\rho_N \le \rho_N^{(1)} + \rho_N^{(2)} \tag{3.2.10}$$

where

$$\rho_N^{(i)} = \sup\Big\{ \iint_{\{[-N,N]^d\times[-N,N]^d\}} K_N^{(i)}(u-v)f(u)f(v)dudv;$$
$$\int_{[-N,N]^d} |f(u)|^{\frac{2p}{2p-1}}du = 1\Big\} \qquad i = 1, 2.$$

In our following argument, every function f on $[-N,N]^d$ is automatically extended into a periodic function on $\mathbb{R}^d$ with

$$f(x + 2Ny) = f(x) \qquad x \in \mathbb{R}^d \qquad y \in \mathbb{Z}^d.$$

By an argument similar to the proof of Lemma 3.1.1, we get

$$\rho_N^{(2)} \le \Big\{ \int_{[-N,N]^d} \big[K_N^{(2)}(x)\big]^p dx \Big\}^{1/p} = \Big\{ \int_{\{\mathbb{R}^d\setminus[-M,M]^d\}} G^p(x)dx \Big\}^{1/p}. \tag{3.2.11}$$

Observe that $K_N^{(1)}(x)$ is periodic and that

$$K_N^{(1)}(x) = G^{(1)}(x) \qquad x \in [-2(N-M), 2(N-M)]^d. \tag{3.2.12}$$

Write $E_N = [-N,N]^d \setminus [-(N-M), N-M]^d$ and let $f(x) \ge 0$ be a non-negative function on $[-N,N]^d$ such that

$$\int_{[-N,N]^d} f^{\frac{2p}{2p-1}}(x)dx = 1.$$

We claim that there is $a \in [-N,N]^d$ such that

$$\int_{E_N} f^{\frac{2p}{2p-1}}(x+a)dx \le \frac{|E_N|}{(2N)^d} \tag{3.2.13}$$

where $|E_N|$ is the volume of E_N.

If not, we would have

$$\int_{[-N,N]^d}\int_{E_N} f^{\frac{2p}{2p-1}}(x+y)dxdy > \int_{[-N,N]^d} \frac{|E_N|}{(2N)^d}dy = |E_N|.$$

This cannot be true since

$$\begin{aligned}&\int_{[-N,N]^d}\int_{E_N} f^{\frac{2p}{2p-1}}(x+y)dxdy\\&=\int_{E_N}\int_{[-N,N]^d} f^{\frac{2p}{2p-1}}(x+y)dydx\\&=\int_{E_N} dx = |E_N|.\end{aligned}$$

Define $\bar{f}(x)=f(x+a)$. Then

$$\int_{[-N,N]^d}\bar{f}^{\frac{2p}{2p-1}}(x)dx=\int_{[-N,N]^d} f^{\frac{2p}{2p-1}}(x)dx=1.$$

Set

$$h(x,y)=K_N^{(1)}(x-y)f(x)f(y).$$

By the translation invariance led by periodicity, we get

$$\begin{aligned}&\iint_{[-N,N]^d\times[-N,N]^d} K_N^{(1)}(x-y)f(x)f(y)dxdy\\&=\iint_{[-N,N]^d\times[-N,N]^d} h(x,y)dxdy\\&=\iint_{[-N,N]^d\times[-N,N]^d} h(x+a,y+a)dxdy\\&=\iint_{[-N,N]^d\times[-N,N]^d} K_N^{(1)}(x-y)\bar{f}(x)\bar{f}(y)dxdy\\&\le\iint_{[-(N-M),\ N-M]^d\times[-(N-M),\ N-M]^d} K_N^{(1)}(x-y)\bar{f}(x)\bar{f}(y)dxdy\\&+2\iint_{E_N\times[-N,N]^d} K_N^{(1)}(x-y)\bar{f}(x)\bar{f}(y)dxdy.\end{aligned}$$

Notice that as $x,y\in[-(N-M),\ N-M]^d$, $x-y\in[-2(N-M),2(N-M)]^d$. By (3.2.12), we get

$$\begin{aligned}&\iint_{[-(N-M),\ N-M]^d\times[-(N-M),\ N-M]^d} K_N^{(1)}(x-y)\bar{f}(x)\bar{f}(y)dxdy\\&=\iint_{[-(N-M),\ N-M]^d\times[-(N-M),\ N-M]^d} G^{(1)}(x-y)\bar{f}(x)\bar{f}(y)dxdy\\&\le\iint_{[-N,N]^d\times[-N,N]^d} G(x-y)\bar{f}(x)\bar{f}(y)dxdy\le\rho.\end{aligned}$$

In addition,

$$\begin{aligned}&\iint_{E_N\times[-N,N]^d} K_N^{(1)}(x-y)\bar f(x)\bar f(y)dxdy\\&=\iint_{E_N\times[-N,N]^d} K_N^{(1)}(x-y)\big(\bar f(x)\bar f(y)\big)^{\frac{1}{2p-1}}\big(\bar f(x)\bar f(y)\big)^{\frac{2(p-1)}{2p-1}}dxdy\\&\le\Big\{\iint_{[-N,N]^d\times[-N,N]^d}\big[K_N^{(1)}(x-y)\big]^p\big(\bar f(x)\bar f(y)\big)^{\frac{p}{2p-1}}dxdy\Big\}^{1/p}\\&\times\Big\{\iint_{E_N\times[-N,N]^d}\big(\bar f(x)\bar f(y)\big)^{\frac{2p}{2p-1}}dxdy\Big\}^{\frac{p-1}{p}}.\end{aligned}$$

In view of (3.2.13),

$$\iint_{E_N\times[-N,N]^d}\big(\bar f(x)\bar f(y)\big)^{\frac{2p}{2p-1}}dxdy\le\frac{|E_N|}{(2N)^d}.$$

By periodicity, we get

$$\begin{aligned}&\iint_{[-N,N]^d\times[-N,N]^d}\big[K_N^{(1)}(x-y)\big]^p\big(\bar f(x)\bar f(y)\big)^{\frac{p}{2p-1}}dxdy\\&=\iint_{[-N,N]^d\times[-N,N]^d}\big[K_N^{(1)}(x)\big]^p\big(\bar f(x+y)\bar f(y)\big)^{\frac{p}{2p-1}}dxdy\\&=\int_{[-N,N]^d}\big[K_N^{(1)}(x)\big]^p\Big[\int_{[-N,N]^d}\big(\bar f(x+y)\bar f(y)\big)^{\frac{p}{2p-1}}dy\Big]dx.\end{aligned}$$

By periodicity again, we get

$$\begin{aligned}&\int_{[-N,N]^d}\big(\bar f(x+y)\bar f(y)\big)^{\frac{p}{2p-1}}dy\\&\le\Big\{\int_{[-N,N]^d}\bar f^{\frac{2p}{2p-1}}(x+y)dy\Big\}^{1/2}\Big\{\int_{[-N,N]^d}\bar f^{\frac{2p}{2p-1}}(y)dy\Big\}^{1/2}=1.\end{aligned}$$

Therefore,

$$\begin{aligned}&\iint_{[-N,N]^d\times[-N,N]^d}\big[K_N^{(1)}(x-y)\big]^p\big(\bar f(x)\bar f(y)\big)^{\frac{p}{2p-1}}dxdy\\&\le\int_{[-N,N]^d}\big[K_N^{(1)}(x)\big]^pdx=\int_{\mathbb{R}^d}\big[G^{(1)}(x)\big]^pdx\le\int_{\mathbb{R}^d}G^p(x)dx.\end{aligned}$$

Summarizing our argument, we have

$$\begin{aligned}&\iint_{[-N,N]^d\times[-N,N]^d}K_N^{(1)}(x-y)f(x)f(y)dxdy\\&\le\rho+2\Big(\frac{|E_N|}{(2N)^d}\Big)^{\frac{p-1}{p}}\Big\{\int_{\mathbb{R}^d}G^p(x)dx\Big\}^{1/p}.\end{aligned}$$

Taking supremum over f on the right hand side gives us

$$\rho_N^{(1)}\le\rho+2\Big(\frac{|E_N|}{(2N)^d}\Big)^{\frac{p-1}{p}}\Big\{\int_{\mathbb{R}^d}G^p(x)dx\Big\}^{1/p}. \tag{3.2.14}$$

By the obvious fact that

$$\frac{|E_N|}{(2N)^d} \longrightarrow 0 \quad (N \to \infty),$$

we have

$$\limsup_{N\to\infty} \rho_N^{(1)} \le \rho.$$

In view of (3.2.10) and (3.2.11), therefore,

$$\limsup_{N\to\infty} \rho_N \le \rho + \left\{ \int_{\{\mathbb{R}^d \setminus [-M,M]^d\}} G^p(x)dx \right\}^{1/p}.$$

Letting $M \to \infty$ on the right hand side yields the desired conclusion. □

3.3. Large deviation for $\alpha([0,1]^p)$

By Theorem 1.2.8, an immediate consequence of Theorem 3.2.2 is the following result in large deviations.

THEOREM 3.3.1. *Under $p(d-2) < d$, we get*

$$\lim_{t\to\infty} t^{-1/p} \log \mathbb{P}\Big\{\alpha([0,\tau_1]\times\cdots\times[0,\tau_p]) \ge t\Big\} = -\frac{p}{\rho}.$$

We are more interested in the tail of the intersection local time up to a deterministic time. In view of the scaling property given in Proposition 2.2.6, we may let the Brownian motions run up to time 1.

THEOREM 3.3.2. *Under $p(d-2) < d$, we get*

$$\begin{aligned}(3.3.1)\qquad &\lim_{t\to\infty} t^{-\frac{2}{d(p-1)}} \log \mathbb{P}\Big\{\alpha\big([0,1]^p\big) \ge t\Big\} \\ &= -\frac{d(p-1)}{2}\Big(\frac{2p-d(p-1)}{2p}\Big)^{\frac{2p-d(p-1)}{d(p-1)}} \rho^{-\frac{2p}{d(p-1)}},\end{aligned}$$

where ρ is defined by (3.2.1).

REMARK 3.3.3. *Let $\kappa(d,p) > 0$ be the best constant of the Gagliardo-Nirenberg inequality (Section C, Appendix)*

$$(3.3.2)\qquad ||f||_{2p} \le C||\nabla f||_2^{\frac{d(p-1)}{2p}} ||f||_2^{1-\frac{d(p-1)}{2p}} \qquad f \in W^{1,2}(\mathbb{R}^d)$$

where

$$W^{1,2}(\mathbb{R}^d) = \big\{f \in \mathcal{L}^2(\mathbb{R}^d);\ \ \nabla f \in \mathcal{L}^2(\mathbb{R}^d)\big\}.$$

According to Theorem C.2 in the Appendix,

$$(3.3.3)\qquad \rho = \Big(\frac{2p-d(p-1)}{2p}\Big)^{\frac{2p-d(p-1)}{2p}} \Big(\frac{d(p-1)}{p}\Big)^{\frac{d(p-1)}{2p}} \kappa(d,p)^2.$$

Consequently, (3.3.1) can be rewritten as

$$(3.3.4)\qquad \lim_{t\to\infty} t^{-\frac{2}{d(p-1)}} \log \mathbb{P}\Big\{\alpha\big([0,1]^p\big) \ge t\Big\} = -\frac{p}{2}\kappa(d,p)^{-\frac{4p}{d(p-1)}}.$$

The relation (3.3.3) serves as a bridge between two very important and very different approaches used in this book: The large deviations by high moment asymptotics and the large deviations by Feynman-Kac formula (the latter will be introduced in Chapter 4).

The problem on the exact value of Gagliardo-Nirenberg constant has some significance in physics but remains largely open. The case $d = 1$ is an exception in which

$$\kappa(1,p) = 2^{\frac{1}{2p}}(p-1)^{\frac{p-1}{4p}}(p+1)^{\frac{p-3}{4p}}B\Big(\frac{1}{p-1},\frac{1}{2}\Big)^{-\frac{p-1}{2p}}$$

(Theorem C.4, Appendix), where $B(\cdot,\cdot)$ is the beta function. By (2.2.18),

$$\alpha\big([0,1]^p\big) = \int_{-\infty}^{\infty}\Big[\prod_{j=1}^p L_j(1,x)\Big]dx$$

when $d = 1$, where $L_j(t,x)$ is the local time of $W_j(t)$ $(1 \le j \le p)$. The above leads to the following corollary.

COROLLARY 3.3.4. *When $d = 1$, for any $p \ge 2$,*

$$\lim_{t\to\infty} t^{-\frac{2}{p-1}}\log\mathbb{P}\Big\{\int_{-\infty}^{\infty}\Big[\prod_{j=1}^p L_j(1,x)\Big]dx \ge t\Big\}$$
$$= -\frac{p}{4(p-1)}\Big(\frac{p+1}{2}\Big)^{\frac{3-p}{p-1}}B\Big(\frac{1}{p-1},\frac{1}{2}\Big)^2.$$

REMARK 3.3.5. *The large deviation stated in (3.3.4) implies that*

$$\mathbb{E}\exp\Big\{\theta\Big(\alpha\big([0,1]^p\big)\Big)^{\frac{2}{d(p-1)}}\Big\}\begin{cases} < \infty & \theta < \dfrac{p}{2}\kappa(d,p)^{-\frac{4p}{d(p-1)}}, \\ = \infty & \theta > \dfrac{p}{2}\kappa(d,p)^{-\frac{4p}{d(p-1)}}, \end{cases}$$

which appears to be an improved version of Theorem 2.2.9.

Proof of Theorem 3.3.2. By Theorem 1.2.8, we need only to show that

$$\lim_{m\to\infty}\frac{1}{m}\log(m!)^{-\frac{d(p-1)}{2}}\mathbb{E}\Big[\alpha\big([0,1]^p\big)^m\Big] \tag{3.3.5}$$
$$= p\log\rho + \frac{2p-d(p-1)}{2}\log\frac{2p}{2p-d(p-1)}.$$

Let $\tau_1,\cdots,\tau_p$ be independent exponential times with parameter 1. We assume the independence between $\{\tau_1,\cdots,\tau_p)$ and $(W_1,\cdots,W_p)$. In view of (2.2.20), Stirling formula and Theorem 3.2.2 lead to the requested upper bound

$$\limsup_{m\to\infty}\frac{1}{m}\log(m!)^{-\frac{d(p-1)}{2}}\mathbb{E}\Big[\alpha\big([0,1]^p\big)^m\Big] \tag{3.3.6}$$
$$\le p\log\rho + \frac{2p-d(p-1)}{2}\log\frac{2p}{2p-d(p-1)}.$$

We now establish the lower bound for (3.3.5). By the moment inequality given in (2.2.12) and by the scaling property given in Proposition 2.2.6, we get

$$\mathbb{E}\Big[\alpha\big([0,t_1]\times\cdots\times[0,t_p]\big)^m\Big] \le \prod_{j=1}^p \Big\{\mathbb{E}\Big[\alpha\big([0,t_j]^p\big)^m\Big]\Big\}^{1/p}$$

$$= \mathbb{E}\Big[\alpha\big([0,1]^p\big)^m\Big]\prod_{j=1}^p t_j^{\frac{2p-d(p-1)}{2p}m}$$

for any $t_1,\cdots,t_p>0$ and for any integer $m\ge 1$.

Thus,

$$\mathbb{E}\Big[\alpha\big([0,\tau_1]\times\cdots\times[0,\tau_p]\big)^m\Big]$$

$$= \int_0^\infty\cdots\int_0^\infty dt_1\cdots dt_p \exp\Big\{-\sum_{j=1}^p t_j\Big\}\mathbb{E}\Big[\alpha\big([0,t_1]\times\cdots\times[0,t_p]\big)^m\Big]$$

$$\le \mathbb{E}\Big[\alpha\big([0,1]^p\big)^m\Big]\Big\{\int_0^\infty t^{\frac{2p-d(p-1)}{2p}m}e^{-t}dt\Big\}^p$$

$$= \mathbb{E}\Big[\alpha\big([0,1]^p\big)^m\Big]\Big[\Gamma\Big(\frac{2p-d(p-1)}{2p}m+1\Big)\Big]^p.$$

Consequently,

$$\mathbb{E}\Big[\alpha\big([0,1]^p\big)^m\Big] \ge \Big[\Gamma\Big(\frac{2p-d(p-1)}{2p}m+1\Big)\Big]^{-p}\mathbb{E}\Big[\alpha\big([0,\tau_1]\times\cdots\times[0,\tau_p]\big)^m\Big].$$

By Stirling formula and by Theorem 3.2.2,

$$\liminf_{m\to\infty}\frac{1}{m}\log(m!)^{-\frac{d(p-1)}{2}}\mathbb{E}\Big[\alpha\big([0,1]^p\big)^m\Big] \ge p\log\rho+\frac{2p-d(p-1)}{2}\log\frac{2p}{2p-d(p-1)}. \tag{3.3.7}$$

Finally, (3.3.5) follows from (3.3.6) and (3.3.7). □

A direct application of the above result is the law of the iterated logarithm. To establish it, we recall that for $\bar{y}=(y_1,\cdots,y_p)$, $\mathbb{P}_{\bar{y}}$ is the probability distribution of the random path $\overline{W}(t)=\big(W_1(t),\cdots,W_p(t)\big)$ started at $W_1(0)=y_1,\cdots,W_p(0)=y_p$.

LEMMA 3.3.6. *Under $p(d-2)<d$, for any $C>0$, we have*

$$\lim_{t\to\infty} t^{\frac{2}{d(p-1)}}\log\inf_{|\bar{y}|\le C}\mathbb{P}_{\bar{y}}\Big\{\alpha\big([0,1]^p\big)\ge t\Big\} = -\frac{p}{2}\kappa(d,p)^{-\frac{4p}{d(p-1)}}.$$

Proof. By (2.2.24) and (3.3.5),

$$\lim_{m\to\infty}\frac{1}{m}\log(m!)^{-\frac{d(p-1)}{2}}\sup_{\bar{y}}\mathbb{E}_{\bar{y}}\Big[\alpha\big([0,1]^p\big)^m\Big]$$

$$= p\log\rho+\frac{2p-d(p-1)}{2}\log\frac{2p}{2p-d(p-1)}.$$

If we can prove that

$$\liminf_{m\to\infty}\frac{1}{m}\log(m!)^{-\frac{d(p-1)}{2}}\inf_{|\bar{y}|\le C}\mathbb{E}_{\bar{y}}\Big[\alpha\big([0,1]^p\big)^m\Big] \tag{3.3.8}$$
$$\ge p\log\rho+\frac{2p-d(p-1)}{2}\log\frac{2p}{2p-d(p-1)},$$

Lemma 3.3.6 will follow from the proof of Theorem 1.2.8 and from Remark 3.3.3.

By (2.2.23), we get

$$\mathbb{E}_{\bar{y}}\Big[\big([0,1]^p\big)^m\Big]$$
$$=\int_{(\mathbb{R}^d)^m}dx_1\cdots dx_m\prod_{j=1}^p\sum_{\sigma\in\Sigma_m}\int_{[0,1]^m_<}ds_1\cdots ds_m$$
$$\times\, p_{s_1}(x_{\sigma(1)}-y_j)\prod_{k=2}^m p_{s_k-s_{k-1}}(x_{\sigma(k)}-x_{\sigma(k-1)}).$$

Let the small number $\delta>0$ be fixed for a while. A straightforward verification gives that for $0<\epsilon<\delta/2$, there is $c>0$ such that

$$\inf_{|y|\le C}p_s(x-y)\ge cp_{s-\epsilon}(x)\qquad\forall s\ge\delta,\quad x\in\mathbb{R}^d.$$

Thus,

$$\int_{[0,1]^m_<}ds_1\cdots ds_m p_{s_1}(x_{\sigma(1)}-y_j)\prod_{k=2}^m p_{s_k-s_{k-1}}(x_{\sigma(k)}-x_{\sigma(k-1)})$$
$$\ge c\int_{[\delta,1]^m_<}ds_1\cdots ds_m p_{s_1-\epsilon}(x_{\sigma(1)})\prod_{k=2}^m p_{s_k-s_{k-1}}(x_{\sigma(k)}-x_{\sigma(k-1)})$$
$$=c\int_{[\delta-\epsilon,1-\epsilon]^m_<}ds_1\cdots ds_m p_{s_1}(x_{\sigma(1)})\prod_{k=2}^m p_{s_k-s_{k-1}}(x_{\sigma(k)}-x_{\sigma(k-1)})$$

where the last step follows from time translation.

Summarizing our argument, we have obtained that

$$\inf_{|\bar{y}|\le C}\mathbb{E}_{\bar{y}}\Big[\alpha\big([0,1]^p\big)^m\Big]\ge c^p\mathbb{E}\Big[\alpha\big([\delta-\epsilon,1-\epsilon]^p\big)^m\Big]$$
$$\ge c^p\mathbb{E}\Big[\alpha\big([\delta,1-\delta]^p\big)^m\Big].$$

It remains to show that

$$\lim_{\delta\to0^+}\liminf_{m\to\infty}\frac{1}{m}\log(m!)^{-\frac{d(p-1)}{2}}\mathbb{E}\Big[\alpha\big([\delta,1-\delta]^p\big)^m\Big]\ge L \tag{3.3.9}$$

where L denotes the right hand side of the inequality (3.3.8).

Write

$$\alpha\big([0,1-\delta]^p\big)=\alpha\big([0,\delta]\times[0,1-\delta]^{p-1}\big)+\alpha\big([\delta,1-\delta]\times[0,1-\delta]^{p-1}\big).$$

By triangular inequality, we get

$$\begin{aligned}
&\Big\{\mathbb{E}\Big[\alpha\big([0,1-\delta]^p\big)^m\Big]\Big\}^{1/m}\\
&\le \Big\{\mathbb{E}\Big[\alpha\big([0,\delta]\times[0,1-\delta]^{p-1}\big)^m\Big]\Big\}^{1/m} + \Big\{\mathbb{E}\Big[\alpha\big([\delta,1-\delta]\times[0,1-\delta]^{p-1}\big)^m\Big]\Big\}^{1/m}\\
&\le \Big\{\mathbb{E}\Big[\alpha\big([0,\delta]^p\big)^m\Big]\Big\}^{\frac{1}{mp}}\Big\{\mathbb{E}\Big[\alpha\big([0,1-\delta]^p\big)^m\Big]\Big\}^{\frac{p-1}{pm}}\\
&+\Big\{\mathbb{E}\Big[\alpha\big([\delta,1-\delta]^p\big)^m\Big]\Big\}^{\frac{1}{mp}}\Big\{\mathbb{E}\Big[\alpha\big([0,1-\delta]^p\big)^m\Big]\Big\}^{\frac{p-1}{pm}}
\end{aligned}$$

where the last step follows from (2.2.12).

Hence,

$$\Big\{\mathbb{E}\Big[\alpha\big([0,1-\delta]^p\big)^m\Big]\Big\}^{\frac{1}{mp}} \le \Big\{\mathbb{E}\Big[\alpha\big([0,\delta]^p\big)^m\Big]\Big\}^{\frac{1}{mp}} + \Big\{\mathbb{E}\Big[\alpha\big([\delta,1-\delta]^p\big)^m\Big]\Big\}^{\frac{1}{mp}}.$$

On the other hand, the scaling property stated in Proposition 2.2.6 gives that

$$\Big\{\mathbb{E}\Big[\alpha\big([0,1-\delta]^p\big)^m\Big]\Big\}^{\frac{1}{mp}} = (1-\delta)^{\frac{2p-p(d-1)}{2p}}\Big\{\mathbb{E}\Big[\alpha\big([0,1]^p\big)^m\Big]\Big\}^{\frac{1}{mp}},$$

$$\Big\{\mathbb{E}\Big[\alpha\big([0,\delta]^p\big)^m\Big]\Big\}^{\frac{1}{mp}} = \delta^{\frac{2p-p(d-1)}{2p}}\Big\{\mathbb{E}\Big[\alpha\big([0,1]^p\big)^m\Big]\Big\}^{\frac{1}{mp}}.$$

So we obtain that

$$\begin{aligned}
&\mathbb{E}\Big[\alpha\big([\delta,1-\delta]^p\big)^m\Big]\\
&\ge \Big[(1-\delta)^{\frac{2p-p(d-1)}{2p}} - \delta^{\frac{2p-p(d-1)}{2p}}\Big]^{mp}\mathbb{E}\Big[\alpha\big([0,1]^p\big)^m\Big].
\end{aligned}$$

By (3.3.5), therefore,

$$\begin{aligned}
&\liminf_{m\to\infty}\frac{1}{m}\log(m!)^{-\frac{d(p-1)}{2}}\mathbb{E}\Big[\alpha\big([\delta,1-\delta]^p\big)^m\Big]\\
&\ge L + p\log\Big[(1-\delta)^{\frac{2p-p(d-1)}{2p}} - \delta^{\frac{2p-p(d-1)}{2p}}\Big].
\end{aligned}$$

Letting $\delta\to 0^+$ on the right hand side gives us (3.3.9). □

We now state the *law of the iterated logarithm (LIL)* for $\alpha\big([0,t]^p\big)$.

THEOREM 3.3.7. *Under $p(d-2)<d$, we have*

$$\limsup_{t\to\infty} t^{-\frac{2p-d(p-1)}{2}}(\log\log t)^{-\frac{d(p-1)}{2}}\alpha\big([0,t]^p\big) = \Big(\frac{2}{p}\Big)^{-\frac{d(p-1)}{2}}\kappa(d,p)^{2p} \quad a.s.$$

Proof. By the scaling property described in Proposition 2.2.6, we get

$$\begin{aligned}
&\mathbb{P}\Big\{\alpha\big([0,t]^p\big)\ge \lambda t^{\frac{2p-d(p-1)}{2}}(\log\log t)^{\frac{d(p-1)}{2}}\Big\}\\
&=\mathbb{P}\Big\{\alpha\big([0,1]^p\big)\ge \lambda(\log\log t)^{\frac{d(p-1)}{2}}\Big\}.
\end{aligned}$$

Replacing t by $\lambda(\log\log t)^{\frac{d(p-1)}{2}}$ in (3.3.4), we have that for any $\lambda > 0$,

$$\lim_{t\to\infty}\frac{1}{\log\log t}\log\mathbb{P}\Big\{\alpha\big([0,t]^p\big)\geq \lambda t^{\frac{2p-d(p-1)}{2}}(\log\log t)^{\frac{d(p-1)}{2}}\Big\}$$
$$= -\lambda^{\frac{2}{d(p-1)}}\frac{p}{2}\kappa(d,p)^{-\frac{4p}{d(p-1)}}.$$

Notice that

$$\lambda_0 = \Big(\frac{2}{p}\Big)^{-\frac{d(p-1)}{2}}\kappa(d,p)^{2p}$$

is the solution of the equation

$$\lambda^{\frac{2}{d(p-1)}}\frac{d(p-1)}{2}\frac{p}{2}\kappa(d,p)^{-\frac{4p}{d(p-1)}} = 1.$$

For any $\lambda_1 > \lambda_0$, there is a $\epsilon > 0$ such that

$$\mathbb{P}\Big\{\alpha\big([0,t]^p\big)\geq \lambda_1 t^{\frac{2p-d(p-1)}{2}}(\log\log t)^{\frac{d(p-1)}{2}}\Big\}$$
$$\leq \exp\{-(1+\epsilon)\log\log t\} = (\log t)^{-(1+\epsilon)}$$

as t is sufficiently large. Let $\theta > 1$ be fixed but arbitrary and define $t_k = \theta^k$ $k = 1, 2, \cdots$. We have

$$\sum_{k=1}^\infty \mathbb{P}\Big\{\alpha\big([0,t_k]^p\big)\geq \lambda_1 t_k^{\frac{2p-d(p-1)}{2}}(\log\log t_k)^{\frac{d(p-1)}{2}}\Big\} < \infty.$$

By the Borel-Cantelli lemma,

$$\limsup_{k\to\infty} t_k^{-\frac{2p-d(p-1)}{2}}(\log\log t_k)^{-\frac{d(p-1)}{2}}\alpha\big([0,t_k]^p\big)\leq \lambda_1 \qquad a.s.$$

For any large t there is a $k \geq 1$ such that $t_k \leq t \leq t_{k+1}$. So

$$t^{-\frac{2p-d(p-1)}{2}}(\log\log t)^{-\frac{d(p-1)}{2}}\alpha\big([0,t]^p\big)$$
$$\leq t_k^{-\frac{2p-d(p-1)}{2}}(\log\log t_k)^{-\frac{d(p-1)}{2}}\alpha\big([0,t_{k+1}]^p\big)$$
$$\leq \Big(\theta + o(1)\Big)^{\frac{2p-d(p-1)}{2}} t_{k+1}^{-\frac{2p-d(p-1)}{2}}(\log\log t_{k+1})^{-\frac{d(p-1)}{2}}\alpha\big([0,t_{k+1}]^p\big)$$

as $k \to \infty$. Consequently,

$$\limsup_{t\to\infty} t^{-\frac{2p-d(p-1)}{2}}(\log\log t)^{-\frac{d(p-1)}{2}}\alpha\big([0,t]^p\big)\leq \theta\lambda_1 \qquad a.s.$$

Letting $\theta \to 1^+$ and $\lambda_2 \to \lambda_0^+$ on the right hand side gives the desired upper bound

$$\limsup_{t\to\infty} t^{-\frac{2p-d(p-1)}{2}}(\log\log t)^{-\frac{d(p-1)}{2}}\alpha\big([0,t]^p\big)\leq \lambda_0 \qquad a.s. \tag{3.3.10}$$

An obstacle for the lower bound is that the needed part of the classic Borel-Cantelli lemma requires the independence of the event sequence, which does not hold in our case. To fix this problem we introduce an *extended Borel-Cantelli lemma*.

LEMMA 3.3.8. *Let* $\{A_k\}_{k\geq 1}$ *be a sequence of events that is adaptive to the filtration* $\{\mathcal{A}_k\}_{k\geq 1}$. *Then almost surely,*

$$\{A_k \textit{ occur infinitely often}\} = \Big\{\sum_{k=1}^{\infty}\mathbb{P}(A_{k+1}|\mathcal{A}_k) = \infty\Big\}.$$

This lemma was established in the book by Breiman (Corollary 5.29, p. 96, [**18**]), where the proof appears as an elegant application of the Martingale convergence theorem.

We now return to our work on the lower bound. Take $s_k = k^k$ $(k = 1, 2, \cdots)$. By Lemma 3.3.6 and (2.2.22), we get

$$\lim_{t\to\infty}\frac{1}{\log\log t}\log\inf_{|\bar{x}|\leq\sqrt{t}}\mathbb{P}_{\bar{x}}\Big\{\alpha([0,t]^p)\geq\lambda t^{\frac{2p-d(p-1)}{2}}(\log\log t)^{\frac{d(p-1)}{2}}\Big\}$$
$$= -\lambda^{\frac{2}{d(p-1)}}\frac{p}{2}\kappa(d,p)^{-\frac{4p}{d(p-1)}}.$$

Consequently, for any $\lambda_2 < \lambda_0$, the series

$$\sum_{k=1}^{\infty}\inf_{|\bar{x}|\leq\sqrt{s_{k+1}}}\mathbb{P}_{\bar{x}}\Big\{\alpha([0,s_{k+1}-s_k]^p)\geq\lambda_2 s_{k+1}^{\frac{2p-d(p-1)}{2}}(\log\log s_{k+1})^{\frac{d(p-1)}{2}}\Big\}$$

diverges.

On the other hand, recall the classic law of the iterated logarithm for Brownian motions which states that

$$\limsup_{t\to\infty}\frac{|\overline{W}(t)|}{\sqrt{2t\log\log t}} = 1 \qquad a.s.$$

Consequently, for any $u > 0$,

$$\mathbb{P}\Big\{|\overline{W}(s_k)|\leq u\sqrt{s_{k+1}}\ \text{ eventually}\Big\} = 1.$$

Hence, with probability 1,

$$\sum_{k=1}^{\infty}\mathbb{P}_{\overline{W}(s_k)}\Big\{\alpha([0,s_{k+1}-s_k]^p)\geq\lambda_2 s_{k+1}^{\frac{2p-d(p-1)}{2}}(\log\log s_{k+1})^{\frac{d(p-1)}{2}}\Big\} = \infty.$$

By Markov property of Brownian motions, this is equivalent to

$$\sum_{k=1}^{\infty}\mathbb{P}\Big\{\alpha([s_{k+1},s_k]^p)\geq\lambda_2 s_{k+1}^{\frac{2p-d(p-1)}{2}}(\log\log s_{k+1})^{\frac{d(p-1)}{2}}\Big|\mathcal{A}_k\Big\} = \infty \qquad a.s.$$

where $\mathcal{A}_k$ is the σ-field generated by $\{\overline{W}(s);\ \ 0\leq s\leq s_k\}$. By Lemma 3.3.8,

$$\limsup_{k\to\infty}s_{k+1}^{-\frac{2p-d(p-1)}{2}}(\log\log s_{k+1})^{-\frac{d(p-1)}{2}}\alpha([s_k,s_{k+1}]^p)\geq\lambda_2 \qquad a.s.$$

This leads to

$$\limsup_{t\to\infty}t^{-\frac{2p-d(p-1)}{2}}(\log\log t)^{-\frac{d(p-1)}{2}}\alpha([0,t]^p)\geq\lambda_2 \qquad a.s.$$

Letting $\lambda_2\to\lambda_0^-$ on the right hand side gives the lower bound

$$\limsup_{t\to\infty}t^{-\frac{2p-d(p-1)}{2}}(\log\log t)^{-\frac{d(p-1)}{2}}\alpha([0,t]^p)\geq\lambda_0 \qquad a.s. \tag{3.3.11}$$

Finally, Theorem 3.3.7 follows from (3.3.10) and (3.3.11). □

3.4. Notes and comments

Section 3.1

The study of the high moment asymptotics was initiated by König and Mörters [**114**]. Their investigation leads to the large deviation for Brownian intersection local time limited to a finite domain and killed upon exit from the domain. Their approach relies on some sophisticated combinatorial estimate (given in II.2, [**97**]) for the tail of the empirical measure

$$L_n = \frac{1}{n}\sum_{k=1}^{n}\delta_{(Y_k, Y_{k+1})}$$

where $\{Y_k\}$ is an i.i.d. sequence taking values in a finite catalog. The treatment given here appeared in Chen [**31**] and in Chen and Mörters [**37**].

At its beginning stage, the study of high moment asymptotics has shown its power of solving hard problems in large deviations which are out of reach of the conventional methods. We shall see later in the proof of Theorem 7.5.3 how this powerful tool is applied to the super-critical dimensions. In addition, we point out the references Chen [**29**], [**31**], [**32**], Bass, Chen and Rosen [**9**] for the key role played by the method of high moment asymptotics in the recent progress on the large deviations for the local times, self-intersection local times and Riesz potentials of additive processes, where the asymptotics of the form

$$\int_{(\mathbb{R}^d)^m}\pi(d\lambda_1)\cdots\pi(d\lambda_m)\Big[\sum_{\sigma\in\Sigma_m}\prod_{k=1}^{m}Q\Big(\sum_{j=1}^{k}\lambda_{\sigma(j)}\Big)\Big]^p$$

is investigated. For further application, a challenge we face is how to extend the upper bound established in Theorem 3.1.2 to the settings of reasonable generality where, for example, the state space E can be non-compact or non-discrete.

EXERCISE 3.4.1. *In addition to the assumptions in Theorem 3.1.1, assume that E is a compact metric space, $\mathcal{E}$ is the Borel sets on E, $\pi(E) < \infty$ and $K(x, y)$ is continuous in (x, y) with respect to the product topology on $E \times E$. Prove that*

$$\lim_{m\to\infty}\frac{1}{m}\log\frac{1}{(m!)^p}\log\int_{E^m}\pi(dx_1)\cdots\pi(dx_m)\Big[\sum_{\sigma\in\Sigma_m}\prod_{k=1}^{m}K(x_{\sigma(k-1)}, x_{\sigma(k)})\Big]^p = p\log\rho.$$

EXERCISE 3.4.2. *In connection to Exercise 2.5.7, we show that the high moment asymptotics developed in this section has a potential going beyond the processes of independent increments. Let the intersection local time $I\big([0,t_1]\times\cdots\times[0,t_p]\big)$ be defined in (2.5.9). Here we make the following assumptions:*

(1) *The Markov processes are irreducible: For any $x, y \in E$,*

$$G_\lambda(x, y) \equiv \int_0^\infty e^{-\lambda t}p_t(x, y)dt > 0 \qquad \lambda > 0.$$

(2) *The Markov process is symmetric:* $p_t(x,y) = p_t(y,x)$ *for all* $x, y \in E$ *and* $t > 0$.
(3) *The state space* E *is finite.*

The intuition of the irreducibility is that with positive probability, the Markov process moves from one state to another in a finite period; while the symmetry says that the chance of moving from x *to* y *is the same as moving from* y *to* x.

Let $\tau_1, \cdots, \tau_p$ *be independent exponential times of parameter* $\lambda > 0$. *Compute the limit*

$$\lim_{m\to\infty} \frac{1}{m} \log \frac{1}{(m!)^p} \mathbb{E}\Big[I\big([0,\tau_1] \times \cdots \times [0,\tau_p]\big)^m\Big]. \tag{3.4.1}$$

Section 3.2

Concerning developing the general theory of high moment asymptotics, the main result Theorem 3.2.2 serves as a good example supporting generalization of Theorem 3.1.2. The folding technique leading to (3.2.3) is partially inspired by a similar idea going back to Donsker and Varadhan [**57**].

We propose a shorter proof for a weaker version of Theorem 3.2.2 in the following exercise.

EXERCISE 3.4.3 (hard). *Assume that* $d(p-2) < d$. *Write*

$$H_m = \int_{(\mathbb{R}^d)^m} dx_1 \cdots dx_m \bigg[\sum_{\sigma\in\Sigma_m} \prod_{k=1}^m G(x_{\sigma(k)} - x_{\sigma(k-1)})\bigg]^p \qquad m = 1, 2, \cdots.$$

(1) *Prove for any integers* $m, n \geq 1$,

$$H_{m+n} \leq \binom{m+n}{n}^p H_m H_n.$$

Hint: Let Σ'_{m+n} *be the sub-group of* Σ_{m+n} *consisting of the permutations that map* $\{1, \cdots, m\}$ *to* $\{1, \cdots, m\}$ *(and therefore map* $\{m+1, \cdots, m+n\}$ *to* $\{m+1, \cdots, m+n\}$*). First prove*

$$H_{m+n} \leq \binom{m+n}{n}^p \int_{(\mathbb{R}^d)^{m+n}} dx_1 \cdots dx_{m+n}$$
$$\times \bigg[\sum_{\pi\in\Sigma'_{m+n}} \prod_{k=1}^{m+n} G(x_{\pi(k)} - x_{\pi(k-1)})\bigg]^p.$$

(2) *Prove the weaker version of Theorem 3.2.2: There is a constant* $-\infty < c(d,p) < \infty$ *such that*

$$\lim_{m\to\infty} \frac{1}{m} \log \frac{1}{(m!)^p} \mathbb{E}\Big[\alpha([0,\tau_1] \times \cdots \times [0,\tau_p])^m\Big] = c(d,p).$$

Section 3.3

Theorem 3.3.2 was obtained in [**27**] by a very different method. Prior to that, Le Gall ([**124**], [**130**]), Brydges and Slade ([**19**]) had estimated the moment of

$\alpha([0,1]^p)$. Together with Chebyshev inequality, some of their results give bounds to the tail probability of $\alpha([0,1]^p)$ that are sharp up to constant.

In addition to their pioneering paper on the high moment asymptotics, König and Mörters ([**115**]) applied their method to the exponential moments for a functional of intersection local times of independent Brownian motions.

The Gagliardo-Nirenberg inequality has a close tie to Sobolev inequality (see Section C). The full class of Gagliardo-Nirenberg inequalities defined in the analysis literature is wider than the one given in (3.3.2). We point out the books by Ziemer ([**176**]) and by Saloff-Coste ([**154**]) for the general reference. Finding the best constants for Gagliardo-Nirenberg inequality appears to be a difficult problem when $d \geq 2$ and still remains open in general. It has attracted considerable attention partially due to its connection to some problems in physics. The best constant for Nash's inequality, which is a special case of Gagliardo-Nirenberg inequality, was found by Carlen and Loss ([**23**]). See also papers by Cordero-Erausquin ([**40**]) and by Del Pino and Dolbeault ([**45**]) for recent progress on the best constants for a class of Gagliardo-Nirenberg inequalities and logarithmic Sobolev inequalities. For a connection between the best constants of Gagliardo-Nirenberg inequalities and logarithmic Sobolev inequalities, the interested reader is referred to Del Pino and Dolbeault [**46**]. In the critical case $p(d-2) = d$, (3.3.2) becomes the Sobolev inequality

$$||f||_{\frac{2d}{d-2}} \leq C||\nabla f||_2.$$

The best constant in this case is already known (see, e.g., Theorem 1.3.2, p. 20, [**154**]).

Unfortunately, none of this progress was made for the kind of Gagliardo-Nirenberg inequalities given in (3.3.2) as $p(d-2) < d$ and $d \geq 2$.

The papers directly related to $\kappa(d,p)$ for $p(d-2) < d$ are [**132**] and [**164**]. In [**132**], Levine obtained a sharp estimate for $\kappa(2,3)$, which claims that

$$\sqrt[3]{\frac{1}{4.6016}} < \kappa(2,3) < \sqrt[3]{\frac{1}{4.5981}}.$$

Based on that, he conjectured that $\kappa(2,3) = \pi^{-4/9}$.

Weinstein ([**164**]) studied the problem of the best constants for the Gagliardo-Nirenberg inequalities given in (3.3.2). It was shown (Theorem B, [**164**]) that under $p(d-2) < d$, the best constant in (3.3.2) is attained at an infinitely smooth, positive and spherically symmetric function f_0, which solves the non-linear equation

$$\frac{d(p-1)}{2}\Delta f - \frac{2p-d(p-1)}{2} f + f^{2p-1} = 0. \tag{3.4.2}$$

In addition, f_0 has the smallest L^2-norm among all solutions of the above equation (such a solution is called a *ground state solution*). Furthermore,

$$\kappa(d,p) = \left(p||f_0||_{2p}^{-2(p-1)}\right)^{\frac{1}{2p}}. \tag{3.4.3}$$

Using this result, Weinstein obtained the following numerical approximation in the case $d = p = 2$:

$$\kappa(2,2) = \sqrt[4]{\frac{1}{\pi \times 1.86225...}}. \tag{3.4.4}$$

EXERCISE 3.4.4. *Our starting point is (3.4.1) and we assume everything assumed there. Let τ be an independent exponential time of parameter $\lambda > 0$.*

(1) *Prove that*

$$\lim_{m\to\infty} \frac{1}{m} \log \frac{1}{(m!)^p} \mathbb{E}\Big[I\big([0,\tau]^p\big)^m\Big] = p \log \rho_{\lambda/p}, \tag{3.4.5}$$

where for any $\lambda > 0$,

$$\rho_\lambda = \sup\Big\{ \sum_{x,y\in E} G_\lambda(x,y) f(x) f(y); \ \sum_{x\in E} |f(x)|^{\frac{2p}{2p-1}} = 1\Big\}$$

and

$$G_\lambda(x,y) = \int_0^\infty e^{-\lambda t} p_t(x,y) dt \quad x, y \in E.$$

(2) *Prove that*

$$\lim_{t\to\infty} \frac{1}{t} \log \mathbb{P}\Big\{I\big([0,\tau]^p\big) \geq t^p\Big\} = -\frac{p}{\rho_{\lambda/p}}.$$

Under the same assumptions as in the last exercise, we are concerned about the LDP related to the probabilities

$$\mathbb{P}\Big\{I\big([0,\theta t]^p\big) \geq t^p\Big\} \quad \text{and} \quad \mathbb{P}\Big\{I\big([0,t]^p\big) \geq \theta t^p\Big\}.$$

We discuss this topic in the next two exercises.

EXERCISE 3.4.5 (hard). *Let $q > 1$ be the conjugate of p. Prove that for any $\theta > \#(E)^{1/q}$ there is a $u_\theta > 0$ independent of t such that*

$$\mathbb{P}\Big\{I\big([0,\ \theta t]^p\big) \geq t^p\Big\} \geq u_\theta.$$

Consequently,

$$\lim_{t\to\infty} \frac{1}{t} \log \mathbb{P}\Big\{I\big([0,\ \theta t]^p\big) \geq t^p\Big\} = 0 \qquad \theta > \#(E)^{1/q}.$$

Hint: Write

$$\eta_j(t,x) = \int_0^t 1_{\{X_j(s)=x\}} ds \qquad x \in E, \ \ t \geq 0, \ \ j = 1, \cdots, p$$

and establish the deterministic relation

$$\sum_{x\in E} \eta_j^p(\theta t, x) \geq \theta^p t^p \#(E)^{-\frac{1}{p-1}} \qquad j = 1, \cdots, p.$$

Then argue that when the tuple

$$\Big(\frac{\eta_1(\theta t, \cdot)}{t}, \cdots, \frac{\eta_p(\theta t, \cdot)}{t}\Big)$$

is in a neighborhood of the diagonal in $\bigotimes_{j=1}^{p}\mathcal{L}^p(E)$, *the quantities*

$$\sum_{x\in E}\prod_{j=1}^{p}\frac{\eta_j(\theta t,x)}{t}\quad\text{and}\quad\prod_{j=1}^{p}\Big\{\sum_{x\in E}\Big[\frac{\eta_j(\theta t,x)}{t}\Big]^p\Big\}^{1/p}$$

are close to each other.

EXERCISE 3.4.6 (hard). *We continue our discussion of the model* $I\big([0,\theta t]^p\big)$. *In addition to the case* $\theta > \#(E)^{\frac{p-1}{p}}$, *another trivial situation happens when* $\theta < 1$. *By the fact that* $I\big([0,\ \theta t]^p\big) \le (\theta t)^p$ *we have that*

$$\lim_{t\to\infty}\frac{1}{t}\log\mathbb{P}\Big\{I\big([0,\ \theta t]^p\big)\ge t^p\Big\} = -\infty \qquad \forall\theta < 1.$$

In this exercise we mainly focus on the case $1 < \theta < \#(E)^{\frac{p-1}{p}}$.

(1) *Prove that*
   (a) ρ_λ *is differentiable on* $(0,\infty)$ *as the function of* λ.
   (b) $\lim_{\lambda\to 0^+}\lambda G_\lambda(x,y) = \#(E)^{-1}$ *and* $\lim_{\lambda\to\infty}\lambda G_\lambda(x,y) = 1$ *for* $x,y\in E$.
   (c) *Write*
$$\Lambda(\lambda) = \frac{p}{\rho_{\lambda/p}}.$$
   Derive that $\Lambda(\lambda)$ *is differentiable on* $(0,\infty)$ *and that*
$$\lim_{\lambda\to 0^+}\frac{\Lambda(\lambda)}{\lambda} = \#(E)^{\frac{p-1}{p}}\quad\text{and}\quad\lim_{\lambda\to\infty}\frac{\Lambda(\lambda)}{\lambda} = 1.$$
   Hint: You may need the following two facts:

   First, for any $\lambda > 0$ *there is a unique non-negative function* $f_\lambda(x)$ *on* E *such that*
$$\sum_{x\in E} f_\lambda^{\frac{2p}{2p-1}}(x) = 1,$$
$$\rho_\lambda = \sum_{x,y\in E} G_\lambda(x,y)f_\lambda(x)f_\lambda(y).$$
   Second, by a well-known ergodic theorem for Markov processes,
$$\lim_{t\to\infty}p_t(x,y) = \#(E)^{-1}\qquad x,y\in E.$$

(2) *Define*
$$\xi(\theta) = \sup_{\lambda>0}\Big\{\Lambda(\lambda)-\lambda\theta\Big\}\qquad \theta\ge 0.$$
   Prove that
   (a) $\xi(\cdot)$ *is non-negative and lower semi-continuous on* $\mathbb{R}^+$.
   (b) $0 < \xi(\theta) < \infty$ *as* $1 < \theta < \#(E)^{\frac{p-1}{p}}$, *and that for each* $1 < \theta_0 < \#(E)^{\frac{p-1}{p}}$ *there is a* $0 < \lambda_0 < \infty$ *such that*
$$\xi(\theta_0) = \Lambda(\lambda_0) - \lambda_0\theta_0.$$
   Finally, $\xi(\theta)$ *is continuous on* $\Big(1,\ \#(E)^{\frac{p-1}{p}}\Big)$.

(3) *Prove that for any* $1 < \theta_0 < \#(E)^{\frac{p-1}{p}}$,

$$\lim_{t\to\infty} \frac{1}{t} \log \mathbb{P}\Big\{ I\big([0, \theta_0 t]^p\big) \ge t^p \Big\} = -\xi(\theta_0).$$

Hint: Notice that for an independent exponential time with parameter $\lambda > 0$,

$$\mathbb{P}\Big\{ I\big([0, \tau]^p\big) \ge t^p \Big\} = \lambda \int_0^\infty e^{-\lambda s} \mathbb{P}\Big\{ I\big([0, s]^p\big) \ge t^p \Big\} ds.$$

Use this and a Chebyshev type argument to establish the upper bound.

As for the lower bound, use an argument similar to the one used for (1.1.6) to show that for any interval $J \subset \mathbb{R}^+$,

$$\limsup_{t\to\infty} \frac{1}{t} \log \int_{tJ} e^{-\lambda s} \mathbb{P}\Big\{ I\big([0, s]^p\big) \ge t^p \Big\} ds \le - \inf_{\theta\in J} \Big\{ \lambda\theta + \xi(\theta) \Big\}.$$

(4) *Prove that for any* $\#(E)^{-(p-1)} < \theta < 1$,

$$\lim_{t\to\infty} \frac{1}{t} \log \mathbb{P}\Big\{ I\big([0, t]^p\big) \ge \theta t^p \Big\} = -\theta^{1/p} \xi\Big(\frac{1}{\theta^{1/p}}\Big).$$

CHAPTER 4

Self-intersection: large deviations

According to the discussion in Section 2.3 and Section 2.4, self-intersection local time $\beta([0,t]^p_<)$ exists in the case $d=1$ (Proposition 2.3.5) and is renormalizable in the case $d=2$ (Theorem 2.4.1). The goals of this chapter are to establish the large deviation principles for the self-intersection local times in 1-dimension and for the renormalized self-intersection local time in 2-dimension. Our approach consists of the following three main ingredients:

1. Feynman-Kac formula combined with some ideas developed from the area of probability in Banach spaces. The method is used for the p-square integral of the local time given in (2.3.5).

2. Comparison between $\gamma\big([0,t]^2_<\big)$ and $\alpha\big([0,1]^2\big)$. The connections between the self-intersection local times and mutual intersection local times have been seen in the discussion in Section 2.3 and Section 2.4. We shall explore this link in a much more quantitative way.

3. Argument by sub-additivity. This method has been developed into an effective tool in large deviation theory. Even so, one might still be surprised to see throughout this book how often the relations such as (4.3.15) below hold in our models. Naturally, the sub-additivity method becomes an important part of the methodology adopted in this book.

4.1. Feynman-Kac formula

Let $W(t)$ be a d-dimensional Brownian motion and let $f(x)$ be a bounded continuous function on $\mathbb{R}^d$. Write, for each $g\in\mathcal{L}^2(\mathbb{R}^d)$,

$$T_tg(x)=\mathbb{E}\left(\exp\left\{\int_0^t f\big(x+W(s)\big)ds\right\}g\big(x+W(t)\big)\right)\qquad t\geq 0. \tag{4.1.1}$$

One can see that when $t=0$, $T_0=I$ is an identical operator. For $t>0$,

$$\begin{aligned}\int_{\mathbb{R}^d}\big[T_tg(x)\big]^2dx&=\int_{\mathbb{R}^d}\left[\mathbb{E}\left(\exp\left\{\int_0^t f\big(x+W(s)\big)ds\right\}g\big(x+W(t)\big)\right)\right]^2dx\\&\leq\int_{\mathbb{R}^d}\mathbb{E}\left(\exp\left\{2\int_0^t f\big(x+W(s)\big)ds\right\}g^2\big(x+W(t)\big)\right)dx\\&=\int_{\mathbb{R}^d}\exp\left\{2\int_0^t f(x)ds\right\}g^2(x)dx\leq e^{tC}\int_{\mathbb{R}^d}g^2(x)dx\end{aligned}$$

where the second step follows from Cauchy-Schwarz inequality, the third step from translation invariance, and the last step from the boundedness of f.

It shows that for each $t \geq 0$, T_t defines a continuous linear operator on the Hilbert space $\mathcal{L}^2(\mathbb{R}^d)$. For any $t > 0$, $g, h \in \mathcal{L}^2(\mathbb{R}^d)$,

$$\begin{aligned}
&\int_{\mathbb{R}^d} T_t g(x) \cdot h(x) dx \\
&= \mathbb{E}\bigg(\int_{\mathbb{R}^d} \exp\bigg\{ \int_0^t f\big(x + W(s)\big) ds \bigg\} g\big(x + W(t)\big) h(x) dx \bigg) \\
&= \mathbb{E}\bigg(\int_{\mathbb{R}^d} \exp\bigg\{ \int_0^t f\big(x + W(s) - W(t)\big) ds \bigg\} g(x) h\big(x - W(t)\big) dx \bigg) \\
&= \mathbb{E}\bigg(\int_{\mathbb{R}^d} \exp\bigg\{ \int_0^t f\big(x + W(t-s) - W(t)\big) ds \bigg\} g(x) h\big(x - W(t)\big) dx \bigg) \\
&= \int_{\mathbb{R}^d} g(x) \cdot T_t h(x) dx
\end{aligned}$$

where the last step follows from the identity in law that for any fixed $t > 0$,

$$\Big\{ W(t-s) - W(t);\ \ 0 \leq s \leq t \Big\} \stackrel{d}{=} \Big\{ W(s);\ \ 0 \leq s \leq t \Big\}.$$

The established equation shows that T_t is symmetric. By the boundedness of T_t established earlier, T_t is a self-adjoint operator on $\mathcal{L}^2(\mathbb{R}^d)$ (see the discussion in Section E, Appendix).

LEMMA 4.1.1. *$\{T_t;\ t \geq 0\}$ is a semi-group in the sense that for any $s, t \geq 0$,*

$$T_{s+t} = T_s \circ T_t.$$

Proof. For any $g \in \mathcal{L}^(\mathbb{R}^d)$,

$$\begin{aligned}
T_{s+t} g(x) &= \mathbb{E}\bigg(\exp\bigg\{ \int_0^{s+t} f\big(x + W(u)\big) du \bigg\} g\big(x + W(s+t)\big) \bigg) \\
&= \mathbb{E}\bigg(\exp\bigg\{ \int_0^s f\big(x + W(u)\big) du \bigg\} \\
&\times \exp\bigg\{ \int_0^t f\big(x + W(s) + W(s+u) - W(s)\big) du \bigg\} \\
&\times g\big(x + W(s) + W(s+t) - W(s)\big) \bigg) \\
&= \mathbb{E}\bigg(\exp\bigg\{ \int_0^s f\big(x + W(u)\big) du \bigg\} (T_t g)\big(x + W(s)\big) \bigg) \\
&= (T_s \circ T_t)(g)(x)
\end{aligned}$$

where the third step follows from independence of increment and the Fubini theorem. □

A self-adjoint operator T is said to be *non-negative* if $\langle g, Tg\rangle \geq 0$ for any g in the domain $\mathcal{D}(T)$ of T (In the case of bounded operator, of course, $\mathcal{D}(T) = \mathcal{L}^2(\mathbb{R}^d)$). By semi-group property and self-adjointness,

$$\langle g, T_t g\rangle = \langle g,\ T_{t/2}\circ T_{t/2}g\rangle = ||T_{t/2}g||_2^2 \geq 0 \qquad g \in \mathcal{L}^2(\mathbb{R}^d).$$

Therefore, for each $t \geq 0$, T_t is non-negative.

An important aspect of a semi-group is its infinitesimal generator A, which is a linear operator satisfying

$$\lim_{t\to 0^+} \frac{T_t g - g}{t} = Ag$$

on a subspace of $\mathcal{L}^2(\mathbb{R}^d)$, where the limit is taken in an appropriate sense. To make the matter less complicated, we first consider the case when $g(x)$ is a rapidly decreasing function. An infinitely smooth function $g(x)$ is said to be *rapidly decreasing*, if $g(x)$ and its derivatives of all orders decay at infinity faster than any negative power of $|x|$.

We now state the famous *Feynman-Kac formula.*

THEOREM 4.1.2. *For any rapidly decreasing function $g(x)$ on $\mathbb{R}^d$,*

$$\lim_{t\to 0^+} \frac{T_t g(x) - g(x)}{t} = \frac{1}{2}\Delta g(x) + f(x)g(x) \qquad x \in \mathbb{R}^d. \tag{4.1.2}$$

Proof. First we consider the special case when $f \equiv 0$, in which

$$T_t^0 g(x) \equiv T_t g(x) = \int_{\mathbb{R}^d} p_t(y)g(x+y)dy = \int_{\mathbb{R}^d} p_t(y-x)g(y)dy.$$

By Fourier transform,

$$\widehat{T_t^0 g}(\lambda) \equiv \int_{\mathbb{R}^d} e^{i\lambda\cdot x} T_t^0 g(x)dx = \widehat{g}(\lambda)\exp\Big\{-\frac{t|\lambda|^2}{2}\Big\} \qquad \lambda \in \mathbb{R}^d$$

where

$$\widehat{g}(\lambda) = \int_{\mathbb{R}^d} e^{i\lambda\cdot x} g(x)dx.$$

Taking Fourier inversion (Theorem B.1, Appendix),

$$T_t^0 g(x) - g(x) = \frac{1}{(2\pi)^d}\int_{\mathbb{R}^d} e^{-i\lambda\cdot x}\widehat{g}(\lambda)\Big[\exp\Big\{-\frac{t|\lambda|^2}{2}\Big\} - 1\Big]d\lambda.$$

Consequently, by dominated convergence theorem, and by the fact that the Fourier transform of a rapidly decreasing function is rapidly decreasing,

$$\lim_{t\to 0^+} \frac{T_t^0 g(x) - g(x)}{t} = -\frac{1}{2}\frac{1}{(2\pi)^d}\int_{\mathbb{R}^d} e^{-i\lambda\cdot x}|\lambda|^2\widehat{g}(\lambda)d\lambda = \frac{1}{2}\Delta g(x) \tag{4.1.3}$$

where the last step follows from (B.8), Appendix. So we have (4.1.2) when $f \equiv 0$.

We now prove (4.1.2) with full generality. First notice that

$$\begin{aligned}
&\exp\Big\{\int_0^t f\big(x + W(s)\big)ds\Big\}\\
&= 1 + \int_0^t f\big(x+W(s)\big)\exp\Big\{\int_s^t f\big(x+W(u)\big)du\Big\}ds.
\end{aligned}$$

Multiplying both sides by $g\big(x+W(t)\big)$, taking expectation and applying increment independence on the right hand side, we get

$$T_t g(x) = T_t^0 g(x) + \int_0^t \mathbb{E}\Big\{f\big(x+W(s)\big)T_{t-s}g\big(x+W(s)\big)\Big\}ds.$$

Therefore, (4.1.2) follows from (4.1.3) and the fact that

$$\lim_{t\to 0^+}\frac{1}{t}\int_0^t \mathbb{E}\Big\{f\big(x+W(s)\big)T_{t-s}g\big(x+W(s)\big)\Big\}ds = f(x)g(x).$$

□

Let $\mathcal{S}(\mathbb{R}^d)$ be the set of all rapidly decreasing functions on $\mathbb{R}^d$, which is also known as *Schwartz space.* Write

$$Ag(x) = \frac{1}{2}\Delta g(x) + f(x)g(x) \qquad g\in\mathcal{S}(\mathbb{R}^d). \tag{4.1.4}$$

The linear operator A is called the *infinitesimal generator* of the semi-group $\{T_t,\ t\ge 0\}$. In (4.1.4) we momentarily limit the domain of A to $\mathcal{S}(\mathbb{R}^d)$. By integration by parts one can see that

$$\langle g, Ah\rangle = \langle Ag, h\rangle \qquad g,h\in\mathcal{S}(\mathbb{R}^d). \tag{4.1.5}$$

This means that A is a symmetric operator (see Section E, Appendix for the definition of symmetric operator on the general Hilbert space).

As pointed out in Section E of the Appendix, an unbounded symmetric operator does not have to be self-adjoint. It appears to be extremely difficult to examine the self-adjointness of the unbounded operator A. On the other hand, the self-adjointness is crucially needed for later development. Our treatment is to extend A into a self-adjoint operator. Notice that for each $g\in\mathcal{S}(\mathbb{R}^d)$,

$$\langle Ag, g\rangle = \int_{\mathbb{R}^d} f(x)g^2(x)dx - \frac{1}{2}\int_{\mathbb{R}^d}|\nabla g(x)|^2dx \le C||g||_2^2$$

where $C = \sup_{x\in\mathbb{R}^d}|f(x)|$. This means that the operator A is upper semi-bounded. According to Friedrichs' extension theorem (Theorem E.1, Appendix), A admits a self-adjoint extension known as Friedrichs' extension. We still use the same notation for the Friedrichs' extension of A and still call it infinitesimal generator of the semi-group T_t. We denote $\mathcal{D}(A)$ for the domain of the self-adjoint operator A. Clearly, $\mathcal{S}(\mathbb{R}^d)\subset\mathcal{D}(A)\subset\mathcal{L}^2(\mathbb{R}^d)$.

Recall our notation

$$W^{1,2}(\mathbb{R}^d) = \Big\{g\in\mathcal{L}^2(\mathbb{R}^d);\ \ \nabla g\in\mathcal{L}^2(\mathbb{R}^d)\Big\}$$

and set

$$\mathcal{F}_d = \Big\{g\in\mathcal{L}^2(\mathbb{R}^d);\ \ ||g||_2 = 1 \ \text{ and } \ ||\nabla g||_2<\infty\Big\}.$$

In the following lemma we determine some basic structure of the operator A.

LEMMA 4.1.3. *We have that* $\mathcal{D}(A)\subset W^{1,2}(\mathbb{R}^d)$, *and that*

$$\langle g, Ag\rangle = \int_{\mathbb{R}^d} f(x)g^2(x)dx - \frac{1}{2}\int_{\mathbb{R}^d}|\nabla g(x)|^2dx \qquad g\in\mathcal{D}(A), \tag{4.1.6}$$

$$(4.1.7) \qquad \sup_{\substack{g\in\mathcal{D}(A)\\ ||g||_2=1}} \langle g, Ag\rangle = \sup_{g\in\mathcal{F}_d}\left\{\int_{\mathbb{R}^d} f(x)g^2(x)dx - \frac{1}{2}\int_{\mathbb{R}^d}|\nabla g(x)|^2dx\right\}.$$

Proof. Let $g \in \mathcal{D}(A)$ and write, for each $\epsilon > 0$,

$$g_\epsilon(x) = \int_{\mathbb{R}^d} g(y)p_\epsilon(y-x)dx$$

where $p_\epsilon(x)$ is the density of $W(\epsilon)$. It is easy to see that $g_\epsilon \in \mathcal{S}(\mathbb{R}^d)$. By (4.1.4),

$$\langle Ag_\epsilon, g\rangle = \int_{\mathbb{R}^d} f(x)g_\epsilon(x)g(x)dx + \frac{1}{2}\int_{\mathbb{R}^d}\Delta g_\epsilon(x)g(x)dx.$$

We shall let $\epsilon \to 0^+$ on the both sides. For the left-hand side, observe that $g_\epsilon \to g$ in $\mathcal{L}^2(\mathbb{R}^d)$. Consequently,

$$\langle Ag_\epsilon, g\rangle = \langle g_\epsilon, Ag\rangle \longrightarrow \langle g, Ag\rangle \qquad (\epsilon \to 0^+).$$

By the boundedness of $f(x)$, the first term on the right hand side of the previous equality approaches

$$\int_{\mathbb{R}^d} f(x)g^2(x)dx.$$

To treat the second integral on the right hand side, we apply the tool of Fourier transformation. By Parseval identity ((B.9), Appendix), the convolution identity ((B.11), Appendix), and the fact ((B.8), Appendix) that

$$\widehat{\Delta g_\epsilon}(\lambda) = |\lambda|^2\widehat{g_\epsilon}(\lambda) = |\lambda|^2\widehat{g}(\lambda)\exp\left\{-\frac{\epsilon}{2}|\lambda|^2\right\},$$

$$\int_{\mathbb{R}^d}\Delta g_\epsilon(x)g(x)dx = \frac{1}{(2\pi)^d}\int_{\mathbb{R}^d}|\lambda|^2\exp\left\{-\frac{\epsilon}{2}|\lambda|^2\right\}|\widehat{g}(\lambda)|^2d\lambda.$$

Therefore, we must have

$$\int_{\mathbb{R}^d}|\lambda|^2|\widehat{g}(\lambda)|^2d\lambda < \infty.$$

By Theorem B.2, Appendix, $g \in W^{1,2}(\mathbb{R}^d)$ and

$$\begin{aligned}\langle g, Ag\rangle &= \int_{\mathbb{R}^d} f(x)g^2(x)dx - \frac{1}{2}\frac{1}{(2\pi)^d}\int_{\mathbb{R}^d}|\lambda|^2|\widehat{g}(\lambda)|^2d\lambda\\ &= \int_{\mathbb{R}^d} f(x)g^2(x)dx - \frac{1}{2}\int_{\mathbb{R}^d}|\nabla g(x)|^2dx.\end{aligned}$$

Finally, (4.1.7) follows from the relation $\mathcal{S}(\mathbb{R}^d) \subset D(A) \subset W^{1,2}(\mathbb{R}^d)$, the fact that $\mathcal{S}(\mathbb{R}^d)$ is dense in $W^{1,2}(\mathbb{R}^d)$ under the Sobolev norm $\sqrt{||g||_2^2 + ||\nabla g||_2^2}$, and the continuity (in g) of the quadratic form on the right-hand side of (4.1.6) under the Sobolev norm. □

The quadratic form $\langle g, Ag\rangle$ given in (4.1.6) is often known as *the Dirichlet form* of the semi-group $\{T_t\}$.

By the fact that $||T_t|| \le e^{Ct}$ ($t \ge 0$) for some constant $C > 0$, the linear operator

$$G_\lambda g(x) = \int_0^\infty e^{-\lambda t} T_t g(x) dt \qquad g \in \mathcal{L}^2(\mathbb{R}^d) \tag{4.1.8}$$

is well defined and is a bounded self-adjoint operator on $\mathcal{L}^2(\mathbb{R}^d)$.

LEMMA 4.1.4. *For any $g \in \mathcal{S}(\mathbb{R}^d)$ and $x \in \mathbb{R}^d$, $T_t g(x)$ is differentiable on $(0, \infty)$ as a function of t and*

$$\frac{d}{dt}\Big(T_t g(x)\Big) = (A \circ T_t) g(x) = (T_t \circ A) g(x). \tag{4.1.9}$$

In addition, there is a $C > 0$ such that for any $\lambda > C$,

$$(\lambda I - A) \circ G_\lambda g = G_\lambda \circ (\lambda I - A) g = g \qquad g \in \mathcal{S}(\mathbb{R}^d). \tag{4.1.10}$$

Proof. We first point out the following analytic lemma (the proof can be found in p. 239, Section 3, Chapter IX, [**175**]): Let $u(t)$ be a continuous function on $(0, \infty)$ such that its right derivative

$$D^+ u(t) \equiv \lim_{\Delta t \to 0^+} \frac{u(t + \Delta t) - u(t)}{\Delta t}$$

is finite and continuous on $(0, \infty)$. Then $u(t)$ is differentiable on $(0, \infty)$.

To establish (4.1.9), therefore, it suffices to show that

$$\lim_{\Delta t \to 0^+} \frac{T_{t+\Delta t} g(x) - T_t g(x)}{\Delta t} = (A \circ T_t) g(x) = (T_t \circ A) g(x).$$

Indeed, this follows from the equality

$$\frac{T_{t+\Delta t} g(x) - T_t g(x)}{\Delta t} = T_t \circ \frac{T_{\Delta t} - I}{\Delta t} g(x) = \frac{T_{\Delta t} - I}{\Delta t} \circ T_t g(x)$$

and (4.1.2) in Theorem 4.1.2.

By (4.1.9), we get

$$\begin{aligned} (\lambda I - A) \circ G_\lambda g(x) &= \lambda G_\lambda g(x) - \int_0^\infty e^{-\lambda t} (A \circ T_t) g(x) dt \\ &= \lambda G_\lambda g(x) - \int_0^\infty e^{-\lambda t} \frac{d}{dt}\Big(T_t g(x)\Big) dt. \end{aligned}$$

Thus, (4.1.10) follows from integration by parts. □

An important fact is that the semi-group $\{T_t;\ t \ge 0\}$ can be written in terms of exponential functions of A. Our discussion here is based on the material introduced in Section E, Appendix. Generally speaking, a function $\xi(A)$ of the self-adjoint operator is another self-adjoint operator. Being precise, we consider the spectral integral representation (Theorem E.2, Appendix)

$$A = \int_{-\infty}^\infty \lambda E(d\lambda) \tag{4.1.11}$$

of the self-adjoint operator A, where $\{E(\lambda);\ -\infty < \lambda < \infty\}$ is a resolution of identity, a family of projection operators on $\mathcal{L}^2(\mathbb{R}^d)$ with distribution-function-type properties. In addition,

$$\mathcal{D}(A) = \left\{g \in \mathcal{L}^2(\mathbb{R}^d);\ \int_{-\infty}^{\infty} |\lambda|^2 \mu_g(d\lambda) < \infty\right\} \tag{4.1.12}$$

where μ_g is a finite measure (known as spectral measure) on $\mathbb{R}$ induced by the distribution function $F(\lambda) \equiv \langle g, E(\lambda)g\rangle$ with

$$\mu_g(\mathbb{R}) = ||g||_2^2 \qquad g \in \mathcal{L}^2(\mathbb{R}^d). \tag{4.1.13}$$

Given a Borel function $\xi(\lambda)$ on $\mathbb{R}$, the function $\xi(A)$ of A is a self-adjoint operator defined by the spectral integral

$$\xi(A) = \int_{-\infty}^{\infty} \xi(\lambda) E(d\lambda) \tag{4.1.14}$$

with the domain $\mathcal{D}\big(\xi(A)\big)$ given by

$$\mathcal{D}\big(\xi(A)\big) = \left\{g \in \mathcal{L}^2(\mathbb{R}^d);\ \int_{-\infty}^{\infty} |\xi(\lambda)|^2 \mu_g(d\lambda) < \infty\right\}. \tag{4.1.15}$$

Given $\lambda > 0$, for example, we take $\xi(\gamma) = (\lambda - \gamma)^{-1}$ and define the following operator:

$$(\lambda I - A)^{-1} = \int_{-\infty}^{\infty} \frac{1}{\lambda - \gamma} E(d\gamma). \tag{4.1.16}$$

Here we use the fact ((4.1.7)) that

$$\lambda_0 \equiv \sup_{\substack{g\in\mathcal{D}(A)\\ ||g||_2=1}} \langle g, Ag\rangle < \infty. \tag{4.1.17}$$

By Theorem E.3, Appendix, μ_g is supported by $(-\infty, \lambda_0]$ for every $g \in \mathcal{L}^2(\mathbb{R}^d)$. Consequently, by examining the integrability listed in (4.1.15) the operator $(\lambda I - A)^{-1}$ is defined on the whole space $\mathcal{L}^2(\mathbb{R}^d)$ as $\lambda > \lambda_0$. Further, by (E.14) in the Appendix, we get

$$\Big|\Big|(\lambda I - A)^{-1} g\Big|\Big|_2^2 = \int_{-\infty}^{\infty} \frac{1}{(\lambda - \gamma)^2} \mu_g(d\lambda) \le \frac{1}{(\lambda - \lambda_0)^2} ||g||_2^2 \qquad g \in \mathcal{L}^2(\mathbb{R}^d).$$

This implies that $(\lambda I - A)^{-1}$ is a bounded operator. As its notation suggested, $(\lambda I - A)^{-1}$ is the inverse operator of $\lambda I - A$ in the sense that $(\lambda I - A)^{-1} \circ (\lambda I - A)$ and $(\lambda I - A) \circ (\lambda I - A)^{-1}$ are identical operators on $\mathcal{D}(A)$ and on $\mathcal{L}^2(\mathbb{R}^d)$, respectively. To see this, first notice that by (4.1.11)

$$(\lambda I - A) = \int_{-\infty}^{\infty} (\lambda - \gamma) E(d\gamma).$$

Our claim follows immediately from Theorem E.4, Appendix.

Further, applying $(\lambda I - A)^{-1}$ to both sides of (4.1.10) we see that $(\lambda I - A)^{-1}$ and G_λ agree on $\mathcal{S}(\mathbb{R}^d)$ when $\lambda > 0$ is sufficiently large. By the fact that $\mathcal{S}(\mathbb{R}^d)$ is dense in $\mathcal{L}^2(\mathbb{R}^d)$, and by the continuity of the operators $(\lambda I - A)^{-1}$ and G_λ, we have that for large $\lambda > 0$,

$$(\lambda I - A)^{-1} = G_\lambda. \tag{4.1.18}$$

For given $t \geq 0$, we now take $\xi(\lambda) = \exp\{t\lambda\}$ and consider the self-adjoint operator $\exp\{tA\}$ defined by

$$\exp\{tA\} = \int_{-\infty}^{\infty} \exp\{t\lambda\}E(d\lambda). \tag{4.1.19}$$

By Theorem E.3 the support of the spectral measure μ_g is bounded from above by λ_0 given in (4.1.17) for all $g \in \mathcal{L}^2(\mathbb{R}^d)$. Consequently, $\exp\{tA\}$ is defined on the whole space $\mathcal{L}^2(\mathbb{R}^d)$. Moreover, by (E.14) in the Appendix, we get

$$\big|\big|\exp\{tA\}g\big|\big|^2 = \int_{-\infty}^{\infty} \exp\{2t\lambda\}\mu_g(d\lambda) \leq \exp\{2t\lambda_0\}||g||_2^2.$$

Hence, $\exp\{tA\}$ is a bounded operator. Further, by (E.15) in the Appendix, we get

$$\Big\langle g, \exp\{tA\}g\Big\rangle = \int_{-\infty}^{\infty} \exp\{t\lambda\}\mu_g(d\lambda) \geq 0 \quad g \in \mathcal{L}^2(\mathbb{R}^d).$$

Consequently, $\exp\{tA\}$ is non-negative.

Now, we state the following representation theorem.

THEOREM 4.1.5. *For each* $t \geq 0$, $T_t = e^{tA}$.

Proof. First, we reduce Theorem 4.1.5 to the equality

$$\Big\langle g,\ \exp\{tA\}g\Big\rangle = \langle g, T_t g\rangle \quad g \in \mathcal{L}^2(\mathbb{R}^d). \tag{4.1.20}$$

Indeed, for any $g, h \in \mathcal{L}^2(\mathbb{R}^d)$,

$$\Big\langle h,\ \exp\{tA\}g\Big\rangle = \frac{1}{4}\Big\{\Big\langle g+h,\ \exp\{tA\}(g+h)\Big\rangle + \Big\langle g-h,\ \exp\{tA\}(g-h)\Big\rangle\Big\},$$

$$\langle h, T_t g\rangle = \frac{1}{4}\Big\{\langle g+h,\ T_t(g+h)\rangle + \langle g-h,\ T_t(g-h)\rangle\Big\}.$$

If (4.1.20) holds, then

$$\Big\langle h,\ \exp\{tA\}g\Big\rangle = \langle h, T_t g\rangle.$$

Since g and h are arbitrary, this leads to the conclusion of Theorem 4.1.5.

To establish (4.1.20), we begin with the operator equation (4.1.18), which is rewritten as

$$\Big(I - \frac{A}{\lambda}\Big)^{-1} = \lambda\int_0^{\infty} e^{-\lambda s}T_s ds.$$

In our argument given below, we keep $\lambda > 0$ sufficiently large so the operators on both sides of the above equation are well defined and bounded.

Iterating n times (in the sense of operator composition) on both sides of the equation, we get

$$\begin{aligned}
&\Big(I - \frac{A}{\lambda}\Big)^{-n} \\
&= \lambda^n \int_{(\mathbb{R}^+)^n} \exp\{-\lambda(s_1 + \cdots + s_n)\}T_{s_1} \circ \cdots \circ T_{s_n} ds_1 \cdots ds_n \\
&= \lambda^n \int_{(\mathbb{R}^+)^n} \exp\{-\lambda(s_1 + \cdots + s_n)\}T_{s_1+\cdots+s_n} ds_1 \cdots ds_n.
\end{aligned}$$

By the spectral integral representation (4.1.16) and by Theorem E.4, Appendix, the left hand side can be represented as

$$\int_{-\infty}^{\infty}\Big(1-\frac{\gamma}{\lambda}\Big)^{-n}E(d\gamma).$$

By (E.15) in the Appendix, therefore,

$$\int_{-\infty}^{\infty}\Big(1-\frac{\gamma}{\lambda}\Big)^{-n}\mu_g(d\gamma)=\Big\langle g,\ \Big(I-\frac{A}{\lambda}\Big)^{-n}g\Big\rangle$$

$$=\lambda^n\int_{(\mathbb{R}^+)^n}\exp\{-\lambda(s_1+\cdots+s_n)\}\langle g,\ T_{s_1+\cdots+s_n}g\rangle ds_1\cdots ds_n$$

$$=\int_{(\mathbb{R}^+)^n}\exp\{-(s_1+\cdots+s_n)\}\langle g,\ T_{\lambda^{-1}(s_1+\cdots+s_n)}g\rangle ds_1\cdots ds_n.$$

Replacing λ by n/t, we get

$$\int_{-\infty}^{\infty}\Big(1-\frac{t\gamma}{n}\Big)^{-n}\mu_g(d\gamma)=\mathbb{E}\langle g,T_{t\zeta_n}g\rangle \tag{4.1.21}$$

where the expectation on the right hand side is taken with respect to the random variable

$$\zeta_n=\frac{\tau_1+\cdots+\tau_n}{n}$$

given as the sample average of i.i.d. exponential random variables $\tau_1,\cdots,\tau_n$ with parameter 1.

According to the classic law of large numbers, $\zeta_n\stackrel{d}{\longrightarrow}1$ as $n\to\infty$. By the continuity of $\langle g,T_sg\rangle$ as a function of s and by the bound $\langle g,T_sg\rangle\le C_1e^{C_2s}$,

$$\lim_{n\to\infty}\mathbb{E}\langle g,T_{t\zeta_n}g\rangle=\langle g,T_tg\rangle.$$

By the fact that the supporting set of the measure μ_g is bounded from above, and by the dominated convergence theorem,

$$\lim_{n\to\infty}\int_{-\infty}^{\infty}\Big(1-\frac{t\gamma}{n}\Big)^{-n}\mu_g(d\gamma)=\int_{-\infty}^{\infty}e^{t\gamma}\mu_g(d\gamma)=\Big\langle g,\ \exp\{tA\}g\Big\rangle$$

where the second equality follows from (E.15), Appendix.

Finally, letting $n\to\infty$ on both sides of (4.1.21) leads to (4.1.20). □

An interesting application of the Feynman-Kac formula is the following large deviation result.

THEOREM 4.1.6. *For any bounded, continuous function f on $\mathbb{R}^d$,*

$$\lim_{t\to\infty}\frac{1}{t}\log\mathbb{E}\exp\Big\{\int_0^t f\big(W(s)\big)ds\Big\}$$

$$=\sup_{g\in\mathcal{F}_d}\Big\{\int_{\mathbb{R}^d}f(x)g^2(x)dx-\frac{1}{2}\int_{\mathbb{R}^d}|\nabla g(x)|^2dx\Big\},$$

where

$$\mathcal{F}_d=\Big\{g\in\mathcal{L}^2(\mathbb{R}^d);\ \int_{\mathbb{R}^d}g^2(x)dx\ \textit{and}\ \int_{\mathbb{R}^d}|\nabla g(x)|^2dx<\infty\Big\}. \tag{4.1.22}$$

Proof. By the boundedness of f there is $C>0$ such that

$$C^{-1}\le\exp\Big\{\int_0^1 f\big(W(s)\big)ds\Big\}\le C. \tag{4.1.23}$$

Let $g\in\mathcal{S}(\mathbb{R}^d)$ be compactly supported and satisfy

$$\int_{\mathbb{R}^d} g^2(x)dx=1.$$

We have

$$\begin{aligned}
&\mathbb{E}\exp\Big\{\int_1^t f\big(W(s)\big)ds\Big\}\\
&\ge \|g\|_\infty^{-2}\mathbb{E}\bigg(g\big(W(1)\big)\exp\Big\{\int_1^t f\big(W(s)\big)ds\Big\}g\big(W(t)\big)\bigg)\\
&= \|g\|_\infty^{-2}\mathbb{E}\bigg(g\big(W(1)\big)\exp\Big\{\int_1^{t-1} f\big(W(1)+W(1+s)-W(1)\big)ds\Big\}\\
&\qquad\times g\big(W(1)+W(t)-W(1)\big)\bigg)\\
&= \|g\|_\infty^{-2}\mathbb{E}\Big(g\big(W(1)\big)T_{t-1}g\big(W(1)\big)\Big)\\
&= \|g\|_\infty^{-2}\frac{1}{(2\pi)^{d/2}}\int_{\mathbb{R}^d} e^{-|x|^2/2}g(x)T_{t-1}g(x)dx.
\end{aligned}$$

On the supporting set of g, $e^{-|x|^2/2}$ is bounded from below by a positive number. Therefore, we have established that

$$\mathbb{E}\exp\Big\{\int_0^t f\big(W(s)\big)ds\Big\}\ge\delta\int_{\mathbb{R}^d} g(x)T_{t-1}g(x)dx=\delta\langle g,T_{t-1}g\rangle. \tag{4.1.24}$$

By Theorem 4.1.5 and by (E.15) in the Appendix (with $\xi(\lambda)=\exp\{t\lambda\}$), we get

$$\begin{aligned}
\langle g,T_tg\rangle&=\int_{-\infty}^\infty e^{t\lambda}\mu_g(d\lambda)\\
&\ge\exp\Big\{t\int_{-\infty}^\infty \lambda\mu_g(d\lambda)\Big\}=\exp\Big\{t\langle g,Ag\rangle\Big\}\\
&=\exp\Big\{t\Big(\int_{\mathbb{R}^d} f(x)g^2(x)dx-\frac{1}{2}\int_{\mathbb{R}^d}|\nabla g(x)|^2dx\Big)\Big\}
\end{aligned} \tag{4.1.25}$$

where the inequality follows from Jensen inequality and the fact ((4.1.13)) that $\mu_g(\mathbb{R}^+)=1$, which shows that the spectral measure μ_g is a probability measures; and where the third step follows from (E.15) in the Appendix (with $\xi(\lambda)=\lambda$).

By (4.1.23) and (4.1.24), therefore, we get

$$\liminf_{t\to\infty}\frac{1}{t}\log\mathbb{E}\exp\Big\{\int_0^t f\big(W(s)\big)ds\Big\}\ge\int_{\mathbb{R}^d} f(x)g^2(x)dx-\frac{1}{2}\int_{\mathbb{R}^d}|\nabla g(x)|^2dx.$$

Taking supremum over g gives the desired lower bound

$$
\begin{aligned}
(4.1.26)\qquad &\liminf_{t\to\infty}\frac{1}{t}\log\mathbb{E}\exp\left\{\int_0^t f\big(W(s)\big)ds\right\}\\
&\ge \sup_{g\in\mathcal{F}_d}\left\{\int_{\mathbb{R}^d} f(x)g^2(x)dx-\frac{1}{2}\int_{\mathbb{R}^d}|\nabla g(x)|^2dx\right\}.
\end{aligned}
$$

We now establish the upper bound. For each $t>0$,

$$
\begin{aligned}
&\mathbb{E}\exp\left\{\int_1^t f\big(W(s)\big)ds\right\}\\
&\le \mathbb{E}\left(1_{\{|W(1)|\le t^2\}}\exp\left\{\int_1^t f\big(W(s)\big)ds\right\}1_{\{|W(t)|\le t^2\}}\right)\\
&+\mathbb{E}\left(1_{\{|W(1)|> t^2\}}\exp\left\{\int_1^t f\big(W(s)\big)ds\right\}\right)\\
&+\mathbb{E}\left(\exp\left\{\int_1^t f\big(W(s)\big)ds\right\}1_{\{|W(t)|> t^2\}}\right).
\end{aligned}
$$

By boundedness of f, there is $C>0$ such that

$$
\begin{aligned}
&\mathbb{E}\left(1_{\{|W(1)|> t^2\}}\exp\left\{\int_1^t f\big(W(s)\big)ds\right\}\right)\\
&\le e^{Ct}\mathbb{P}\{|W(1)|>t^2\}\le Ce^{Ct}e^{-t^4/2}
\end{aligned}
$$

where the second estimate follows from the estimation of Gaussian tail. Similarly,

$$
\begin{aligned}
&\mathbb{E}\left(\exp\left\{\int_1^t f\big(W(s)\big)ds\right\}1_{\{|W(t)|> t^2\}}\right)\\
&\le e^{Ct}\mathbb{P}\{|W(t)|>t^2\}\le Ce^{Ct}e^{-t^3/2}.
\end{aligned}
$$

Thus,

$$
\begin{aligned}
(4.1.27)\qquad &\limsup_{t\to\infty}\frac{1}{t}\log\mathbb{E}\exp\left\{\int_1^t f\big(W(s)\big)ds\right\}\\
&\le \limsup_{t\to\infty}\frac{1}{t}\log\mathbb{E}\left(1_{\{|W(1)|\le t^2\}}\exp\left\{\int_1^t f\big(W(s)\big)ds\right\}1_{\{|W(t)|\le t^2\}}\right).
\end{aligned}
$$

For each $t>0$, let V_t be the volume of a dimensional ball of the radius t^2, and define

$$
g_t(x)=\begin{cases}\dfrac{1}{\sqrt{V_t}} & |x|\le t^2,\\[2ex] 0 & |x|>t^2.\end{cases}
$$

Then $g_t \in \mathcal{L}^2(\mathbb{R}^d)$ with $||g_t|| = 1$ and

$$\begin{aligned}(4.1.28)\qquad &\mathbb{E}\biggl(1_{\{|W(1)|\le t^2\}}\exp\Bigl\{\int_1^t f\bigl(W(s)\bigr)ds\Bigr\}1_{\{|W(t)|\le t^2\}}\biggr)\\ &= V_t\mathbb{E}\biggl(g_t\bigl(W(1)\bigr)\exp\Bigl\{\int_1^t f\bigl(W(s)\bigr)ds\Bigr\}g_t\bigl(W(t)\bigr)\biggr)\\ &= \frac{1}{(2\pi)^{d/2}}V_t\int_{\mathbb{R}^d}\exp\Bigl\{-\frac{|x|^2}{2}\Bigr\}g_t(x)(T_{t-1}g_t)(x)dx\\ &\le Ct^{2d}\langle g_t, T_{t-1}g_t\rangle.\end{aligned}$$

Applying Theorem 4.1.5 and (E.15) in the Appendix, we get

$$\langle g_t, T_{t-1}g_t\rangle = \int_{-\infty}^{\infty} e^{(t-1)\lambda}\mu_{g_t}(d\lambda).$$

By (4.1.13) and the fact that $||g_t||_2 = 1$, μ_{g_t} is a probability measure. In view of (4.1.17) and (4.1.7) and by Theorem E.3 in the Appendix, the smallest supporting set of μ_{g_t} is bounded from above by

$$\sup_{\substack{g\in\mathcal{D}(A)\\ ||g||_2=1}}\langle g, Ag\rangle = \sup_{g\in\mathcal{F}_d}\Bigl\{\int_{\mathbb{R}^d} f(x)g^2(x)dx - \frac{1}{2}\int_{\mathbb{R}^d}|\nabla g(x)|^2dx\Bigr\}.$$

Hence,

$$\begin{aligned}(4.1.29)\qquad &\langle g_t, T_{t-1}g_t\rangle\\ &\le \exp\biggl((t-1)\sup_{g\in\mathcal{F}_d}\Bigl\{\int_{\mathbb{R}^d} f(x)g^2(x)dx - \frac{1}{2}\int_{\mathbb{R}^d}|\nabla g(x)|^2dx\Bigr\}\biggr).\end{aligned}$$

Combining (4.1.23), (4.1.27), (4.1.28) and (4.1.29) we have

$$\begin{aligned}(4.1.30)\qquad &\limsup_{t\to\infty}\frac{1}{t}\log\mathbb{E}\exp\Bigl\{\int_0^t f\bigl(W(s)\bigr)ds\Bigr\}\\ &\le \sup_{g\in\mathcal{F}_d}\Bigl\{\int_{\mathbb{R}^d} f(x)g^2(x)dx - \frac{1}{2}\int_{\mathbb{R}^d}|\nabla g(x)|^2dx\Bigr\}.\end{aligned}$$

Finally, the desired conclusion follows from (4.1.26) and (4.1.30). □

4.2. One-dimensional case

The local time $L(t,x)$ of a 1-dimensional Brownian motion $W(t)$ exists. According to Proposition 2.3.5, for any integer $p \ge 2$ the self-intersection local time $\beta\bigl([0,t]^p_<\bigr)$ is equal to

$$\frac{1}{p!}\int_{-\infty}^{\infty} L^p(t,x)dx.$$

By (2.3.8) in Proposition 2.3.3, we get

$$(4.2.1)\qquad \int_{-\infty}^{\infty} L^p(t,x)dx \overset{d}{=} t^{\frac{p+1}{2}}\int_{-\infty}^{\infty} L^p(1,x)dx.$$

Thus, we focus on the large deviation principle for the L_p-norm of the Brownian local times up to the time 1. The following theorem is the main result of this section.

THEOREM 4.2.1. *When $d = 1$, for any real number $p > 1$,*

$$\lim_{a\to\infty} a^{-\frac{2}{p-1}} \log \mathbb{P}\Big\{ \int_{-\infty}^{\infty} L^p(1,x)dx \geq a \Big\}$$
$$= -\frac{1}{4(p-1)}\Big(\frac{p+1}{2}\Big)^{\frac{3-p}{p-1}} B\Big(\frac{1}{p-1}, \frac{1}{2}\Big)$$

where $B(\cdot,\cdot)$ is beta function.

Proof. By Corollary 1.2.5, Theorem 4.2.1 holds if we can prove for each $\theta > 0$ that

$$\lim_{a\to\infty} a^{-\frac{2}{p-1}} \log \mathbb{E}\exp\left\{ a^{\frac{p+1}{p(p-1)}}\theta \Big(\int_{-\infty}^{\infty} L^p(1,x)dx\Big)^{1/p}\right\} \tag{4.2.2}$$
$$= \Big(\frac{\theta}{p}\Big)^{\frac{2p}{p+1}} \Big(\frac{\sqrt{2}}{(p-1)(p+1)} B\Big(\frac{1}{p-1},\frac{1}{2}\Big)\Big)^{-\frac{2(p-1)}{p+1}}.$$

By (4.2.1) and (C.8) in Theorem C.4, Appendix, this is equivalent to

$$\lim_{t\to\infty} \frac{1}{t}\log \mathbb{E}\exp\left\{ \Big(\int_{-\infty}^{\infty} L^p(t,x)dx\Big)^{1/p}\right\} \tag{4.2.3}$$
$$= \sup_{g\in\mathcal{F}} \left\{ \Big(\int_{-\infty}^{\infty} |g(x)|^{2p}dx\Big)^{1/p} - \frac{1}{2}\int_{-\infty}^{\infty} |g'(x)|^2 dx\right\},$$

where

$$\mathcal{F} = \left\{ g;\ \int_{-\infty}^{\infty} |g(x)|^2 dx = 1 \text{ and } \int_{-\infty}^{\infty} |g'(x)|^2 dx < \infty\right\}.$$

The proof of (4.2.3) is a combination of the Feynman-Kac formula and some methods developed along the line of the probability in Banach space. Let $q > 1$ be the conjugate number and f a bounded continuous function on $\mathbb{R}$ such that

$$\int_{-\infty}^{\infty} |f(x)|^q dx = 1. \tag{4.2.4}$$

Using Hölder inequality, we get

$$\Big(\int_{-\infty}^{\infty} L^p(t,x)dx\Big)^{1/p} \geq \int_{-\infty}^{\infty} f(x)L(t,x)dx = \int_0^t f(W(s))ds.$$

By Theorem 4.1.6, therefore, we get

$$\liminf_{t\to\infty} \frac{1}{t}\log \mathbb{E}\exp\left\{ \Big(\int_{-\infty}^{\infty} L^p(t,x)dx\Big)^{1/p}\right\}$$
$$\geq \sup_{g\in\mathcal{F}} \left\{ \int_{-\infty}^{\infty} f(x)g^2(x)dx - \frac{1}{2}\int_{-\infty}^{\infty} |g'(x)|^2 dx\right\}.$$

Notice that the functions f satisfying our conditions are dense in the unit sphere of $\mathcal{L}^q(\mathbb{R})$. Taking supremum over f on the right hand side gives

$$\liminf_{t\to\infty}\frac{1}{t}\log\mathbb{E}\exp\left\{\left(\int_{-\infty}^{\infty}L^p(t,x)dx\right)^{1/p}\right\}$$
$$\geq\sup_{g\in\mathcal{F}}\left\{\left(\int_{-\infty}^{\infty}|g(x)|^{2p}dx\right)^{1/p}-\frac{1}{2}\int_{-\infty}^{\infty}|g'(x)|^2dx\right\}. \tag{4.2.5}$$

The upper bound is more delicate. The self-intersection outside any compact set of $\mathbb{R}$ is not negligible. Indeed, for any $M>0$, by the argument for the lower bound (4.2.5),

$$\liminf_{t\to\infty}\frac{1}{t}\log\mathbb{E}\exp\left\{\left(\int_{M}^{\infty}L^p(t,x)dx\right)^{1/p}\right\}$$
$$\geq\sup_{g\in\mathcal{F}}\left\{\left(\int_{M}^{\infty}|g(x)|^{2p}dx\right)^{1/p}-\frac{1}{2}\int_{-\infty}^{\infty}|g'(x)|^2dx\right\}.$$

By shifting the space variable x, one can see that the variation on the right hand side is equal to the right hand side of (4.2.3). This indicates that $L(t,x)$ is not exponentially tight when embedded into $\mathcal{L}^p(\mathbb{R})$.

To resolve this problem, we wrap the real line $\mathbb{R}$ into a compact circle. Let $M>0$ be large but fixed. We have

$$\int_{-\infty}^{\infty}L^p(t,x)dx=\sum_{k\in\mathbb{Z}}\int_0^M L^p(t,x+kM)dx\leq\int_0^M\tilde{L}^p(t,x)dx \tag{4.2.6}$$

where

$$\tilde{L}(t,x)=\sum_{k\in\mathbb{Z}}L(t,x+kM).$$

For any $s>0$, the process

$$W_s(u)=W(s+u)-W(s)\qquad u\geq 0$$

is a 1-dimensional Brownian motion. If we write $L_s(t,x)$ for the local time of $W_s(u)$ and write

$$\tilde{L}_s(t,x)=\sum_{k\in\mathbb{Z}}L_s(t,kM+x),$$

then for any $t>0$, we get

$$\int_0^M\left[\tilde{L}(s+t,x)-\tilde{L}(s,x)\right]^p dx=\int_0^M\tilde{L}_s^p(t,x-W(t))dx=\int_0^M\tilde{L}_s^p(t,x)dx$$

where the last step follows from periodicity of $\tilde{L}_s^p(t,\cdot)$. By independence and stationarity of the Brownian increments, therefore, the quantity

$$\int_0^M\left[\tilde{L}(s+t,x)-\tilde{L}(s,x)\right]^p dx$$

is independent of $\{W(u);\ \ 0\leq u\leq s\}$ and has the same distribution as

$$\int_0^M\tilde{L}^p(t,x)dx.$$

By triangular inequality, for any $s, t > 0$,

$$\begin{aligned}(4.2.7)\qquad &\left(\int_0^M \tilde{L}^p(s+t,x)dx\right)^{1/p}\\ &\le \left(\int_0^M \tilde{L}^p(s,x)dx\right)^{1/p} + \left(\int_0^M \big[\tilde{L}(s+t,x)-\tilde{L}(s,x)\big]^p dx\right)^{1/p}.\end{aligned}$$

Based on all properties listed above, we conclude that the non-negative process

$$\left(\int_0^M \tilde{L}^p(t,x)dx\right)^{1/p} \qquad t \ge 0$$

is sub-additive. Using Theorem 1.3.5, we get

$$(4.2.8)\qquad \mathbb{E}\exp\left\{\theta\left(\int_0^M \tilde{L}^p(1,x)dx\right)^{1/p}\right\} < \infty,$$

$$(4.2.9)\qquad \lim_{t\to\infty}\frac{1}{t}\log\mathbb{E}\exp\left\{\theta\left(\int_0^M \tilde{L}^p(t,x)dx\right)^{1/p}\right\} < \infty$$

for every $\theta > 0$.

We view $\tilde{L}(t,\cdot)$ as a stochastic process taking values in the separable Banach space $\mathcal{L}^p[0,M]$. For any $x\in\mathbb{R}$ we adopt the notations $\mathbb{P}_x$ and $\mathbb{E}_x$ for the distribution and expectation, respectively, of the Brownian motion $W(t)$ starting at x (i.e., $W(0)=x$). We let $t=1$ and claim that the family of distributions

$$\mathcal{L}_x\left\{\tilde{L}(1,\cdot)\right\} \qquad x\in\mathbb{R}$$

is uniformly tight in $\mathcal{L}^p[0,M]$. That is, for any $\epsilon > 0$ there is a compact set $K\subset\mathcal{L}^p[0,M]$ such that

$$(4.2.10)\qquad \inf_{x\in\mathbb{R}}\mathbb{P}_x\big\{\tilde{L}(1,\cdot)\in K\big\} \ge 1-\epsilon.$$

Indeed, by continuity of $L(1,x)$ as a function of x and by the fact that only finitely many $L(1,\cdot+kM)$ are non-zero at any given moment t, $\tilde{L}(1,x)$ is continuous. We have that for any $\gamma > 0$,

$$\begin{aligned}&\sup_{x\in\mathbb{R}}\mathbb{P}_x\Big\{\sup_{|y-z|\le\delta}|\tilde{L}(1,y)-\tilde{L}(1,z)|\ge\gamma\Big\}\\ &=\sup_{x\in\mathbb{R}}\mathbb{P}\Big\{\sup_{|y-z|\le\delta}|\tilde{L}(1,y-x)-\tilde{L}(1,z-x)|\ge\gamma\Big\}\\ &=\mathbb{P}\Big\{\sup_{|y-z|\le\delta}|\tilde{L}(1,y)-\tilde{L}(1,z)|\ge\gamma\Big\}\longrightarrow 0 \qquad (\delta\to 0^+).\end{aligned}$$

By Arzelá-Ascoli theorem (Theorem D.2, Appendix), there is a compact set $\widetilde{K}$ compact in $C[0,M]$ such that

$$\inf_{x\in\mathbb{R}}\mathbb{P}_x\big\{\tilde{L}(1,\cdot)\in\widetilde{K}\big\}\ge 1-\epsilon.$$

Notice that

$$\int_0^M \tilde{L}(1,x)dx = \int_{-\infty}^{\infty} L(1,x)dx = 1.$$

Hence, the set

$$A \equiv \widetilde{K} \cap \left\{ f \in C[0, M]; \quad f \geq 0 \text{ and } \int_0^M f(x)dx = 1 \right\}$$

is relatively compact in $\mathcal{L}^p[0, M]$. Let K be the closure (in $\mathcal{L}^p[0, M]$) of A. We have confirmed our claim on the uniform tightness.

Notice that for any $x \in \mathbb{R}$ and $\lambda > 0$,

$$\begin{aligned}
&\mathbb{E}_x \exp\left\{ \lambda \Big(\int_0^M \tilde{L}^p(1, y)dy \Big)^{1/p} \right\} \\
&= \mathbb{E} \exp\left\{ \lambda \Big(\int_0^M \tilde{L}^p(1, y - x)dy \Big)^{1/p} \right\} \\
&= \mathbb{E} \exp\left\{ \lambda \Big(\int_0^M \tilde{L}^p(1, y)dy \Big)^{1/p} \right\}
\end{aligned}$$

where the last step follows from periodicity. By (4.2.8), the right hand side is finite. Thus,

$$\sup_{x \in \mathbb{R}} \mathbb{E}_x \exp\left\{ \lambda \Big(\int_0^M \tilde{L}^p(1, y)dy \Big)^{1/p} \right\} < \infty \qquad \forall \lambda > 0. \tag{4.2.11}$$

By Theorem 1.1.7, the uniform tightness of the family

$$\left\{ \mathcal{L}_x\big(\tilde{L}(1, \cdot)\big); \quad x \in \mathbb{R} \right\}$$

and (4.2.11) imply that there is a convexed, positively balanced and compact set $K \subset \mathcal{L}^p[0, M]$ such that

$$\sup_{x \in \mathbb{R}} \mathbb{E}_x \exp\left\{ q_K\big(\tilde{L}(1, \cdot)\big) \right\} < \infty. \tag{4.2.12}$$

For any $s, t > 0$ and $x \in \mathbb{R}$, by sub-additivity of the Minkowski functional given in (1.1.10) and by Markov property we get

$$\begin{aligned}
&\mathbb{E}_x \exp\left\{ q_K\big(\tilde{L}(s + t, \cdot)\big) \right\} \\
&\leq \mathbb{E}_x \exp\left\{ q_K\big(\tilde{L}(s, \cdot)\big) + q_K\big(\tilde{L}(s + t, \cdot) - \tilde{L}(s, \cdot)\big) \right\} \\
&= \mathbb{E}_x \Big(\exp\left\{ q_K\big(\tilde{L}(s, \cdot)\big) \right\} \mathbb{E}_{W(s)} \exp\left\{ q_K\big(\tilde{L}(t, \cdot)\big) \right\} \Big) \\
&\leq \mathbb{E}_x \exp\left\{ q_K\big(\tilde{L}(s, \cdot)\big) \right\} \sup_{y \in \mathbb{R}} \mathbb{E}_y \exp\left\{ q_K\big(\tilde{L}(t, \cdot)\big) \right\}.
\end{aligned}$$

Thus,

$$\begin{aligned}
&\sup_{x \in \mathbb{R}} \mathbb{E}_x \exp\left\{ q_K\big(\tilde{L}(s + t, \cdot)\big) \right\} \\
&\leq \sup_{x \in \mathbb{R}} \mathbb{E}_x \exp\left\{ q_K\big(\tilde{L}(s, \cdot)\big) \right\} \sup_{x \in \mathbb{R}} \mathbb{E}_x \exp\left\{ q_K\big(\tilde{L}(t, \cdot)\big) \right\}.
\end{aligned}$$

The sub-additivity established here implies (Lemma 1.3.1) that

$$\lim_{t \to \infty} \frac{1}{t} \log \sup_{x \in \mathbb{R}} \mathbb{E}_x \exp\left\{ q_K\big(\tilde{L}(t, \cdot)\big) \right\} = \inf_{s > 0} \frac{1}{s} \log \sup_{x \in \mathbb{R}} \mathbb{E}_x \exp\left\{ q_K\big(\tilde{L}(s, \cdot)\big) \right\}.$$

By (4.2.12), the right hand side is finite. Hence,

$$\lim_{t\to\infty}\frac{1}{t}\log\mathbb{E}\exp\Big\{q_K\big(\tilde{L}(t,\cdot)\big)\Big\}<\infty. \tag{4.2.13}$$

Let $\lambda>0$ be arbitrary and write

$$\begin{aligned}\mathbb{E}\exp&\bigg\{\Big(\int_0^M\tilde{L}^p(t,x)dx\Big)^{1/p}\bigg\} \\ &=\mathbb{E}\bigg[\exp\bigg\{\Big(\int_0^M\tilde{L}^p(t,x)dx\Big)^{1/p}\bigg\}1_{\{\frac{1}{t}\tilde{L}(t,\cdot)\in\lambda K\}}\bigg] \\ &+\mathbb{E}\bigg[\exp\bigg\{\Big(\int_0^M\tilde{L}^p(t,x)dx\Big)^{1/p}\bigg\}1_{\{\frac{1}{t}\tilde{L}(t,\cdot)\notin\lambda K\}}\bigg].\end{aligned} \tag{4.2.14}$$

For each $g\in\lambda K$, by Hahn-Banach theorem (Corollary 2, p. 108, Section 6, Chapter IV, [**175**]) and by the facts that $\mathcal{L}^q[0,M]$ is the *topological dual space* of $\mathcal{L}^p[0,M]$ and that bounded and continuous functions are dense in $\mathcal{L}^q[0,M]$, there is a bounded and continuous function f such that

$$f(0)=f(M)\quad\text{and}\quad\int_0^M|f(x)|^qdx=1 \tag{4.2.15}$$

and

$$\Big(\int_0^M|g(x)|^pdx\Big)^{1/p}<\epsilon+\int_0^M f(x)g(x)dx.$$

By the finite cover theorem, there are bounded and continuous functions $f_1,\cdots,f_m$ on $[0,M]$ satisfying (4.2.15) such that

$$\Big(\int_0^M|g(x)|^pdx\Big)^{1/p}<\epsilon+\max_{1\le i\le m}\int_0^M f_i(x)g(x)dx\qquad\forall g\in\lambda K.$$

Consequently,

$$\begin{aligned}&\mathbb{E}\bigg[\exp\bigg\{\Big(\int_0^M\tilde{L}^p(t,x)dx\Big)^{1/p}\bigg\}1_{\{\frac{1}{t}\tilde{L}(t,\cdot)\in\lambda K\}}\bigg] \\ &\le e^{\epsilon t}\sum_{i=1}^m\mathbb{E}\exp\bigg\{\int_0^M f_i(x)\tilde{L}(t,x)dx\bigg\}.\end{aligned}$$

In view of the first equation in (4.2.15), f_i can be extended into a bounded, continuous and periodic function (with period M) on $\mathbb{R}$. By periodicity,

$$\begin{aligned}&\int_0^M f_i(x)\tilde{L}(t,x)dx=\sum_{k\in\mathbb{Z}}\int_0^M f_i(x)L(t,kM+x)dx \\ &=\int_{-\infty}^{\infty}f_i(x)L(t,x)dx=\int_0^t f_i\big(W(s)\big)ds.\end{aligned}$$

According to Theorem 4.1.6, therefore,

$$\lim_{t\to\infty}\frac{1}{t}\log\mathbb{E}\exp\left\{\int_0^M f_i(x)\tilde{L}(t,x)dx\right\}$$
$$=\sup_{g\in\mathcal{F}}\left\{\int_{-\infty}^{\infty} f_i(x)g^2(x)dx-\frac{1}{2}\int_{-\infty}^{\infty}|g'(x)|^2dx\right\}.$$

Using periodicity again,

$$\int_{-\infty}^{\infty} f_i(x)g^2(x)dx=\int_0^M f_i(x)\sum_{k\in\mathbb{Z}}g^2(kM+x)dx$$
$$\le\left(\int_0^M\Big(\sum_{k\in\mathbb{Z}}g^2(kM+x)\Big)^p dx\right)^{1/p}$$

where the last step follows from Hölder inequality. Summarizing our argument, we have

$$\limsup_{t\to\infty}\frac{1}{t}\log\mathbb{E}\left[\exp\left\{\left(\int_0^M\tilde{L}^p(t,x)dx\right)^{1/p}\right\}1_{\{\frac{1}{t}\tilde{L}(t,\cdot)\in\lambda K\}}\right]$$
$$\le\epsilon+\sup_{g\in\mathcal{F}}\left\{\left(\int_0^M\Big(\sum_{k\in\mathbb{Z}}g^2(kM+x)\Big)^p dx\right)^{1/p}-\frac{1}{2}\int_{-\infty}^{\infty}|g'(x)|^2dx\right\}.$$

Letting $\epsilon\to 0^+$ on the right hand side gives

$$\limsup_{t\to\infty}\frac{1}{t}\log\mathbb{E}\left[\exp\left\{\left(\int_0^M\tilde{L}^p(t,x)dx\right)^{1/p}\right\}1_{\{\frac{1}{t}\tilde{L}(t,\cdot)\in\lambda K\}}\right]$$
$$\le\sup_{g\in\mathcal{F}}\left\{\left(\int_0^M\Big(\sum_{k\in\mathbb{Z}}g^2(kM+x)\Big)^p dx\right)^{1/p}-\frac{1}{2}\int_{-\infty}^{\infty}|g'(x)|^2dx\right\}. \tag{4.2.16}$$

We now return to (4.2.14). By Cauchy-Schwarz inequality, we get

$$\mathbb{E}\left[\exp\left\{\left(\int_0^M\tilde{L}^p(t,x)dx\right)^{1/p}\right\}1_{\{\frac{1}{t}\tilde{L}(t,\cdot)\notin\lambda K\}}\right]$$
$$\le\left[\mathbb{E}\exp\left\{2\left(\int_0^M\tilde{L}^p(t,x)dx\right)^{1/p}\right\}\right]^{1/2}\left[\mathbb{P}\Big\{\frac{1}{t}\tilde{L}(t,\cdot)\notin\lambda K\Big\}\right]^{1/2}.$$

Observe that

$$\mathbb{P}\Big\{\frac{1}{t}\tilde{L}(t,\cdot)\notin\lambda K\Big\}\le\mathbb{P}\Big\{q_K\big((\lambda t)^{-1}\tilde{L}(t,\cdot)\big)\ge 1\Big\}=\mathbb{P}\Big\{q_K\big(\tilde{L}(t,\cdot)\big)\ge\lambda t\Big\}$$

where the second step follows from the positive homogeneity of the Minkowski functional given in (1.1.10).

By (4.2.12) and Chebyshev inequality, therefore, we get

$$\lim_{\lambda\to\infty}\limsup_{t\to\infty}\frac{1}{t}\log\mathbb{P}\Big\{\frac{1}{t}\tilde{L}(t,\cdot)\notin\lambda K\Big\}=-\infty.$$

This, combined with (4.2.9), leads to

$$\lim_{\lambda\to\infty}\limsup_{t\to\infty}\frac{1}{t}\log\mathbb{E}\left[\exp\left\{\left(\int_0^M\tilde{L}^p(t,x)dx\right)^{1/p}\right\}1_{\{\frac{1}{t}\tilde{L}(t,\cdot)\notin\lambda K\}}\right]=-\infty.$$

By (4.2.14) and (4.2.16), therefore, we get

$$\limsup_{t\to\infty}\frac{1}{t}\log\mathbb{E}\exp\left\{\left(\int_0^M \tilde{L}^p(t,x)dx\right)^{1/p}\right\}$$
$$\le \sup_{g\in\mathcal{F}}\left\{\left(\int_0^M\Big(\sum_{k\in\mathbb{Z}}g^2(kM+x)\Big)^p dx\right)^{1/p}-\frac{1}{2}\int_{-\infty}^{\infty}|g'(x)|^2dx\right\}.$$

From (4.2.6) we have

$$\limsup_{t\to\infty}\frac{1}{t}\log\mathbb{E}\exp\left\{\left(\int_{-\infty}^{\infty} L^p(t,x)dx\right)^{1/p}\right\}$$
$$\le \sup_{g\in\mathcal{F}}\left\{\left(\int_0^M\Big(\sum_{k\in\mathbb{Z}}g^2(kM+x)\Big)^p dx\right)^{1/p}-\frac{1}{2}\int_{-\infty}^{\infty}|g'(x)|^2dx\right\}.$$

By Lemma 4.2.2 below, letting $M\to\infty$ on the right hand side gives the upper bound

$$\limsup_{t\to\infty}\frac{1}{t}\log\mathbb{E}\exp\left\{\left(\int_{-\infty}^{\infty} L^p(t,x)dx\right)^{1/p}\right\}$$
$$\le \sup_{g\in\mathcal{F}}\left\{\left(\int_{-\infty}^{\infty}|g(x)|^{2p}dx\right)^{1/p}-\frac{1}{2}\int_{-\infty}^{\infty}|g'(x)|^2dx\right\}. \tag{4.2.17}$$

Finally, (4.2.3) follows from (4.2.5) and (4.2.17). □

LEMMA 4.2.2.

$$\limsup_{M\to\infty}\sup_{g\in\mathcal{F}}\left\{\left(\int_0^M\Big(\sum_{k\in\mathbb{Z}}g^2(kM+x)\Big)^p dx\right)^{1/p}-\frac{1}{2}\int_{-\infty}^{\infty}|g'(x)|^2dx\right\}$$
$$\le \sup_{f\in\mathcal{F}}\left\{\left(\int_{-\infty}^{\infty}|f(x)|^{2p}dx\right)^{1/p}-\frac{1}{2}\int_{-\infty}^{\infty}|f'(x)|^2dx\right\}.$$

Proof. Let $g\in\mathcal{F}$ be fixed but arbitrary and write

$$\bar{g}(x)=\left(\sum_{k\in\mathbb{Z}}g^2(kM+x)\right)^{1/2}.$$

Then $\bar{g}$ is absolutely continuous with

$$\int_0^M \bar{g}^2(x)dx=1 \quad\text{and}\quad |\bar{g}'(x)|^2\le\sum_{k\in\mathbb{Z}}|g'(kM+x)|^2 \tag{4.2.18}$$

where the second fact follows from direct computation of $\bar{g}'(x)$ and Cauchy-Schwarz inequality. By continuity there is a $0\le a\le M$ such that $\bar{g}(a)=\min_{0\le x\le M}\bar{g}(x)$. In view of the equality in (4.2.18), we must have $\bar{g}(a)\le M^{-1}$. We may take $a=0$, for otherwise we replace $g(x)$ by $g_a(x)=g(x+a)$.

We henceforth assume that

$$\bar{g}(0)=\min_{0\le x\le M}\bar{g}(x)\le M^{-1}. \tag{4.2.19}$$

Write $\tilde{g}(x) = \bar{g}(x) - \bar{g}(0)$. The function $\tilde{g}(\cdot)$ is absolutely continuous with $\tilde{g}(0) = \tilde{g}(M) = 0$ and $\tilde{g}'(x) = \bar{g}'(x)$. By triangular inequality and by (4.2.19),

$$\begin{aligned}
&\left(\int_0^M |\bar{g}(x)|^{2p}dx\right)^{1/p} \leq \left[\left(\int_0^M |\tilde{g}(x)|^{2p}dx\right)^{\frac{1}{2p}} + \bar{g}(0)M^{\frac{1}{2p}}\right]^2 \\
&\leq \left(\int_0^M |\tilde{g}(x)|^{2p}dx\right)^{1/p} + 2M^{-\frac{2p-1}{2p}}\left(\int_0^M |\tilde{g}(x)|^{2p}dx\right)^{\frac{1}{2p}} + M^{-\frac{2p-1}{p}} \\
&\leq \left(1 + M^{-\frac{2p-1}{2p}}\right)\left(\int_0^M |\tilde{g}(x)|^{2p}dx\right)^{1/p} + \left(M^{-\frac{2p-1}{p}} + M^{-\frac{2p-1}{2p}}\right)
\end{aligned}$$

where the last step follows from the inequality $2u \leq u^2 + 1$.

By the inequality in (4.2.18), therefore, we get

$$\begin{aligned}
(4.2.20) \qquad &\left(\int_0^M |\bar{g}(x)|^{2p}dx\right)^{1/p} - \frac{1}{2}\int_{-\infty}^{\infty} |g'(x)|^2 dx \\
&\leq \theta_M\left(\int_0^M |\tilde{g}(x)|^{2p}dx\right)^{1/p} - \frac{1}{2}\int_{-\infty}^{\infty} |\tilde{g}'(x)|^2 dx + \epsilon_M
\end{aligned}$$

where $\theta_M = 1 + M^{-\frac{2p-1}{2p}}$ and $\epsilon_M = M^{-\frac{2p-1}{p}} + M^{-\frac{2p-1}{2p}}$.

Write

$$C_g = \left\{\int_0^M |\tilde{g}(x)|^2 dx\right\}^{1/2}$$

and define

$$f_M(x) = \begin{cases} C_g^{-1}\tilde{g}(x) & x \in [0, M], \\ 0 & \text{else}. \end{cases}$$

Partially, by the fact that $\tilde{g}(0) = \tilde{g}(M) = 0$, $f_M \in \mathcal{F}$. Consequently, (4.2.20) implies that

$$\begin{aligned}
&\left(\int_0^M |\bar{g}(x)|^{2p}dx\right)^{1/p} - \frac{1}{2}\int_{-\infty}^{\infty} |g'(x)|^2 dx \\
&\leq C_g^2 \sup_{f\in\mathcal{F}}\left\{\theta_M\left(\int_{-\infty}^{\infty} |f(x)|^{2p}dx\right)^{1/p} - \frac{1}{2}\int_{-\infty}^{\infty} |f'(x)|^2 dx\right\} + \epsilon_M \\
&= C_g^2\theta_M^{\frac{2p}{p+1}} \sup_{f\in\mathcal{F}}\left\{\left(\int_{-\infty}^{\infty} |f(x)|^{2p}dx\right)^{1/p} - \frac{1}{2}\int_{-\infty}^{\infty} |f'(x)|^2 dx\right\} + \epsilon_M
\end{aligned}$$

where the last step follows from substituting $f(x)$ by $\theta_M^{\frac{p}{2(p+1)}} f\big(\theta_M^{\frac{p}{p+1}} x\big)$.

Notice that (4.2.19) implies that

$$C_g \leq M^{-1/2} + \left\{\int_0^M |\bar{g}(x)|^2 dx\right\}^{1/2} = M^{-1/2} + 1.$$

We finally reach the bound

$$\Big(\int_0^M |\bar{g}(x)|^{2p}dx\Big)^{1/p} - \frac{1}{2}\int_{-\infty}^{\infty}|g'(x)|^2dx$$
$$\le \epsilon_M + (M^{-1/2}+1)^2\theta_M^{\frac{2p}{p+1}} \sup_{f\in\mathcal{F}}\Big\{\Big(\int_{-\infty}^{\infty}|f(x)|^{2p}dx\Big)^{1/p} - \frac{1}{2}\int_{-\infty}^{\infty}|f'(x)|^2dx\Big\}$$

uniformly over $g\in\mathcal{F}$, which leads to the desired conclusion. □

In the case $d\ge 2$, the local time $L(t,x)$ no longer exists. We may introduce

$$L(t,x,\epsilon) = \int_0^t p_\epsilon\big(W(s)-x\big)ds$$

instead. With some slight modification on the proof of (4.2.3), we can prove the following.

THEOREM 4.2.3. *For each $d\ge 1$, $p>1$ and $\epsilon,\theta>0$,*

$$\lim_{t\to\infty}\frac{1}{t}\log\mathbb{E}\exp\Big\{\theta\Big(\int_{\mathbb{R}^d}L^p(t,x,\epsilon)dx\Big)^{1/p}\Big\}$$
$$=\sup_{g\in\mathcal{F}_d}\Big\{\theta\Big(\int_{\mathbb{R}^d}\big[g(x,\epsilon)\big]^{2p}dx\Big)^{1/p} - \frac{1}{2}\int_{\mathbb{R}^d}|\nabla g(x)|^2dx\Big\}$$

where $\mathcal{F}_d$ is defined in (4.1.22) and

$$g(x,\epsilon) = \Big\{\int_{\mathbb{R}^d}p_\epsilon(y)g^2(x-y)dx\Big\}^{1/2} \qquad x\in\mathbb{R}^d.$$

4.3. Two-dimensional case

Recall that in Section 2.4 we constructed the renormalized self-intersection local time $\gamma\big([0,t]^2_<\big)$ (formally written in (2.4.1)) run by a 2-dimensional Brownian motion $W(t)$. The discussion naturally leads to the "renormalized" polymer models

$$\widehat{\mathbb{P}}_\lambda(A) = \widehat{C}_\lambda^{-1}\mathbb{E}\Big(\exp\Big\{\lambda\gamma\big([0,1]^2_<\big)\Big\}1_{\{W(\cdot)\in A\}}\Big) \qquad A\subset C\big\{[0,1],\mathbb{R}^2\big\}, \tag{4.3.1}$$

$$\widetilde{\mathbb{P}}_\lambda(A) = \widetilde{C}_\lambda^{-1}\mathbb{E}\Big(\exp\Big\{-\lambda\gamma\big([0,1]^2_<\big)\Big\}1_{\{W(\cdot)\in A\}}\Big) \quad A\subset C\big\{[0,1],\mathbb{R}^2\big\} \tag{4.3.2}$$

where

$$\widehat{C}_\lambda = \mathbb{E}\exp\Big\{\lambda\gamma\big([0,1]^2_<\big)\Big\} \quad\text{and}\quad \widetilde{C}_\lambda = \mathbb{E}\exp\Big\{-\lambda\gamma\big([0,1]^2_<\big)\Big\}$$

are normalizers.

In view of Theorem 2.4.2, $\widetilde{C}_\lambda<\infty$ for all $\lambda>0$. In the term of physics, it shows that there is no phase transition in the self-repelling polymer model given in (4.3.2). On the other hand, we shall show that there is a $\lambda_0>0$ such that $\widehat{C}_\lambda=\infty$ for sufficiently large $\lambda>0$. An important question is to find the $\lambda_0>0$ such that

$$\mathbb{E}\exp\Big\{\lambda\gamma\big([0,1]^2_<\big)\Big\}\begin{cases} <\infty & \lambda<\lambda_0,\\ =\infty & \lambda>\lambda_0.\end{cases} \tag{4.3.3}$$

In physics, the value λ_0 is critical to the "melt-down" of a self-attracting polymer. We shall identify λ_0 in terms of the Gagliardo-Nirenberg constant.

THEOREM 4.3.1. *Let $\kappa(2,2)>0$ be the best constant of the Gagliardo-Nirenberg inequality*

$$||f||_4 \le C\sqrt{||\nabla f||_2}\sqrt{||f||_2} \qquad f\in W^{1,2}(\mathbb{R}^2) \tag{4.3.4}$$

where

$$W^{1,2}(\mathbb{R}^2)=\big\{f\in\mathcal{L}^2(\mathbb{R}^2);\ \ \nabla f\in\mathcal{L}^2(\mathbb{R}^d)\big\}.$$

Then

$$\lim_{t\to\infty}\frac{1}{t}\log\mathbb{P}\Big\{\gamma\big([0,1]^2_<\big)\ge t\Big\}=-\kappa(2,2)^{-4}.$$

Proof. Let $0\le\lambda_0\le\infty$ be defined by (4.3.3). Then

$$\limsup_{t\to\infty}\frac{1}{t}\log\mathbb{P}\Big\{\gamma\big([0,1]^2_<\big)\ge t\Big\}=-\lambda_0.$$

Theorem 2.4.2 implies that $\lambda_0>0$. To establish the upper bound, we need to show that $\lambda_0\ge\kappa(2,2)^{-4}$. For this purpose we may assume that $\lambda_0<\infty$.

Consider the decomposition

$$\gamma\big([0,1]^2_<\big)=\gamma\big([0,1/2]^2_<\big)+\gamma\big([1/2,1]^2_<\big)+\gamma\big([0,1/2]\times[1/2,1]\big).$$

It gives that for each $\epsilon>0$,

$$\begin{aligned}-\lambda_0\le\max\bigg\{&\limsup_{t\to\infty}\frac{1}{t}\log\mathbb{P}\Big\{\gamma\big([0,1/2]^2_<\big)+\gamma\big([1/2,1]^2_<\big)\ge\frac{1+\epsilon}{2}t\Big\},\\ &\limsup_{t\to\infty}\frac{1}{t}\log\mathbb{P}\Big\{\gamma\big([0,1/2]\times[1/2,1]\big)\ge\frac{1-\epsilon}{2}t\Big\}\bigg\}.\end{aligned}$$

By Proposition 2.3.4,

$$\gamma\big([0,1/2]\times[1/2,1]\big)\overset{d}{=}\frac{1}{2}\Big\{\alpha\big([0,1]^2\big)-\mathbb{E}\alpha\big([0,1]^2\big)\Big\}.$$

Taking $d=p=2$ in (3.3.4),

$$\lim_{t\to\infty}\frac{1}{t}\log\mathbb{P}\Big\{\alpha\big([0,1]^2\big)\ge t\Big\}=-\kappa(2,2)^{-4}.$$

Consequently,

$$\lim_{t\to\infty}\frac{1}{t}\log\mathbb{P}\Big\{\gamma\big([0,1/2]\times[1/2,1]\big)\ge\frac{1-\epsilon}{2}t\Big\}=-(1-\epsilon)\kappa(2,2)^{-4}.$$

Observe that $\gamma\big([0,1/2]^2_<\big)$ and $\gamma\big([1/2,1]^2_<\big)$ are independent and have the same law as $(1/2)\gamma\big([0,1]^2_<\big)$. According to the definition of λ_0,

$$\mathbb{E}\exp\Big\{2\lambda\Big[\gamma\big([0,1/2]^2_<\big)+\gamma\big([1/2,1]^2_<\big)\Big]\Big\}=\Big(\mathbb{E}\exp\Big\{\lambda\gamma\big([0,1]^2_<\big)\Big\}\Big)^2<\infty$$

for every $\lambda<\lambda_0$. A standard application of Chebyshev inequality leads to

$$\limsup_{t\to\infty}\frac{1}{t}\log\mathbb{P}\Big\{\gamma\big([0,1/2]^2_<\big)+\gamma\big([1/2,1]^2_<\big)\ge\frac{1+\epsilon}{2}t\Big\}\le-(1+\epsilon)\lambda_0.$$

Summarizing our argument, we have

$$-\lambda_0 \le \max\Big\{-(1+\epsilon)\lambda_0, \quad -(1-\epsilon)\kappa(2,2)^{-4}\Big\}.$$

By the fact that $\lambda_0 > 0$, we must have $\lambda_0 \ge (1-\epsilon)\kappa(2,2)^{-4}$. Letting $\epsilon \to 0^+$ we have proved the upper bound

$$\limsup_{t\to\infty} \frac{1}{t} \log \mathbb{P}\Big\{\gamma\big([0,1]^2_<\big) \ge t\Big\} \le -\kappa(2,2)^{-4}. \tag{4.3.5}$$

To establish the lower bound, recall that

$$L(t,x,\epsilon) = \int_0^t p_\epsilon\big(W(s)-x\big)ds$$

and notice that

$$\int_{\mathbb{R}^2} L^2(t,x,\epsilon)dx = 2\iint_{\{0\le r<s\le t\}} p_{2\epsilon}\big(W(r)-W(t)\big)drds.$$

By Theorem 4.2.7 (with $p=2$) for any $\theta > 0$,

$$\begin{aligned}
&\lim_{t\to\infty} \frac{1}{t} \log \mathbb{E}\exp\left\{\theta\left(\iint_{\{0\le r<s\le t\}} p_{2\epsilon}\big(W(r)-W(t)\big)drds\right)^{1/2}\right\}\\
&= \sup_{g\in\mathcal{F}_2}\left\{\frac{\theta}{\sqrt{2}}\left(\int_{\mathbb{R}^2} g^4(x,\epsilon)dx\right)^{1/2} - \frac{1}{2}\int_{\mathbb{R}^d} |\nabla g(x)|^2dx\right\}.
\end{aligned}$$

Let $t=n$ be a positive integer and write

$$\begin{aligned}
&\iint_{\{0\le r<s\le n\}} p_{2\epsilon}\big(W(r)-W(t)\big)drds\\
&= \sum_{k=1}^{n-1} \beta_\epsilon\big([0,k]\times[k,k+1]\big) + \sum_{k=1}^{n} \beta_\epsilon\big([k-1,k]^2_<\big).
\end{aligned} \tag{4.3.6}$$

Here we recall that

$$\beta_\epsilon(A) = \int_A p_{2\epsilon}\big(W(r)-W(s)\big)drds,$$

$$\beta(A) = \int_A \delta_0\big(W(r)-W(s)\big)drds$$

are the random measures discussed in Section 2.3 (with $p=d=2$). By Theorem 2.3.2 and by a treatment similar to the one used in Proposition 2.3.4 we conclude that for any fixed n,

$$\sum_{k=1}^{n-1} \beta_\epsilon\big([0,k]\times[k,k+1]\big) \overset{\epsilon\to 0^+}{\longrightarrow} \sum_{k=1}^{n-1} \beta\big([0,k]\times[k,k+1]\big)$$

in $\mathcal{L}^m(\Omega,\mathcal{A},\mathbb{P})$ for all $m > 0$.

Notice that the second term on the right hand side of (4.3.6) is bounded by $C_\epsilon n$ for some constant $C_\epsilon > 0$. So we have

$$\lim_{n\to\infty}\frac{1}{n}\log\mathbb{E}\exp\left\{\theta\Big(\sum_{k=1}^{n-1}\beta_\epsilon\big([0,k]\times[k,k+1]\big)\Big)^{1/2}\right\}$$
$$=\sup_{g\in\mathcal{F}_2}\left\{\frac{\theta}{\sqrt{2}}\Big(\int_{\mathbb{R}^2}g^4(x,\epsilon)dx\Big)^{1/2}-\frac{1}{2}\int_{\mathbb{R}^d}|\nabla g(x)|^2dx\right\}.$$

By Lemma 1.2.6, this leads to

$$\lim_{n\to\infty}\frac{1}{n}\log\sum_{m=0}^\infty\frac{\theta^m}{m!}\left\{\mathbb{E}\Big[\sum_{k=1}^{n-1}\beta_\epsilon\big([0,k]\times[k,k+1]\big)\Big]^m\right\}^{1/2}\tag{4.3.7}$$
$$=\frac{1}{2}\sup_{g\in\mathcal{F}_2}\left\{\sqrt{2}\theta\Big(\int_{\mathbb{R}^2}g^4(x,\epsilon)dx\Big)^{1/2}-\frac{1}{2}\int_{\mathbb{R}^d}|\nabla g(x)|^2dx\right\}.$$

We now claim that for any $\epsilon > 0$, integers $m, n \geq 1$,

$$\mathbb{E}\Big[\sum_{k=1}^{n-1}\beta_\epsilon\big([0,k]\times[k,k+1]\big)\Big]^m\leq\mathbb{E}\Big[\sum_{k=1}^{n-1}\beta\big([0,k]\times[k,k+1]\big)\Big]^m.\tag{4.3.8}$$

Indeed, write

$$D=\bigcup_{k=1}^{n-1}[0,k]\times[k,k+1].$$

By Fourier inversion (Theorem B.1, Appendix), we get

$$p_{2\epsilon}(x)=(2\pi)^{-2}\int_{\mathbb{R}^2}e^{-i\lambda\cdot x}\exp\big\{-\epsilon|\lambda|^2\big\}d\lambda.$$

Thus,

$$\sum_{k=1}^{n-1}\beta_\epsilon\big([0,k]\times[k,k+1]\big)=\int_D drdsp_{2\epsilon}\big(W(r)-W(s)\big)drds$$
$$=\int_{\mathbb{R}^2}d\lambda\exp\big\{-\epsilon|\lambda|^2\big\}\int_D drds\exp\big\{-i\lambda\cdot\big(W(r)-W(s)\big)\big\}.$$

Consequently,

$$\mathbb{E}\Big[\sum_{k=1}^{n-1}\beta_\epsilon\big([0,k]\times[k,k+1]\big)\Big]^m$$
$$=\int_{(\mathbb{R}^2)^m}d\lambda_1\cdots d\lambda_m\exp\Big\{-\epsilon\sum_{k=1}^m|\lambda_k|^2\Big\}$$
$$\times\int_{D^m}dr_1ds_1\cdots dr_mds_m\exp\Big\{-\frac{1}{2}\mathrm{Var}\Big(\sum_{k=1}^m\lambda_k\cdot\big(W(r_k)-W(s_k)\big)\Big)\Big\}.$$

Therefore, for any $0<\epsilon'<\epsilon$,

$$\mathbb{E}\Big[\sum_{k=1}^{n-1}\beta_\epsilon\big([0,k]\times[k,k+1]\big)\Big]^m\leq\mathbb{E}\Big[\sum_{k=1}^{n-1}\beta_{\epsilon'}\big([0,k]\times[k,k+1]\big)\Big]^m,$$

which leads to (4.3.8) by letting $\epsilon'\to 0$.

By (4.3.7) and (4.3.8),

$$\liminf_{n\to\infty}\frac{1}{n}\log\sum_{m=0}^{\infty}\frac{\theta^m}{m!}\Big\{\mathbb{E}\Big[\sum_{k=1}^{n-1}\beta\big([0,k]\times[k,k+1]\big)\Big]^m\Big\}^{1/2}$$
$$\ge\frac{1}{2}\sup_{g\in\mathcal{F}_2}\Big\{\sqrt{2}\theta\Big(\int_{\mathbb{R}^2}g^4(x,\epsilon)dx\Big)^{1/2}-\frac{1}{2}\int_{\mathbb{R}^d}|\nabla g(x)|^2dx\Big\}.$$

When $\epsilon\to 0^+$, the right hand side converges to

$$\frac{1}{2}\sup_{g\in\mathcal{F}_2}\Big\{\sqrt{2}\theta\Big(\int_{\mathbb{R}^2}|g(x)|^4dx\Big)^{1/2}-\frac{1}{2}\int_{\mathbb{R}^d}|\nabla g(x)|^2dx\Big\}=\frac{1}{2}\theta^2\kappa(2,2)^4$$

where the equality follows from Theorem C.1, Appendix (with $d=p=2$). Thus,

$$\liminf_{n\to\infty}\frac{1}{n}\log\sum_{m=0}^{\infty}\frac{\theta^m}{m!}\Big\{\mathbb{E}\Big[\sum_{k=1}^{n-1}\beta\big([0,k]\times[k,k+1]\big)\Big]^m\Big\}^{1/2}\ge\frac{1}{2}\theta^2\kappa(2,2)^4\qquad(\theta>0).\tag{4.3.9}$$

Notice that (4.3.9) alone is not enough for the lower bound of the large deviation. In the following effort, we strengthen (4.3.9) into an equality.

By the scaling property, (4.3.5) can be rewritten as

$$\limsup_{n\to\infty}\frac{1}{n}\log\mathbb{P}\Big\{\gamma\big([0,n]_<^2\big)\ge\lambda n^2\Big\}\le-\lambda\kappa(2,2)^{-4}\qquad(\lambda>0).$$

Similarly to the decomposition (4.3.6), we have

$$\gamma\big([0,n]_<^2\big)=\sum_{k=1}^{n-1}\beta\big([0,k]\times[k,k+1]\big)-\sum_{k=1}^{n-1}\mathbb{E}\beta\big([0,k]\times[k,k+1]\big)+\sum_{k=1}^{n}\gamma\big([k-1,k]_<^2\big).\tag{4.3.10}$$

Notice that

$$\sum_{k=1}^{n-1}\mathbb{E}\beta\big([0,k]\times[k,k+1]\big)=\frac{1}{2\pi}\sum_{k=1}^{n-1}\int_0^k\int_k^{k+1}\frac{1}{s-r}dsdr=\frac{1}{2\pi}\sum_{k=1}^{n-1}\big\{\log(k+1)-\log k\big\}=\frac{1}{2\pi}n\log n.\tag{4.3.11}$$

The random variables $\gamma\big([k-1,k]_<^2\big)$ $(k=1,2\cdots)$ form an i.i.d. exponentially integrable sequence. By a standard application of Chebyshev inequality, we get

$$\lim_{n\to\infty}\frac{1}{n}\log\mathbb{P}\Big\{\Big|\sum_{k=1}^{n}\gamma\big([k-1,k]_<^2\big)\Big|\ge\epsilon n^2\Big\}=\infty.\tag{4.3.12}$$

Summarizing our argument, we have

$$\limsup_{n\to\infty}\frac{1}{n}\log\mathbb{P}\Big\{\sum_{k=1}^{n-1}\beta\big([0,k]\times[k,k+1]\big)\ge\lambda n^2\Big\}\le-\lambda\kappa(2,2)^{-4}.$$

By Lemma 1.2.9 (part (2)) and Lemma 1.2.10, this leads to

$$\limsup_{n\to\infty}\frac{1}{n}\log\sum_{m=0}^{\infty}\frac{\theta^m}{m!}\left\{\mathbb{E}\left[\sum_{k=1}^{n-1}\beta\big([0,k]\times[k,k+1]\big)\right]^m\right\}^{1/2}$$
$$\le\sup_{\lambda>0}\left\{\theta\lambda^{1/2}-\frac{1}{2}\lambda\kappa(2,2)^{-4}\right\}=\frac{1}{2}\theta^2\kappa(2,2)^4.$$

Combining this with (4.3.9) gives us

$$\lim_{n\to\infty}\frac{1}{n}\log\sum_{m=0}^{\infty}\frac{\theta^m}{m!}\left\{\mathbb{E}\left[\sum_{k=1}^{n-1}\beta\big([0,k]\times[k,k+1]\big)\right]^m\right\}^{1/2}=\frac{1}{2}\theta^2\kappa(2,2)^4.$$

Applying Theorem 1.2.7, we get

$$\lim_{n\to\infty}\frac{1}{n}\log\mathbb{P}\left\{\sum_{k=1}^{n-1}\beta\big([0,k]\times[k,k+1]\big)\ge\lambda n^2\right\}=-\lambda\kappa(2,2)^{-4}.$$

We bring this back to the decomposition (4.3.10). By (4.3.11) and (4.3.12),

$$\lim_{n\to\infty}\frac{1}{n}\log\mathbb{P}\left\{\gamma\big([0,n]^2_<\big)\ge\lambda n^2\right\}=-\lambda\kappa(2,2)^{-4},$$

which ends the proof. □

Comparing Theorem 4.3.1 with (3.3.4) (with $d=p=2$), one can see that the intersection local time $\alpha\big([0,1/2]^2\big)$ obeys the same large deviation principle as $\gamma\big([0,1]^2_<\big)$. This observation is crucial in our proof of the upper bound part of Theorem 4.3.1, since $\alpha\big([0,1/2]^2\big)\overset{d}{=}2^{-1}\alpha\big([0,1]^2\big)$; so the tail of $\alpha\big([0,1/2]^2\big)$ is given by Theorem 3.3.2. We try to turn our mathematical argument into the following intuitive explanation:

Cut the Brownian path $\big\{W(t);\ 0\le t\le 1\big\}$ into two paths at the middle point $W(1/2)$ and shift the whole system so the cutting point $W(1/2)$ becomes the origin after shifting. If we reverse the direction of the path before the time $t=1/2$, then the resulted two paths are trajectories of two independent Brownian motions running up to time $1/2$.

The total self-intersection of the original path $\big\{W(t);\ 0\le t\le 1\big\}$ is the sum of the intersection within each sub-path and the intersection between two sub-paths. When $d=2$, the first type intersection out-numbers the second type intersection (that is the main reason why $\beta\big([0,1]^2_<\big)=\infty$ (Proposition 2.3.6)). The renormalization subdues the short range intersection so that these two intersections are comparable. By chance, the ratio is 1 to 1 here in the sense of large deviations. That is to say, about half of $\gamma\big([0,1]^2_<\big)$ is made of a random quantity distributed as $\alpha\big([0,1/2]^2\big)$. As for another half, it is equal to the sum of the (renormalized) self-intersection local times of two independent paths and each of them can be analyzed in the way proposed above.

The proportion between these two types of intersection varies in different settings. From later development, we shall see that finding the ratio is an important part of establishing the large deviations related to self intersections.

The following theorem shows a completely different tail asymptotic behavior.

THEOREM 4.3.2. *There is a constant* $0 < L < \infty$ *such that for any* $\theta > 0$,

$$\lim_{t\to\infty} t^{-2\pi\theta} \log \mathbb{P}\Big\{ -\gamma\big([0,1]^2_<\big) \ge \theta \log t \Big\} = -L.$$

Proof. One needs only to prove Theorem 4.3.2 in the case $\theta = (2\pi)^{-1}$. That is,

$$\lim_{t\to\infty} \frac{1}{t} \log \mathbb{P}\Big\{ -\gamma\big([0,1]^2_<\big) \ge (2\pi)^{-1} \log t \Big\} = -L \tag{4.3.13}$$

for some $0 < L < \infty$. Indeed, the general statement will follows from (4.3.13) if we substitute t by $t^{2\pi\theta}$.

The argument for (4.3.13) is based on sub-additivity. Define

$$Z_t = -\frac{1}{2\pi} t \log t - \gamma\big([0,t]^2_<\big) \qquad t \ge 0. \tag{4.3.14}$$

For any $s, t > 0$,

$$\begin{aligned}
Z_{s+t} &= -\frac{1}{2\pi}(s+t)\log(s+t) - \gamma\big([0,s]^2_<\big) - \gamma\big([s,s+t]^2_<\big) \\
&\quad - \beta\big([0,s]\times[s,s+t]\big) + \mathbb{E}\beta\big([0,s]\times[s,s+t]\big) \\
&\le -\frac{1}{2\pi}(s+t)\log(s+t) - \gamma\big([0,s]^2_<\big) \\
&\quad - \gamma\big([s,s+t]^2_<\big) + \mathbb{E}\beta\big([0,s]\times[s,s+t]\big).
\end{aligned}$$

By Proposition 2.3.4, we get

$$\begin{aligned}
&\mathbb{E}\beta\big([0,s]\times[s,s+t]\big) = \mathbb{E}\alpha\big([0,s]\times[0,t]\big) \\
&= \int_{\mathbb{R}^2} dx \Big[\int_0^s p_u(x)du\Big]\Big[\int_0^t p_u(x)du\Big] = \int_0^s\int_0^t p_{u+v}(0)dudv
\end{aligned}$$

where the second equality follows from Le Gall's moment formula in the special case $d = p = 2$ and $m = 1$. By a straightforward calculation, we have

$$\begin{aligned}
&\int_0^s\int_0^t p_{u+v}(0)dudv = \frac{1}{2\pi}\int_0^s\int_0^t \frac{1}{u+v}dudv \\
&= \frac{1}{2\pi}\Big[(s+t)\log(s+t) - s\log s - t\log t\Big].
\end{aligned}$$

Summarizing our argument gives us

$$Z_{s+t} \le Z_s + Z'_t \tag{4.3.15}$$

where

$$Z'_t = -\frac{1}{2\pi} t \log t - \gamma\big([s,s+t]^2_<\big)$$

is independent of Z_s and $Z'_t \overset{d}{=} Z_t$. This means that Z_t is sub-additive.

By Theorem 2.4.7, and the fact $\gamma\big([0,\delta]^2_<\big) \overset{d}{=} \delta\gamma\big([0,1]^2_<\big)$, for any $\lambda > 0$, one can take $\delta > 0$ sufficiently small so that

$$\mathbb{E}\exp\Big\{\lambda Z_\delta\Big\} < \infty.$$

By sub-additivity given in (4.3.15), therefore, we conclude that[1]

$$\mathbb{E}\exp\Big\{\lambda Z_t\Big\} < \infty \qquad \forall \lambda, t > 0. \tag{4.3.16}$$

By (4.3.15) again, for any $s, t > 0$,

$$\mathbb{E}\exp\Big\{2\pi Z_{s+t}\Big\} \le \mathbb{E}\exp\Big\{2\pi Z_s\Big\}\mathbb{E}\exp\Big\{2\pi Z_t\Big\}$$

or,

$$\begin{aligned}&(s+t)^{-(s+t)}\mathbb{E}\exp\Big\{-2\pi\gamma\big([0,s+t]^2_<\big)\Big\}\\ &\le \Big(s^{-s}\mathbb{E}\exp\Big\{-2\pi\gamma\big([0,s]^2_<\big)\Big\}\Big)\Big(t^{-t}\mathbb{E}\exp\Big\{-2\pi\gamma\big([0,t]^2_<\big)\Big\}\Big).\end{aligned}$$

By Lemma 1.3.1, the limit

$$\lim_{t\to\infty}\frac{1}{t}\log\Big(t^{-t}\mathbb{E}\exp\Big\{-2\pi\gamma\big([0,t]^2_<\big)\Big\}\Big) = A \tag{4.3.17}$$

exists with $A < \infty$. By Lemma 4.3.3 below, $A > -\infty$. Let $t = n$ be integer. By Stirling formula and by scaling, we get

$$\lim_{n\to\infty}\frac{1}{n}\log\frac{1}{n!}\mathbb{E}\exp\Big\{-2\pi n\gamma\big([0,1]^2_<\big)\Big\} = A+1.$$

Finally, applying Theorem 1.2.8 to the non-negative random variable

$$Y = \exp\Big\{-2\pi\gamma\big([0,1]^2_<\big)\Big\}$$

leads to (4.3.13) with

$$L = \exp\{-1-A\}. \tag{4.3.18}$$

□

We end this section by establishing the following lemma.

Lemma 4.3.3.

$$\liminf_{n\to\infty}\frac{1}{n}\log\Big(n^{-n}\mathbb{E}\exp\Big\{-2\pi\gamma\big([0,n]^2_<\big)\Big\}\Big) > -\infty.$$

Proof. By (4.3.10) and (4.3.11)

$$\begin{aligned}&n^{-n}\mathbb{E}\exp\Big\{-2\pi\gamma\big([0,n]^2_<\big)\Big\}\\ &= \exp\Big\{-2\pi\sum_{k=1}^{n}\gamma\big([k-1,k]^2_<\big) - 2\pi\sum_{k=1}^{n-1}\beta\big([0,k]\times[k,k+1]\big)\Big\}\\ &\ge e^{-2\pi Mn}\mathbb{E}\exp\Big\{-2\pi\sum_{k=1}^{n-1}\beta\big([0,k]\times[k,k+1]\big)\Big\}\\ &\quad - e^{-2\pi Mn}\mathbb{P}\Big\{\sum_{k=1}^{n}\gamma\big([k-1,k]^2_<\big) \ge Mn\Big\}.\end{aligned}$$

---

[1]Here we are not allowed to apply Theorem 1.3.5 directly for lack of monotonicity.

By Chebyshev inequality, for any $0 < \lambda < \kappa(2,2)^{-4}$,

$$\mathbb{P}\Big\{\sum_{k=1}^{n}\gamma\big([k-1,k]_<^2\big) \geq Mn\Big\} \leq e^{-n\lambda M}\Big(\mathbb{E}\exp\Big\{\lambda\gamma\big([0,1]_<^2\big)\Big\}\Big)^n$$

and the right hand side can be exponentially small to a requested level by choosing sufficiently large M. To complete the proof, therefore, we need only to establish the lower bound

$$\mathbb{E}\exp\Big\{-2\pi\sum_{k=1}^{n-1}\beta\big([0,k]\times[k,k+1]\big)\Big\} \geq c_1^n \tag{4.3.19}$$

for some constant $c_1 > 0$ and sufficiently large n.

Let the 1-dimensional Brownian motion $W_0(t)$ be the first component of $W(t)$ and write

$$D_n = \Big\{\sup_{0\leq s\leq n}|W_0(s) - s| \leq \delta\Big\}$$

where $0 < \delta < 1/2$ is fixed. Notice that on D_n,

$$\sum_{k=1}^{n-1}\beta\big([0,k]\times[k,k+1]\big) = \sum_{k=1}^{n-1}\beta\big([k-1,k]\times[k,k+1]\big).$$

Consequently, for any $N > 0$,

$$\begin{aligned}
&\mathbb{E}\exp\Big\{-2\pi\sum_{k=1}^{n-1}\beta\big([0,k]\times[k,k+1]\big)\Big\} \\
&\geq \mathbb{E}\Big[\exp\Big\{-2\pi\sum_{k=1}^{n-1}\beta\big([k-1,k]\times[k,k+1]\big)\Big\}1_{D_n}\Big] \\
&\geq e^{-2\pi Nn}\Big\{\mathbb{P}(D_n) - \mathbb{P}\Big\{\sum_{k=1}^{n-1}\beta\big([k-1,k]\times[k,k+1]\big) \geq Nn\Big\}\Big\}.
\end{aligned} \tag{4.3.20}$$

Write

$$\begin{aligned}
&\sum_{k=1}^{n-1}\beta\big([k-1,k]\times[k,k+1]\big) \\
&= \sum_k \beta\big([2(k-1),2k-1]\times[2k-1,2k]\big) \\
&+ \sum_k \beta\big([2k-1,2k]\times[2k,2k+1]\big).
\end{aligned}$$

Observe that

$$\begin{aligned}
&\mathbb{P}\Big\{\sum_{k=1}^{n-1}\beta\big([k-1,k]\times[k,k+1]\big) \geq Nn\Big\} \\
&\leq \mathbb{P}\Big\{\sum_k \beta\big([2(k-1),2k-1]\times[2k-1,2k]\big) \geq 2^{-1}Nn\Big\} \\
&+ \mathbb{P}\Big\{\sum_k \beta\big([2k-1,2k]\times[2k,2k+1]\big)\big) \geq 2^{-1}Nn\Big\}.
\end{aligned}$$

By the fact that the sequence

$$\beta\big([2(k-1),2k-1]\times[2k-1,2k]\big) \qquad k=1,2,\cdots$$

is an i.i.d. with common distribution the same as $\alpha([0,1]^2)$,

$$\mathbb{P}\Big\{\sum_k \beta\big([2(k-1),2k-1]\times[2k-1,2k]\big)\geq 2^{-1}Nn\Big\}$$
$$\leq e^{-2^{-1}n\lambda N}\left(\mathbb{E}\exp\left\{\lambda\alpha\big([0,1]^2\big)\right\}\right)^n$$

where $\lambda>0$ is chosen in order that

$$\mathbb{E}\exp\left\{\lambda\alpha\big([0,1]^2\big)\right\}<\infty.$$

Similarly,

$$\mathbb{P}\Big\{\sum_k \beta\big([2k-1,2k]\times[2k,2k+1]\big)\geq 2^{-1}Nn\Big\}$$
$$\leq e^{-2^{-1}n\lambda N}\left(\mathbb{E}\exp\left\{\lambda\alpha\big([0,1]^2\big)\right\}\right)^n.$$

Thus, by triangular inequality, we get

$$\mathbb{P}\Big\{\sum_{k=1}^{n-1} \beta\big([k-1,k]\times[k,k+1]\big)\geq Nn\Big\}$$
$$\leq 2e^{-2^{-1}n\lambda N}\left(\mathbb{E}\exp\left\{\lambda\alpha\big([0,1]^2\big)\right\}\right)^n.$$

On the other hand,

$$D_n=\Big\{\sup_{0\leq s\leq n-1}|W_0(s)-s|\leq\delta\Big\}$$
$$\cap\Big\{\sup_{0\leq s\leq 1}\big|\big(W_0(n-1+s)-W_0(n-1)\big)-s+\big(W_0(n-1)-(n-1)\big)\big|\leq\delta\Big\}.$$

By independence of increments,

$$\mathbb{P}(D_n)\geq\mathbb{P}(D_{n-1})\inf_{|x|\leq\delta}\mathbb{P}_x\Big\{\sup_{0\leq s\leq 1}\big|W_0(s)-s\big|\leq\delta\Big\}.$$

Repeating our argument,

$$\mathbb{P}(D_n)\geq\left(\inf_{|x|\leq\delta}\mathbb{P}_x\Big\{\sup_{0\leq s\leq 1}\big|W_0(s)-s\big|\leq\delta\Big\}\right)^n=c_2^n,$$

where

$$c_2\equiv\inf_{|x|\leq\delta}\mathbb{P}_x\Big\{\sup_{0\leq s\leq 1}\big|W_0(s)-s\big|\leq\delta\Big\}>0.$$

Thus, one can take $N>0$ sufficiently large so that

$$\mathbb{P}\Big\{\sum_{k=1}^{n-1} \beta\big([k-1,k]\times[k,k+1]\big)\geq Nn\Big\}\leq\frac{1}{2}\mathbb{P}(D_n).$$

Hence, (4.3.19) follows from (4.3.20). □

4.4. Applications to LIL

In this section we apply the large deviations established in Section 4.2 and Section 4.3 to the laws of the iterated logarithm for self-intersection local times.

THEOREM 4.4.1. *Let $L(t,x)$ be the local time of a 1-dimensional Brownian motion. For any $p > 1$,*

$$\limsup_{t\to\infty} t^{-\frac{p+1}{2}}(\log\log t)^{-\frac{p-1}{2}}\int_{-\infty}^{\infty} L^p(t,x)dx$$
$$= 2^{\frac{p+1}{2}}(p-1)^{\frac{p-1}{2}}(p+1)^{\frac{p-3}{2}}B\Big(\frac{1}{p-1},\frac{1}{2}\Big)^{-(p-1)} \qquad a.s.$$

Proof. By the scaling (4.2.1), Theorem 4.2.1 can be rewritten as

$$\lim_{t\to\infty}\frac{1}{\log\log t}\log\mathbb{P}\Big\{\int_{-\infty}^{\infty}L^p(t,x)dx \ge \lambda t^{\frac{p+1}{2}}(\log\log t)^{\frac{p-1}{2}}\Big\}$$
$$= -\lambda^{\frac{2}{p-1}}\frac{1}{4(p-1)}\Big(\frac{p+1}{2}\Big)^{\frac{3-p}{p-1}}B\Big(\frac{1}{p-1},\frac{1}{2}\Big) \qquad (\lambda>0).$$

Observe that the constant

$$\lambda_0 \equiv 2^{\frac{p+1}{2}}(p-1)^{\frac{p-1}{2}}(p+1)^{\frac{p-3}{2}}B\Big(\frac{1}{p-1},\frac{1}{2}\Big)^{-(p-1)}$$

appearing in Theorem 4.4.1 solves the equation

$$\lambda^{\frac{2}{p-1}}\frac{1}{4(p-1)}\Big(\frac{p+1}{2}\Big)^{\frac{3-p}{p-1}}B\Big(\frac{1}{p-1},\frac{1}{2}\Big) = 1.$$

Consequently,

$$\sum_k \mathbb{P}\Big\{\int_{-\infty}^{\infty}L^p(\theta^k,x)dx \ge \lambda_1(\theta^k)^{\frac{p+1}{2}}(\log\log\theta_k)^{\frac{p-1}{2}}\Big\} < \infty$$

for every $\theta > 1$ and every $\lambda_1 > \lambda_0$. By the Borel-Cantelli lemma, we get

$$\limsup_{k\to\infty}(\theta^k)^{-\frac{p+1}{2}}(\log\log\theta_k)^{-\frac{p-1}{2}}\int_{-\infty}^{\infty}L^p(\theta^k,x)dx \le \lambda_1 \qquad a.s.$$

For any t between θ^k and θ^{k+1}, the bound

$$t^{-\frac{p+1}{2}}(\log\log t)^{-\frac{p-1}{2}}\int_{-\infty}^{\infty}L^p(t,x)dx$$
$$\le \Big(\theta^{\frac{p+1}{2}}+o(1)\Big)(\theta^{k+1})^{-\frac{p+1}{2}}(\log\log\theta_{k+1})^{-\frac{p-1}{2}}\int_{-\infty}^{\infty}L^p(\theta^{k+1},x)dx$$

implies that

$$\limsup_{t\to\infty} t^{-\frac{p+1}{2}}(\log\log t)^{-\frac{p-1}{2}}\int_{-\infty}^{\infty}L^p(t,x)dx \le \theta^{\frac{p+1}{2}}\lambda_1 \qquad a.s.$$

Letting $\theta\to 1^+$ and $\lambda_1\to\lambda_0^+$ on the right hand side leads to the upper bound for Theorem 4.4.1.

In the argument for the lower bound, we make $\theta > 1$ large. For any $\lambda_2 < \lambda_0$,

$$\sum_k \mathbb{P}\Big\{\int_{-\infty}^{\infty}\big[L(\theta^{k+1},x)-L(\theta^k,x)\big]^p dx \ge \lambda_2(\theta^{k+1})^{\frac{p+1}{2}}(\log\log\theta^{k+1})^{\frac{p-1}{2}}\Big\}$$
$$=\sum_k \mathbb{P}\Big\{\int_{-\infty}^{\infty} L^p(\theta^{k+1}-\theta^k,x)dx \ge \lambda_2(\theta^{k+1})^{\frac{p+1}{2}}(\log\log\theta^{k+1})^{\frac{p-1}{2}}\Big\}=\infty$$

for sufficiently large θ. By the independence among the random variables

$$\int_{-\infty}^{\infty}\big[L(\theta^{k+1},x)-L(\theta^k,x)\big]^p dx \qquad k=1,2,\cdots$$

and by the Borel-Cantelli lemma, with probability 1, we get

$$\limsup_{k\to\infty}(\theta^{k+1})^{-\frac{p+1}{2}}(\log\log\theta^{k+1})^{-\frac{p-1}{2}}\int_{-\infty}^{\infty}\big[L(\theta^{k+1},x)-L(\theta^k,x)\big]^p dx \ge \lambda_2.$$

Consequently,

$$\limsup_{k\to\infty}(\theta^{k+1})^{-\frac{p+1}{2}}(\log\log\theta^{k+1})^{-\frac{p-1}{2}}\int_{-\infty}^{\infty} L^p(\theta^{k+1},x)dx \ge \lambda_2 \qquad a.s.$$

This clearly implies that

$$\limsup_{t\to\infty} t^{-\frac{p+1}{2}}(\log\log t)^{-\frac{p-1}{2}}\int_{-\infty}^{\infty} L^p(t,x)dx \ge \lambda_2 \qquad a.s.$$

Letting $\lambda_2 \to \lambda_0^-$ leads to the requested lower bound. □

THEOREM 4.4.2. *Let $\gamma([0,t]_<^2)$ be the renormalized self-intersection local time of a planar Brownian motion $W(t)$.*

$$\limsup_{t\to\infty}\frac{1}{t\log\log t}\gamma\big([0,t]_<^2\big)=\kappa(2,2)^4 \qquad a.s., \tag{4.4.1}$$

$$\liminf_{t\to\infty}\frac{1}{t\log\log\log t}\gamma\big([0,t]_<^2\big)=-\frac{1}{2\pi} \qquad a.s. \tag{4.4.2}$$

Proof. Let $\theta > 1$ be fixed but arbitrary and write $t_n = \theta^n$. By Theorem 4.3.1 and the Borel-Cantelli lemma, we get

$$\limsup_{n\to\infty}\frac{1}{t_n\log\log t_n}\gamma\big([0,t_n]_<^2\big)\le\kappa(2,2)^4 \qquad a.s. \tag{4.4.3}$$

Given $\epsilon > 0$, we have

$$\mathbb{P}\Big\{\sup_{t_{n-1}\le t\le t_n}\big|\gamma\big([0,t]_<^2\big)-\gamma\big([0,t_{n-1}]_<^2\big)\big|\ge\epsilon t_n\log\log t_n\Big\}$$
$$\le\mathbb{P}\Big\{\sup_{\substack{|t-s|\le(\theta-1)t_n\\ s,t\in[0,t_n]}}\big|\gamma\big([0,t]_<^2\big)-\gamma\big([0,s]_<^2\big)\big|\ge\epsilon t_n\log\log t_n\Big\}$$
$$=\mathbb{P}\Big\{\sup_{\substack{|t-s|\le(\theta-1)\\ s,t\in[0,1]}}\big|\gamma\big([0,t]_<^2\big)-\gamma\big([0,s]_<^2\big)\big|\ge\epsilon\log\log t_n\Big\},$$

where the last step follows from scaling. By (2.4.8), therefore, one can have

$$\sum_n \mathbb{P}\Big\{\sup_{t_{n-1}\le t\le t_n}\big|\gamma\big([0,t]_<^2\big)-\gamma\big([0,t_{n-1}]_<^2\big)\big|\ge \epsilon t_n\log\log t_n\Big\}<\infty$$

by making θ close to 1 enough. By the Borel-Cantelli lemma, we get

$$\limsup_{n\to\infty}\frac{1}{t_n\log\log t_n}\sup_{t_{n-1}\le t\le t_n}\big|\gamma\big([0,t]_<^2\big)-\gamma\big([0,t_{n-1}]_<^2\big)\big|\le\epsilon\qquad a.s.$$

Combining this with (4.4.3) we establish the upper bound

$$\limsup_{t\to\infty}\frac{1}{t\log\log t}\gamma\big([0,t]_<^2\big)\le\kappa(2,2)^4\qquad a.s. \tag{4.4.4}$$

Write $K_n=[\log\log t_n]$ and $s_n=t_n/K_n$. For $(l-1)s_n\le t\le ls_n$,

$$\begin{aligned}-\gamma\big([0,t]_<^2\big)&\le-\sum_{k=1}^{l-1}\gamma\big([(k-1)s_n,ks_n]_<^2\big)-\gamma\big([(l-1)s_n,s]_<^2\big)\\&\quad+\sum_{k=1}^{K_n-1}\mathbb{E}\beta\big([0,(k-1)s_n]\times[ks_n,(k+1)s_n]\big)\\&\le\sum_{k=1}^{K_n}\sup_{(k-1)s_n\le s\le ks_n}\Big\{-\gamma\big([(k-1)s_n,s]_<^2\big)\Big\}\\&\quad+\sum_{k=1}^{K_n-1}\mathbb{E}\beta\big([0,(k-1)s_n]\times[ks_n,(k+1)s_n]\big).\end{aligned}$$

The second term on the right hand side is equal to

$$\begin{aligned}&\frac{1}{2\pi}\sum_{k=1}^{K_n-1}\Big\{(k+1)s_n\log\big((k+1)s_n\big)-ks_n\log(ks_n)-s_n\log s_n\Big\}\\&=\frac{1}{2\pi}K_ns_n\log K_n\le\frac{1}{2\pi}t_n\log\log t_n.\end{aligned}$$

Thus, for any $\epsilon>0$,

$$\begin{aligned}&\mathbb{P}\Big\{\sup_{t\le t_n}\Big\{-\gamma\big([0,t]_<^2\big)\Big\}\ge(1+\epsilon)t_n\log\log\log t_n\Big\}\\&\le\mathbb{P}\Big\{\sum_{k=1}^{K_n}\sup_{(k-1)s_n\le s\le ks_n}\Big\{-\gamma\big([(k-1)s_n,s]_<^2\big)\Big\}\ge\epsilon t_n\log\log\log t_n\Big\}.\end{aligned}$$

The random sequence

$$\sup_{(k-1)s_n\le s\le ks_n}\Big\{-\gamma\big([(k-1)s_n,s]_<^2\big)\Big\}\qquad k=1,2,\cdots$$

is an i.i.d. with the common distribution the same as

$$s_n\sup_{0\le s\le1}\Big\{-\gamma\big([0,s]_<^2\big)\Big\}.$$

By Chebyshev inequality, therefore, we get

$$\mathbb{P}\Big\{\sup_{t\le t_n}\Big\{-\gamma\big([0,t]_<^2\big)\Big\}\ge(1+\epsilon)t_n\log\log\log t_n\Big\}$$
$$\le\exp\Big\{-\lambda\epsilon K_n\log K_n\Big\}\Big(\mathbb{E}\exp\Big\{\lambda\sup_{0\le s\le 1}\Big\{-\gamma\big([0,s]_<^2\big)\Big\}\Big\}\Big)^{K_n}$$

where, by (2.4.9), we can take $\lambda>0$ small enough so that

$$\mathbb{E}\exp\Big\{\lambda\sup_{0\le s\le 1}\Big\{-\gamma\big([0,s]_<^2\big)\Big\}\Big\}<\infty.$$

Therefore,

$$\sum_n\mathbb{P}\Big\{\sup_{t\le t_n}\Big\{-\gamma\big([0,t]_<^2\big)\Big\}\ge(1+\epsilon)t_n\log\log\log t_n\Big\}<\infty.$$

A standard procedure via the Borel-Cantelli lemma gives the upper bound

$$\limsup_{t\to\infty}\frac{1}{t\log\log\log t}\Big(-\gamma\big([0,t]_<^2\big)\Big)\le\frac{1}{2\pi}\quad a.s. \tag{4.4.5}$$

We now work on the lower bound. Recall that $t_n=\theta^n$. This time $\theta>1$ stands for a large number. For any $\lambda_1<\kappa(2,2)^4$, by Theorem 4.3.1 and by scaling, we get

$$\sum_n\mathbb{P}\Big\{\gamma\big([t_{n-1},t_n]_<^2\big)\ge\lambda_1t_n\log\log t_n\Big\}$$
$$=\sum_n\mathbb{P}\Big\{\gamma\big([0,1]_<^2\big)\ge\lambda_1\frac{t_n}{t_n-t_{n-1}}\log\log t_n\Big\}=\infty$$

for sufficiently large θ. Notice that

$$\gamma\big([t_{n-1},t_n]_<^2\big)\qquad n=1,2,\cdots$$

is an independent sequence. By the Borel-Cantelli lemma

$$\limsup_{n\to\infty}\frac{1}{t_n\log\log t_n}\gamma\big([t_{n-1},t_n]_<^2\big)\ge\lambda_1\qquad a.s. \tag{4.4.6}$$

In addition,

$$\begin{aligned}\gamma\big([0,t_n]_<^2\big)&=\gamma\big([0,t_{n-1}]_<^2\big)+\gamma\big([t_{n-1},t_n]_<^2\big)\\&\quad+\beta\big([0,t_{n-1}]\times[t_{n-1},t_n]\big)-\mathbb{E}\beta\big([0,t_{n-1}]\times[t_{n-1},t_n]\big)\\&\ge\gamma\big([0,t_{n-1}]_<^2\big)+\gamma\big([t_{n-1},t_n]_<^2\big)+\mathbb{E}\beta\big([0,t_{n-1}]\times[t_{n-1},t_n]\big)\end{aligned}$$

and the last term on the right hand side is equal to

$$\frac{1}{2\pi}\Big\{t_n\log t_n-t_{n-1}\log t_{n-1}-(t_n-t_{n-1})\log(t_n-t_{n-1})\Big\}=O(t_n).$$

By (4.4.5), we get

$$\liminf_{n\to\infty}\frac{1}{t_n\log\log t_n}\gamma\big([0,t_{n-1}]_<^2\big)=0\qquad a.s.$$

Hence, it follows from (4.4.6) that

$$\limsup_{t\to\infty}\frac{1}{t\log\log t}\gamma\big([0,t]_<^2\big)\ge\kappa(2,2)^4\qquad a.s. \tag{4.4.7}$$

Write $u_n = \exp\{n^2\}$ $(n = 1, 2, \cdots)$. By Theorem 4.3.2, for any $\lambda_2 < (2\pi)^{-1}$,

$$\sum_n \mathbb{P}\Big\{ -\gamma\big([u_{n-1}, u_n]_<^2\big) \ge \lambda_2 u_n \log\log\log u_n \Big\}$$
$$= \sum_n \mathbb{P}\Big\{ -\gamma\big([0,1]_<^2\big) \ge \lambda_2 \frac{u_n}{u_n - u_{n-1}} \log\log\log u_n \Big\} = \infty.$$

By the Borel-Cantelli lemma, we get

$$\limsup_{n\to\infty} \frac{1}{u_n \log\log\log u_n}\Big(-\gamma\big([u_{n-1}, u_n]_<^2\big)\Big) \ge \frac{1}{2\pi} \qquad a.s. \tag{4.4.8}$$

Notice that

$$-\gamma\big([0, u_n]_<^2\big) \ge -\gamma\big([0, u_{n-1}]_<^2\big) - \beta\big([0, u_{n-1}] \times [u_{n-1}, u_n]\big) - \gamma\big([u_{n-1}, u_n]_<^2\big).$$

To control the first two terms on the right hand side, we use the fact that as $n \to \infty$,

$$u_n \log\log\log u_n \gg \sqrt{u_n u_{n-1}} \log\log u_n \gg u_{n-1} \log\log u_{n-1}.$$

Indeed, by (4.4.4) and (4.4.5),

$$\lim_{n\to\infty} \frac{1}{u_n \log\log\log u_n} \gamma\big([0, u_{n-1}]_<^2\big) = 0 \qquad a.s.$$

According to Proposition 2.3.4,

$$\beta\big([0, u_{n-1}] \times [u_{n-1}, u_n]\big) \overset{d}{=} \alpha\big([0, u_{n-1}] \times [0, u_n - u_{n-1}]\big).$$

For any integer $m \ge 1$, by (2.2.12), we get

$$\mathbb{E}\Big[\beta\big([0, u_{n-1}] \times [u_{n-1}, u_n]\big)^m\Big]$$
$$\le \Big\{\mathbb{E}\Big[\alpha\big([0, u_{n-1}]^2\big)^m\Big]\Big\}^{1/2} \Big\{\mathbb{E}\Big[\alpha\big([0, u_n - u_{n-1}]^2\big)^m\Big]\Big\}^{1/2}$$
$$= \big(u_{n-1}(u_n - u_{n-1})\big)^{m/2} \mathbb{E}\Big[\alpha\big([0,1]^2\big)^m\Big].$$

By Theorem 2.2.9 (with $d = p = 2$), therefore, there is a $c > 0$ such that

$$\sup_n \mathbb{E}\exp\Big\{c(u_{n-1}u_n)^{-1/2} \beta\big([0, u_{n-1}] \times [u_{n-1}, u_n]\big)\Big\} < \infty.$$

Using Chebyshev inequality one can see that for any $\epsilon > 0$,

$$\sum_n \mathbb{P}\Big\{\beta\big([0, u_{n-1}] \times [u_{n-1}, u_n]\big) \ge \epsilon u_n \log\log\log u_n\Big\} < \infty.$$

By the Borel-Cantelli lemma, we get

$$\lim_{n\to\infty} \frac{1}{u_n \log\log\log u_n} \beta\big([0, u_{n-1}] \times [u_{n-1}, u_n]\big) = 0 \qquad a.s.$$

Therefore, it follows from (4.4.8) that

$$\limsup_{t\to\infty} \frac{1}{t \log\log\log t}\Big(-\gamma\big([0, t]_<^2\big)\Big) \ge \frac{1}{2\pi} \qquad a.s. \tag{4.4.9}$$

Finally, Theorem 4.4.2 follows from (4.4.4), (4.4.5), (4.4.7) and (4.4.9). □

4.5. Notes and comments

Section 4.1

The Feynman-Kac formula goes back to Kac [**105**] and exists in literature in several different forms. It has become a popular tool in the study of Markov processes. We refer the interested reader to the books by Fukushima ([**81**]), Rogers and Williams ([**144**], [**145**]), Sznitman (Chapter 1, [**158**]) for the Feynman-Kac formula in the general context. The well-known Donsker-Varadhan large deviations ([**56**], [**58**], [**61**]) for empirical measures of Markov processes are closely related to the Feynman-Kac formula. Some idea in the proof of Theorem 4.1.6 comes from Remillard [**140**]. We also mention the book by Pinsky [**139**] for the applications of the Feynman-Kac formula to the logarithmic moment generating functions of the occupation times of the diffusion processes.

The following exercise further addresses the relation between Brownian motion and Laplacian operator. Viewing Brownian motion as a Markov process, we allow $W(t)$ to start at $x \in \mathbb{R}$ and use "$\mathbb{E}_x$" to denote the expectation correspondent to such a path.

EXERCISE 4.5.1. *Let $g(x)$: $\mathbb{R}^d \longrightarrow \mathbb{R}$ be twice continuously differentiable on $\mathbb{R}^d$. Show that under $\mathbb{E}_x$, the process*

$$X(t) = g\big(W(t)\big) - \frac{1}{2}\int_0^t \Delta g\big(W(s)\big)ds \qquad t \ge 0 \tag{4.5.1}$$

is a Martingale with respect to the Brownian filtration $\mathcal{F}_t = \sigma\Big\{W(s);\quad 0 \le s \le t\Big\}$, provided

$$\mathbb{E}\big|g\big(W(t)\big)\big| < \infty \quad \textit{and} \quad \int_0^t \mathbb{E}\big|\Delta g\big(W(s)\big)\big|ds < \infty$$

for all $t > 0$.

Hint: You may need the fact that

$$\frac{\partial p_t}{\partial t}(x) = \frac{1}{2}\Delta p_t(x) \qquad t > 0 \;\; x \in \mathbb{R}^d.$$

EXERCISE 4.5.2. *Let $X(t)$ be an irreducible, symmetric Markov chain with finite state space E. It is a classic fact (section 3.1, [**92**]) that there is a non-positive matrix $\mathcal{A} = (q_{x,y})_{x,y\in E}$, known as infinitesimal matrix (operator), such that*

$$\frac{d}{dt}p_t(x,y) = \sum_{z\in E} q_{x,z}p_t(z,y) \qquad x,y \in E, \quad t \ge 0.$$

(1) *Follow the discuss in section 4.1 to show that for any function $f(x)$ on E,*

$$\begin{aligned}&\lim_{t\to\infty}\frac{1}{t}\log\mathbb{E}\exp\left\{\int_0^t f\big(X(s)\big)ds\right\}\\ &= \sup_{||g||_2=1}\left\{\sum_{x\in E} f(x)g^2(x) + \sum_{x\in E} g(x)\mathcal{A}g(x)\right\}\end{aligned}$$

where

$$\mathcal{A}g(x) = \sum_{z\in E} q_{x,z} g(z) \qquad x \in E.$$

(2) *Write*

$$SI_t = \int_{[0,t]^p} 1_{\{X(s_1)=\cdots=X(s_p)\}} ds_1 \cdots ds_p.$$

Prove that

$$\lim_{t\to\infty} \frac{1}{t} \log \mathbb{E} \exp\Big\{\theta SI_t\Big\} = \sup_{||g||_2=1} \bigg\{\theta\Big(\sum_{x\in E} |g(x)|^{2p}\Big)^{1/p} + \sum_{x\in E} g(x)\mathcal{A}g(x)\bigg\}$$

(you may need to read the proof of Theorem 4.2.1 before doing part (2)).

In the following exercises we present the applications of the Feynman-Kac formula to the small ball probabilities.

EXERCISE 4.5.3. *Let $d = 1$. The only reason why we assume the boundedness of f in this section is to make the linear operator A defined in (4.1.4) upper semi-bounded. A careful reader may realize that this assumption can be relaxed to the upper boundedness of f. We may take $f(x) = -x^2$ in Theorem 4.1.6 that leads to*

$$\begin{aligned}(4.5.2)\qquad &\lim_{t\to\infty} \frac{1}{t} \log \mathbb{E} \exp\bigg\{ - \int_0^t |W(s)|^2 ds\bigg\} \\ &= - \inf_{g\in\mathcal{F}_1} \bigg\{ \int_{-\infty}^{\infty} x^2 g^2(x) dx + \frac{1}{2} \int_{-\infty}^{\infty} |g'(x)|^2 dx\bigg\}.\end{aligned}$$

(1) *Based on (4.5.2), prove that*

$$(4.5.3)\qquad \lim_{\epsilon\to 0^+} \epsilon \log \mathbb{P}\bigg\{ \int_0^1 |W(s)|^2 ds \le \epsilon\bigg\} = -\frac{1}{8}.$$

Hint: For the computation of the variation on the right hand side of (4.5.2),

$$\begin{aligned}\frac{1}{2} &= \frac{1}{2} \int_{-\infty}^{\infty} g^2(x) dx = - \int_{-\infty}^{\infty} x g(x) g'(x) dx \\ &\le \bigg\{ \int_{-\infty}^{\infty} x^2 g^2(x) dx\bigg\}^{1/2} \bigg\{ \int_{-\infty}^{\infty} |g'(x)|^2 dx\bigg\}^{1/2} \\ &\le \frac{1}{\sqrt{2}} \bigg\{ \int_{-\infty}^{\infty} x^2 g^2(x) dx + \frac{1}{2} \int_{-\infty}^{\infty} |g'(x)|^2 dx\bigg\}.\end{aligned}$$

Find the function g which turns the inequalities into equalities.

(2) *Extend the small ball result to the case $d \ge 2$.*

Comment. The precise Laplace transform of the L^2-norm of 1-dimensional Brownian motion is given as the following (p. 414, [**141**]):

$$\mathbb{E} \exp\bigg\{ - \lambda \int_0^1 |W(s)|^2 ds\bigg\} = \big(\cosh \sqrt{2\lambda}\big)^{-1/2} \qquad \lambda > 0.$$

The above identity is known as the *Cameron-Martin formula.*

EXERCISE 4.5.4. *We compute the small ball probability for the 1-dimensional Brownian motion $W(t)$ under the supremum norm. Define the exit time T as*

$$T = \inf\big\{t \ge 0;\quad W(t) \notin [-1, 1]\big\}.$$

(1) *Let $g(x)$ be a twice continuously differentiable function on $\mathbb{R}$ with $g(-1) = g(1) = 0$. Prove that*

$$\mathbb{E}\Big\{g\big(W(t)\big)1_{\{T\ge t\}}\Big\} = g(0) + \frac{1}{2}\int_0^t \mathbb{E}\Big\{g''\big(W(s)\big)1_{\{T\ge s\}}\Big\}ds.$$

Hint: Stop the martingale $X(t)$ defined in (4.5.1) at $T \wedge t$.

(2) *Prove that*

$$\mathbb{E}\Big\{\cos\Big(\frac{\pi}{2}W(t)\Big)1_{\{T\ge t\}}\Big\} = \exp\Big\{-\frac{\pi^2}{8}t\Big\}.$$

(3) *Prove that*

$$\lim_{\epsilon\to 0^+}\epsilon^2\log\mathbb{P}\Big\{\sup_{0\le s\le 1}|W(s)| \le \epsilon\Big\} = -\frac{\pi^2}{8}.$$

Comment. By the method of reflection principle, the precise distribution of $\sup_{0\le s\le 1}|W(s)|$ is given as (p. 342, [**76**])

$$\mathbb{P}\Big\{\sup_{0\le s\le 1}|W(s)| \le x\Big\} = \sum_{k=0}^\infty(-1)^k\frac{4}{(2k+1)\pi}\exp\Big\{-\frac{(2k+1)\pi^2}{8x^2}\Big\}\qquad x > 0.$$

On the other hand, the idea appearing in the above exercise is adaptable to a wider class of models.

Section 4.2

Theorem 4.2.1 is philosophically linked to the famous *Donsker-Varadhan's large deviations* for the empirical measure

$$\mathcal{L}_t(A) = \frac{1}{t}\int_0^t 1_{\{W(s)\in A\}}ds\qquad A \in \mathcal{B}(\mathbb{R}).$$

Viewing $\mathcal{L}_t(\cdot)$ as a stochastic process taking values in $\mathcal{P}(\mathbb{R})$, the space of the probability measures on $\mathbb{R}$ equipped with the topology of weak convergence, Donsker and Varadhan ([**55**]) establish a weak version of the large deviation principle claiming

$$\lim_{t\to\infty}\frac{1}{t}\log\mathbb{P}\big\{\mathcal{L}_t(\cdot) \in \mathcal{U}\big\} = -\inf_{\mu\in\mathcal{U}} I(\mu)$$

for a class of $\mathcal{U} \subset \mathcal{P}(\mathbb{R})$, where the rate function

$$I(\mu) = \frac{1}{8}\int_{-\infty}^\infty \frac{|f'(x)|^2}{f(x)}dx$$

for absolute continuous probability measure μ with the density f.

Define the function Ψ on (a subset of) $\mathcal{P}(\mathbb{R})$ by

$$\Psi(\mu) = \Big(\int_{-\infty}^\infty f^p(x)dx\Big)^{1/p}.$$

Varadhan's integral lemma (a functional version of Theorem 1.1.6, see Theorem 4.3.1 in [**47**] for full details) morally supports (4.2.3) in the following way:

$$\lim_{t\to\infty}\frac{1}{t}\log\mathbb{E}\exp\left\{\left(\int_{-\infty}^{\infty}L^p(t,x)dx\right)^{1/p}\right\}$$
$$=\lim_{t\to\infty}\frac{1}{t}\log\mathbb{E}\exp\Big\{t\Psi(\mathcal{L}_t)\Big\} = \sup_{\mu\in\mathcal{P}(\mathbb{R}^d)}\Big\{\Psi(\mu)-I(\mu)\Big\}$$
$$=\sup_f\left\{\left(\int_{-\infty}^{\infty}f^p(x)dx\right)^{1/p}-\frac{1}{8}\int_{-\infty}^{\infty}\frac{|f'(x)|^2}{f(x)}dx\right\}$$

where the supremum on the right hand side is taken for the probability density functions f on $\mathbb{R}$. It becomes (4.2.3) under the substitution $f(x)=g^2(x)$.

To make a mathematical argument out of the above heuristics, one has to overcome two major obstacles: the singularity of the function $\Psi(\mu)$ and the absence of the full large deviation for $\mathcal{L}_t$. Theorem 4.2.1 in the special case $p=2$ was first obtained by Mansmann [**135**] in this approach. To make it work, some very delicate approximation was adopted.

Prior to Mansmann's work, Donsker and Varadhan ([**55**], [**56**], [**57**], [**58**], [**59**] [**60**], [**61**]) had applied their large deviations for empirical measures of Markov processes to a variety of the models which can be viewed as the functionals of the empirical measure. In addition, see Bolthausen ([**17**]), van den Berg, Bolthausen and den Hollander ([**11**], [**12**]) for the applications of this approach to some related models.

By excursion theory, Csáki, König and Shi ([**43**]) established the large deviations for self-intersection local times of Brownian bridges that take a form exactly the same as for Theorem 4.2.1. The result given in Theorem 4.2.1 was obtained by Chen and Li [**23**] in a more general form and the argument used here essentially comes from that paper.

EXERCISE 4.5.5. *Prove Theorem 4.2.3.*

The main idea in the proof of Theorem 4.2.1 is to view $L(t,\cdot)$ as a stochastic process taking values in $\mathcal{L}^p(\mathbb{R})$. In the next two exercises we try to extended it to the case $p=\infty$. The fact that the Banach space of the bounded functions on an infinite set is not separable (under the supremum norm) brings some technical difficulties. Thanks to the continuity of the Brownian local time $L(t,x)$ with respect to x, we may embed the local time into the space of the continuous functions.

EXERCISE 4.5.6 (hard). *Prove that for any $M>0$,*

$$\mathbb{E}\exp\Big\{\sup_{x\in\mathbb{R}}L(t,x)\Big\}$$
$$\le\left(\mathbb{P}\Big\{\sup_{0\le s\le t}|W(s)|\le M\Big\}\right)^{-1}\Big\{1+\mathbb{E}\exp\Big\{\sup_{|x|\le M}L(t,x)\Big\}\Big\}.$$

Hint: What makes the case $p = \infty$ different from $1 < p < \infty$ is the inequality

$$\mathbb{E}\exp\Big\{\sup_{x\in\mathbb{R}} L(t,x)\Big\} \le \mathbb{E}\exp\Big\{\sup_{|x|\le M} L(t,x)\Big\} + \mathbb{E}\exp\Big\{\sup_{|x|>M} L(t,x)\Big\}.$$

Work on the second term on the right hand side.

EXERCISE 4.5.7. *Again, $L(t,x)$ is the local time of a 1-dimensional Brownian motion.*

(1) *Prove that*

$$\lim_{t\to\infty}\frac{1}{t}\log\mathbb{E}\exp\Big\{\sup_{x\in\mathbb{R}} L(t,x)\Big\} = \sup_{g\in\mathcal{F}}\Big\{||g||_\infty^2 - \frac{1}{2}\int_{-\infty}^{\infty}|g'(x)|^2dx\Big\}.$$

(2) *Show that*

$$\sup_{g\in\mathcal{F}}\Big\{||g||_\infty^2 - \frac{1}{2}\int_{-\infty}^{\infty}|g'(x)|^2dx\Big\} = \frac{1}{2}.$$

Hint: For one side, take $g(x) = e^{-|x|}$. For another side, apply Cauchy-Schwarz inequality to the representations

$$g^2(x_0) = -2\int_{x_0}^{\infty} g(x)g'(x)dx = 2\int_{-\infty}^{x_0} g(x)g'(x)dx$$

where $x_0\in\mathbb{R}$ can be chosen as a maximizer of $g^2(x)$.

(3) *Conclude that*

$$\lim_{t\to\infty}\frac{1}{t}\log\mathbb{P}\Big\{\sup_{x\in\mathbb{R}} L(1,x) \ge \sqrt{t}\Big\} = -\frac{1}{2}. \tag{4.5.4}$$

Comment. In comparison to Exercise 2.5.6, the tail probabilities of $L(1,0)$ and $\sup_{x\in\mathbb{R}} L(1,x)$ are asymptotically the same, despite the obvious relation

$$L(1,0) \le \sup_{x\in\mathbb{R}} L(1,x).$$

EXERCISE 4.5.8. *In this exercise we compute the small ball probability of p-multiple self-intersection local time of 1-dimensional Brownian motion.*

(1) *Prove that*

$$\lim_{t\to\infty}\frac{1}{t}\log\mathbb{E}\exp\Big\{-t^{\frac{p+1}{2}}\int_{-\infty}^{\infty} L^p(1,x)dx\Big\} = -C_p$$

where $0 < C_p \le \infty$.

Hint: Sub-additivity.

(2) *It can be proved that $0 < C_p < \infty$. Accept this fact and prove that*

$$\lim_{\epsilon\to 0^+}\epsilon^{\frac{2}{p-1}}\log\mathbb{P}\Big\{\int_{-\infty}^{\infty} L^p(1,x)dx \le \epsilon\Big\} = -2^{\frac{2}{p-1}}(p-1)(p+1)^{-\frac{p+1}{p-1}}C_p^{\frac{p+1}{p-1}}. \tag{4.5.5}$$

The value of C_p remains unknown.

Section 4.3

To explain Theorem 4.3.1 by Donsker-Varadhan large deviation ([**55**]), we hypothetically assume the existence of the local time $L(t,x)$ of a planar Brownian motion. The second term in the decomposition

$$\gamma([0,1]^2_<) = \frac{1}{2}\int_{\mathbb{R}^2} L^2(1,x)dx - \frac{1}{2}\mathbb{E}\int_{\mathbb{R}^2} L^2(1,x)dx$$

does not play any role in the LDP given in Theorem 4.3.1. So Donsker-Varadhan large deviation suggests Theorem 4.3.1 in the same way as it suggests Theorem 4.2.1.

Duplantier ([**64**]) points out that in quantum field theory, "Families of universal critical exponents are associated with intersection properties of sets of random walks". The large deviation in Theorem 4.3.1 and the related exponential integrability problem in (4.3.3) correspond to the continuum limit of the physical model known as Edwards model. We refer the reader to the papers by Westwater ([**165**], [**166**], [**167**]) and by Duplantier ([**63**], [**64**]) for details. The result stated in Theorem 4.3.1 comes from the paper by Bass and Chen ([**5**]). Prior to [**5**], the critical exponential integrability given in (4.3.3) was studied by Yor ([**174**]) and by Le Gall ([**130**]) without identifying the constant λ_0. In the context of the random walks with discrete time and state space, the same type of exponential integrability was obtained by Brydges and Slade ([**19**]). Further, Brydges and Slade provided a heuristic argument (p. 294, [**19**]) by using Donsker-Varadhan large deviation suggesting the relevance of Gagliardo-Nirenberg constant $\kappa(2,2)$. For the random walks of lattice values but continuous time, Bolthausen ([**17**]) obtained a large deviation with a form essentially the same as Theorem 4.3.1, except that in his model, the self-intersection local time does not have to be renormalized.

Among all the works prior to ([**5**]), a private letter from Duplantier to Yor captures some of the idea about the critical value λ_0 given in (4.3.3). In that letter, Duplantier pointed out (without proof) that

$$\lambda_0 = \frac{\pi}{2}\int_0^\infty rg^4(r)dr$$

where the function g solves the differential equation

$$g''(r) + \frac{1}{r}g'(r) = g(r) - g^3(r).$$

His observation is accurate. Indeed, let $f_0(x)$ be the positive, infinitely smooth, spherically symmetric ground state solution of the non-linear partial differential equation (3.4.2). By spherical symmetry $f(x) = g(|x|)$ for some function g on $\mathbb{R}^+$. One can see that when $d = p = 2$, g satisfies the ordinary differential equation proposed by Duplantier and that by (3.4.3),

$$\lambda_0 = \kappa(2,2)^{-4} = \frac{1}{4}\int_{\mathbb{R}^2} f^4(x)dx = \frac{\pi}{2}\int_0^\infty rg^4(r)dr.$$

In addition, from (3.4.4), we get

$$\lambda_0 \approx \pi \times 1.86225\cdots \approx 5.85043\cdots .$$

A striking feature of the renormalized self-intersection local time is its asymptotically asymmetric tail, which was first observed by Le Gall ([**130**]). Theorem 4.3.2 sharpens Le Gall's observation. This result was obtained in Bass and Chen [**5**]. Finally, we mention a non-trivial extension of Theorem 4.3.1 and Theorem 4.3.2 to the setting of stable processes ([**6**]), where the stable counterpart of Theorem 4.3.2 takes a drastically different form.

A natural question remains: What can we say about the LDP for the renormalized p-multiple self-intersection local time $\gamma\big([0,t]_<^p\big)$ heuristically given in (2.5.13) as $p \geq 3$? Based on an earlier explanation of Theorem 4.3.1 in terms of Donsker-Varadhan large deviation, we post the following conjecture.

CONJECTURE 4.5.9. *As $d = 2$ and $p \geq 2$,*

$$\lim_{t\to\infty} t^{-\frac{1}{p-1}} \log \mathbb{P}\Big\{\gamma\big([0,1]_<^p\big) \geq t\Big\} = -\frac{1}{2}(p!)^{\frac{1}{p-1}}\kappa(2,p)^{-\frac{2p}{p-1}}.$$

A further open problem is the LDP for $\gamma\big([0,t]_<^p\big)$ in the case $p \geq 3$, which corresponds to the lower tail asymptotics given in Theorem 4.3.2.

Section 4.4

The application of large deviations to the laws of the iterated logarithm has become nearly standard: The large deviations provide the estimates needed for the Borel-Cantelli type of argument. The results in this section are expected consequences of the large deviations established in the previous sections.

EXERCISE 4.5.10. *Let $L(t,x)$ be the local time of a 1-dimensional Brownian motion. Use the LDP established in (4.5.4) to prove the following law of the iterated logarithm:*

$$\limsup_{t\to\infty}(2t\log\log t)^{-1}\sup_{x\in\mathbb{R}} L(t,x) = 1 \quad a.s. \tag{4.5.6}$$

Hint: Read the proof of Theorem 4.4.1.

EXERCISE 4.5.11. *Let $L(t,x)$ be the local time of a 1-dimensional Brownian motion. Prove the following law of the iterated logarithm:*

$$\begin{aligned}\liminf_{t\to\infty}\frac{1}{t}\Big(\frac{\log\log t}{t}\Big)^{\frac{p-2}{2}}\int_{-\infty}^{\infty} L^p(t,x)dx \\ = \frac{1}{2}(p-1)^{-\frac{p-1}{2}}(p+1)^{\frac{p+1}{2}}C_p^{-\frac{p+1}{2}} \quad a.s.\end{aligned} \tag{4.5.7}$$

where the constant C_p is given in the small ball probability computed in (4.5.5).

EXERCISE 4.5.12. *Let $W(t)$ be a 1-dimensional Brownian motion. Prove Chung's law of the iterated logarithm*

$$\liminf_{t\to\infty}\sqrt{\frac{\log\log t}{t}}\sup_{0\leq s\leq t}|W(s)| = \frac{\pi}{\sqrt{8}} \quad a.s.$$

CHAPTER 5

Intersections on lattices: weak convergence

A $\mathbb{Z}^d$-valued *random walk* $\{S(n)\}$ is defined by

$$S(n) = X_1 + \cdots + X_n \quad n = 1, 2, \cdots$$

where $\{X_n\}$ is an i.i.d. sequence of Z^d-valued random variables. Unless claiming otherwise, we define $S(0) = 0$. Throughout we assume that X_1 is symmetric and has finite second moment and we write Γ for the covariance matrix of X_1. In the special case $d = 1$, we like to use σ^2 for the variance of X_1. We assume that the smallest sub-group of $\mathbb{Z}^d$ that supports these random walks is $\mathbb{Z}^d$ itself.

Let $S_1(n), \cdots, S_p(n)$ be independent copies of $S(n)$. For any $\Delta \subset \mathbb{Z}^+$, set

$$S_j(\Delta) = \{S_j(k); \quad k \in \Delta\}.$$

Here is a list of the models of interest:

1. *Self-intersection local time*:

$$Q_n^{(p)} = \sum_{1 \le k_1 < \cdots < k_p \le n} 1_{\{S(k_1) = \cdots = S(k_p)\}}.$$

An important special case is when $p = 2$. In this case we write $Q_n = Q_n^{(2)}$.

2. *Range of the random walk*:

$$R_n = \#\{S[1, n]\} = \#\{S(1), \cdots, S(n)\}.$$

3. *Mutual intersection local time*:

$$I_n = \sum_{k_1, \cdots, k_p = 1}^{n} 1_{\{S_1(k_1) = \cdots = S_p(k_p)\}}.$$

4. *Intersection of the independent ranges*:

$$J_n = \#\{S_1[1, n] \cap \cdots \cap S_p[1, n]\}.$$

The goal of this chapter is to establish weak laws of convergence for these models.

5.1. Preliminary on random walks

Let

$$\varphi(\lambda) = \sum_{x \in \mathbb{Z}^d} e^{i\lambda \cdot x} \mathbb{P}\{X_1 = x\}$$

be the characteristic function of X_1. Then $\varphi(\lambda)$ is a real, *periodic function* with period 2π:

$$\varphi(\lambda + 2\pi x) = \varphi(\lambda) \qquad x \in \mathbb{Z}^d.$$

We now discuss another (unrelated) kind of (a)periodicity. The *period* of the random walk $\{S(n)\}$ is defined as the greatest common factor of the set

$$\big\{n \geq 1;\ \ \mathbb{P}\{S_n = 0\} > 0\big\}.$$

$\{S(n)\}$ is said to be *aperiodic* if its period is 1. It is known that $\{S(n)\}$ is aperiodic if and only if

$$|\varphi(\lambda)| < 1 \qquad \forall \lambda \in [-\pi, \pi]^d \setminus \{0\}.$$

Aperiodicity is a good property under which things usually become easier. On the other hand, aperiodicity is not satisfied even in some important special cases. A random walk on $\mathbb{Z}^d$ is called *simple random walk* if the distribution of its increment is given by

$$\mathbb{P}\{X_1 = \pm e_j\} = \frac{1}{2d} \qquad j = 1, \cdots d$$

where $\{e_1, \cdots, e_d\} \subset \mathbb{Z}^d$ is the canonical orthogonal basis in $\mathbb{Z}^d$. It is easy to see that the period of a simple random walk is 2.

LEMMA 5.1.1. *Assume that $\{S(n)\}$ is aperiodic.*

(1) *There is $\delta > 0$ such that*

$$|\varphi(\lambda)| \leq \exp\Big\{-\delta|\lambda|^2\Big\} \qquad \lambda \in [-\pi, \pi]^d. \tag{5.1.1}$$

(2)

$$\sup_{x \in \mathbb{Z}^d} \left| n^{d/2}\mathbb{P}\{S(n) = x\} - \frac{1}{(2\pi)^{d/2}\det(\Gamma)^{1/2}} \exp\Big\{-\frac{\langle x, \Gamma^{-1}x\rangle}{2n}\Big\}\right| \longrightarrow 0 \quad (n \to \infty). \tag{5.1.2}$$

In particular,

$$\mathbb{P}\{S_n = 0\} \sim \frac{1}{(2\pi)^{d/2}\sqrt{\det(\Gamma)}} n^{-d/2} \qquad (n \to \infty). \tag{5.1.3}$$

(3) *There is a constant $C > 0$ such that*

$$\Big|\mathbb{P}\{S(n) = x\} - \mathbb{P}\{S(n) = y\}\Big| \leq C n^{-\frac{d+1}{2}} |x - y| \tag{5.1.4}$$

for any $x, y \in \mathbb{Z}^d$ and any $n = 1, 2, \cdots$.

Proof. To prove the first part, notice that by Taylor expansion

$$0 \leq \varphi(\lambda) = \mathbb{E}e^{i\lambda\cdot X_1} \sim 1 - \frac{1}{2}\lambda\cdot\Gamma\lambda \qquad (|\lambda| \to 0).$$

Observe that there is a $c > 0$ such that $\lambda \cdot \Gamma\lambda \geq c|\lambda|^2$. This concludes that there are $\epsilon > 0$ and $c_1 > 0$ such that

$$|\varphi(\lambda)| \leq \exp\Big\{-c_1|\lambda|^2\Big\} \qquad |\lambda| \leq \epsilon.$$

Write $T_\epsilon = \{\lambda \in [-\pi, \pi]^d;\ \ |\lambda| \geq \epsilon\}$. By aperiodicity we have that

$$\theta \equiv \sup_{\lambda \in T_\epsilon} |\varphi(\lambda)| < 1.$$

Consequently, there is $c_2 > 0$ such that

$$\theta \le \exp\Big\{ - c_2 \sup_{\lambda \in [-\pi,\pi]^d} |\lambda|^2 \Big\}.$$

Letting $\delta = \min\{c_1, c_2\}$ give the conclusion in the first part.

The proof of (5.1.2), (5.1.3) and (5.1.4) relies on (5.1.1) and Fourier transformation. For similarity, we only prove (5.1.3). By Fourier inversion ((B.4), Appendix),

$$\begin{aligned}\mathbb{P}\{S_n = x\} &= \frac{1}{(2\pi)^d} \int_{[-\pi,\pi]^d} e^{-i\lambda\cdot x} \mathbb{E} e^{i\lambda\cdot S_n} d\lambda \\ &= \frac{1}{(2\pi)^d} \int_{[-\pi,\pi]^d} e^{-i\lambda\cdot x} \varphi(\lambda)^n d\lambda.\end{aligned}$$

Take $x = 0$. By variable substitution, we get

$$n^{d/2}\mathbb{P}\{S_n = 0\} = \frac{1}{(2\pi)^d} \int_{[-\sqrt{n}\pi,\sqrt{n}\pi]^d} \varphi(\lambda/\sqrt{n})^n d\lambda.$$

By the classic central limit theorem, we get

$$\lim_{n\to\infty} \varphi(\lambda/\sqrt{n})^n = \exp\Big\{ - \frac{1}{2}\lambda\cdot\Gamma\lambda\Big\}$$

and by (5.1.1),

$$1_{[-\sqrt{n}\pi,\sqrt{n}\pi]^d}(\lambda)\varphi(\lambda/\sqrt{n})^n \le \exp\{ - \delta|\lambda|^2\}.$$

Applying the dominated convergence theorem gives us

$$\lim_{n\to\infty} n^{d/2}\mathbb{P}\{S_n = 0\} = \frac{1}{(2\pi)^d} \int_{\mathbb{R}^d} \exp\Big\{ - \frac{1}{2}\langle\lambda,\Gamma\lambda\rangle\Big\} d\lambda = \frac{1}{(2\pi)^{d/2}\sqrt{\det(\Gamma)}}.$$

□

It is easy to see that the aperiodicity is necessary in Lemma 5.1.1. In some other situations, however, the result is first proved under aperiodicity and is then extended to the general case. Define, for example,

$$g_n = \sum_{k=0}^{n} \mathbb{P}\{S(n) = 0\} \qquad n = 0, 1, 2, \cdots. \tag{5.1.5}$$

We call g_n the *partial Green's function* of the random walk.

LEMMA 5.1.2. *When $n \to \infty$,*

$$g_n \sim \begin{cases} \dfrac{1}{\sqrt{2\pi}\sigma}\sqrt{n} & d = 1, \\[2ex] \dfrac{1}{2\pi\sqrt{\det(\Gamma)}} \log n & d = 2, \\[2ex] O(1) & d \ge 3. \end{cases} \tag{5.1.6}$$

Proof. With aperiodicity, this lemma holds immediately from (5.1.3). We now extend it to the general case. For similarity we only consider the case $d = 1$ in the following argument.

The approach we shall adopt is called *resolvent approximation* . Let $0 < \theta < 1$ be fixed and let $\{\delta_n\}$ be a Bernoulli sequence independent of $\{S(n)\}$ such that

$$\mathbb{P}\{\delta_n = 1\} = 1 - \mathbb{P}\{\delta_n = 0\} = \theta.$$

Write $\tau_0 = 0$ and

$$\tau_n = \min\{k > \tau_{n-1};\ \ \delta_k = 1\} \qquad n = 1, 2, \cdots .$$

Then $\{\tau_n - \tau_{n-1}\}_{n\ge 1}$ is an i.i.d. sequence with common (geometric) distribution

$$\mathbb{P}\{\tau_1 = k\} = \theta(1-\theta)^{k-1} \qquad k = 1, 2, \cdots .$$

Notice that $\{S(\tau_n)\}$ is a random walk whose increment has the distribution

$$\mathbb{P}\{S(\tau_1) = x\} = \sum_{k=1}^{\infty} \theta(1-\theta)^{k-1}\mathbb{P}\{S(k) = x\}.$$

In particular,

$$\mathrm{Var}\,\big(S(\tau_1)\big) = \mathbb{E}\tau_1 \cdot \sigma^2 = \theta^{-1}\sigma^2$$

and, $\mathbb{P}\{S(\tau_1) = 0\} > 0$ implies that $\{S(\tau_n)\}$ is aperiodic. $\{S(\tau_n)\}$ is called the *resolvent random walk* of $\{S(n)\}$ in literature.

Applying what we know under aperiodicity to $\{S(\tau_n)\}$,

$$\sum_{k=1}^{n} \mathbb{P}\{S(\tau_k) = 0\} \sim \frac{\sqrt{\theta}}{\sqrt{2\pi}\sigma}\sqrt{n} \qquad (n \to \infty).$$

On the other hand, by definition one can see that

$$\sum_{k=1}^{n} 1_{\{S(\tau_k)=0\}} = \sum_{k=1}^{\tau_n} \delta_k 1_{\{S(k)=0\}} = \theta \sum_{k=1}^{\tau_n} 1_{\{S(k)=0\}} + \sum_{k=1}^{\tau_n} (\delta_k - \theta) 1_{\{S(k)=0\}}.$$

Conditionally (on $\{S_k\}$), applying Doob's stopping rule for martingales to the second term on the right hand side, we get

$$\mathbb{E}\sum_{k=1}^{\tau_n} (\delta_k - \theta) 1_{\{S(k)=0\}} = 0.$$

By the classic law of large numbers, $\tau_n/n \to \theta^{-1}$ in probability.

$$\mathbb{E}\sum_{k=1}^{\tau_n} 1_{\{S(k)=0\}} = \mathbb{E}g_{\tau_n} \sim g_{[n\theta^{-1}]} \qquad (n \to \infty).$$

Summarizing our argument, we have

$$g_{[n\theta^{-1}]} \sim \frac{1}{\sqrt{2\pi}\sigma}\sqrt{n\theta} \qquad (n \to \infty),$$

which leads to the desired conclusion. □

Write

(5.1.7) $$T_x = \inf\{k \ge 1;\ \ S(k) = x\} \qquad x \in \mathbb{Z}^d.$$

Then for each x, T_x is a stopping time and is called the *hitting time* in literature. The equation (5.1.8) given below is called the *first entry formula*.

LEMMA 5.1.3. *For any* $x \in \mathbb{Z}^d$,

$$\mathbb{P}\{S(k) = x\} = \sum_{j=1}^{k} \mathbb{P}\{T_x = j\}\mathbb{P}\{S(k-j) = 0\}. \tag{5.1.8}$$

Consequently,

$$\sum_{k=1}^{n} \mathbb{P}\{S(k) = x\} = \sum_{k=1}^{n} \mathbb{P}\{T_x = k\} g_{n-k} \tag{5.1.9}$$

where $g_0 = 0$ *and* $\mathbb{P}\{S(0) = 0\} = 1$.

Proof. The first identity follows from the following computation:

$$\mathbb{P}\{S(k) = x\} = \mathbb{P}\{S(k) = x,\ T_x \le k\} = \sum_{j=1}^{k} \mathbb{P}\{S(k) = x,\ T_x = j\}$$

$$= \sum_{j=1}^{k} \mathbb{P}\{S(k) - S(j) = 0,\ T_x = j\} = \sum_{j=1}^{k} \mathbb{P}\{T_x = j\}\mathbb{P}\{S(k-j) = 0\}.$$

Summing up on the both side of (5.1.8) gives (5.1.9). □

THEOREM 5.1.4.

$$\mathbb{P}\{T_0 < \infty\} = \begin{cases} 1 & d = 1, 2, \\ \dfrac{\sum_{k=1}^{\infty} \mathbb{P}\{S(k) = 0\}}{1 + \sum_{k=1}^{\infty} \mathbb{P}\{S(k) = 0\}} & d \ge 3. \end{cases} \tag{5.1.10}$$

In addition,

$$\mathbb{P}\{S(k) = 0 \text{ infinitely often}\} = \begin{cases} 1 & d \le 2, \\ 0 & d \ge 3. \end{cases} \tag{5.1.11}$$

Proof. We first establish (5.1.10). Due to similarity, we only consider the case $d \ge 3$. Taking $x = 0$ in (5.1.8) and summing up both sides

$$\sum_{k=1}^{\infty} \mathbb{P}\{S(k) = 0\} = \sum_{k=1}^{\infty} \sum_{j=1}^{k} \mathbb{P}\{T_0 = j\}\mathbb{P}\{S(k-j) = 0\}$$

$$= \sum_{j=1}^{\infty} \mathbb{P}\{T_0 = j\} \sum_{k=j}^{\infty} \mathbb{P}\{S(k-j) = 0\}$$

$$= \mathbb{P}\{T_0 < \infty\}\Big\{1 + \sum_{k=1}^{\infty} \mathbb{P}\{S(k) = 0\}\Big\}.$$

We now prove (5.1.11). As $d \geq 3$, (5.1.11) follows from the Borel-Cantelli lemma and the fact that

$$\sum_{k=1}^{\infty} \mathbb{P}\{S(k) = 0\} < \infty.$$

Let $d \leq 2$. For each $m \geq 1$, let $T_0(m)$ be the time when $\{S(k)\}$ has mth hit of 0. We take $T_0(m) = \infty$ if the mth hit never comes. Under our notation, $T_0(1) = T_0$. Clearly,

$$\mathbb{P}\Big\{\sum_{k=1}^{\infty} 1_{\{S(k)=0\}} \geq m\Big\} = \mathbb{P}\big\{T_0(m) < \infty\big\}.$$

By (5.1.10), we get

$$\begin{aligned}
\mathbb{P}\{T_0(m) < \infty\} &= \mathbb{P}\big\{T_0(m-1) < \infty,\ T_0(m) < \infty\big\} \\
&= \sum_{1 \leq j < k < \infty} \mathbb{P}\{T_0(m-1) = j,\ T_0(m) = k\} \\
&= \sum_{1 \leq j < k < \infty} \mathbb{P}\{T_0(m-1) = j\}\mathbb{P}\{T_0 = k - j\} \\
&= \mathbb{P}\{T_0 < \infty\} \sum_{j=1}^{\infty} \mathbb{P}\{T_0(m-1) = j\} \\
&= \mathbb{P}\{T_0(m-1) < \infty\}.
\end{aligned}$$

Repeating the procedure gives us

$$\mathbb{P}\Big\{\sum_{k=1}^{\infty} 1_{\{S(k)=0\}} \geq m\Big\} = \mathbb{P}\{T_0(m) < \infty\} = \mathbb{P}\{T_0 < 1\} = 1$$

for $m = 1, 2, \cdots$. Consequently,

$$\mathbb{P}\Big\{\sum_{k=1}^{\infty} 1_{\{S(k)=0\}} = \infty\Big\} = \lim_{m \to \infty} \mathbb{P}\Big\{\sum_{k=1}^{\infty} 1_{\{S(k)=0\}} \geq m\Big\} = 1.$$

□

Theorem 5.1.4 shows that the random walk is *recurrent* when $d = 1, 2$ and *transient* when $d \geq 3$. A quantity we use very often later (mostly in the case $d \geq 3$) is *the probability of no return* defined as

(5.1.12) $$\gamma_S = \mathbb{P}\big\{S(k) \neq 0 \text{ for all } n \geq 1\big\}.$$

Another way to write (5.1.10) is

(5.1.13) $$\gamma_S = \begin{cases} 0 & d = 1, 2, \\ \Big(\sum_{k=0}^{\infty} \mathbb{P}\{S(k) = 0\}\Big)^{-1} & d \geq 3. \end{cases}$$

5.2. Intersection in 1-dimension

The local time of the random walk is given as

(5.2.1) $$l(n,x)=\sum_{k=1}^{n}1_{\{S(k)=x\}}\qquad x\in\mathbb{Z}^d.$$

In the rest of this section we let $d=1$ and $p\geq 2$. By a combinatorial argument,

(5.2.2) $$\sum_{x\in\mathbb{Z}}l^p(n,x)=\sum_{k_1,\cdots,k_p=1}^{n}1_{\{S(k_1)=\cdots=S(k_p)\}}$$
$$=n+\sum_{j=2}^{p}\binom{p-1}{j-1}j!\sum_{1\leq k_1<\cdots<k_j\leq n}1_{\{S(k_1)=\cdots=S(k_j)\}}.$$

We shall see that the typical order of the left hand side is $n^{\frac{p+1}{2}}$ as $n\to\infty$. Consequently,

(5.2.3) $$\sum_{x\in\mathbb{Z}}l^p(n,x)-p!\sum_{1\leq k_1<\cdots<k_p\leq n}1_{\{S(k_1)=\cdots=S(k_p)\}}$$
$$=O\big(n^{p/2}\big)\qquad(n\to\infty)$$

in the sense of the weak convergence. Therefore, the weak law for the p-multiple self-intersection is reduced to the study of the weak law for the p-square

$$\sum_{x\in\mathbb{Z}}l^p(n,x).$$

Define

$$\zeta_n(x)=\begin{cases}n^{-1/2}l(n,k) & x=\dfrac{k}{\sqrt{n}}\text{ for all }k\in\mathbb{Z},\\ \text{linear} & \text{else};\end{cases}$$

or,

$$\zeta_n(x)=\frac{1}{\sqrt{n}}l(n,k)+\Big(x-\frac{k}{\sqrt{n}}\Big)\Big[l(n,k+1)-l(n,k)\Big]$$

for $x\in\Big[\dfrac{k}{\sqrt{n}},\dfrac{k+1}{\sqrt{n}}\Big]$ and for all $k\in\mathbb{Z}$.

Recall that for any $t>0$, the local time $L(t,x)$ of a 1-dimensional Brownian motion $W(t)$ is defined as the density of the occupation measure

$$\mu_t(A)=\int_0^t 1_B\big(W(s)\big)ds.$$

Here we view $\zeta_n(\cdot)$ and $L(t,\cdot)$ as stochastic processes taking values in the Banach space $\mathcal{L}^p(\mathbb{R})$.

THEOREM 5.2.1. *When $n\to\infty$,*

(5.2.4) $$\zeta_n(\cdot)\xrightarrow{d}\sigma^{-1}L\big(1,\sigma^{-1}(\cdot)\big)$$

where $\sigma^2>0$ is the variance of the random walk $\{S(n)\}$.

The hard part of the proof is to show that the sequence $\{\zeta_n\}$ is uniformly tight. For this we need the following moment estimate.

LEMMA 5.2.2. *Assume that the random walk $S(n)$ is aperiodic. There is a $C>0$ such that*

$$\mathbb{E}\big|l(n,x)-l(n,y)\big|^{2l}\le(2l)!C^l n^{\frac{3l}{4}}|x-y|^{l/2} \tag{5.2.5}$$

for any integers $l,n\ge 1$ and for any $x,y\in\mathbb{Z}$.

Let $\zeta_n(\cdot)$ be defined as in Theorem 5.2.1. Then

$$\mathbb{E}\big|\zeta_n(x)-\zeta_n(y)\big|^{2l}\le(2l)!C^l|x-y|^{l/2} \tag{5.2.6}$$

for any integers $l,n\ge 1$ and for any $x,y\in\mathbb{R}$.

Proof. We first establish the bound

$$\sum_{k=0}^n\big|\mathbb{P}\{S(k)=x\}-\mathbb{P}\{S(k)=0\}\big|\le Cn^{1/4}|x|^{1/2} \tag{5.2.7}$$

where $C>0$ represents a constant independent of n and $x\in\mathbb{Z}$ (to simplify the notation, we allow C to be different from place to place in our proof). This is obvious as $x=0$. We assume $x\ne 0$. By Lemma 5.1.1, we get

$$\begin{aligned}&\sum_{k=0}^{[\sqrt{n}]|x|}\big|\mathbb{P}\{S(k)=0\}-\mathbb{P}\{S(k)=x\}\big|\le 1+C\sum_{k=1}^{[\sqrt{n}]|x|}k^{-1/2}\\&\le 1+Cn^{1/4}|x|^{1/2}\le Cn^{1/4}|x|^{1/2}\end{aligned}$$

where the last step follows from the fact that $|x|\ge 1$.

By (5.1.4), we get

$$\begin{aligned}&\sum_{k=[\sqrt{n}]|x|+1}^n\big|\mathbb{P}\{S(k)=0\}-\mathbb{P}\{S(k)=x\}\big|\\&\le C\sum_{k=[\sqrt{n}]|x|+1}^n\frac{|x|}{k^{3/2}}\le C\frac{|x|}{n^{1/4}\sqrt{|x|}}\sum_{k=1}^n k^{-1/2}\\&\le Cn^{1/4}|x|^{1/2}.\end{aligned}$$

We now come to the proof of (5.2.5). Write $\eta(z)=1_x(z)-1_y(z)$. Then

$$\mathbb{E}\big[l(n,x)-l(n,y)\big]^m=\mathbb{E}\Big[\sum_{k=1}^n\eta(S(k))\Big]^m=\sum_{l_1,\cdots,l_m=1}^n\mathbb{E}\prod_{k=1}^m\eta(S(l_k)).$$

Break the summation on the right hand side into $m!$ parts. In each part, the index $l_1,\cdots l_m$ is in non-decreasing order. For fixed $l_1\le\cdots\le l_m$,

$$\begin{aligned}\mathbb{E}\prod_{k=1}^m\eta(S(l_k))&=\mathbb{E}\Big\{\Big(\prod_{k=1}^{m-2}\eta(S(l_k))\Big)1_x(S(l_{m-1}))\eta(S(l_m)-S(l_{m-1})+x)\Big\}\\&-\mathbb{E}\Big\{\Big(\prod_{k=1}^{m-2}\eta(S(l_k))\Big)1_y(S(l_{m-1}))\eta(S(l_m)-S(l_{m-1})+y)\Big\}.\end{aligned}$$

By Markov property, the right hand side is equal to

$$\mathbb{E}\Big\{\Big(\prod_{k=1}^{m-2}\eta\big(S(l_k)\big)\Big)1_x\big(S(l_{m-1})\big)\Big\}\mathbb{E}\eta\big(S(l_m-l_{m-1})+x\big)$$
$$-\mathbb{E}\Big\{\Big(\prod_{k=1}^{m-2}\eta\big(S(l_k)\big)\Big)1_y\big(S(l_{m-1})\big)\Big\}\mathbb{E}\eta\big(S(l_m-l_{m-1})+y\big).$$

Notice that

$$-\mathbb{E}\eta\big(S(l_m-l_{m-1})+y\big)=\mathbb{E}\eta\big(S(l_m-l_{m-1})+x\big)$$
$$=\mathbb{P}\{S(l_m-l_{m-1})=0\}-\mathbb{P}\{S(l_m-l_{m-1})=x-y\}.$$

Hence,

$$\mathbb{E}\prod_{k=1}^{m}\eta\big(S(l_k)\big)=\mathbb{E}\eta\big(S(l_m-l_{m-1})+x\big)\mathbb{E}\Big\{\Big(\prod_{k=1}^{m-2}\eta\big(S(l_k)\big)\Big)\tilde{\eta}\big(S(l_{m-1})\big)\Big\}$$

where $\tilde{\eta}(z)=1_x(z)+1_y(z)$. Similarly,

$$\mathbb{E}\Big\{\Big(\prod_{k=1}^{m-2}\eta\big(S(l_k)\big)\Big)\tilde{\eta}\big(S(l_{m-1})\big)\Big\}$$
$$=\mathbb{E}\Big\{\Big(\prod_{k=1}^{m-3}\eta\big(S(l_k)\big)\Big)1_x\big(S(l_{m-2})\big)\Big\}\mathbb{E}\tilde{\eta}\big(S(l_{m-1}-l_{m-2})+x\big)$$
$$-\mathbb{E}\Big\{\Big(\prod_{k=1}^{m-3}\eta\big(S(l_k)\big)\Big)1_y\big(S(l_{m-2})\big)\Big\}\mathbb{E}\tilde{\eta}\big(S(l_{m-1}-l_{m-2})+y\big)$$
$$=\mathbb{E}\tilde{\eta}\big(S(l_{m-1}-l_{m-2})+x\big)\mathbb{E}\Big(\prod_{k=1}^{m-2}\eta\big(S(l_k)\big)\Big)$$

where the last step follows from the fact that

$$\mathbb{E}\tilde{\eta}\big(S(l_{m-1}-l_{m-2})+x\big)=\mathbb{E}\tilde{\eta}\big(S(l_{m-1}-l_{m-2})+y\big)$$
$$=\mathbb{P}\{S(l_m-l_{m-1})=0\}+\mathbb{P}\{S(l_m-l_{m-1})=x-y\}.$$

Summarizing our argument we have

$$\mathbb{E}\prod_{k=1}^{m}\eta\big(S(l_k)\big)$$
$$=\mathbb{E}\tilde{\eta}\big(S(l_{m-1}-l_{m-2})+x\big)\mathbb{E}\eta\big(S(l_m-l_{m-1})+x\big)\mathbb{E}\Big(\prod_{k=1}^{m-2}\eta\big(S(l_k)\big)\Big).$$

Therefore, when $m=2l$ is even,

$$\mathbb{E}\prod_{k=1}^{2l}\eta\big(S(l_k)\big)=\prod_{k=1}^{l}\Big(\mathbb{E}\tilde{\eta}\big(S(l_{2k-1}-l_{2k-2})+x\big)\mathbb{E}\eta\big(S(l_{2k}-l_{2k-1})+x\big)\Big).$$

Thus,

$$\begin{aligned}
&\mathbb{E}\big[l(n,x) - l(n,y)\big]^{2l} \\
&\le (2l)!\Big\{\sum_{k=0}^{n} \mathbb{E}\tilde{\eta}\big(S(k)+x\big)\Big\}^l \Big\{\sum_{k=0}^{n} \big|\mathbb{E}\eta\big(S(k)+x\big)\big|\Big\}^l \\
&\le (2l)!\Big\{\sum_{k=0}^{n} \Big(\mathbb{P}\{S(k)=0\} + \mathbb{P}\{S(k)=x-y\}\Big)\Big\}^l \\
&\times \Big\{\sum_{k=0}^{n} \big|\mathbb{P}\{S(k)=0\} - \mathbb{P}\{S(k)=x-y\}\big|\Big\}^l.
\end{aligned}$$

Thus, (5.2.5) follows from Lemma 5.1.1 and (5.2.7).

We come to prove (5.2.6). When $|x-y| \ge 1/\sqrt{n}$, by (5.2.5), we get

$$\mathbb{E}\Big[\zeta_n(x) - \zeta_n(y)\Big]^{2l} \le (2l)! C^l n^{-l/4} |[\sqrt{n}x] - [\sqrt{n}y]|^{l/2}.$$

Thus, (5.2.6) follows from the simple fact that

$$|[\sqrt{n}x] - [\sqrt{n}y]| \le 3\sqrt{n}|x-y|.$$

Assume that $|x-y| < 1/\sqrt{n}$. If x, y is in the same interval $\Big[\frac{k}{\sqrt{n}}, \frac{k+1}{\sqrt{n}}\Big]$, then we have

$$\zeta_n(x) - \zeta_n(y) = (y-x)\Big[l(n,k) - l(n,k+1)\Big].$$

By (5.2.5),

$$\mathbb{E}\Big[\zeta_n(x) - \zeta_n(y)\Big]^{2l} \le |y-x|^{2l}(2l)! C^l n^{3l/4} \le (2l)! C^l |y-x|^{l/2}.$$

If x, y is not in the same interval, we may assume that $x < y$. In this case, we must have k such that $x \in \Big[\frac{k-1}{\sqrt{n}}, \frac{k}{\sqrt{n}}\Big]$ and $y \in \Big[\frac{k}{\sqrt{n}}, \frac{k+1}{\sqrt{n}}\Big]$. By definition of ζ_n, we get

$$\begin{aligned}
\zeta_n(y) - \zeta_n(x) &= \Big(y - \frac{k}{\sqrt{n}}\Big)\Big[l(n,k+1) - l(n,k)\Big] \\
&+ \Big(\frac{k}{\sqrt{n}} - x\Big)\Big[l(n,k) - l(n,k-1)\Big].
\end{aligned}$$

Notice that

$$0 \le \Big(y - \frac{k}{\sqrt{n}}\Big),\ \Big(\frac{k}{\sqrt{n}} - x\Big) \le |x-y|.$$

By (5.2.5), we get

$$\mathbb{E}\Big[\zeta_n(x) - \zeta_n(y)\Big]^{2l} \le |y-x|^{2l}(2l)! C^l n^{3l/4} \le (2l)! C^l |y-x|^{l/2}.$$

□

Proof of Theorem 5.2.1. By the method of resolvent approximation introduced in Section 5.1, we may assume that the random walk $\{S(n)\}$ is aperiodic. Let $q > 1$

be the conjugate of p and let $f \in \mathcal{L}^q(\mathbb{R})$ be continuous:

$$\begin{aligned}
&\int_{\mathbb{R}} f(x)\zeta_n(x)dx \\
&= \sum_{k\in\mathbb{Z}} \int_{\frac{k}{\sqrt{n}}}^{\frac{k+1}{\sqrt{n}}} f(x)\Big\{\frac{1}{\sqrt{n}}l(n,k) + \Big(x - \frac{k}{\sqrt{n}}\Big)\Big[l(n,k+1) - l(n,k)\Big]\Big\}dx \\
&= \frac{1}{n}\sum_{k\in\mathbb{Z}} \int_k^{k+1} f\Big(\frac{x}{\sqrt{n}}\Big)\Big\{l(n,k) + \Big(\frac{x-k}{\sqrt{n}}\Big)\Big[l(n,k+1) - l(n,k)\Big]\Big\}dx \\
&= o(1) + \frac{1}{n}\sum_{k\in\mathbb{Z}} f\Big(\frac{k}{\sqrt{n}}\Big)l(n,k) = o(1) + \frac{1}{n}\sum_{l=1}^n f\Big(\frac{S(l)}{\sqrt{n}}\Big).
\end{aligned}$$

By invariance principle

$$\begin{aligned}
\int_{\mathbb{R}} f(x)\zeta_n(x) &\xrightarrow{d} \int_0^1 f\big(\sigma W(1)\big)dx \\
&= \int_{\mathbb{R}} f(\sigma x)L(1,x)dx = \sigma^{-1}\int_{\mathbb{R}} f(x)L(1,\sigma^{-1}x)dx.
\end{aligned}$$

It remains to show that $\zeta_n(\cdot)$ is uniformly tight in $\mathcal{L}^p(\mathbb{R})$. Given $\epsilon > 0$ there is a $a > 0$ such that

$$\mathbb{P}\Big\{\max_{k\le n}|S(k)| \ge a\sqrt{n}\Big\} < \epsilon/2.$$

Recall the classic fact that (Theorem 6.5, [**26**])

$$l(n,0)/\sqrt{n} \xrightarrow{d} \sigma^{-1}|U|$$

where U is a standard normal random variable. In particular, the real random sequence $\zeta_n(0) = l(n,0)/\sqrt{n}$ is uniformly tight in $\mathbb{R}$. By (5.2.6) and the Kolmogorov continuity theorem (Theorem D.7, Appendix), conditioning on the event $\Big\{\max_{k\le n}|S(k)| \le a\sqrt{n}\Big\}$, the sequence $\{\zeta_n(\cdot)\}$ vanishes outside $[-a,a]$ and is uniformly tight in $C[-a,a]$. Consequently, there is a compact set $K \subset C[-a,a]$ such that

$$\mathbb{P}\Big\{\zeta_n(\cdot) \notin K,\ \max_{k\le n}|S(k)| \le a\sqrt{n}\Big\} < \epsilon/2.$$

Notice that

$$\int_{\mathbb{R}} \zeta_n(x)dx = 1 \qquad n = 1,2,\cdots.$$

Hence, $\mathbb{P}\{\zeta_n(\cdot) \notin A\} < \epsilon$, where

$$\begin{aligned}
A \equiv \Big\{ & f \in C(\mathbb{R});\ \ f \ge 0,\ f(x) = 0 \text{ for } |x| > a, \text{ the restriction of} \\
& f(\cdot) \text{ on } [-a,a] \text{ is in } K, \text{ and } \int_{\mathbb{R}} f(x)dx = 1\Big\}
\end{aligned}$$

is relatively compact in $\mathcal{L}^p(\mathbb{R})$.

Take $\widetilde{K}$ as the closure of A in $\mathcal{L}^p(\mathbb{R})$. Then $\widetilde{K}$ is compact in $\mathcal{L}^p(\mathbb{R})$ and $\mathbb{P}\{\zeta_n(\cdot) \notin \widetilde{K}\} < \epsilon$. □

Here is how we apply Theorem 5.2.1: A straightforward computation leads to

$$\begin{aligned}&\int_{\mathbb{R}} \zeta_n^p(x)dx\\ &= \frac{1}{n^{\frac{p+1}{2}}}\sum_{k\in\mathbb{Z}}\int_k^{k+1}\Big\{l(n,k)+(x-k)\big[l(n,k+1)-l(n,k)\big]\Big\}dx\\ &= \frac{1}{n^{\frac{p+1}{2}}}\Big\{o(1)+\sum_{x\in\mathbb{Z}} l^p(n,x)\Big\}\end{aligned}$$

where the second step follows from (5.2.5). By Theorem 5.2.1, on the other hand, we get

$$\int_{\mathbb{R}} \zeta_n^p(x)dx \xrightarrow{d} \sigma^{-p}\int_{\mathbb{R}} L^p(1,\sigma^{-1}x)dx = \sigma^{-(p-1)}\int_{\mathbb{R}} L^p(1,x)dx.$$

Moreover, let $S_1(n),\cdots,S_p(n)$ be independent copies of $S(n)$ and consider the mutual intersection local time

$$I_n = \#\Big\{(k_1,\cdots,k_p)\in[1,n]^p;\ \ S_1(k_1)=\cdots=S_p(k_p)\Big\}.$$

Then we have

$$I_n = \sum_{x\in\mathbb{Z}}\prod_{j=1}^p l_j(n,x) = n^{\frac{p+1}{2}}\Big\{o(1)+\int_{\mathbb{R}}\prod_{j=1}^p \zeta_n^{(j)}(x)dx\Big\}$$

where $l_j(n,x)$ is the local time of $S_j(n)$ and the $\mathcal{L}^p(\mathbb{R})$-valued random variable $\zeta_n^{(j)}(\cdot)$ $(j=1,\cdots,p)$ is generated from $l_j(n,x)$ in the same way as we generate $\zeta_n(\cdot)$. By Theorem 5.2.1, we get

$$\Big(\zeta_n^{(1)}(\cdot),\cdots,\zeta_n^{(p)}(\cdot)\Big) \xrightarrow{d} \sigma^{-1}\Big(L_1\big(1,\sigma^{-1}(\cdot)\big),\cdots,L_1\big(1,\sigma^{-1}(\cdot)\big)\Big)$$

in the product space $\bigotimes_{j=1}^p \mathcal{L}^p(\mathbb{R})$, where $L_1(t,x),\cdots,L_p(t,x)$ are the local times of the independent 1-dimensional Brownian motions $W_1(t),\cdots,W_p(t)$, respectively. This leads to

$$\begin{aligned}\int_{\mathbb{R}}\prod_{j=1}^p \zeta_n^{(j)}(x)dx &\xrightarrow{d} \sigma^{-p}\int_{\mathbb{R}}\prod_{j=1}^p L_j(1,\sigma^{-1}x)dx\\ &= \sigma^{-(p-1)}\int_{\mathbb{R}}\prod_{j=1}^p L_j(1,x)dx = \sigma^{-(p-1)}\alpha\big([0,1]^p\big)\end{aligned}$$

where the last equality follows from (2.2.18).

We summarize our discussion in the following theorem.

THEOREM 5.2.3. *When* $d=1$,

$$n^{-\frac{p+1}{2}}\sum_{x\in\mathbb{Z}} l^p(n,x) \xrightarrow{d} \sigma^{-(p-1)}\int_{\mathbb{R}} L^p(1,x)dx, \tag{5.2.8}$$

$$n^{-\frac{p+1}{2}} I_n \xrightarrow{d} \sigma^{-(p-1)}\alpha\big([0,1]^p\big). \tag{5.2.9}$$

The weak law in (5.2.9), together with its multi-dimensional version, will be established in the next section by an alternative approach.

We now come to the weak laws for

$$R_n = \#\{S[1,n]\} \quad \textit{and} \quad J_n = \#\{S_1[1,n]\cap\cdots\cap S_p[1,n]\}.$$

Notice that in the special case of simple random walk in which

$$\mathbb{P}\{S(k)-S(k-1)=\pm 1\}=\frac{1}{2} \qquad k=1,2,\cdots$$

we have that

$$R_n = \max_{1\le k\le n} S(k) - \min_{1\le k\le n} S(k).$$

By the classic invariance principle we have that

$$R_n/\sqrt{n} \xrightarrow{d} \max_{0\le t\le 1} W(t) - \min_{0\le t\le 1} W(t).$$

Things get a little complicated in the general case, because of the possible "holes" of unvisited sites in the integer interval $[\min_{1\le k\le n} S(k), \max_{1\le k\le n} S(k)]$. The following theorem indicates that these holes are asymptotically insignificant. We include it without proof, for which the reader is referred to [**124**].

THEOREM 5.2.4. *When* $d=1$,

$$R_n/\sqrt{n} \xrightarrow{d} \sigma\Big(\max_{0\le t\le 1} W(t) - \min_{0\le t\le 1} W(t)\Big), \tag{5.2.10}$$

$$J_n/\sqrt{n} \xrightarrow{d} \sigma\Big(\min_{1\le j\le p}\max_{0\le t\le 1} W_j(t) - \max_{0\le j\le p}\min_{0\le t\le 1} W_j(t)\Big). \tag{5.2.11}$$

Compared to the weak laws in the case $d\ge 2$ (Theorem 5.3.4, Theorem 5.4.3 below), the behavior of J_n and R_n in $d=1$ follows a completely different pattern. The asymptotic pattern given in Theorem 5.2.4 is not the main focus of this book. In addition to [**124**], the interested reader is referred to [**30**] for the large and small deviations for R_n and J_n in the case $d=1$.

5.3. Mutual intersection in sub-critical dimensions

The objectives of this section are the weak laws for the mutual intersection local time

$$I_n = \#\Big\{(k_1,\cdots,k_p)\in[1,n]^p;\ \ S_1(k_1)=\cdots=S_p(k_p)\Big\}$$

and for the intersection of independent ranges

$$J_n = \#\{S_1[1,n]\cap\cdots\cap S_p[1,n]\}$$

generated by the independent, identically distributed symmetric random walks $S_1(n),\cdots,S_p(n)$ on $\mathbb{Z}^d$ with $d\ge 2$. It has been known for more than fifty years (Dvoretsky-Erdös [**65**] and Dvoretsky-Erdös-Kakutani [**66**], [**67**]) that the trajectories of $S_1(n),\cdots,S_p(n)$ intersect infinitely often, i.e.,

$$\#\Big\{S_1[1,\infty)\cap\cdots\cap S_p[1,\infty)\Big\}=\infty \qquad a.s.$$

if and only if $p(d-2)\le d$. Naturally, all combinations of d and p are classified into *sub-critical dimensions*, *critical dimensions* and *super-critical dimensions*, defined by the inequalities $p(d-2) < d$, $p(d-2) = d$ and $p(d-2) > d$, respectively. Our discussion in this section is focused on the sub-critical dimensions $p(d-2) < d$.

The difference also appears between cases $d = 1$ and $d \geq 2$. Indeed, the limiting random variable in Theorem 5.2.1 is not well defined in the case $d \geq 2$. As consequence, the argument for (5.2.9) is no longer valid in the multi-dimensions.

Fix $\epsilon > 0$ and define

$$I_{n,\epsilon} = C_{n,\epsilon}^{-1} \sum_{x\in\mathbb{Z}^d} \prod_{j=1}^p \sum_{k=1}^n p_\epsilon\Big(\frac{\Gamma^{-1/2}\big(S_j(k) - x\big)}{\sqrt{n}}\Big) \tag{5.3.1}$$

where the $d \times d$ positive matrix $\Gamma^{-1/2}$ solves the matrix equation $A^2 = \Gamma^{-1}$ (the role of $\Gamma^{-1/2}$ is to canonicalize the random walks), and where

$$C_{n,\epsilon} = \Big(\sum_{x\in\mathbb{Z}^d} p_\epsilon\Big(\frac{\Gamma^{-1/2}x}{\sqrt{n}}\Big)\Big)^p \sim n^{\frac{dp}{2}} \det(\Gamma)^{p/2} \qquad (n \to \infty). \tag{5.3.2}$$

Applying the invariance principle to the continuous function

$$\widetilde{\Psi}(f_1, \cdots, f_p) = \int_{\mathbb{R}^d} \Big[\prod_{j=1}^p \int_0^1 p_\epsilon\Big(\Gamma^{-1/2} f_j(s) - x\Big)\Big] dx$$

on the functional space $\bigotimes_{j=1}^p C\big\{[0,1], \mathbb{R}^d\big\}$ we obtain that

$$\begin{aligned} & n^{-p} \int_{\mathbb{R}^d} \Big[\prod_{j=1}^p \sum_{k=1}^n p_\epsilon\Big(\frac{\Gamma^{-1/2} S_j(k)}{\sqrt{n}} - x\Big)\Big] dx \\ & \xrightarrow{d} \int_{\mathbb{R}^d} \Big[\prod_{j=1}^p \int_0^1 p_\epsilon\big(W_j(s) - x\big) ds\Big] dx \\ & = \alpha_\epsilon\big([0,1]^p\big) \end{aligned}$$

as $n \to \infty$. Here we recall that the process $\alpha_\epsilon\big([0,t]^p\big)$ is introduced in Section 2.2.

By continuity of $p_\epsilon(\cdot)$, on the other hand, we have

$$\begin{aligned} & \int_{\mathbb{R}^d} \Big[\prod_{j=1}^p \sum_{k=1}^n p_\epsilon\Big(\frac{\Gamma^{-1/2} S_j(k)}{\sqrt{n}} - x\Big)\Big] dx \\ & = n^{-d/2} \det(\Gamma)^{-1} \int_{\mathbb{R}^d} \Big[\prod_{j=1}^p \sum_{k=1}^n p_\epsilon\Big(\frac{\Gamma^{-1/2}\big(S_j(k) - x\big)}{\sqrt{n}}\Big)\Big] dx \\ & = n^{-d/2} \det(\Gamma)^{-1} \Big\{o(1) + \sum_{x\in\mathbb{Z}^d} \prod_{j=1}^p \sum_{k=1}^n p_\epsilon\Big(\frac{\Gamma^{-1/2}\big(S_j(k) - x\big)}{\sqrt{n}}\Big)\Big\}. \end{aligned}$$

Therefore, for each $\epsilon > 0$,

$$n^{-\frac{2p-d(p-1)}{2}} I_{n,\epsilon} \xrightarrow{d} \det(\Gamma)^{-\frac{p-1}{2}} \alpha_\epsilon\big([0,1]^p\big), \tag{5.3.3}$$

as $n \to \infty$.

The key step toward establishing the weak law for I_n is the following lemma.

LEMMA 5.3.1. *Under $p(d-2) < d$, we get*

$$\lim_{\epsilon\to 0^+}\limsup_{n\to\infty} n^{-\big(2p-d(p-1)\big)}\mathbb{E}\big[I_n - I_{n,\epsilon}\big]^2 = 0.$$

Proof. By resolvent approximation (see an exercise in "Notes and comments" section), we may assume that the random walks are aperiodic.

The main idea is Fourier transformation. Write

$$I_n(x_1,\cdots,x_{p-1}) = \sum_{k_1,\cdots,k_p=1}^n \prod_{j=1}^{p-1} 1_{\{S_j(k_j)-S_p(k_p)=x_j\}} \qquad x_1,\cdots,x_{p-1}\in\mathbb{Z}^d.$$

We have that

$$\begin{aligned}(5.3.4)\qquad &\widehat{I}_n(\lambda_1,\cdots,\lambda_{p-1})\\ &\equiv \sum_{x_1,\cdots,x_{p-1}\in\mathbb{Z}^d} I_n(x_1,\cdots,x_{p-1})\exp\Big\{i\sum_{j=1}^{p-1}\lambda_j\cdot x_j\Big\}\\ &= \sum_{k_1,\cdots,k_p=1}^n \prod_{j=1}^{p-1}\exp\big\{i\lambda_j\cdot\big(S_j(k_j)-S_p(p)\big)\big\}\\ &= \prod_{j=1}^p\sum_{k=1}^n e^{i\lambda_j\cdot S_j(k)}.\end{aligned}$$

Here, and elsewhere in the proof, we adopt the convention that

$$\lambda_p = -(\lambda_1+\cdots+\lambda_{p-1}).$$

Taking Fourier inversion ((B.4), Appendix),

$$(5.3.5)\qquad I_n = I_n(0) = \frac{1}{(2\pi)^{d(p-1)}}\int_{([-\pi,\pi]^d)^{p-1}}\prod_{j=1}^p\sum_{k=1}^n e^{i\lambda_j\cdot S_j(k)}d\lambda_1\cdots d\lambda_{p-1}.$$

Write

$$(5.3.6)\qquad g_{n,\epsilon}(x_1,\cdots,x_{p-1}) = C_{n,\epsilon}^{-1}\sum_{x\in\mathbb{Z}^d} p_\epsilon\Big(\frac{\Gamma^{-1/2}x}{\sqrt{n}}\Big)\prod_{j=1}^{p-1}p_\epsilon\Big(\frac{\Gamma^{-1/2}(x_j-x)}{\sqrt{n}}\Big)$$

where $x_1,\cdots,x_{p-1}\in\mathbb{Z}^d$ and $C_{n,\epsilon}>0$ is given in (5.3.2). One can see that $g_{n,\epsilon}$ is a probability density on $\mathbb{Z}^{d(p-1)}$ and that

$$\begin{aligned}I_{n,\epsilon} &= \sum_{x_1,\cdots,x_p\in\mathbb{Z}^d} g_{n,\epsilon}(x_1,\cdots,x_{p-1})I_n(x_1,\cdots,x_{p-1})\\ &= \frac{1}{(2\pi)^{d(p-1)}}\int_{([-\pi,\pi]^d)^{p-1}}\widehat{g}_{n,\epsilon}(\lambda_1,\cdots,\lambda_{p-1})\prod_{j=1}^p\sum_{k=1}^n e^{i\lambda_j\cdot S_j(k)}d\lambda_1\cdots d\lambda_{p-1}\end{aligned}$$

where the second step follows from (5.3.4) and the Parseval identity ((B.3), Appendix).

Together with (5.3.5), this leads to

$$\begin{aligned}&\mathbb{E}\big[I_n - I_{n,\epsilon}\big]^2\\&= \frac{1}{(2\pi)^{2d(p-1)}}\int_{([-\pi,\pi]^d)^{2(p-1)}} d\lambda_1 d\gamma_1\cdots d\lambda_{p-1}d\gamma_{p-1}\\&\quad\times F_{n,\epsilon}(\lambda_1,\gamma_1,\cdots,\lambda_{p-1},\gamma_{p-1})\prod_{j=1}^p H_n(\lambda_j,\gamma_j)\end{aligned}$$

where

$$\begin{aligned}&F_{n,\epsilon}(\lambda_1,\gamma_1,\cdots,\lambda_{p-1},\gamma_{p-1})\\&= \Big[1-\widehat{g}_{n,\epsilon}(\lambda_1,\cdots,\lambda_{p-1})\Big]\Big[1-\widehat{g}_{n,\epsilon}(\gamma_1,\cdots,\gamma_{p-1})\Big],\end{aligned}$$

$$H_n(\lambda,\gamma) = \sum_{j,k=1}^n \mathbb{E}\exp\Big\{i\big(\lambda\cdot S(j)+\gamma\cdot S(k)\big)\Big\}.$$

From the fact that $\widehat{g}_{n,\epsilon}$ is real with $|\widehat{g}_{n,\epsilon}|\le 1$, we have

$$0\le F_{n,\epsilon}(\lambda_1,\gamma_1,\cdots,\lambda_{p-1},\gamma_{p-1})\le 4.$$

Write

$$\prod_{j=1}^p \big|H_n(\lambda_j,\gamma_j)\big| = \prod_{j=1}^p\prod_{\substack{1\le k\le p\\ k\ne j}}\big|H_n(\lambda_k,\gamma_k)\big|^{\frac{1}{p-1}}.$$

By Hölder inequality, we have

$$\begin{aligned}&\mathbb{E}\big[I_n - I_{n,\epsilon}\big]^2\\&\le \frac{1}{(2\pi)^{2d(p-1)}}\prod_{j=1}^p\bigg\{\int_{([-\pi,\pi]^d)^{2(p-1)}} d\lambda_1 d\gamma_1\cdots d\lambda_{p-1}d\gamma_{p-1}\\&\qquad\times F_{n,\epsilon}(\lambda_1,\gamma_1,\cdots,\lambda_{p-1},\gamma_{p-1})\prod_{\substack{1\le k\le p\\ k\ne j}}\big|H_n(\lambda_k,\gamma_k)\big|^q\bigg\}^{1/p}\end{aligned}$$

where $q=p(p-1)^{-1}$ is the conjugate of p.

The factor with the index $j=p$ is nice enough so we leave it alone at this step. As for $1\le j\le p-1$,

$$\begin{aligned}&\int_{([-\pi,\pi]^d)^{2(p-1)}} d\lambda_1 d\gamma_1\cdots d\lambda_{p-1}d\gamma_{p-1}\\&\qquad\times F_{n,\epsilon}(\lambda_1,\gamma_1,\cdots,\lambda_{p-1},\gamma_{p-1})\prod_{\substack{1\le k\le p\\ k\ne j}}\big|H_n(\lambda_k,\gamma_k)\big|^q\\&\le 4\int_{([-\pi,\pi]^d)^{2(p-1)}} d\lambda_1 d\gamma_1\cdots d\lambda_{p-1}d\gamma_{p-1}\prod_{\substack{1\le k\le p\\ k\ne j}}\big|H_n(\lambda_k,\gamma_k)\big|^q\\&= 4\bigg(\int_{([-\pi,\pi]^d)^2}\big|H_n(\lambda,\gamma)\big|^q d\lambda d\gamma\bigg)^{p-1}\end{aligned}$$

where the last step follows from the translation invariance

$$\begin{aligned}
&\int_{([-\pi,\pi]^d)^2} d\lambda_j d\gamma_j \big|H_n(\lambda_p,\gamma_p)\big|^q \\
&= \int_{([-\pi,\pi]^d)^2} d\lambda_j d\gamma_j \big|H_n(\lambda_1+\cdots+\lambda_{p-1},\gamma_1+\cdots+\gamma_{p-1})\big|^q \\
&= \int_{([-\pi,\pi]^d)^2} d\lambda_j d\gamma_j \big|H_n(\lambda_j,\gamma_j)\big|^q
\end{aligned}$$

and Fubini theorem.

Summarizing our argument, we have

$$\begin{aligned}
(5.3.7)\qquad & \mathbb{E}\big[I_n - I_{n,\epsilon}\big]^2 \\
&\le \frac{4^{p-1}}{(2\pi)^{2d(p-1)}}\bigg(\int_{([-\pi,\pi]^d)^2} |H_n(\lambda,\gamma)|^q d\lambda d\gamma\bigg)^{\frac{(p-1)^2}{p}} \\
&\times \bigg\{\int_{([-\pi,\pi]^d)^{2(p-1)}} d\lambda_1 d\gamma_1\cdots d\lambda_{p-1}d\gamma_{p-1} \\
&\qquad \times F_{n,\epsilon}(\lambda_1,\gamma_1,\cdots,\lambda_{p-1},\gamma_{p-1})\prod_{j=1}^{p-1}\big|H_n(\lambda_j,\gamma_j)\big|^q\bigg\}^{1/p}.
\end{aligned}$$

We now bound $H_n(\lambda,\gamma)$. Let $\varphi(\lambda)$ be the characteristic function of the i.i.d. increments of the random walks. For $j\le k$,

$$\begin{aligned}
&\mathbb{E}\exp\Big\{i\lambda\cdot S(j) + i\gamma\cdot S(k)\Big\} \\
&= \mathbb{E}\exp\Big\{i(\lambda-\gamma)\cdot S(j) + i\gamma\cdot\big(S(k)-S(j)\big)\Big\} \\
&= \varphi(\lambda-\gamma)^j\varphi(\gamma)^{k-j}.
\end{aligned}$$

Therefore,

$$\begin{aligned}
(5.3.8)\qquad \big|H_n(\lambda,\gamma)\big| &\le \bigg(\sum_{k=0}^n|\varphi(\lambda-\gamma)|^k\bigg)\bigg(\sum_{l=0}^n|\varphi(\gamma)|^l\bigg) \\
&+ \bigg(\sum_{l=0}^n|\varphi(\gamma-\lambda)|^l\bigg)\bigg(\sum_{k=0}^n|\varphi(\lambda)|^l\bigg) \\
&= \xi_1(\lambda,\gamma)+\xi_2(\lambda,\gamma)\qquad \text{(say)}.
\end{aligned}$$

Consequently, by triangular inequality and translation invariance, we get

$$(5.3.9)\qquad \int_{([-\pi,\pi]^d)^2}|H_n(\lambda,\gamma)|^q d\lambda d\gamma \le 2^q\bigg\{\int_{[-\pi,\pi]^d}\bigg(\sum_{l=0}^n|\varphi(\lambda)|^l\bigg)^q d\lambda\bigg\}^2.$$

To continue, we claim that there is a constant $C>0$ such that for any $n\ge 1$,

$$(5.3.10)\qquad \sum_{k=0}^n\Big|\varphi\Big(\frac{\lambda}{\sqrt{n}}\Big)\Big|^k \le Cn(1+|\lambda|^2)^{-1}\qquad \lambda\in[-\sqrt{n}\pi,\sqrt{n}\pi]^d.$$

Indeed, For $|\lambda| \le 1$, we use the obvious bound

$$\sum_{k=0}^{n}\Big|\varphi\Big(\frac{\lambda}{\sqrt{n}}\Big)\Big|^{k} \le n \le 2n(1+|\lambda|^2)^{-1}.$$

For $|\lambda| \ge 1$, by (5.1.1) in Lemma 5.1.1,[1]

$$\sum_{k=0}^{n}\Big|\varphi\Big(\frac{\lambda}{\sqrt{n}}\Big)\Big|^{k} \le \Big\{1-\Big|\varphi\Big(\frac{\lambda}{\sqrt{n}}\Big)\Big|\Big\}^{-1} \le \Big(1-\exp\Big\{-\frac{\delta|\lambda|^2}{n}\Big\}\Big)^{-1}$$
$$\le Cn|\lambda|^{-2} \le 2Cn(1+|\lambda|^2)^{-1}.$$

Apply the variable substitution $\lambda \mapsto \lambda/\sqrt{n}$ to the integral on the right hand side of (5.3.9). By (5.3.10) and by the fact that $(1+|\lambda|)^{-1} \in \mathcal{L}^q(\mathbb{R}^d)$,

$$\int_{([-\pi,\pi]^d)^2} |H_n(\lambda,\gamma)|^q d\lambda d\gamma = O\Big(n^{2q-d}\Big) \quad (n\to\infty). \tag{5.3.11}$$

In view of (5.3.7), it remains to show that

$$\lim_{\epsilon\to\infty}\limsup_{n\to\infty} n^{-(2p-d(p-1))}\int_{([-\pi,\pi]^d)^{2(p-1)}} d\lambda_1 d\gamma_1\cdots d\lambda_{p-1}d\gamma_{p-1}$$
$$\times F_{n,\epsilon}(\lambda_1,\gamma_1,\cdots,\lambda_{p-1},\gamma_{p-1})\prod_{k=1}^{p-1}|H_n(\lambda_k,\gamma_k)|^q = 0.$$

From (5.3.8),

$$\prod_{k=1}^{p-1}|H_n(\lambda_k,\gamma_k)|^q \le 2^q \sum_{j_1,\cdots,j_k=1}^{2}\prod_{k=1}^{p-1}\xi_{j_k}^q(\lambda_k,\gamma_k).$$

Hence, it suffices to show that for any $(j_1,\cdots,j_{p-1}) \in \{1,2\}^{p-1}$,

$$\lim_{\epsilon\to\infty}\limsup_{n\to\infty} n^{-(2p-d(p-1))}\int_{([-\pi,\pi]^d)^{2(p-1)}} d\lambda_1 d\gamma_1\cdots d\lambda_{p-1}d\gamma_{p-1}$$
$$\times F_{n,\epsilon}(\lambda_1,\gamma_1,\cdots,\lambda_{p-1},\gamma_{p-1})\prod_{j=1}^{p-1}\xi_{j_k}^q(\lambda_k,\gamma_k) = 0. \tag{5.3.12}$$

Indeed, by the periodicity of the integrand, and therefore by translation invariance, we have

$$\int_{[-\pi,\pi]^{2d(p-1)}} d\lambda_1 d\gamma_1\cdots d\lambda_{p-1}d\gamma_{p-1}$$
$$\times F_{n,\epsilon}(\lambda_1,\gamma_1,\cdots,\lambda_{p-1},\gamma_{p-1})\prod_{k=1}^{p-1}\xi_{j_k}(\lambda_k,\gamma_k)$$
$$= \int_{[-\pi,\pi]^{2d(p-1)}} d\lambda_1 d\gamma_1\cdots d\lambda_{p-1}d\gamma_{p-1}F_{n,\epsilon}(\lambda_1^*,\gamma_1^*,\cdots,\lambda_{p-1}^*\gamma_{p-1}^*)$$
$$\times\prod_{k=1}^{p-1}\Big(\sum_{l=0}^{n}|\varphi(\lambda_k)|^l\Big)\Big(\sum_{l=0}^{n}|\varphi(\gamma_k)|^l\Big)$$

---

[1] This is the only place in the entire proof where we use aperiodicity

where for each $1 \le k \le p-1$, λ_k^* and γ_k^* are linear combinations of λ_k and γ_k. By the maps $\lambda_k \mapsto \lambda_k/\sqrt{n}$ and $\gamma_k \mapsto \gamma_k/\sqrt{n}$ $(1 \le k \le p-1)$, and by (5.3.10), the right hand side is bounded by a constant multiple of

$$n^{2p-d(p-1)} \int_{\mathbb{R}^{2d(p-1)}} d\lambda_1 d\gamma_1 \cdots d\lambda_{p-1} d\gamma_{p-1}$$
$$\times F_{n,\epsilon}\Big(\frac{\lambda_1^*}{\sqrt{n}}, \frac{\gamma_1^*}{\sqrt{n}}, \cdots, \frac{\lambda_{p-1}^*}{\sqrt{n}}, \frac{\gamma_{p-1}^*}{\sqrt{n}}\Big) \prod_{k=1}^{p-1} (1+|\lambda_k|)^{-q}(1+|\gamma_k|)^{-q}.$$

Using the fact that

$$\widehat{h}_{n,\epsilon}\Big(\frac{\lambda_1}{\sqrt{n}}, \cdots, \frac{\lambda_{p-1}}{\sqrt{n}}\Big) \longrightarrow \int_{\mathbb{R}^d} dx p_\epsilon(x) \prod_{j=1}^{p-1} \int_{\mathbb{R}^d} e^{-i\lambda_j \cdot y} p_t(x,y)_\epsilon(y-x) dy$$
$$= \exp\Big\{-\frac{\epsilon}{2}\Big(\sum_{j=1}^{p-1} |\lambda_j|^2 + \Big|\sum_{j=1}^{p-1} \lambda_j\Big|^2\Big)\Big\} \qquad (n \to \infty)$$

and using dominated convergence theorem, we get

$$\lim_{n\to\infty} \int_{\mathbb{R}^{2d(p-1)}} d\lambda_1 d\gamma_1 \cdots d\lambda_{p-1} d\gamma_{p-1}$$
$$\times F_{n,\epsilon}\Big(\frac{\lambda_1^*}{\sqrt{n}}, \frac{\gamma_1^*}{\sqrt{n}}, \cdots, \frac{\lambda_{p-1}^*}{\sqrt{n}}, \frac{\gamma_{p-1}^*}{\sqrt{n}}\Big) \prod_{k=1}^{p-1} (1+|\lambda_k|)^{-q}(1+|\gamma_k|)^{-q}$$
$$= \int_{\mathbb{R}^{2d(p-1)}} d\lambda_1 d\gamma_1 \cdots d\lambda_{p-1} d\gamma_{p-1} \Big[1 - \exp\Big\{-\frac{\epsilon}{2}\Big(\sum_{j=1}^{p-1} |\lambda_j^*|^2 + \Big|\sum_{j=1}^{p-1} \lambda_j^*\Big|^2\Big)\Big\}\Big]$$
$$\times \Big[1 - \exp\Big\{-\frac{\epsilon}{2}\Big(\sum_{j=1}^{p-1} |\gamma_j^*|^2 + \Big|\sum_{j=1}^{p-1} \gamma_j^*\Big|^2\Big)\Big\}\Big] \prod_{k=1}^{p-1} (1+|\lambda_k|)^{-q}(1+|\gamma_k|)^{-q}.$$

Letting $\epsilon \to 0^+$ on both sides completes the proof of (5.3.12). □

THEOREM 5.3.2. *Under $p(d-2) < d$, we get*

$$n^{-\frac{2p-d(p-1)}{2}} I_n \xrightarrow{d} \det(\Gamma)^{-\frac{p-1}{2}} \alpha\big([0,1]^p\big), \tag{5.3.13}$$

$$\lim_{n\to\infty} n^{-(2p-d(p-1))} \mathbb{E} I_n^2 = \det(\Gamma)^{-(p-1)} \mathbb{E}\Big[\alpha\big([0,1]^p\big)^2\Big]. \tag{5.3.14}$$

Proof. According to Theorem 2.2.3, $\alpha_\epsilon\big([0,1]^p\big)$ converges to $\alpha\big([0,1]^p\big)$ in all positive moments. Therefore, (5.3.13) and (5.3.14) are direct consequences of (5.3.3) and Lemma 5.3.1. □

We now establish the weak laws for the range intersection J_n. Our approach is to approximate J_n by a multiple of I_n.

LEMMA 5.3.3. *When $d = 2$ and $p \ge 2$,*

$$\lim_{n\to\infty} \frac{1}{n^2} \mathbb{E}\Big\{I_n - \Big(\frac{\log n}{2\pi\sqrt{\det(\Gamma)}}\Big)^p J_n\Big\}^2 = 0. \tag{5.3.15}$$

When $d = 3$ and $p = 2$,

$$\lim_{n\to\infty}\frac{1}{n}\mathbb{E}\Big\{I_n - \gamma_S^{-2}J_n\Big\}^2 = 0 \tag{5.3.16}$$

where γ_S is the probability of no return defined in (5.1.12).

Proof. Due to similarity, we only prove (5.3.15). In view of (5.3.14), it is sufficient to show that

$$\limsup_{n\to\infty}\Big(\frac{(\log n)^p}{n}\Big)^2\mathbb{E}J_n^2 \le (2\pi)^{2p}\det(\Gamma)\mathbb{E}\Big[\alpha\big([0,1]^p\big)^2\Big], \tag{5.3.17}$$

$$\liminf_{n\to\infty}\frac{(\log n)^p}{n^2}\mathbb{E}\big(I_nJ_n\big) \ge (2\pi)^p\det(\Gamma)^{-\frac{p-2}{2}}\mathbb{E}\Big[\alpha\big([0,1]^p\big)^2\Big]. \tag{5.3.18}$$

Notice that

$$I_n = \sum_{x\in\mathbb{Z}^2}\prod_{j=1}^p l_j(n,x), \qquad J_n = \sum_{x\in\mathbb{Z}^2}\prod_{j=1}^p 1_{\{T_x^{(j)}\le n\}}. \tag{5.3.19}$$

where $l_j(n,x)$ and $T_x^{(j)}$ are the local time and the hitting time generated by $\{S_j(k)\}_{k\ge1}$ in the way same as in (5.2.1) and (5.1.7), respectively. Thus,

$$\begin{aligned}\mathbb{E}J_n^2 &= \sum_{x,y\in\mathbb{Z}^2}\Big[\mathbb{P}\{T_x\le n,\ T_y\le n\}\Big]^p\\ &= \sum_{x\in\mathbb{Z}^2}\Big[\mathbb{P}\{T_x\le n\}\Big]^p + \sum_{x\ne y}\Big[\mathbb{P}\{T_x\le n,\ T_y\le n\}\Big]^p.\end{aligned}$$

Observe that

$$\sum_{x\in\mathbb{Z}^2}\Big[\mathbb{P}\{T_x\le n\}\Big]^p = \mathbb{E}J_n \le \mathbb{E}I_n.$$

In view of (5.3.14), this quantity is negligible. Thus, to prove (5.3.17) it is sufficient to show that

$$\begin{aligned}&\limsup_{n\to\infty}\Big(\frac{(\log n)^p}{n}\Big)^2\sum_{x\ne y}\Big[\mathbb{P}\{T_x\le n,\ T_y\le n\}\Big]^p\\ &\le (2\pi)^{2p}\det(\Gamma)\mathbb{E}\Big[\alpha\big([0,1]^p\big)^2\Big].\end{aligned} \tag{5.3.20}$$

For $x\ne y$, write

$$\begin{aligned}&\mathbb{P}\{T_x\le n,\ T_y\le n\}\\ &= \sum_{1\le j<k\le n}\mathbb{P}\{T_x=j,\ T_y=k\} + \sum_{1\le k<j\le n}\mathbb{P}\{T_x=j,\ T_y=k\}.\end{aligned}$$

Notice that for $j<k$,

$$\begin{aligned}&\mathbb{P}\{T_x=j,\ T_y=k\}\\ &\le \mathbb{P}\Big\{T_x=j, S(j+1)\ne y,\cdots,S(k-1)\ne y, S(k)=y\Big\}\\ &= \mathbb{P}\{T_x=j\}\mathbb{P}\{T_{y-x}=k-j\}.\end{aligned}$$

Therefore,

$$\sum_{1\le j<k\le n} \mathbb{P}\{T_x = j,\ T_y = k\} \le \sum_{j=1}^{n-1} \mathbb{P}\{T_x = j\}\mathbb{P}\{T_{y-x} \le n - j\}.$$

On the other hand, recall ((5.1.5) and (5.1.6)) that

$$g_n \equiv \sum_{k=0}^{n} \mathbb{P}\{S(k) = 0\} \sim \frac{1}{2\pi\sqrt{\det(\Gamma)}} \log n \qquad (n \to \infty). \tag{5.3.21}$$

The key observation here is the slow varying property of g_n: $g_{[\epsilon n]} \sim g_n$ for any $\epsilon > 0$.

Given $\epsilon > 0$ and $1 \le j < [(1+2\epsilon)n]$, by (5.1.9) in Lemma 5.1.3, we have

$$\sum_{k=1}^{[(1+2\epsilon)n]-j} \mathbb{P}\{S(k) = y - x\} = \sum_{k=1}^{[(1+2\epsilon)n]-j} \mathbb{P}\{T_{y-x} = k\} g_{[(1+2\epsilon)n]-k}$$
$$\ge g_{[\epsilon n]} \mathbb{P}\{T_{y-x} \le [(1+\epsilon)n] - j\}.$$

Consequently,

$$\sum_{1\le j<k\le[(1+2\epsilon)n]} \mathbb{P}\{S(j) = x\}\mathbb{P}\{S(k-j) = y - x\}$$
$$\ge g_{[n\epsilon]} \sum_{j=1}^{[(1+2\epsilon)n]-1} \mathbb{P}\{S(j) = x\}\mathbb{P}\{T_{y-x} \le [(1+\epsilon)n] - j\}$$
$$\ge g_{[n\epsilon]} \sum_{j=1}^{[(1+\epsilon)n]-1} \mathbb{P}\{T_{x-y} = j\} \sum_{k=1}^{[(1+\epsilon)n]-j} \mathbb{P}\{S(k) = x\}$$
$$\ge g_{[n\epsilon]}^2 \sum_{k=1}^{n-j} \mathbb{P}\{T_{x-y} = j\}\mathbb{P}\{T_x \le n - j\}$$

where the last step again follows from (5.1.9) in Lemma 5.1.3.

Summarizing our argument, we have

$$\begin{aligned} &\sum_{1\le j<k\le n} \mathbb{P}\{T_x = j,\ T_y = k\} \\ &\le g_{[n\epsilon]}^{-2} \sum_{1\le j<k\le[(1+2\epsilon)n]} \mathbb{P}\{S(j) = x\}\mathbb{P}\{S(k-j) = y - x\} \\ &= g_{[n\epsilon]}^{-2} \sum_{1\le j<k\le[(1+2\epsilon)n]} \mathbb{P}\{S(j) = x,\ S(k) = y\}. \end{aligned} \tag{5.3.22}$$

Therefore,

$$\sum_{x\ne y} \Big[\mathbb{P}\{T_x \le n,\ T_y \le n\}\Big]^p$$
$$\le g_{[\epsilon n]}^{-2} \sum_{x,\in\mathbb{Z}^2} \Bigg[\sum_{j,k=1}^{[(1+2\epsilon)n]} \mathbb{P}\{S(j) = x,\ S(k) = y\}\Bigg]^p$$
$$= g_{[\epsilon n]}^{-2} \mathbb{E} I_{[(1+2\epsilon)n]}^2.$$

Hence, (5.3.20) follows from (5.3.14) in Theorem 5.3.2, (5.3.21) and the fact that ϵ can be arbitrarily small.

We now prove (5.3.18). By (5.3.19), we have

$$\begin{aligned}
\mathbb{E}(I_n J_n) &= \sum_{x,y\in\mathbb{Z}^2} \Big[\mathbb{E}\Big\{1_{\{T_x\le n\}} l(n,y)\Big\}\Big]^p \\
&\ge \sum_{x\neq y}\Big[\sum_{j,k=1}^{n} \mathbb{P}\{T_x=j,\ S(k)=y\}\Big]^p.
\end{aligned} \tag{5.3.23}$$

By independence of increments and by (5.1.9) in Lemma 5.1.3, we have

$$\begin{aligned}
&\sum_{1\le j<k\le n} \mathbb{P}\{T_x=j,\ S(k)=y\} \\
&= \sum_{1\le j<k\le n} \mathbb{P}\{T_x=j\}\mathbb{P}\{S(k-j)=y-x\} \\
&= \sum_{k=1}^{n-1} \mathbb{P}\{S(k)=y-x\}\mathbb{P}\{T_x\le n-k\} \\
&\ge g_n^{-1}\sum_{k=1}^{n-1} \mathbb{P}\{S(k)=y-x\}\sum_{j=1}^{n-k}\mathbb{P}\{S(j)=x\} \\
&= g_n^{-1}\sum_{1\le j<k\le n} \mathbb{P}\{S(j)=x,\ S(k)=y\}.
\end{aligned}$$

In addition,

$$\begin{aligned}
&\sum_{1\le k<j\le n} \mathbb{P}\{T_x=j,\ S(k)=y\} \\
&= \sum_{1\le k<j\le n} \mathbb{P}\{T_x\ge k,\ S(k)=y,\ S(k+1)\neq x, \\
&\qquad\qquad \cdots, S(j-1)\neq x,\ S(j)=x\} \\
&= \sum_{1\le k<j\le n} \mathbb{P}\{T_x\ge k,\ S(k)=y\}\mathbb{P}\{T_{x-y}=j-k\} \\
&= \sum_{1\le k<j\le n} \mathbb{P}\{S(k)=y\}\mathbb{P}\{T_{x-y}=j-k\} \\
&- \sum_{1\le k<j\le n} \mathbb{P}\{T_x<k,\ S(k)=y\}\mathbb{P}\{T_{x-y}=j-k\}.
\end{aligned}$$

Using (5.1.9) one can see that

$$\begin{aligned}
&\sum_{1\le k<j\le n} \mathbb{P}\{S(k)=y\}\mathbb{P}\{T_{x-y}=j-k\} \\
&\ge g_n^{-1}\sum_{1\le k<j\le n} \mathbb{P}\{S(k)=y\}\mathbb{P}\{S(j-k)=x-y\} \\
&= g_n^{-1}\sum_{1\le k<j\le n} \mathbb{P}\{S(j)=x,\ S(k)=y\}.
\end{aligned}$$

Therefore, we conclude that

$$\begin{aligned}&\Big\{\sum_{x\neq y}\Big[\sum_{j,k=1}^{n}\mathbb{P}\{T_x=j,\ S(k)=y\}\Big]^p\Big\}^{1/p}\\&\geq g_n^{-1}\Big\{\sum_{x\neq y}\Big[\sum_{j,k=1}^{n}\mathbb{P}\{S(j)=x,\ S(k)=y\}\Big]^p\Big\}^{1/p}\\&-\Big\{\sum_{x\neq y}\Big[\sum_{1\leq k<j\leq n}\mathbb{P}\{T_x<k,\ S(k)=y\}\mathbb{P}\{T_{x-y}=j-k\}\Big]^p\Big\}^{1/p}\end{aligned}$$

using triangular inequality. Notice that

$$\sum_{x\neq y}\Big[\sum_{j,k=1}^{n}\mathbb{P}\{S(j)=x,\ S(k)=y\}\Big]^p=\mathbb{E}I_n^2-\mathbb{E}I_n.$$

In view of (5.3.14), (5.3.21) and (5.3.23), we will have (5.3.18) if

(5.3.24)
$$\begin{aligned}\lim_{n\to\infty}\frac{(\log n)^p}{n^2}\sum_{x,y\in\mathbb{Z}^2}\Big[&\sum_{1\leq k<j\leq n}\mathbb{P}\{T_x<k,\ S(k)=y\}\\&\times\mathbb{P}\{T_{x-y}=j-k\}\Big]^p=0.\end{aligned}$$

By independence of random walk increments, we get

$$\begin{aligned}&\sum_{1\leq k<j\leq n}\mathbb{P}\{T_x<k,\ S(k)=y\}\mathbb{P}\{T_{x-y}=j-k\}\\&=\sum_{1\leq i<k<j\leq n}\mathbb{P}\{T_x=i,\ S(k)=y\}\mathbb{P}\{T_{x-y}=j-k\}\\&=\sum_{1\leq i<k<j\leq n}\mathbb{P}\{T_x=i\}\mathbb{P}\{S(k-i)=y-x\}\mathbb{P}\{T_{x-y}=j-k\}\\&\leq\mathbb{P}\{T_x\leq n\}\mathbb{P}\{T_{x-y}\leq n\}\sum_{k=1}^{n}\mathbb{P}\{S(k)=y-x\}.\end{aligned}$$

Noticing that $T_{x-y}\stackrel{d}{=}T_{y-x}$, we have

$$\begin{aligned}&\sum_{x,y\in\mathbb{Z}^2}\Big[\sum_{1\leq k<j\leq n}\mathbb{P}\{T_x<k,\ S(k)=y\}\mathbb{P}\{T_{x-y}=j-k\}\Big]^p\\&\leq\Big\{\sum_{x\in\mathbb{Z}^2}\Big[\mathbb{P}\{T_x<n\}\Big]^p\Big\}\sum_{y\in\mathbb{Z}^2}\Big[\mathbb{P}\{T_y\leq n\}\sum_{k=1}^{n}\mathbb{P}\{S(k)=y\}\Big]^p.\end{aligned}$$

Observe that

(5.3.25)
$$\sum_{x\in\mathbb{Z}^2}\Big[\mathbb{P}\{T_x\leq n\}\Big]^p=\mathbb{E}J_n=O\Big(\frac{n}{(\log n)^p}\Big)$$

where the second step follows from (5.3.17). By Cauchy-Schwarz inequality

$$\sum_{y\in\mathbb{Z}^2}\Big[\mathbb{P}\{T_y\le n\}\sum_{k=1}^n\mathbb{P}\{S(k)=y\}\Big]^p$$
$$\le\Big\{\sum_{y\in\mathbb{Z}^2}\Big[\mathbb{P}\{T_y\le n\}\Big]^{2p}\Big\}^{1/2}\Big\{\sum_{y\in\mathbb{Z}^2}\Big[\sum_{k=1}^n\mathbb{P}\{S(k)=y\}\Big]^{2p}\Big\}^{1/2}.$$

Replacing p by $2p$ in (5.3.25) and in (5.3.14) gives, respectively,

$$\sum_{y\in\mathbb{Z}^2}\Big[\mathbb{P}\{T_y\le n\}\Big]^{2p}=O\Big(\frac{n}{(\log n)^{2p}}\Big)$$

and

$$\sum_{y\in\mathbb{Z}^2}\Big[\sum_{k=1}^n\mathbb{P}\{S(k)=y\}\Big]^{2p}=O(n).$$

We have proved the claim (5.3.24). □

In view of Theorem 5.3.2, an immediate consequence of Lemma 5.3.3 is the following weak law for J_n.

THEOREM 5.3.4. *When $d=2$ and $p\ge 2$,*

$$\frac{(\log n)^p}{n}J_n\xrightarrow{d}(2\pi)^p\sqrt{\det(\Gamma)}\alpha\big([0,1]^p\big).\tag{5.3.26}$$

When $d=3$ and $p=2$,

$$J_n/\sqrt{n}\xrightarrow{d}\det(\Gamma)^{-1/2}\gamma_S^2\alpha\big([0,1]^p\big).\tag{5.3.27}$$

Write

$$F_n^{(p)}=\begin{cases}\displaystyle\sum_{x\in\mathbb{Z}^2}\prod_{j=1}^p\Big[l_j(n,x)-\frac{\log n}{2\pi\sqrt{\det(\Gamma)}}1_{\{x\in S_j[1,n]\}}\Big] & d=2,\ p\ge 2,\\ \displaystyle\sum_{x\in\mathbb{Z}^3}\prod_{j=1}^2\Big[l_j(n,x)-\gamma_S^{-1}1_{\{x\in S_j[1,n]\}}\Big] & d=3,\ p=2.\end{cases}\tag{5.3.28}$$

By a slight modification of the proof of Lemma 5.3.3, one can show that

$$\lim_{\epsilon\to 0^+}\limsup_{n\to\infty}\frac{1}{n^2}\mathbb{E}\big[F_n^{(p)}\big]^2=0\tag{5.3.29}$$

when $d=2$, $p\ge 2$, and that

$$\lim_{\epsilon\to 0^+}\limsup_{n\to\infty}\frac{1}{n}\mathbb{E}\big[F_n^{(2)}\big]^2=0\tag{5.3.30}$$

when $d=3$ and $p=2$.

Corresponding to $I_{n,\epsilon}$ defined in (5.3.1), set

$$J_{n,\epsilon}=C_{n,\epsilon}^{-1}\sum_{x\in\mathbb{Z}^d}\prod_{j=1}^p\sum_{y\in S_j[1,n]}p_\epsilon\Big(\frac{\Gamma^{-1/2}(y-x)}{\sqrt{n}}\Big).\tag{5.3.31}$$

A natural question is whether J_n can be approximated by $J_{n,\epsilon}$ in a manner similar to Lemma 5.3.1. First, the treatment for Lemma 5.3.1 fails when we move from I_n, and $I_{n,\epsilon}$ to J_n and $J_{n,\epsilon}$. On the other hand, given what we already know now, J_n can be approximated by $J_{n,\epsilon}$.

Set

$$D_{n,\epsilon}^{(p)} = \sum_{x\in\mathbb{Z}^d}\prod_{j=1}^p \Big[1_{\{x\in S_j[1,n]\}} - C_{n,\epsilon}^{1/p}\sum_{z\in S_j[1,n]} p_\epsilon\Big(\frac{\Gamma^{-1/2}(z-x)}{\sqrt{n}}\Big)\Big]. \tag{5.3.32}$$

The next lemma will be used in Chapter 8.

LEMMA 5.3.5. (1) *When $d=2$ and $p\ge 2$,*

$$\lim_{\epsilon\to 0^+}\limsup_{n\to\infty}\frac{(\log n)^{2p}}{n^2}\mathbb{E}\big[D_{n,\epsilon}^{(p)}\big]^2 = 0, \tag{5.3.33}$$

$$\lim_{\epsilon\to 0^+}\limsup_{n\to\infty}\frac{(\log n)^{2p}}{n^2}\mathbb{E}\big[J_n - J_{n,\epsilon}\big]^2 = 0. \tag{5.3.34}$$

(2) *When $d=3$ and $p=2$,*

$$\lim_{\epsilon\to 0^+}\limsup_{n\to\infty}\frac{1}{n}\mathbb{E}\big[D_{n,\epsilon}^{(2)}\big]^2 = 0, \tag{5.3.35}$$

$$\lim_{\epsilon\to 0^+}\limsup_{n\to\infty}\frac{1}{n}\mathbb{E}\big[J_n - J_{n,\epsilon}\big]^2 = 0. \tag{5.3.36}$$

Proof. Due to similarity, we only consider the case $d=2$, $p\ge 2$. Write

$$F_{n,\epsilon}^{(p)} = C_{n,\epsilon}^{-1}\sum_{x\in\mathbb{Z}^2}\prod_{j=1}^p\Big[\sum_{k=1}^n p_\epsilon\Big(\frac{\Gamma^{-1/2}(S_j(k)-x)}{\sqrt{n}}\Big) - \frac{\log n}{2\pi\sqrt{\det(\Gamma)}}\sum_{z\in S_j[1,n]} p_\epsilon\Big(\frac{\Gamma^{-1/2}(z-x)}{\sqrt{n}}\Big)\Big]. \tag{5.3.37}$$

By Lemma 5.3.1, Lemma 5.3.3 and (5.3.29), it suffices to show that

$$\lim_{\epsilon\to 0^+}\limsup_{n\to\infty}\frac{1}{n^2}\mathbb{E}\big[F_{n,\epsilon}^{(p)}\big]^2 = 0. \tag{5.3.38}$$

By independence, we have

$$\begin{aligned}
&\mathbb{E}\big[F_{n,\epsilon}^{(p)}\big]^2 \\
&= C_{n,\epsilon}^{-2}\sum_{x,y\in\mathbb{Z}^2}\bigg[\mathbb{E}\bigg\{\sum_{k=1}^n p_\epsilon\Big(\frac{\Gamma^{-1/2}(S(k)-x)}{\sqrt{n}}\Big)-\frac{\log n}{2\pi\sqrt{\det(\Gamma)}}\\
&\times\sum_{z\in S[1,n]}p_\epsilon\Big(\frac{\Gamma^{-1/2}(z-x)}{\sqrt{n}}\Big)\bigg\}\bigg\{\sum_{k=1}^n p_\epsilon\Big(\frac{\Gamma^{-1/2}(S(k)-y)}{\sqrt{n}}\Big)\\
&-\frac{\log n}{2\pi\sqrt{\det(\Gamma)}}\sum_{z\in S[1,n]}p_\epsilon\Big(\frac{\Gamma^{-1/2}(z-y)}{\sqrt{n}}\Big)\bigg\}\bigg]^p\\
&= C_{n,\epsilon}^{-2}\sum_{x,y\in\mathbb{Z}^2}\bigg[\sum_{z_1,z_2\in\mathbb{Z}^2}p_\epsilon\Big(\frac{\Gamma^{-1/2}(z_1-x)}{\sqrt{n}}\Big)p_\epsilon\Big(\frac{\Gamma^{-1/2}(z_2-y)}{\sqrt{n}}\Big)\\
&\times\mathbb{E}\bigg\{l(n,z_1)-\frac{\log n}{2\pi\sqrt{\det(\Gamma)}}1_{\{T_{z_1}\le n\}}\bigg\}\bigg\{l(n,z_2)-\frac{\log n}{2\pi\sqrt{\det(\Gamma)}}1_{\{T_{z_2}\le n\}}\bigg\}\bigg]^p.
\end{aligned}$$

Applying Jensen inequality, the right hand side is less than or equal to

$$\begin{aligned}
&C_{n,\epsilon}^{-2/p}\sum_{x,y\in\mathbb{Z}^2}\sum_{z_1,z_2\in\mathbb{Z}^2}p_\epsilon\Big(\frac{\Gamma^{-1/2}(z_1-x)}{\sqrt{n}}\Big)p_\epsilon\Big(\frac{\Gamma^{-1/2}(z_2-y)}{\sqrt{n}}\Big)\\
&\times\bigg|\mathbb{E}\bigg\{l(n,z_1)-\frac{\log n}{2\pi\sqrt{\det(\Gamma)}}1_{\{T_{z_1}\le n\}}\bigg\}\bigg\{l(n,z_2)-\frac{\log n}{2\pi\sqrt{\det(\Gamma)}}1_{\{T_{z_2}\le n\}}\bigg\}\bigg|^p\\
&= C_{n,\epsilon}^{-2/p}\sum_{z_1,z_2\in\mathbb{Z}^2}\bigg|\mathbb{E}\bigg\{l(n,z_1)-\frac{\log n}{2\pi\sqrt{\det(\Gamma)}}1_{\{T_{z_1}\le n\}}\bigg\}\\
&\qquad\times\bigg\{l(n,z_2)-\frac{\log n}{2\pi\sqrt{\det(\Gamma)}}1_{\{T_{z_2}\le n\}}\bigg\}\bigg|^p.
\end{aligned}$$

When p is even, this is equal to $\mathbb{E}\big[F_n^{(p)}\big]^2$. Consequently, (5.3.38) follows from (5.3.29).

When p is odd, by the inequality

$$\sum_{z_1,z_2\in\mathbb{Z}^2}|\cdots|^p\le\Big\{\sum_{z_1,z_2\in\mathbb{Z}^2}[\cdots]^{p-1}\Big\}^{1/2}\Big\{\sum_{z_1,z_2\in\mathbb{Z}^2}[\cdots]^{p+1}\Big\}^{1/2}$$

we obtain

$$\mathbb{E}\big[F_{n,\epsilon}^{(p)}\big]^2\le\Big\{\mathbb{E}\big[F_n^{(p-1)}\big]^2\Big\}^{1/2}\Big\{\mathbb{E}\big[F_n^{(p+1)}\big]^2\Big\}^{1/2}.$$

Again, (5.3.38) follows from (5.3.29) with p being replaced by $p\pm 1$. $\square$

In the rest of the section, $S(k)$ is a symmetric random walk on $\mathbb{Z}^d$ and we view $S_1(k),\cdots,S_p(k)$ as its independent copies. Recall our notation

$$R_n=\#\{S(1),\cdots,S(n)\}.$$

By (5.1.9) in Lemma 5.1.3 and by (5.1.6), (5.1.10), when $d \ge 2$ (in which case g_n is slowly increasing), we have

$$\mathbb{E}R_n = \sum_{x\in\mathbb{Z}^2} \mathbb{P}\{T_x \le n\} \sim g_n^{-1} \sum_{x\in\mathbb{Z}^2} \sum_{k=1}^n \mathbb{P}\{S(k) = x\} \tag{5.3.39}$$

$$= g_n^{-1} n = \begin{cases} 2\pi\sqrt{\det(\Gamma)}\dfrac{n}{\log n} & d = 2, \\ \gamma_S n & d \ge 3. \end{cases}$$

THEOREM 5.3.6. *Let $f(x)$ be a bounded continuous function on $\mathbb{R}^d$.*

(1) *When $d = 2$,*

$$\frac{1}{n}\Big\{\sum_{k=1}^n f\Big(\frac{S(k)}{\sqrt{n}}\Big) - \frac{\log n}{2\pi\sqrt{\det(\Gamma)}} \sum_{x\in S[1,n]} f\Big(\frac{x}{\sqrt{n}}\Big)\Big\} \xrightarrow{P} 0. \tag{5.3.40}$$

Consequently,

$$\Big(\frac{\log n}{2\pi n\sqrt{\det(\Gamma)}} \sum_{x\in S[1,n]} f\Big(\frac{x}{\sqrt{n}}\Big),\ \frac{S(n)}{\sqrt{n}}\Big) \tag{5.3.41}$$

$$\xrightarrow{d} \Big(\int_0^1 f\big(\Gamma^{1/2}W(t)\big)dt,\ W(1)\Big) \qquad (n\to\infty).$$

(2) *When $d = 3$,*

$$\frac{1}{n}\Big\{\sum_{k=1}^n f\Big(\frac{S(k)}{\sqrt{n}}\Big) - \gamma_S^{-1} \sum_{x\in S[1,n]} f\Big(\frac{x}{\sqrt{n}}\Big)\Big\} \xrightarrow{P} 0. \tag{5.3.42}$$

Consequently,

$$\Big(\frac{1}{\gamma_S n} \sum_{x\in S[1,n]} f\Big(\frac{x}{\sqrt{n}}\Big),\ \frac{S(n)}{\sqrt{n}}\Big) \xrightarrow{d} \Big(\int_0^1 f\big(\Gamma^{1/2}W(t)\big)dt,\ W(1)\Big), \tag{5.3.43}$$

as $n\to\infty$.

Proof. Due to similarity we only deal with $d = 2$. By invariance principle, (5.3.41) is a direct consequence of (5.3.40).

For each $M > 0$, define $f_M(x) = f(x)$ if $|x| \le M$, and $f_M(x) = 0$ if $|x| > M$. By the classic fact that the random sequence $\max_{1\le k\le n} |S(k)|/\sqrt{n}$ is bounded in probability, we only need to show that for any fixed $M > 0$,

$$\lim_{n\to\infty} \frac{1}{n^2}\mathbb{E}\Big[\sum_{k=1}^n f_M\Big(\frac{S(k)}{\sqrt{n}}\Big) - \frac{\log n}{2\pi\sqrt{\det(\Gamma)}} \sum_{x\in S[1,n]} f_M\Big(\frac{x}{\sqrt{n}}\Big)\Big]^2 = 0. \tag{5.3.44}$$

Take $d = p = 2$ and let $F_n^{(2)}$ be defined in (5.3.28), then

$$\mathbb{E}\big[F_n^{(2)}\big]^2 = \sum_{x,y\in\mathbb{Z}^2} \Big[\mathbb{E}\Big\{l(n,x) - \frac{\log n}{2\pi\sqrt{\det(\Gamma)}} 1_{\{T_x\le n\}}\Big\}$$

$$\times \Big\{l(n,y) - \frac{\log n}{2\pi\sqrt{\det(\Gamma)}} 1_{\{T_y\le n\}}\Big\}\Big]^2.$$

By Cauchy-Schwarz inequality,

$$\begin{aligned}&\Big\{\sum_{x\in\mathbb{Z}^2}\Big|f_M\Big(\frac{x}{\sqrt{n}}\Big)\Big|^2\Big\}\Big\{\mathbb{E}[F_n^{(2)}]^2\Big\}^{1/2}\\ &\ge\sum_{x,y\in\mathbb{Z}^2}f_M\Big(\frac{x}{\sqrt{n}}\Big)f_M\Big(\frac{y}{\sqrt{n}}\Big)\mathbb{E}\Big\{l(n,x)-\frac{\log n}{2\pi\sqrt{\det(\Gamma)}}1_{\{T_x\le n\}}\Big\}\\ &\qquad\times\Big\{l(n,y)-\frac{\log n}{2\pi\sqrt{\det(\Gamma)}}1_{\{T_y\le n\}}\Big\}.\end{aligned}$$

Notice that the right hand side is equal to

$$\begin{aligned}&\mathbb{E}\Big[\sum_{x\in\mathbb{Z}^2}f_M\Big(\frac{x}{\sqrt{n}}\Big)l(n,x)-\frac{\log n}{2\pi\sqrt{\det(\Gamma)}}\sum_{x\in\mathbb{Z}^2}f_M\Big(\frac{x}{\sqrt{n}}\Big)1_{\{T_y\le n\}}\Big]^2\\ &=\mathbb{E}\Big[\sum_{k=1}^n f_M\Big(\frac{S(k)}{\sqrt{n}}\Big)-\frac{\log n}{2\pi\sqrt{\det(\Gamma)}}\sum_{x\in S[1,n]}f_M\Big(\frac{x}{\sqrt{n}}\Big)\Big]^2.\end{aligned}$$

Therefore, the requested (5.3.44) follows from (5.3.29) (with $p=2$) and the fact that

$$\sum_{x\in\mathbb{Z}^2}\Big|f_M\Big(\frac{x}{\sqrt{n}}\Big)\Big|^2\sim n\int_{\mathbb{R}^2}|f_M(x)|^2dx,$$

as $n\to\infty$. □

Theorem 5.3.6 will play an important role in Chapter 7.

5.4. Self-intersection in dimension two

The objective of this section is to establish the weak laws for the double self-intersection local time

$$Q_n=\sum_{1\le j<k\le n}1_{\{S(j)=S(k)\}}$$

and for the range

$$R_n=\#\{S[1,n]\}=\#\{S(1),\cdots,S(n)\}$$

of the planar symmetric random walk $S(k)$.

By (5.1.6), it is straightforward to see that

$$\mathbb{E}Q_n=\sum_{1\le j<k\le n}\mathbb{P}\{S(k-j)=0\}\sim\frac{1}{2\pi\sqrt{\det(\Gamma)}}n\log n. \tag{5.4.1}$$

The asymptotic expansion of $\mathbb{E}R_n$ is given in (5.3.39). Taking $f\equiv 1$ in Theorem 5.3.6 leads to

$$\lim_{n\to\infty}R_n/\mathbb{E}R_n=1$$

in probability.

Concentration phenomena also happen to Q_n in multi-dimensions. The right question is to ask what can be said about the limit laws for $Q_n-\mathbb{E}Q_n$ and for $R_n-\mathbb{E}R_n$. We start our investigation by estimating the variance of Q_n and R_n.

LEMMA 5.4.1.

$$\text{Var}\,(Q_n) = O(n^2) \quad (n \to \infty), \tag{5.4.2}$$

$$\text{Var}\,(R_n) = O\Big(\frac{n^2}{(\log n)^4}\Big) \quad (n \to \infty). \tag{5.4.3}$$

Proof. Due to similarity, we only prove (5.4.3). For each integer $k \geq 1$, write

$$a_k = \sup\Big\{\text{Var}\,(R_n)^{1/2};\ \ 2^k < n \leq 2^{k+1}\Big\}.$$

For each $n \in (2^k, 2^{k+1}]$ let $n_1 = [n/2]$ and $n_2 = n - n_1$. Consider the decomposition

$$R_n = R_{n_1} + \widetilde{R}_{n_2} - \#\{S[1, n_1] \cap S[n_1 + 1, n]\}$$

where $\widetilde{R}_{n_2} = \#\big\{S[n_1 + 1, n]\big\}$ is independent of R_{n_1} and of the distribution the same as R_{n_2}, and

$$\#\big\{S[1, n_1] \cap S[n_1, n]\big\} \stackrel{d}{=} \#\big\{S[0, n_1 - 1] \cap \widetilde{S}[1, n_2]\big\}$$

where $\widetilde{S}(n)$ is an independent copy of $S(n)$. By Cauchy-Schwarz inequality, we have

$$\begin{aligned}
&\text{Var}\,(R_n)^{1/2}\\
&\leq \Big(\text{Var}\,(R_{n_1}) + \text{Var}\,(R_{n_2})\Big)^{1/2} + \text{Var}\,\Big(\#\big\{S[0, n_1 - 1] \cap \widetilde{S}[1, n_2]\big\}\Big)^{1/2}.
\end{aligned}$$

Taking $p = 2$ in (5.3.17) gives that the second term on the right hand side is bounded by $C_1 k^{-2} 2^k$ for some constant independent of k. Thus,

$$a_k \leq 2^{1/2} a_{k-1} + C_1 k^{-2} 2^k \qquad k = 2, 3, \cdots.$$

Write $d_k = k^2 2^{-k} a_k$. Then

$$d_k \leq \theta d_{k-1} + C_1$$

for any $\theta > 2^{-1/2}$ and sufficiently large k. Consequently, the sequence d_k is bounded. □

Recall the renormalized self-intersection local time

$$\begin{aligned}
\gamma\big([0, t]^2_<\big) = &\iint_{\{0 \leq r < s \leq t\}} \delta_0\big(W(r) - W(s)\big) dr ds\\
&- \mathbb{E} \iint_{\{0 \leq r < s \leq t\}} \delta_0\big(W(r) - W(s)\big) dr ds
\end{aligned}$$

run by a planar Brownian motion $W(t)$.

THEOREM 5.4.2.

$$n^{-1}\Big\{Q_n - \mathbb{E}Q_n\Big\} \xrightarrow{d} \det(\Gamma)^{-1/2} \gamma\big([0, 1]^2_<\big).$$

Proof. The proof is based on the same type of triangular approximation used in Theorem 2.4.1.

Let $N \geq 1$ be a large integer and write

$$\mathcal{A}_l^k(n) = \Big(\frac{2l}{2^{k+1}}n, \frac{2l+1}{2^{k+1}}n\Big] \times \Big(\frac{2l+1}{2^{k+1}}n, \frac{2l+2}{2^{k+1}}n\Big]$$
$$l = 0, 1, \cdots, 2^k - 1 \quad k = 0, 1, \cdots N-1,$$

$$\mathcal{B}_l(n) = \Big\{(i,j);\ \frac{l-1}{2^N}n < i < j \leq \frac{l}{2^N}n\Big\} \qquad l = 1, \cdots, 2^N.$$

The reader is advised to check Figure 1 in Section 2.4 for visualization of these sets.

We have

$$(5.4.4) \qquad Q_n = \sum_{k=1}^{N-1}\sum_{l=1}^{2^k-1}\sum_{(i,j)\in\mathcal{A}_l^k(n)} 1_{\{S(i)=S(j)\}} + \sum_{l=1}^{2^N}\sum_{(i,j)\in\mathcal{B}_l(n)} 1_{\{S(i)=S(j)\}}.$$

Observe that the sequence

$$\sum_{(i,j)\in\mathcal{B}_l(n)} 1_{\{S(i)=S(j)\}} \qquad l = 1, \cdots, 2^N$$

is independent. Therefore,

$$\begin{aligned}
&\mathrm{Var}\Big\{\sum_{l=1}^{2^N}\sum_{(i,j)\in\mathcal{B}_l(n)} 1_{\{S(i)=S(j)\}}\Big\}\\
&= \sum_{l=1}^{2^N}\mathrm{Var}\Big\{\sum_{(i,j)\in\mathcal{B}_l(n)} 1_{\{S(i)=S(j)\}}\Big\}\\
&= O\Big(2^{-N}n^2\Big).
\end{aligned}$$

In particular,

$$(5.4.5) \qquad \lim_{N\to\infty}\limsup_{n\to\infty} n^{-2}\mathrm{Var}\Big\{\sum_{l=1}^{2^N}\sum_{(i,j)\in\mathcal{B}_l(n)} 1_{\{S(i)=S(j)\}}\Big\} = 0.$$

Assume that we can prove that as $n \to \infty$,

$$(5.4.6) \qquad \frac{1}{n}\sum_{k=1}^{N-1}\sum_{l=1}^{2^k-1}\sum_{(i,j)\in\mathcal{A}_l^k(n)} 1_{\{S(i)=S(j)\}} \xrightarrow{d} \det(\Gamma)^{-1/2}\beta(D_N),$$

$$(5.4.7) \qquad \lim_{n\to\infty}\frac{1}{n}\mathbb{E}\Big\{\sum_{k=1}^{N-1}\sum_{l=1}^{2^k-1}\sum_{(i,j)\in\mathcal{A}_l^k(n)} 1_{\{S(i)=S(j)\}}\Big\} = \det(\Gamma)^{-1/2}\mathbb{E}\beta(D_N)$$

for every N, where $\beta(\cdot)$ is the self-intersection local time (without being renormalized)

$$\beta(A) = \iint_A \delta_0\big(W(r) - W(s)\big)drds \qquad A \subset (\mathbb{R}^+)^2_<$$

discussed in Section 2.3,

$$D_N = \bigcup_{k=1}^{N-1}\bigcup_{l=1}^{2^k-1} A_l^k$$

and the sets A_l^k are defined in (2.4.2). Then the desired conclusion holds, since by Theorem 2.4.1, $\beta(D_N) - \mathbb{E}\beta(D_N)$ converges to $\gamma([0,1]^2_<)$ as $N \to \infty$.

Therefore, it remains to prove (5.4.6) and (5.4.7). Due to similarity in proof, we only deal with (5.4.6). Fix $\epsilon > 0$ and apply the invariance principle to the continuous functional

$$\Psi(f) = \iint_{D_N} p_{2\epsilon}\Big(\Gamma^{-1/2}\big(f(r) - f(s)\big)\Big) drds$$

on $C\{[0,1],\mathbb{R}^2\}$, which gives that

$$\frac{1}{n^2}\sum_{k=1}^{N-1}\sum_{l=1}^{2^k-1}\sum_{(i,j)\in\mathcal{A}_l^k(n)} p_{2\epsilon}\Big(\frac{\Gamma^{-1/2}\big(S(i)-S(j)\big)}{\sqrt{n}}\Big)$$
$$\xrightarrow{d} \iint_{D_N} p_{2\epsilon}\big(W(r) - W(s)\big)drds = \beta_\epsilon(D_N),$$

as $n \to \infty$, where the random measure

$$\beta_\epsilon(A) = \int_{\mathbb{R}^d}\Big[\iint_A p_\epsilon\big(W(r)-x\big)p_\epsilon\big(W(s)-x\big)drds\Big]dx \qquad A \subset (\mathbb{R}+)^2_<$$

was constructed and discussed (in a more general form) in Section 2.3. By Theorem 2.3.2, $\beta_\epsilon(D_N)$ converges to $\beta(D_N)$ in all moments as $\epsilon \to 0^+$.

From the discussion above, the proof of (5.4.6) has been reduced to show that for any fixed $1 \le k \le N-1$ and $l = 0, 1, \cdots, 2^k - 1$,

$$\lim_{\epsilon\to 0^+}\limsup_{n\to\infty}\frac{1}{n^2}\mathbb{E}\Big\{\sum_{(i,j)\in\mathcal{A}_l^k(n)} 1_{\{S(i)=S(j)\}}$$
$$- \frac{1}{\sqrt{\det(\Gamma)}n}\sum_{(i,j)\in\mathcal{A}_l^k(n)} p_{2\epsilon}\Big(\frac{\Gamma^{-1/2}\big(S(i)-S(j)\big)}{\sqrt{n}}\Big)\Big\}^2 = 0. \tag{5.4.8}$$

Indeed, there are $t_n, \tilde{t}_n = \frac{n}{2^k} \pm 1$ such that

$$\sum_{(i,j)\in\mathcal{A}_l^k(n)} 1_{\{S(i)=S(j)\}} - \frac{1}{\sqrt{\det(\Gamma)}n}\sum_{(i,j)\in\mathcal{A}_l^k(n)} p_{2\epsilon}\Big(\frac{\Gamma^{-1/2}\big(S(i)-S(j)\big)}{\sqrt{n}}\Big)$$
$$\stackrel{d}{=} \sum_{i=0}^{t_n}\sum_{j=1}^{\tilde{t}_n} 1_{\{S(i)=\widetilde{S}(j)\}} - \frac{1}{\sqrt{\det(\Gamma)}n}\sum_{i=0}^{t_n}\sum_{j=1}^{\tilde{t}_n} p_{2\epsilon}\Big(\frac{\Gamma^{-1/2}\big(S(i)-\widetilde{S}(j)\big)}{\sqrt{n}}\Big)$$

where $\widetilde{S}(n)$ is an independent copy of $S(n)$. By the continuity of $p_t(x)$ with respect to both t and x, the right hand side has the expansion, in the notation used in (5.3.1) (with $d = p = 2$),

$$o(n) + I_{t_n} - I_{t_n, 2^k\epsilon} \qquad (n \to \infty).$$

Hence, (5.4.8) follows from Lemma 5.3.1. □

THEOREM 5.4.3.

$$\mathbb{E}\Big\{(Q_n - \mathbb{E}Q_n) + \Big(\frac{\log n}{2\pi\sqrt{\det(\Gamma)}}\Big)^2(R_n - \mathbb{E}R_n)\Big\}^2 = o(n^2), \tag{5.4.9}$$

as $n \to \infty$. In particular,

$$\frac{(\log n)^2}{n}(R_n - \mathbb{E}R_n) \xrightarrow{d} -(2\pi)^2\sqrt{\det(\Gamma)}\gamma([0,1]^2_<). \tag{5.4.10}$$

Proof. Correspondent to (5.4.4),

$$R_n = \sum_{l=1}^{2^N} \#\Big\{S\Big(\frac{l-1}{2^N}n, \frac{l}{2^N}n\Big]\Big\}$$
$$- \sum_{k=1}^{N-1}\sum_{l=1}^{2^k-1} \#\Big\{S\Big(\frac{2l}{2^{k+1}}n, \frac{2l+1}{2^{k+1}}n\Big] \cap S\Big(\frac{2l+1}{2^{k+1}}n, \frac{2l+2}{2^{k+1}}n\Big]\Big\}.$$

By an argument similar to the one used for (5.4.5), we have

$$\lim_{N\to\infty}\limsup_{n\to\infty}\frac{(\log n)^4}{n^2}\mathrm{Var}\Big\{\sum_{l=1}^{2^N}\#\Big\{S\Big(\frac{l-1}{2^N}n, \frac{l}{2^N}n\Big]\Big\}\Big\} = 0.$$

To prove (5.4.9), therefore, it is sufficient to prove that for fixed k, l,

$$\lim_{n\to\infty}\frac{1}{n^2}\mathbb{E}\Big\{\sum_{(i,j)\in\mathcal{A}_l^k(n)} 1_{\{S(i)=S(j)\}}$$
$$-\Big(\frac{\log n}{2\pi\sqrt{\det(\Gamma)}}\Big)^2\#\Big\{S\Big(\frac{2l}{2^{k+1}}n, \frac{2l+1}{2^{k+1}}n\Big] \cap S\Big(\frac{2l+1}{2^{k+1}}n, \frac{2l+2}{2^{k+1}}n\Big]\Big\}\Big\}^2$$
$$= 0.$$

This clearly follows from (5.3.15) in Lemma 5.3.3 with $d = p = 2$ and the identity in law

$$\Big(\sum_{(i,j)\in\mathcal{A}_l^k(n)} 1_{\{S(i)=S(j)\}},\ \#\Big\{S\Big(\frac{2l}{2^{k+1}}n, \frac{2l+1}{2^{k+1}}n\Big] \cap S\Big(\frac{2l+1}{2^{k+1}}n, \frac{2l+2}{2^{k+1}}n\Big]\Big\}\Big)$$
$$\stackrel{d}{=} \Big(\sum_{i=0}^{t_n}\sum_{j=1}^{\tilde{t}_n} 1_{\{S(i)=\widetilde{S}(j)\}},\ \#\big\{S[0,t_n] \cap \widetilde{S}[1,\tilde{t}_n]\big\}\Big)$$

where t_n and $\tilde{t}_n$ are equal to $2^{-k}n - 1$, $2^{-k}n$ or $2^{-k}n + 1$.

The statement (5.4.10) is a direct consequence of Theorem 5.4.2 and (5.4.9). □

5.5. Intersection in high dimensions

For the mutual intersection, the phrase "high dimensions" refers to the critical dimensions defined by the equality $p(d-2) = d$ and the super-critical dimensions defined by the inequality $p(d-2) > d$. When $p(d-2) > d$, $\mathbb{E}I_n$ and $\mathbb{E}J_n$ are asymptotically bounded, according to (5.6.4) given in Section 5.6. In the critical dimensions, the Brownian intersection local time $\alpha(A)$ does not exist. Naturally, the weak law for I_n and J_n, if any, will take a form different from those given in Theorem 5.3.2 and Theorem 5.3.4. Here we include, without giving the proof, the following theorem of Le Gall [**125**] on the weak convergence for I_n and J_n.

THEOREM 5.5.1. (1) *When $d = 4$ and $p = 2$,*

$$(\log n)^{-1} I_n \xrightarrow{d} (2\pi)^{-2} \det(\Gamma)^{-1/2} U^2, \tag{5.5.1}$$

$$(\log n)^{-1} J_n \xrightarrow{d} \gamma_S^2 (2\pi)^{-2} \det(\Gamma)^{-1/2} U^2 \tag{5.5.2}$$

where U is a $N(0,1)$ random variable.

(2) *When $d = p = 3$,*

$$(\log n)^{-1} I_n \xrightarrow{d} (2\pi)^{-2} \frac{1}{\det(\Gamma)} V, \tag{5.5.3}$$

$$(\log n)^{-1} J_n \xrightarrow{d} (2\pi)^{-2} \frac{\gamma_S^3}{\det(\Gamma)} V \tag{5.5.4}$$

where the random variable V has the gamma distribution with parameter $1/4$.

As for the self-intersection, the phrase "high dimensions" refers to the dimensions three or above, partially because the Brownian self-intersection local time does not exist (even by renormalization) when $d \geq 3$. To establish the weak laws for $Q_n - \mathbb{E}Q_n$ and $R_n - \mathbb{E}R_n$, we first investigate the asymptotic orders of their variances.

LEMMA 5.5.2. *When $d = 3$,*

$$\mathrm{Var}\,(Q_n) \sim \frac{1}{2\pi^2 \det(\Gamma)} n \log n \qquad (n \to \infty), \tag{5.5.5}$$

$$\mathrm{Var}\,(R_n) \sim \frac{\gamma_S^4}{2\pi^2 \det(\Gamma)} n \log n \qquad (n \to \infty). \tag{5.5.6}$$

When $d \geq 4$,

$$\mathrm{Var}\,(Q_n) \sim \mathfrak{D}^2 n \qquad (n \to \infty), \tag{5.5.7}$$

$$\mathrm{Var}\,(R_n) \sim \widetilde{\mathfrak{D}}^2 n \qquad (n \to \infty). \tag{5.5.8}$$

Here the limiting variances $\mathfrak{D}^2$ and $\widetilde{\mathfrak{D}}^2$ are given by

$$\mathfrak{D}^2 = 3\mathcal{G}^2(0) + \mathcal{G}(0) + 2 \sum_{x \in \mathbb{Z}^d} \mathcal{G}^3(x),$$

$$\widetilde{\mathfrak{D}}^2 = \gamma_S(1 - \gamma_S) + 2\gamma_S^2 \sum_{x \neq 0} \frac{1}{1 + \mathcal{G}(x)} \mathcal{G}^3(x)$$

where $\mathcal{G}(x)$ is Green's function defined as

$$\mathcal{G}(x) = \sum_{k=1}^{\infty} \mathbb{P}\{S(k) = x\} \qquad x \in \mathbb{Z}^d.$$

Proof. By Exercise 5.6.6, $\mathcal{G}(\cdot) \in l^3(\mathbb{Z}^d)$ as $d \geq 4$. Subsequently, the limiting variances $\mathfrak{D}^2$ and $\widetilde{\mathfrak{D}}^2$ are well defined.

We only compute the variance for Q_n. That is, we prove (5.5.5) and (5.5.7). The proof of (5.5.6) and (5.5.8) is similar but harder. We refer the interested reader to [**131**] for proof (Theorem 4.4 for $d \geq 4$ and Theorem 4.6 for $d = 3$).

We introduce the following notations: $P_k(x) = \mathbb{P}\{S(k) = x\}$ and $\{\widetilde{S}(k)\}$ is an independent copy of $\{S(k)\}$. Let

$$\mathcal{G}_k(x) = \sum_{j=1}^{k} P_j(x) \qquad x \in \mathbb{Z}^d, \quad k = 1, 2, \cdots$$

be the partial Green's functions.

We begin with the decomposition

$$\begin{aligned} Q_n &= \sum_{j=1}^{n-1} \sum_{k=j+1}^{\infty} 1_{\{S(k)=S(j)\}} - \sum_{j=1}^{n-1} \sum_{k=n+1}^{\infty} 1_{\{S(k)=S(j)\}} \\ &= \sum_{j=1}^{n-1} Z_j - \sum_{j=1}^{n-1} W_j^n \qquad \text{(say)}. \end{aligned} \tag{5.5.9}$$

We shall show that the second term in the decomposition makes no asymptotic contribution to the variance.

For any $1 \leq i \leq j \leq n-1$,

$$\begin{aligned} \mathbb{E}W_i^n W_j^n &= \sum_{k,l=n+1}^{\infty} \mathbb{P}\{S(l) - S(j) = 0, \ S(k) - S(j) = S(i) - S(j)\} \\ &= \sum_{k,l=n-j+1}^{\infty} \mathbb{P}\{S(l) = 0, \ S(k) = \widetilde{S}(j-i)\} \\ &= \sum_{x \in \mathbb{Z}^d} P_{j-i}(x) \sum_{k,l=n-j+1}^{\infty} \mathbb{P}\{S(l) = 0, \ S(k) = x\} \\ &= \sum_{x \in \mathbb{Z}^d} P_{j-i}(x) \sum_{n-j+1 \leq k \leq l < \infty} \mathbb{P}\{S(l) = 0, \ S(k) = x\} \\ &\quad + \sum_{x \in \mathbb{Z}^d} P_{j-i}(x) \sum_{n-j+1 \leq l < k < \infty} \mathbb{P}\{S(l) = 0, \ S(k) = x\} \\ &= (I) + (II) \qquad \text{(say)}. \end{aligned}$$

For the first term on the right hand side,

$$\begin{aligned} (I) &= \sum_{x \in \mathbb{Z}^d} P_{j-i}(x) \sum_{n-j+1 \leq k \leq l < \infty} \mathbb{P}\{S(l-k) = x\}\mathbb{P}\{S(k) = x\} \\ &\leq C \sum_{x \in \mathbb{Z}^d} P_{j-i}(x) \sum_{n-j+1 \leq k \leq l < \infty} P_{l-k}(x) \frac{1}{k^{d/2}} \\ &= C \sum_{n-j+1 \leq k \leq l < \infty} P_{l-k+j-i}(0) \frac{1}{k^{d/2}} \end{aligned}$$

where the second step follows from the classic fact that $\sup_{x\in\mathbb{Z}^d} P_k(x) = O(k^{-d/2})$ and the third step from Chapman-Kolmogorov equation.

As for the second term, a similar estimate yields that

$$(II) \le C \sum_{n-j+1\le l\le k<\infty} P_{k-l+j-i}(0)\frac{1}{l^{d/2}}.$$

Hence,

$$\begin{aligned}\mathbb{E}W_i^n W_j^n &\le 2C \sum_{n-j+1\le k\le l<\infty} P_{l-k+j-i}(0)\frac{1}{l^{d/2}}\\ &= 2\Big(\sum_{k=n-j+1}^{\infty} P_k(0)\Big)\Big(\sum_{l=j-i}^{\infty}\frac{1}{l^{d/2}}\Big)\\ &= O\Big((n-j)^{1-\frac{d}{2}}(j-i)^{1-\frac{d}{2}}\Big) \qquad (n\to\infty).\end{aligned}$$

Therefore, as $n\to\infty$, we have

$$\sum_{j=1}^{n-1}\sum_{i=1}^{j}\mathbb{E}W_i^n W_j^n = \begin{cases} O(n) & d=3,\\ O\Big((\log n)^2\Big) & d=4,\\ O(1) & d\ge 5.\end{cases} \tag{5.5.10}$$

It shows that the second term in the decomposition (5.5.9) is negligible as far as the variance is concerned.

For $1\le i\le j\le n-1$,

$$\begin{aligned}\operatorname{Cov}(Z_i, Z_j) &= \operatorname{Cov}\Big(\sum_{k=1}^{\infty} 1_{\{S(k)=\widetilde{S}(j-i)\}}, \sum_{k=1}^{\infty} 1_{\{S(k)=0\}}\Big) \qquad (5.5.11)\\ &= \sum_{k,l=1}^{\infty}\Big\{\mathbb{P}\{S(k)=\widetilde{S}(j-i),\ S(l)=0\} - P_{k+j-i}(0)P_l(0)\Big\}\\ &= \sum_{x\in\mathbb{Z}^d} P_{j-i}(x)\sum_{k,l=1}^{\infty}\mathbb{P}\{S(k)=-x,\ S(l)=0\} - \mathcal{G}(0)\sum_{k=j-i+1}^{\infty} P_k(0).\end{aligned}$$

Write

$$\begin{aligned}&\sum_{x\in\mathbb{Z}^d} P_{j-i}(x)\sum_{k,l=1}^{\infty}\mathbb{P}\{S(k)=x,\ S(l)=0\}\\ &= \sum_{x\in\mathbb{Z}^d} P_{j-i}(x)\sum_{1\le k\le l<\infty}\mathbb{P}\{S(k)=x,\ S(l)=0\}\\ &+ \sum_{x\in\mathbb{Z}^d} P_{j-i}(x)\sum_{1\le l<k<\infty}\mathbb{P}\{S(k)=x,\ S(l)=0\}.\end{aligned}$$

The first term on the right hand side is equal to

$$\begin{aligned}&\sum_{x\in\mathbb{Z}^d} P_{j-i}(x)\sum_{1\le k\le l<\infty} P_k(x)P_{l-k}(x)\\ &=\sum_{x\in\mathbb{Z}^d} P_{j-i}(x)\mathcal{G}(x)\sum_{l=0}^{\infty} P_l(x)\\ &= P_{j-i}(0)\mathcal{G}(0)\big(1+\mathcal{G}(0)\big)+\sum_{x\neq 0} P_{j-i}(x)\mathcal{G}^2(x).\end{aligned}$$

As for the second term, it is equal to

$$\begin{aligned}&\sum_{x\in\mathbb{Z}^d} P_{j-i}(x)\sum_{1\le l<k<\infty} P_{k-l}(x)P_l(0)\\ &=\sum_{1\le l<k<\infty} P_{k-l+j-i}(0)P_l(0)=\mathcal{G}(0)\sum_{k=j-i+1}^{\infty} P_k(0).\end{aligned}$$

In summary of the argument since (5.5.11), we have

$$\operatorname{Cov}(Z_i,Z_j)=P_{j-i}(0)\mathcal{G}(0)\big(1+\mathcal{G}(0)\big)+\sum_{x\neq 0} P_{j-i}(x)\mathcal{G}^2(x).$$

Consequently,

$$\begin{aligned}(5.5.12)\qquad \operatorname{Var}\Big(\sum_{i=1}^{n-1} Z_i\Big)&=\sum_{i=1}^{n-1}\operatorname{Var}(Z_i)+2\sum_{1\le i<j\le n-1}\operatorname{Cov}(Z_i,Z_j)\\ &=(n-1)\mathcal{G}(0)\big(1+\mathcal{G}(0)\big)+2\mathcal{G}(0)\big(1+\mathcal{G}(0)\big)\sum_{j=1}^{n-2}\mathcal{G}_{n-j-1}(0)\\ &\quad+2\sum_{x\neq 0}\mathcal{G}^2(x)\sum_{j=1}^{n-2}\mathcal{G}_{n-j-1}(x)\\ &=(n-1)\mathcal{G}(0)\big(1+\mathcal{G}(0)\big)+2\mathcal{G}(0)\sum_{j=1}^{n-2}\mathcal{G}_j(0)+2\sum_{x\in\mathbb{Z}^d}\mathcal{G}^2(x)\sum_{j=1}^{n-2}\mathcal{G}_j(x).\end{aligned}$$

When $d\ge 4$, by (5.6.4), we have

$$\sum_{x\in\mathbb{Z}^d}\mathcal{G}^3(x)<\infty.$$

By (5.5.12), we have

$$\operatorname{Var}\Big(\sum_{i=1}^{n-1} Z_i\Big)\sim n\Big\{\mathcal{G}(0)\big(1+\mathcal{G}(0)\big)+2\mathcal{G}^2(0)+2\sum_{x\in\mathbb{Z}^d}\mathcal{G}^3(x)\Big\},$$

as $n\to\infty$. By (5.5.9) and (5.5.10), this implies (5.5.7).

We now consider the case $d=3$. We use the fact that (p. 308, [**155**])

$$\mathcal{G}(x)\sim(2\pi)^{-1}\det(\Gamma)^{-1/2}\langle x,\Gamma^{-1}x\rangle^{-1/2}\qquad(|x|\to\infty).$$

Thus,

$$\sum_{\langle x,\Gamma^{-1}x\rangle>j}\mathcal{G}^2(x)\mathcal{G}_j(x)=O\Big(\frac{1}{j}\sum_{x\in\mathbb{Z}^3}\mathcal{G}_j(x)\Big)=O(1)\qquad(j\to\infty).\tag{5.5.13}$$

We now claim that uniformly over $x\in\mathbb{Z}^3$,

$$\mathcal{G}(x)-\mathcal{G}_j(x)=O\big(j^{-1/2}\big)\qquad(j\to\infty).\tag{5.5.14}$$

Indeed, by Fourier inversion, we have

$$\mathbb{P}\{S(k)=x\}=\frac{1}{(2\pi)^3}\int_{[-\pi,\pi]^3}e^{-\lambda\cdot x}\varphi(\lambda)^k d\lambda\le\frac{1}{(2\pi)^3}\int_{[-\pi,\pi]^3}|\varphi(\lambda)|^k d\lambda.$$

When the random walk is aperiodic, by the argument used in (5.1.3) in Lemma 5.1.1, the right hand side is asymptotically equivalent to

$$\frac{1}{(2\pi)^{3/2}\sqrt{\det(\Gamma)}}k^{-3/2},$$

as $k\to\infty$. Consequently, (5.5.14) holds under aperiodicity. A standard argument via resolvent approximation extends it to the general case.

By (5.5.14), we have

$$\begin{aligned}&\sum_{\langle x,\Gamma^{-1}x\rangle\le j}\mathcal{G}^2(x)\big(\mathcal{G}(x)-\mathcal{G}_j(x)\big)\\&\le Cj^{-1/2}\sum_{\langle x,\Gamma^{-1}x\rangle\le j}\mathcal{G}^2(x)\\&\le Cj^{-1/2}\sum_{\langle x,\Gamma^{-1}x\rangle\le j}\frac{1}{1+|x|^2}=O(1)\qquad(j\to\infty).\end{aligned}\tag{5.5.15}$$

In addition,

$$\begin{aligned}\sum_{\langle x,\Gamma^{-1}x\rangle\le j}\mathcal{G}^3(x)&\sim\frac{1}{(2\pi)^3\det(\Gamma)^3}\int_{\{1\le\langle x,\Gamma^{-1}x\rangle\le j\}}\langle x,\Gamma^{-1}x\rangle^{-3/2}dx\\&=\frac{1}{(2\pi)^3\det(\Gamma)}\int_{\{1\le|y|\le\sqrt{j}\}}|y|^{-3}dy=\frac{1}{(2\pi)^2\det(\Gamma)}\log j\qquad(j\to\infty).\end{aligned}\tag{5.5.16}$$

Combining (5.5.13), (5.5.15) and (5.5.16) gives us

$$\sum_{x\in\mathbb{Z}^3}\mathcal{G}^2(x)\mathcal{G}_j(x)\sim\frac{1}{(2\pi)^2\det(\Gamma)}\log j\qquad(j\to\infty).$$

By (5.5.11), we have

$$\mathrm{Var}\Big(\sum_{i=1}^{n-1}Z_i\Big)\sim\frac{2}{(2\pi)^2\det(\Gamma)}n\log n\qquad(n\to\infty)$$

which, together with (5.5.9) and (5.5.10), implies (5.5.5). □

The following theorem shows that $Q_n-\mathbb{E}Q_n$ and $R_n-\mathbb{E}R_n$ are attracted by normal distributions as $d\ge3$.

THEOREM 5.5.3. (1) *When $d = 3$,*

$$\frac{1}{\sqrt{n\log n}}\Big(Q_n - \mathbb{E}Q_n\Big) \xrightarrow{d} N\Big(0, \frac{1}{2\pi^2\det(\Gamma)}\Big), \tag{5.5.17}$$

$$\frac{1}{\sqrt{n\log n}}\Big(R_n - \mathbb{E}R_n\Big) \xrightarrow{d} N\Big(0, \frac{\gamma_S^4}{2\pi^2\det(\Gamma)}\Big). \tag{5.5.18}$$

(2) *When $d \geq 4$,*

$$\frac{1}{\sqrt{n}}\Big(Q_n - \mathbb{E}Q_n\Big) \xrightarrow{d} N(0, \mathfrak{D}^2), \tag{5.5.19}$$

$$\frac{1}{\sqrt{n}}\Big(R_n - \mathbb{E}R_n\Big) \xrightarrow{d} N(0, \widetilde{\mathfrak{D}}^2), \tag{5.5.20}$$

where $\mathfrak{D}^2$ and $\widetilde{\mathfrak{D}}^2$ are defined in Lemma 5.5.2.

Proof. We only prove (5.5.17). Let $\{\gamma_n\}$ be a positive sequence such that

$$\gamma_n \longrightarrow \infty \quad \text{and} \quad \gamma_n = o\Big(\sqrt{\log n}\Big) \qquad (n\to\infty).$$

Let $0 = n_0 < n_1 < \cdots < n_{\gamma_n} = n$ be an integer partition of $[0, n]$ such that for each $1 \leq i \leq \gamma_n$, $n - i - n_{i-1} = [n\gamma_n^{-1}]$ or $[n\gamma_n^{-1}] + 1$. Then

$$Q_n = \sum_{i=1}^{\gamma_n}\sum_{n_{i-1}<j<k\leq n_i} 1_{\{S(j)=S(k)\}} + \sum_{i=1}^{\gamma_n-1}\sum_{j=n_{i-1}+1}^{n_i}\sum_{k=n_i+1}^{n} 1_{\{S(j)=S(k)\}}. \tag{5.5.21}$$

For each $1 \leq i \leq \gamma_n - 1$,

$$\sum_{j=n_{i-1}+1}^{n_i}\sum_{k=n_i+1}^{n} 1_{\{S(j)=S(k)\}} \overset{d}{=} \sum_{j=1}^{n_i-n_{i-1}}\sum_{k=1}^{n-n_i} 1_{\{S_j=\widetilde{S}(k)\}} \leq \sum_{j=1}^{n}\sum_{k=1}^{n} 1_{\{S(j)=\widetilde{S}(k)\}}$$

where $\{\widetilde{S}(k)\}$ is an independent copy of $\{S(k)\}$. By (5.3.14) in Theorem 5.3.2 (with $d = 2$ and $p = 2$), we have

$$\mathbb{E}\Big(\sum_{i=1}^{\gamma_n-1}\sum_{j=n_{i-1}+1}^{n_i}\sum_{k=n_i+1}^{n} 1_{\{S(j)=S(k)\}}\Big)^2 \leq \gamma_n^2\mathbb{E}\Big(\sum_{j=1}^{n}\sum_{k=1}^{n} 1_{\{S(j)=\widetilde{S}(k)\}}\Big)^2 = o(n\log n). \tag{5.5.22}$$

In addition, notice that the random variables

$$\sum_{n_{i-1}<j<k\leq n_i} 1_{\{S(j)=S(k)\}} \qquad i = 1, 2, \cdots, \gamma_n$$

are independent with

$$\sum_{n_{i-1}<j<k\leq n_i} 1_{\{S_j=S_k\}} \overset{d}{=} Q_{n_i-n_{i-1}} \qquad i = 1, 2, \cdots, \gamma_n.$$

Comparing (5.5.5) and (5.5.22)

$$\mathrm{Var}\Big(\sum_{i=1}^{\gamma_n}\sum_{n_{i-1}<j<k\le n_i}1_{\{S_j=S_k\}}\Big)\sim\mathfrak{D}^2 n\log n.$$

From Theorem 6.4.1 in the next chapter, we can check the Lindberg condition. Hence,

$$\sum_{i=1}^{\gamma_n}\bigg\{\sum_{n_{i-1}<j<k\le n_i}1_{\{S_j=S_k\}}-\mathbb{E}\Big(\sum_{n_{i-1}<j<k\le n_i}1_{\{S_j=S_k\}}\Big)\bigg\}\Big/\sqrt{n\log n}$$
$$\xrightarrow{d} N\Big(0,\ \frac{1}{2\pi^2\det(\Gamma)}\Big).$$

Finally, (5.5.17) follows from (5.5.21) and (5.5.22). □

From the argument for Theorem 5.5.3, we see that as $d\ge 3$, the short range self-intersection dominates the long-range self-intersection. Consequently, $Q_n-\mathbb{E}Q_n$ (as well as $R_n-\mathbb{E}R_n$) is approximated by independent sums. That is the reason we have Gaussian laws here.

5.6. Notes and comments

Section 5.1

We refer the reader to the book by Spitzer [**155**] for the collection of general theory on random walks. The terminology used by Spitzer is slightly different from what we use here: What we call "aperiodic" is "strongly aperiodic" in Spitzer [**155**].

EXERCISE 5.6.1. *We discuss the relation (see (5.1.10))*

(5.6.1) $$\gamma_S=\Big(\sum_{k=0}^\infty\mathbb{P}\{S(k)=0\}\Big)^{-1}$$

in the case $d\ge 3$.

Let T_S *be the last time that the random walk visits 0. That is,*

$$T_S=\max\big\{k\ge 0;\ \ S(k)=0\big\}.$$

It is easy to see that T_S *is not a stopping time. Since*

$$\sum_{k=0}^\infty\mathbb{P}\{S(k)=0\}<\infty,$$

by the Borel-Cantelli lemma, we have that $\mathbb{P}\{S(k)=0,\ \ i.o.\}=0$. *Equivalently, we have that* $T_S<\infty$ *a.s. Therefore,*

$$\sum_{k=0}^\infty\mathbb{P}\{T_S=k\}=1.$$

(1) *Prove (5.6.1) based on this equation.*

(2) *Prove that*

$$\sum_{k=0}^{\infty}\mathbb{P}\{S(k)=0\}=\frac{1}{(2\pi)^d}\int_{[-\pi,\pi]^d}\frac{1}{1-\varphi(\lambda)}d\lambda$$

and give an alternative representation of γ_S.

EXERCISE 5.6.2 (Chapman-Kolmogorov equation). *Prove that for any* $j,k\geq 1$,

$$\sum_{x\in\mathbb{Z}^d}\mathbb{P}\{S(j)=x\}\mathbb{P}\{S(k)=x\}=\mathbb{P}\{S(j+k)=0\}. \tag{5.6.2}$$

Section 5.2

Theorem 5.2.1, Theorem 5.2.3 and the argument used to their establishment come from the paper by Chen and Li ([**35**]). The proof of Lemma 5.2.2 is essentially adopted from Jain and Pruitt [**104**].

EXERCISE 5.6.3. *Prove that*

$$\sup_{x\in\mathbb{Z}} l(n,x)/\sqrt{n}\xrightarrow{d}\sigma^{-1}\sup_{x\in\mathbb{R}}L(1,x) \tag{5.6.3}$$

when $d=1$, *where* $\sigma^2>0$ *is the variance of the i.i.d. increment of the random walk.*

Section 5.3

The weak laws for I_n and J_n were first studied by Le Gall [**124**] in the sub-critical dimensions. He shows that every integral moment of properly normalized I_n and J_n converges to the same moment of $\alpha([0,1]^p)$. Prior to [**124**], Le Gall had obtained an analogue for the intersection of two independent Brownian sausages ([**123**]). The idea of using Fourier transformation in the proof of the weak law for I_n (Theorem 5.3.2) can be found in the paper by Rosen [**151**]. The treatment of approximating J_n by the multiple of I_n (proof of Lemma 5.3.3) can be seen in the paper by Le Gall and Rosen ([**131**]).

EXERCISE 5.6.4. *Let* $p=2$ *and* $d\leq 3$. *Use the Chapman-Kolmogorov equation to prove that* $\mathbb{E}I_n\sim cn^{\frac{4-d}{2}}$ *and find the constant* c.

EXERCISE 5.6.5. *Complete the proof of Lemma 5.3.1 by extending it from the aperiodic case to the general case.*

Hint: Let $\{\delta_k\}$, $\{\tau_k\}$ *be defined as in the proof of Lemma 5.1.2. Notice that the local time of the aperiodic random walk* $\widetilde{S}(k)=S(\tau_k)$ *can be written as*

$$\tilde{l}(n,x)=\sum_{k=1}^{\tau_n}\delta_k 1_{\{S(k)=x\}},$$

which can be close to $l(n,x)$ *as* θ *is close to 1.*

EXERCISE 5.6.6. *In this exercise we consider the super-critical dimensions defined by the inequality $p(d-2) > d$. Prove that*

$$\sum_{x\in\mathbb{Z}^d}\Big[\sum_{k=1}^\infty \mathbb{P}\{S_k = x\}\Big]^p < \infty. \tag{5.6.4}$$

Then use this fact to show that $I_\infty < \infty$, $J_\infty < \infty$ a.s., where

$$I_\infty = \#\Big\{(k_1,\cdots,k_p)\in(\mathbb{Z}^+)^p;\ \ S_1(k_1) = \cdots = S_p(k_p)\Big\}$$

$$J_\infty = \#\Big\{x\in\mathbb{Z}^d;\ \ x = S_1(k_1) = \cdots = S_p(k_p)\ \textit{for some}\ (k_1,\cdots,k_p)\in(\mathbb{Z}^+)^p\Big\}.$$

Hint: Some idea in the proof of Lemma 5.3.1 may be useful in the proof for the first part.

Section 5.4.

The weak laws for $Q_n - \mathbb{E}Q_n$ and $R_n - \mathbb{E}R_n$ in the case $d = 2$ were first investigated by Le Gall [**124**] and his results were essentially what we stated in Theorem 5.4.2 and in Theorem 5.4.3. Le Gall's work was later extended by Rosen [**151**] and by Le Gall and Rosen [**131**] to more general random walks, for example, to the stable random walks.

The study of the range in $d = 2$ has been sensational and several interesting things happen in this dimension. Let, for example, $R_n^{(p)}$ be the number of distinct sites visited at least p ($p \geq 1$) times by $S(k)$ in the first n steps. Under this notation, $R_n = R_n^{(1)} \geq R_n^{(p)}$ for $p \geq 2$. Hamana ([**87**]) points out that

$$\frac{(\log n)^2}{n}\Big[R_n^{(p)} - \mathbb{E}R_n^{(p)}\Big] \xrightarrow{d} -(2\pi)^2\sqrt{\det(\Gamma)}\gamma\big([0,1]_<^2\big). \tag{5.6.5}$$

A striking fact is that $R_n^{(p)}$ obeys exactly the same weak law as R_n. This seems to suggest that once being visited, it then costs almost nothing to be visited again.

Let $\widehat{R}_n^{(p)}$ be the number of distinct sites visited exactly p ($p \geq 1$) times by the two-dimensional random walk $S(k)$ in the first n steps. In the same work, Hamana ([**87**]) points out that

$$\frac{(\log n)^3}{n}\Big[\widehat{R}_n^{(p)} - \mathbb{E}\widehat{R}_n^{(p)}\Big] \xrightarrow{d} -16\pi^3\det(\Gamma)\gamma\big([0,1]_<^2\big). \tag{5.6.6}$$

Comparing with (5.6.5), the law given in (5.6.6) indicates that there is a price to pay for not being visited ever since. Further, its irrelevance to the parameter p shows that the cost is roughly the same regardless of the number of visits allowed before the visit is stopped.

When p increases with n, the sites counted by $\mathbb{R}_n^{(p)}$ are called *thick points*. A remarkable progress in recent years is the investigation of thick points led by Dembo, Peres, Rosen and Zeitouni ([**48**]) which claims that as $p_n \sim a(\log n)^2$, then

$$\lim_{n\to\infty}(\log n)^{-1}\log R_n^{(p_n)} = 1 - \pi a \qquad a.s.$$

A direct consequence of their work is the confirmation of the long standing Erdös-Taylor's conjecture ([**75**]) on the strong law

$$\lim_{n\to\infty}(\log n)^{-2}\sup_{x\in\mathbb{Z}^2} l(n,x)=\frac{1}{\pi}\qquad a.s.$$

in the setting of 2-dimensional simple random walks.

Other related models are the time needed for the random walk to cover a big disk (cover-time problem) and the waiting time for visiting a fresh site after the first n steps. These models were investigated by Dembo, Peres and Rosen ([**50**]).

Section 5.5.

EXERCISE 5.6.7. *Use the Chapman-Kolmogorov equation to prove that as $d=4$ and $p=2$,*

$$\mathbb{E}I_n\sim\frac{1}{(2\pi)^2\sqrt{\det(\Gamma)}}\log n\qquad(n\to\infty).$$

Then show that

$$\mathbb{E}J_n\sim\frac{\gamma_S^2}{(2\pi)^2\sqrt{\det(\Gamma)}}\log n\qquad(n\to\infty).$$

Section 5.6.

Due to some historic reasons, much of the focus is on the $R_n-\mathbb{E}R_n$ rather than on $Q_n-\mathbb{E}Q_n$ as $d\ge 3$. The central limit theorem given in (5.5.20) was first obtained by Jain and Orey ([**101**]) in the special case $d\ge 5$. Later, it was extended by Jain and Pruitt ([**102**]) to $d=4$. In the same work, Jain and Pruitt established (5.5.18) for $d=3$. The limit laws (5.5.18) and (5.5.20) in their current form come from Le Gall and Rosen ([**131**]), where the parameters are identified and the results are established in a more general setting.

Similar results were obtained by Yor ([**171**]) and by Calais and Yor ([**22**]) in the setting of Brownian sausages. The central limit theorem for $Q_n-\mathbb{E}Q_n$ (given in Theorem 5.5.3) was established in [**33**] much later. The argument used in Lemma 5.5.2 goes back at least to Jain and Orey ([**101**]).

EXERCISE 5.6.8. *Prove (5.5.18) and (5.5.20) in Theorem 5.5.3.*

The next two exercises appear as applications of the weak laws for Q_n stated in Theorem 5.2.3 (with $p=2$), Theorem 5.4.2 and Theorem 5.5.3. They are adopted from Kesten and Spitzer ([**107**]), Bolthausen ([**15**]), Khoshnevisan and Lewis ([**111**]), Chen ([**33**]), Chen and Khoshnevisan ([**34**]).

EXERCISE 5.6.9 (hard). *Let $\{\xi(x);\ x\in\mathbb{Z}^d\}$ be a family of i.i.d.random variables independent of the random walk $\{S(k)\}_{k\ge1}$ such that $\mathbb{E}\xi(0)=0$ and $\mathbb{E}\xi^2(0)=1$. The random sequence*

$$\zeta_n=\sum_{k=1}^n\xi(S(k))\qquad n=1,2,\cdots \tag{5.6.7}$$

is called random walk in random scenery in literature, where the random scenery is referred to as $\{\xi(x);\ x\in\mathbb{Z}^d\}$.

(1) *When* $d=1$,

$$n^{-3/4}\zeta_n \xrightarrow{d} U\sqrt{\frac{1}{\sigma}\int_{-\infty}^{\infty} L^2(1,x)dx} \tag{5.6.8}$$

where $U \sim N(0,1)$ *independent of* $W(t)$.

(2) *When* $d=2$,

$$(n\log n)^{-1/2}\zeta_n \xrightarrow{d} \frac{1}{\sqrt{\pi}\sqrt[4]{\det(\Gamma)}}U. \tag{5.6.9}$$

(3) *When* $d\ge 3$,

$$\zeta_n/\sqrt{n} \xrightarrow{d} U\sqrt{1+2\sum_{k=1}^{\infty}\mathbb{P}\{S(k)=0\}}. \tag{5.6.10}$$

Hint: Conditioning on $\{S(1),\cdots,S(n)\}$, $\zeta_n = \sum_{x\in\mathbb{Z}^d} l(n,x)\xi(x)$ *is a sum of independent random variables.*

EXERCISE 5.6.10 (hard). *Let* $\{\omega_k\}_{k\ge 1}$ *be a sequence of i.i.d. random variables independent of the random walk* $\{S(k)\}_{k\ge 1}$ *such that* $\mathbb{E}\omega_1=0$ *and* $\mathbb{E}\omega_1^2=1$. *Assume, in addition, that* $\mathbb{E}\omega^4<\infty$. *The random quantity*

$$H_n = \sum_{1\le j<k\le n}\omega_j\omega_k 1_{\{S(j)=S(k)\}} \qquad n=1,2,\cdots \tag{5.6.11}$$

represents the interacting energy of the discrete polymer with vertices

$$\{S(1),\cdots,S(n)\},$$

where $\omega_1,\cdots,\omega_n$ *are the random (electrical) charges carried by the polymer.*

(1) *When* $d=1$,

$$n^{-3/4}H_n \xrightarrow{d} U\sqrt{\frac{1}{2\sigma}\int_{-\infty}^{\infty} L^2(1,x)dx} \tag{5.6.12}$$

where $U\sim N(0,1)$ *is independent of* $W(t)$.

(2) *When* $d=2$,

$$(n\log n)^{-1/2}H_n \xrightarrow{d} \frac{1}{\sqrt{2\pi}\sqrt[4]{\det(\Gamma)}}U. \tag{5.6.13}$$

(3) *When* $d\ge 3$,

$$H_n/\sqrt{n} \xrightarrow{d} U\sqrt{\sum_{k=1}^{\infty}\mathbb{P}\{S(k)=0\}}. \tag{5.6.14}$$

Hint: Follow an idea similar to the one in the previous exercise in connection to the representation

$$H_n = \frac{1}{2}\sum_{x\in\mathbb{Z}^d}\bigg\{\Big(\sum_{k=1}^n \omega_k 1_{\{S(k)=x\}}\Big)^2 - \sum_{k=1}^n \omega_k^2 1_{\{S(k)=x\}}\bigg\}.$$

CHAPTER 6

Inequalities and integrabilities

The large deviation principle appears as asymptotically sharp inequalities of the probabilities of rare events. On the other hand, sharp inequalities are usually important tools for studying the large deviation principle. When such inequalities take a clean and symmetric form, they become a focus of interest. Our first goal in this chapter is to establish such inequalities for additive and sub-additive functionals associated with mutual intersections. The key idea in these inequalities is to separate the (mutual) intersections that happen in different time periods.

It is hard to imagine that a practical interesting model satisfying the large deviation principle does not have any basic exponential integrabilities. The second goal of this chapter is to establish the exponential integrabilities for the intersection local times I_n, Q_n, the range R_n and the intersection of independent ranges J_n.

The argument by sub-additivity has been an effective approach for establishing large deviation integrabilities. The multinomial inequalities established in Section 6.1 provide the condition which allows such an argument. In Section 6.2 they are applied directly to the problem of maximal integrabilities for I_n and J_n.

In Section 6.3 and Section 6.4, the integrabilities established for I_n and J_n are extended to the renormalized self-intersection local time $Q_n - \mathbb{E}Q_n$ and to the renormalized range $R_n - \mathbb{E}R_n$.

6.1. Multinomial inequalities

Let $S_1(k), \cdots, S_p(k)$ be independent symmetric random walks in $\mathbb{Z}^d$. For convenience, we view $S_1(k), \cdots, S_p(k)$ as independent copies of the random walk $\{S(k)\}$.

Let the integer $a \geq 2$ and the increasing integers $1 \leq n_1 < \cdots < n_a$ be fixed. We introduce the notations $\mathcal{F}_k^x(S)$ and $\mathcal{G}_k^x(S)$ ($x \in \mathbb{Z}^d$, $k = 1, \cdots, a$) for two measurable functionals of a random walk $S = \{S(k)\}$ on $\mathbb{Z}^d$. The following assumptions occur throughout this section.

(1) Adaptivity of $\mathcal{F}_k^x(S)$: $\mathcal{F}_k^x(S) \in \sigma\{S(j);\ 1 \leq j \leq n_k\}$ for any $x \in \mathbb{Z}^d$, $k = 1, \cdots, a$.

(2) Spatial homogeneity[1] of $\mathcal{G}_k^x(S)$: $\mathcal{G}_k^x(z + S) = \mathcal{G}_k^{x-z}$ for any $x, z \in \mathbb{Z}^d$, $k = 1, \cdots, a$.

[1]Here we allow the random walk to start at the non-zero point. In this way, $z + S$ constitutes a random walk for any $z \in \mathbb{Z}^d$.

The functional $\mathcal{F}_k^x(S)$ $(k = 1, \cdots, a)$ is said to be *sub-additive* on the increasing sequence $n_1 < n_2 < \cdots < n_a$ with respect to the *increment functional* $\mathcal{G}_k^x$, if

$$\mathcal{F}_1(S)^x = \mathcal{G}_1^x(S) \quad \text{and} \quad \mathcal{F}_k^x(S) \le \mathcal{F}_{k-1}^x(S) + \mathcal{G}_k^x(S \circ \theta^{n_{k-1}}) \tag{6.1.1}$$

for $x \in \mathbb{Z}^d$, $k = 1, \cdots, a$, where θ is the usual shift operator of the Markov chain $\{S(k)\}_{k\ge 1}$: $S(n) \circ \theta = S(n+1)$, so that

$$S \circ \theta^{n_{k-1}} = \Big\{S(n_{k-1}+1),\ S(n_{k-1}+2), \cdots\Big\}.$$

Further, $\mathcal{F}_k^x(S)$ is said to be *additive* if the inequality in (6.1.1) is replaced by an equality.

THEOREM 6.1.1. *Assume the sub-additivity defined in (6.1.1) and assume that $\mathcal{F}_k^x(S) \ge 0$, $\mathcal{G}_k^x(S) \ge 0$ for $x \in \mathbb{Z}^d$, $k = 1, \cdots, a$. For any integer $m \ge 1$,*

$$\begin{aligned} &\bigg\{\mathbb{E}\Big[\sum_{x\in\mathbb{Z}^d}\prod_{j=1}^p \mathcal{F}_a^x(S_j)\Big]^m\bigg\}^{1/p} \\ &\le \sum_{\substack{l_1+\cdots+l_a=m\\ l_1,\cdots,l_a\ge 0}} \frac{m!}{l_1!\cdots l_a!}\prod_{k=1}^a\bigg\{\mathbb{E}\Big[\sum_{x\in\mathbb{Z}^d}\prod_{j=1}^p \mathcal{G}_k^x(S_j)\Big]^{l_k}\bigg\}^{1/p}. \end{aligned} \tag{6.1.2}$$

Consequently, for any $u > 0$,

$$\begin{aligned} &\sum_{m=0}^\infty \frac{u^m}{m!}\bigg\{\mathbb{E}\Big[\sum_{x\in\mathbb{Z}^d}\prod_{j=1}^p \mathcal{F}_a^x(S_j)\Big]^m\bigg\}^{1/p} \\ &\le \prod_{k=1}^a\sum_{m=0}^\infty \frac{u^m}{m!}\bigg\{\mathbb{E}\Big[\sum_{x\in\mathbb{Z}^d}\prod_{j=1}^p \mathcal{G}_k^x(S_j)\Big]^m\bigg\}^{1/p}. \end{aligned} \tag{6.1.3}$$

Proof. The inequality (6.1.3) follows directly from (6.1.2). By an induction procedure, we need only to show that for $k = 2, \cdots, a$,

$$\begin{aligned} &\bigg\{\mathbb{E}\Big[\sum_{x\in\mathbb{Z}^d}\prod_{j=1}^p \mathcal{F}_k^x(S_j)\Big]^m\bigg\}^{1/p} \\ &\le \sum_{l=0}^m \binom{m}{l}\bigg\{\mathbb{E}\Big[\sum_{x\in\mathbb{Z}^d}\prod_{j=1}^p \mathcal{F}_{k-1}^x(S_j)\Big]^l\bigg\}^{1/p}\bigg\{\mathbb{E}\Big[\sum_{x\in\mathbb{Z}^d}\prod_{j=1}^p \mathcal{G}_k^x(S_j)\Big]^{m-l}\bigg\}^{1/p}. \end{aligned} \tag{6.1.4}$$

Let k be fixed and write

$$\Delta_1(x) = \mathcal{F}_{k-1}^x(S) \quad \text{and} \quad \Delta_2(x) = \mathcal{G}_k^x(S \circ \theta^{n_{k-1}}).$$

By sub-additivity we have

$$\begin{aligned} \mathbb{E}\Big[\sum_{x\in\mathbb{Z}^d}\prod_{j=1}^p \mathcal{F}_k^x(S_j)\Big]^m &= \sum_{x_1,\cdots,x_m\in\mathbb{Z}^d}\Big[\mathbb{E}\prod_{l=1}^m \mathcal{F}_k^{x_l}(S)\Big]^p \\ &\le \sum_{x_1,\cdots,x_m\in\mathbb{Z}^d}\Big[\sum_{i_1,\cdots,i_m=1}^2 \mathbb{E}\Big(\Delta_{i_1}(x_1)\cdots\Delta_{i_m}(x_m)\Big)\Big]^p. \end{aligned}$$

Applying triangular inequality, we get

$$\Big\{\mathbb{E}\Big[\sum_{x\in\mathbb{Z}^d}\prod_{j=1}^p \mathcal{F}_k^x(S_j)\Big]^m\Big\}^{1/p}$$
$$\le \sum_{i_1,\cdots,i_m=1}^2\Big\{\sum_{x_1,\cdots,x_m\in\mathbb{Z}^d}\Big[\mathbb{E}\Big(\Delta_{i_1}(x_1)\cdots\Delta_{i_m}(x_m)\Big)\Big]^p\Big\}^{1/p}$$
$$=\sum_{l=0}^m\binom{m}{l}\Big\{\sum_{x_1,\cdots,x_m\in\mathbb{Z}^d}\Big[\mathbb{E}\Big(\Delta_1(x_1)\cdots\Delta_1(x_l)\Big)\Big(\Delta_2(x_{l+1})\cdots\Delta_2(x_m)\Big)\Big]^p\Big\}^{1/p}.$$

All we need to prove is that

$$\sum_{x_1,\cdots,x_m\in\mathbb{Z}^d}\Big[\mathbb{E}\Big(\Delta_1(x_1)\cdots\Delta_1(x_l)\Big)\Big(\Delta_2(x_{l+1})\cdots\Delta_2(x_m)\Big)\Big]^p$$
$$\le\mathbb{E}\Big[\sum_{x\in\mathbb{Z}^d}\prod_{j=1}^p\mathcal{F}_{k-1}^x(S_j)\Big]^l\,\mathbb{E}\Big[\sum_{x\in\mathbb{Z}^d}\prod_{j=1}^p\mathcal{G}_k^x(S_j)\Big]^{m-l}. \tag{6.1.5}$$

For fixed $(x_1,\cdots,x_l)$,

$$\sum_{x_{l+1},\cdots,x_m\in\mathbb{Z}^d}\Big[\mathbb{E}\Big(\Delta_1(x_1)\cdots\Delta_1(x_l)\Big)\Big(\Delta_2(x_{l+1})\cdots\Delta_2(x_m)\Big)\Big]^p$$
$$=\mathbb{E}\Big\{\Big(\Delta_1(x_1)\cdots\Delta_1(x_l)\Big)\sum_{x_{l+1},\cdots,x_m\in\mathbb{Z}^d}\Big(\Delta_2(x_{l+1})\cdots\Delta_2(x_m)\Big)$$
$$\times\Big[\mathbb{E}\Big(\Delta_1(x_1)\cdots\Delta_1(x_l)\Big)\Big(\Delta_2(x_{l+1})\cdots\Delta_2(x_m)\Big)\Big]^{p-1}\Big\}$$
$$=\mathbb{E}\Big\{\Big(\prod_{j=1}^l\mathcal{F}_{k-1}^{x_j}(S)\Big)\sum_{x_{l+1},\cdots,x_m\in\mathbb{Z}^d}\prod_{j=l+1}^m\mathcal{G}_k^{x_j}(S\circ\theta^{n_{k-1}})$$
$$\times\Big[\mathbb{E}\Big(\Delta_1(x_1)\cdots\Delta_1(x_l)\Big)\Big(\Delta_2(x_{l+1})\cdots\Delta_2(x_m)\Big)\Big]^{p-1}\Big\}.$$

By spatial homogeneity of $\mathcal{G}_k^x(S)$, we get

$$\prod_{j=l+1}^m\mathcal{G}_k^{x_j}(S\circ\theta^{n_{k-1}})=\prod_{j=l+1}^m\mathcal{G}_k^{x_j-S(n_{k-1})}(S^{(k)})$$

where the random walk

$$S^{(k)}(n)=S(n_{k-1}+n)-S(n)\qquad n=1,2,\cdots$$

is independent of $\{S(1),\cdots,S(n_{k-1})\}$.

Write

$$\xi(x_{l+1},\cdots,x_m)=\mathbb{E}\prod_{j=l+1}^m\mathcal{G}_k^{x_j}(S).$$

We have that

$$\sum_{x_{l+1},\cdots,x_m\in\mathbb{Z}^d}\Big[\mathbb{E}\Big(\Delta_1(x_1)\cdots\Delta_1(x_l)\Big)\Big(\Delta_2(x_{l+1})\cdots\Delta_2(x_m)\Big)\Big]^p$$
$$=\mathbb{E}\bigg\{\Big(\prod_{j=1}^l\mathcal{F}_{k-1}^{x_j}(S)\Big)\sum_{x_{l+1},\cdots,x_m\in\mathbb{Z}^d}\xi\big(x_{l+1}-S(n_{k-1}),\cdots,x_m-S(n_{k-1})\big)$$
$$\times\Big[\mathbb{E}\Big(\Delta_1(x_1)\cdots\Delta_1(x_l)\Big)\Big(\Delta_2(x_{l+1})\cdots\Delta_2(x_m)\Big)\Big]^{p-1}\bigg\}.$$

By Hölder inequality, we get

$$\sum_{x_{l+1},\cdots,x_m\in\mathbb{Z}^d}\xi\big(x_{l+1}-S(n_{k-1}),\cdots,x_m-S(n_{k-1})\big)$$
$$\times\Big[\mathbb{E}\Big(\Delta_1(x_1)\cdots\Delta_1(x_l)\Big)\Big(\Delta_2(x_{l+1})\cdots\Delta_2(x_m)\Big)\Big]^{p-1}$$
$$\le\bigg\{\sum_{x_{l+1},\cdots,x_m\in\mathbb{Z}^d}\xi^p\big(x_{l+1}-S(n_{k-1}),\cdots,x_m-S(n_{k-1})\big)\bigg\}^{1/p}$$
$$\times\bigg\{\sum_{x_{l+1},\cdots,x_m\in\mathbb{Z}^d}\Big[\mathbb{E}\Big(\Delta_1(x_1)\cdots\Delta_1(x_l)\Big)\Big(\Delta_2(x_{l+1})\cdots\Delta_2(x_m)\Big)\Big]^p\bigg\}^{\frac{p-1}{p}}.$$

By translation invariance,

$$\sum_{x_{l+1},\cdots,x_m\in\mathbb{Z}^d}\xi^p\big(x_{l+1}-S(n_{k-1}),\cdots,x_m-S(n_{k-1})\big)$$
$$=\sum_{x_{l+1},\cdots,x_m\in\mathbb{Z}^d}\xi^p\big(x_{l+1},\cdots,x_m\big)=\mathbb{E}\Big[\sum_{x\in\mathbb{Z}^d}\prod_{j=1}^p\mathcal{G}_k^x(S_j)\Big]^{m-l}.$$

Summarizing our argument, for any $x_1,\cdots,x_l\in\mathbb{Z}^d$,

$$\sum_{x_{l+1},\cdots,x_m\in\mathbb{Z}^d}\Big[\mathbb{E}\Big(\Delta_1(x_1)\cdots\Delta_1(x_l)\Big)\Big(\Delta_2(x_{l+1})\cdots\Delta_2(x_m)\Big)\Big]^p$$
$$\le\bigg\{\mathbb{E}\Big[\sum_{x\in\mathbb{Z}^d}\prod_{j=1}^p\mathcal{G}_k^x(S_j)\Big]^{m-l}\bigg\}^{1/p}\Big(\mathbb{E}\prod_{j=1}^l\mathcal{F}_{k-1}^{x_j}(S)\Big)$$
$$\times\bigg\{\sum_{x_{l+1},\cdots,x_m\in\mathbb{Z}^d}\Big[\mathbb{E}\Big(\Delta_1(x_1)\cdots\Delta_1(x_l)\Big)\Big(\Delta_2(x_{l+1})\cdots\Delta_2(x_m)\Big)\Big]^p\bigg\}^{\frac{p-1}{p}}.$$

Hence, we conclude that

$$\sum_{x_{l+1},\cdots,x_m\in\mathbb{Z}^d}\Big[\mathbb{E}\Big(\Delta_1(x_1)\cdots\Delta_1(x_l)\Big)\Big(\Delta_2(x_{l+1})\cdots\Delta_2(x_m)\Big)\Big]^p$$
$$\le\mathbb{E}\Big[\sum_{x\in\mathbb{Z}^d}\prod_{j=1}^p\mathcal{G}_k^x(S_j)\Big]^{m-l}\Big[\mathbb{E}\prod_{j=1}^l\mathcal{F}_{k-1}^{x_j}(S)\Big]^p$$

at least when

$$\mathbb{E}\prod_{j=1}^{l}\mathcal{F}_{k-1}^{x_j}(S)>0.$$

In other case, the left hand side is equal to 0. Therefore, the inequality holds for every $(x_1,\cdots,x_l)$.

Finally, the requested (6.1.5) follows from the fact that

$$\sum_{x_1,\cdots,x_l\in\mathbb{Z}^d}\Big[\mathbb{E}\prod_{j=1}^{l}\mathcal{F}_{k-1}^{x_j}(S)\Big]^p=\mathbb{E}\Big[\sum_{x\in\mathbb{Z}^d}\prod_{j=1}^{p}\mathcal{F}_{k-1}^{x}(S_j)\Big]^l.$$

□

The inequality given in (6.1.2) (and the inequalities in (6.1.13) and (6.1.14) in the next theorem) is called the *multinomial inequality* due to the multinomial form on its right hand side. In this book, the inequalities established in Theorem 6.1.1 are mainly applied to the following three special cases.

Case 1. Recall that $l(n,x)$ is the local time of $\{S(k)\}$. The functional $\mathcal{F}_k^x(S)=l(n_k,x)$ $(k=1,\cdots,a)$ is additive on $n_1<\cdots<n_a$ with increment functional $\mathcal{G}_k^x(S)=l(n_k-n_{k-1},x)$ $(k=1,\cdots,a)$. Here we follow the convention $n_0=0$.

Case 2. The functional $\mathcal{F}_k^x(S)=1_{\{x\in S[1,n_k]\}}$ $(k=1,\cdots,a)$ is sub-additive on $n_1<\cdots<n_a$ with the increment functional

$$\mathcal{G}_k^x(S)=1_{\{x\in S[1,n_k-n_{k-1}]\}}\qquad k=1,\cdots,a.$$

Case 3. An additive version of Case 2 is defined by

$$\mathcal{F}_k^x(S)=\sum_{j=1}^{k}1_{\{x\in S(n_{j-1},n_j]\}}\qquad k=1,\cdots,a; \tag{6.1.6}$$

it is additive on $n_1<\cdots<n_a$ with the same increment functional $\mathcal{G}_k^x(S)=1_{\{x\in S[1,n_k-n_{k-1}]\}}$.

Applying Theorem 6.1.1 to these three cases we obtain the following inequalities:

$$\Big\{\mathbb{E}I_{n_a}^m\Big\}^{1/p}\le\sum_{\substack{l_1+\cdots+l_a=m\\ l_1,\cdots,l_a\ge 0}}\frac{m!}{l_1!\cdots l_a!}\prod_{k=1}^{a}\Big\{\mathbb{E}I_{n_k-n_{k-1}}^{l_k}\Big\}^{1/p}, \tag{6.1.7}$$

$$\sum_{m=0}^{\infty}\frac{u^m}{m!}\big(\mathbb{E}I_{n_a}^m\big)^{1/p}\le\prod_{k=1}^{a}\sum_{m=0}^{\infty}\frac{u^m}{m!}\big(\mathbb{E}I_{n_k-n_{k-1}}^m\big)^{1/p}, \tag{6.1.8}$$

$$\Big\{\mathbb{E}J_{n_a}^m\Big\}^{1/p}\le\sum_{\substack{l_1+\cdots+l_a=m\\ l_1,\cdots,l_a\ge 0}}\frac{m!}{l_1!\cdots l_a!}\prod_{k=1}^{a}\Big\{\mathbb{E}J_{n_k-n_{k-1}}^{l_k}\Big\}^{1/p}, \tag{6.1.9}$$

$$\sum_{m=0}^{\infty}\frac{u^m}{m!}\big(\mathbb{E}J_{n_a}^m\big)^{1/p}\le\prod_{k=1}^{a}\sum_{m=0}^{\infty}\frac{u^m}{m!}\big(\mathbb{E}J_{n_k-n_{k-1}}^m\big)^{1/p}, \tag{6.1.10}$$

$$\left\{\mathbb{E}\mathcal{J}_{n_a}^m\right\}^{1/p} \le \sum_{\substack{l_1+\cdots+l_a=m\\ l_1,\cdots,l_a\ge 0}} \frac{m!}{l_1!\cdots l_a!}\prod_{k=1}^a \left\{\mathbb{E}J_{n_k-n_{k-1}}^{l_k}\right\}^{1/p}, \tag{6.1.11}$$

$$\sum_{m=0}^\infty \frac{u^m}{m!}\left(\mathbb{E}\mathcal{J}_{n_a}^m\right)^{1/p} \le \prod_{k=1}^a \sum_{m=0}^\infty \frac{u^m}{m!}\left(\mathbb{E}J_{n_k-n_{k-1}}^m\right)^{1/p}. \tag{6.1.12}$$

Here we recall that

$$I_n = \sum_{x\in\mathbb{Z}^d}\prod_{j=1}^p l_j(n,x) \quad\text{and}\quad J_n = \sum_{x\in Z^d}\prod_{j=1}^p 1_{\{x\in S_j[1,n]\}}$$

and we define

$$\mathcal{J}_{n_a} = \sum_{x\in Z^d}\prod_{j=1}^p\sum_{k=1}^a 1_{\{x\in S_j(n_{k-1},n_k]\}}.$$

The argument for Theorem 6.1.1 does not work if $\mathcal{F}_k^x(S)$ and $\mathcal{G}_k^x(S)$ are allowed to take negative values. On the other hand, there are some practical cases in which one has to deal with the additive functional with sign flapping. As seen in Section 5.3, an important way to study the limit behavior of the quantity

$$\sum_{x\in\mathbb{Z}^d}\prod_{j=1}^p \mathcal{F}_k^x(S_j)$$

is to approximate it by its smoothed version

$$\sum_{x\in\mathbb{Z}^d}\prod_{j=1}^p (h*\mathcal{F})_k^x(S_j)$$

where $h(x)$ is a suitably chosen probability density function on $\mathbb{Z}^d$ and

$$(h*\mathcal{F})_k^x(S) = \sum_{y\in\mathbb{Z}^d} h(x-y)\mathcal{F}_k^y(S).$$

In this scenario, it is desirable to provide inequalities to bound the error term. The additive (or, sub-additive in general) functional one has to deal with may take form as

$$\mathcal{F}_k^x(S) - (h*\mathcal{F})_k^x(S),$$

which is not constantly non-negative.

In the following discussion, we focus our attention on the case of two independent random walks.

THEOREM 6.1.2. *Let $\mathcal{F}_k^x(S)$ $(k=1,\cdots,a)$ be additive on the increasing sequence $1<n_1<\cdots<n_a$ with the increment $\mathcal{G}_k^x$. For any probability density function $h(x)$*

on $\mathbb{Z}^d$ *and any integer* $m \geq 1$,

(6.1.13)
$$\begin{aligned}&\Big\{\mathbb{E}\Big[\sum_{x\in\mathbb{Z}^d}\prod_{j=1}^2 \mathcal{F}_a^x(S_j) - \sum_{x\in\mathbb{Z}^d}\prod_{j=1}^2 (h*\mathcal{F})_a^x(S_j)\Big]^m\Big\}^{1/2}\\ &\leq \sum_{\substack{l_1+\cdots+l_a=m\\ l_1,\cdots,l_a\geq 0}} \frac{m!}{l_1!\cdots l_a!}\\ &\times\prod_{k=1}^a\Big\{\mathbb{E}\Big|\sum_{x\in\mathbb{Z}^d}\prod_{j=1}^2 \mathcal{G}_k^x(S_j) - \sum_{x\in\mathbb{Z}^d}\prod_{j=1}^2 (h*\mathcal{G})_k^x(S_j)\Big|^{l_k}\Big\}^{1/2},\end{aligned}$$

(6.1.14)
$$\begin{aligned}&\Big\{\mathbb{E}\Big[\sum_{x\in\mathbb{Z}^d}\prod_{j=1}^2 \Big(\mathcal{F}_a^x(S_j) - (h*\mathcal{F})_a^x(S_j)\Big)\Big]^m\Big\}^{1/2}\\ &\leq \sum_{\substack{l_1+\cdots+l_a=m\\ l_1,\cdots,l_a\geq 0}} \frac{m!}{l_1!\cdots l_a!}\prod_{k=1}^a\Big\{\mathbb{E}\Big|\sum_{x\in\mathbb{Z}^d}\prod_{j=1}^2 \Big(\mathcal{G}_k^x(S_j) - (h*\mathcal{G})_k^x(S_j)\Big)\Big|^{l_k}\Big\}^{1/2}.\end{aligned}$$

Consequently, for any $u > 0$,

(6.1.15)
$$\begin{aligned}&\sum_{m=0}^\infty \frac{u^m}{m!}\Big\{\mathbb{E}\Big[\sum_{x\in\mathbb{Z}^d}\prod_{j=1}^2 \mathcal{F}_a^x(S_j) - \sum_{x\in\mathbb{Z}^d}\prod_{j=1}^2 (h*\mathcal{F})_a^x(S_j)\Big]^m\Big\}^{1/2}\\ &\leq \prod_{k=1}^a\sum_{m=0}^\infty \frac{u^m}{m!}\Big\{\mathbb{E}\Big|\sum_{x\in\mathbb{Z}^d}\prod_{j=1}^2 \mathcal{G}_k^x(S_j) - \sum_{x\in\mathbb{Z}^d}\prod_{j=1}^2 (h*\mathcal{G})_k^x(S_j)\Big|^m\Big\}^{1/2},\end{aligned}$$

(6.1.16)
$$\begin{aligned}&\sum_{m=0}^\infty \frac{u^m}{m!}\Big\{\mathbb{E}\Big[\sum_{x\in\mathbb{Z}^d}\prod_{j=1}^2 \Big(\mathcal{F}_a^x(S_j) - (h*\mathcal{F})_a^x(S_j)\Big)\Big]^m\Big\}^{1/2}\\ &\leq \prod_{k=1}^a\sum_{m=0}^\infty \frac{u^m}{m!}\Big\{\mathbb{E}\Big|\sum_{x\in\mathbb{Z}^d}\prod_{j=1}^2 \Big(\mathcal{G}_k^x(S_j) - (h*\mathcal{G})_k^x(S_j)\Big)\Big|^m\Big\}^{1/2}.\end{aligned}$$

Proof. Clearly, (6.1.15) and (6.1.16) are direct consequences of (6.1.13) and (6.1.14), respectively.

By additivity (recall the convention $n_0 = 0$), we have

$$\mathcal{F}_a^x(S) = \sum_{k=1}^a \mathcal{G}_k^x(S\circ\theta^{n_{k-1}}).$$

To prove (6.1.15), by induction procedure, it suffices to show that

$$\begin{aligned}(6.1.17)\qquad &\bigg\{\mathbb{E}\Big[\sum_{x\in\mathbb{Z}^d}\prod_{j=1}^2\mathcal{F}_k^x(S_j)-\sum_{x\in\mathbb{Z}^d}\prod_{j=1}^2(h*\mathcal{F})_k^x(S_j)\Big]^m\bigg\}^{1/2}\\
&\le\sum_{l=0}^m\binom{m}{l}\bigg\{\mathbb{E}\Big|\sum_{x\in\mathbb{Z}^d}\prod_{j=1}^2\mathcal{G}_1^x(S_j)-\sum_{x\in\mathbb{Z}^d}\prod_{j=1}^2(h*\mathcal{G})_1^x(S_j)\Big|^l\bigg\}^{1/2}\\
&\times\bigg\{\mathbb{E}\Big[\sum_{x\in\mathbb{Z}^d}\prod_{j=1}^2\mathcal{J}_{a-1}^x(S_j)-\sum_{x\in\mathbb{Z}^d}\prod_{j=1}^2(h*\mathcal{J}_{a-1})^x(S_j)\Big]^{m-l}\bigg\}^{1/2}\end{aligned}$$

where $\mathcal{J}_{a-1}^x(S)=\sum_{k=2}^a\mathcal{G}_k^x(S\circ\theta^{n_{k-1}-n_1})$ is an additive functional on the increasing sequence $1\le n_2-n_1<\cdots<n_a-n_1$.

The approach here relies on Fourier transformation. By Parseval identity, we get

$$\begin{aligned}&\sum_{x\in\mathbb{Z}^d}\prod_{j=1}^2\mathcal{F}_a^x(S_j)-\sum_{x\in\mathbb{Z}^d}\prod_{j=1}^2(h*\mathcal{F})_a^x(S_j)\\
&=\frac{1}{(2\pi)^d}\int_{[-\pi,\pi]^d}\big[1-|\widehat{h}(\lambda)|^2\big]\Big[\sum_{x\in\mathbb{Z}^d}e^{i\lambda\cdot x}\mathcal{F}_a^x(S_1)\Big]\overline{\Big[\sum_{x\in\mathbb{Z}^d}e^{i\lambda\cdot x}\mathcal{F}_a^x(S_1)\Big]}d\lambda\end{aligned}$$

where

$$\widehat{h}(\lambda)=\sum_{x\in\mathbb{Z}^d}h(x)e^{i\lambda\cdot x}.$$

By independence, we get

$$\begin{aligned}(6.1.18)\qquad &\mathbb{E}\Big[\sum_{x\in\mathbb{Z}^d}\prod_{j=1}^2\mathcal{F}_a^x(S_j)-\sum_{x\in\mathbb{Z}^d}\prod_{j=1}^2(h*\mathcal{F})_a^x(S_j)\Big]^m\\
&=\frac{1}{(2\pi)^{md}}\int_{(\mathbb{R}^d)^m}\mu(d\lambda_1)\cdots\mu(d\lambda_m)\Big|\prod_{k=1}^m\sum_{x\in\mathbb{Z}^d}e^{i\lambda_k\cdot x}\mathcal{F}_a^x(S)\Big|^2\end{aligned}$$

where the measure $\mu(d\lambda)$ is given by

$$\mu(d\lambda)=\big[1-|\widehat{h}(\lambda)|^2\big]d\lambda.$$

In particular, the m-th moment is non-negative. Similarly, one can see that all moments appearing in the statement of Theorem 6.1.2 are non-negative. Consequently, everything we have talked about is well defined.

Write

$$\Delta_1(\lambda)=\sum_{x\in\mathbb{Z}^d}e^{i\lambda_k\cdot x}\mathcal{G}_1^x(S),\qquad \Delta_2(\lambda)=\sum_{x\in\mathbb{Z}^d}e^{i\lambda_k\cdot x}\mathcal{J}_{a-1}^x(S)\circ\theta^{n_1}.$$

From the definition of $\mathcal{J}_{a-1}^x(S)$, we get

$$\sum_{x\in\mathbb{Z}^d}e^{i\lambda\cdot x}\mathcal{F}_a^x(S)=\Delta_1(\lambda)+\Delta_2(\lambda).$$

By triangular inequality, we get

$$\Big\{\int_{(\mathbb{R}^d)^m}\mu(d\lambda_1)\cdots\mu(d\lambda_m)\Big|\prod_{k=1}^m\sum_{x\in\mathbb{Z}^d}e^{i\lambda_k\cdot x}\mathcal{F}_a^x(S)\Big|^2\Big\}^{1/2}$$
$$\le\sum_{j_1,\cdots,j_m=1}^2\Big\{\int_{(\mathbb{R}^d)^m}\mu(d\lambda_1)\cdots\mu(d\lambda_m)\Big|\mathbb{E}\Delta_{j_1}(\lambda_1)\cdots\Delta_{j_m}(\lambda_m)\Big|^2\Big\}^{1/2}$$
$$=\sum_{l=0}^m\binom{m}{l}\Big\{\int_{(\mathbb{R}^d)^m}\mu(d\lambda_1)\cdots\mu(d\lambda_m)\Big|\mathbb{E}\Big(\prod_{k=1}^l\Delta_1(\lambda_k)\Big)\Big(\prod_{k=l+1}^m\Delta_2(\lambda_k)\Big)\Big|^2\Big\}^{1/2}.$$

For (6.1.17) to hold, therefore, it suffices to show that for any $0\le l\le m$,

$$\int_{(\mathbb{R}^d)^m}\mu(d\lambda_1)\cdots\mu(d\lambda_m)\Big|\mathbb{E}\Big(\prod_{k=1}^l\Delta_1(\lambda_k)\Big)\Big(\prod_{k=l+1}^m\Delta_2(\lambda_k)\Big)\Big|^2 \tag{6.1.19}$$
$$\le(2\pi)^{md}\Big\{\mathbb{E}\Big|\sum_{x\in\mathbb{Z}^d}\prod_{j=1}^2\mathcal{G}_1^x(S_j)-\sum_{x\in\mathbb{Z}^d}\prod_{j=1}^2(h*\mathcal{G})_1^x(S_j)\Big|^l\Big\}$$
$$\times\Big\{\mathbb{E}\Big[\sum_{x\in\mathbb{Z}^d}\prod_{j=1}^2\mathcal{J}_{a-1}^x(S_j)-\sum_{x\in\mathbb{Z}^d}\prod_{j=1}^2(h*\mathcal{J}_{a-1})^x(S_j)\Big]^{m-l}\Big\}.$$

By spatial homogeneity, we get

$$\Delta_2(\lambda)=\sum_{x\in\mathbb{Z}^d}e^{i\lambda_k\cdot x}\mathcal{J}_{a-1}^{x-S(n_1)}(S^{(n_1)})$$
$$=e^{-i\lambda\cdot S(n_1)}\sum_{x\in\mathbb{Z}^d}e^{i\lambda_k\cdot x}\mathcal{J}_{a-1}^x(S^{(n_1)})$$
$$=e^{-i\lambda\cdot S(n_1)}\widehat{\mathcal{J}}_{a-1}(S^{(n_1)})(\lambda)$$

where $S^{(n_1)}$ is the random walk defined by

$$S^{(n_1)}(n)=S(n_1+n)-S(n_1)\qquad n=1,2,\cdots.$$

By increment independence, we get

$$\mathbb{E}\Big(\prod_{k=1}^l\Delta_1(\lambda_k)\Big)\Big(\prod_{k=l+1}^m\Delta_2(\lambda_k)\Big)$$
$$=\mathbb{E}\Big[\exp\Big\{-i\Big(\sum_{l+1}^m\lambda_k\Big)\cdot S(n_1)\Big\}\prod_{k=1}^l\Delta_1(\lambda_k)\Big]\mathbb{E}\prod_{k=l+1}^m\widehat{\mathcal{J}}_{a-1}(S)(\lambda_k)$$
$$=\mathbb{E}\Big[\exp\Big\{-i\Big(\sum_{l+1}^m\lambda_k\Big)\cdot S(n_1)\Big\}\prod_{k=1}^l\widehat{\mathcal{G}}_1(S)(\lambda_k)\Big]\mathbb{E}\prod_{k=l+1}^m\widehat{\mathcal{J}}_{a-1}(S)(\lambda_k);$$

so we have the equality

$$\begin{aligned}
&\left|\mathbb{E}\Big(\prod_{k=1}^{l}\Delta_1(\lambda_k)\Big)\Big(\prod_{k=l+1}^{m}\Delta_2(\lambda_k)\Big)\right|^2\\
&=\left|\mathbb{E}\prod_{k=l+1}^{m}\widehat{\mathcal{J}}_{a-1}(S)(\lambda_k)\right|^2\mathbb{E}\Big\{\exp\Big\{i\Big(\sum_{l+1}^{m}\lambda_k\Big)\cdot\big(S_2(n_1)-S_1(n_1)\big)\Big\}\\
&\times\prod_{k=1}^{l}\widehat{\mathcal{G}}_1(S_1)(\lambda_k)\overline{\widehat{\mathcal{G}}_1(S_2)(\lambda_k)}\Big\}.
\end{aligned}$$

Fix $(\lambda_{l+1},\cdots,\lambda_m)$ for a moment and take integration against $\mu(d\lambda_1)\cdots\mu(d\lambda_l)$ on both sides:

$$\begin{aligned}
&\int_{([-\pi,\pi]^d)^l}\mu(d\lambda_1)\cdots\mu(d\lambda_l)\left|\mathbb{E}\Big(\prod_{k=1}^{l}\Delta_1(\lambda_k)\Big)\Big(\prod_{k=l+1}^{m}\Delta_2(\lambda_k)\Big)\right|^2\\
&=\left|\mathbb{E}\prod_{k=l+1}^{m}\widehat{\mathcal{J}}_{a-1}(S)(\lambda_k)\right|^2\mathbb{E}\Big\{\exp\Big\{i\Big(\sum_{l+1}^{m}\lambda_k\Big)\cdot\big(S_2(n_1)-S_1(n_1)\big)\Big\}\\
&\times\left[\int_{[-\pi,\pi]^d}\widehat{\mathcal{G}}_1(S_1)(\lambda)\overline{\widehat{\mathcal{G}}_1(S_2)(\lambda)}d\lambda\right]^l\Big\}.
\end{aligned}$$

By Parseval identity, we get

$$\begin{aligned}
&\int_{[-\pi,\pi]^d}\widehat{\mathcal{G}}_1(S_1)(\lambda)\overline{\widehat{\mathcal{G}}_1(S_2)(\lambda)}d\lambda\\
&=(2\pi)^{ld}\Big\{\sum_{x\in\mathbb{Z}^d}\prod_{j=1}^{2}\mathcal{G}_1^x(S_j)-\sum_{x\in\mathbb{Z}^d}\prod_{j=1}^{2}(h*\mathcal{G})_1^x(S_j)\Big\}.
\end{aligned}$$

Using the fact that

$$\left|\exp\Big\{i(\lambda_{l+1}+\cdots+\lambda_m)\cdot\big(\widetilde{S}(n_1)-S(n_1)\big)\Big\}\right|=1,$$

we conclude that

$$\begin{aligned}
&\int_{([-\pi,\pi]^d)^l}\mu(d\lambda_1)\cdots\mu(d\lambda_l)\left|\mathbb{E}\Big(\prod_{k=1}^{l}\Delta_1(\lambda_k)\Big)\Big(\prod_{k=l+1}^{m}\Delta_2(\lambda_k)\Big)\right|^2\\
&\le(2\pi)^{ld}\mathbb{E}\left|\sum_{x\in\mathbb{Z}^d}\prod_{j=1}^{2}\mathcal{G}_1^x(S_j)-\sum_{x\in\mathbb{Z}^d}\prod_{j=1}^{2}(h*\mathcal{G})_1^x(S_j)\right|^l\left|\mathbb{E}\prod_{k=l+1}^{m}\widehat{\mathcal{J}}_{a-1}(S)(\lambda_k)\right|^2.
\end{aligned}$$

Now take integration against $\mu(d\lambda_{l+1})\cdots\mu(d\lambda_m)$ on both sides. The requested (6.1.19) follows from the equality

$$\begin{aligned}
&\int_{([-\pi,\pi]^d)^{m-l}}\left|\mathbb{E}\prod_{k=l+1}^{m}\widehat{\mathcal{J}}_{a-1}(\lambda_k)\right|^2\mu(d\lambda_{l+1})\cdots\mu(d\lambda_m)\\
&=(2\pi)^{(m-l)d}\mathbb{E}\left[\sum_{x\in\mathbb{Z}^d}\prod_{j=1}^{2}\mathcal{J}_{a-1}^x(S_j)-\sum_{x\in\mathbb{Z}^d}\prod_{j=1}^{2}(h*\mathcal{J}_{a-1})^x(S_j)\right]^{m-l}.
\end{aligned}$$

As for the proof of (6.1.14), observe that correspondent to (6.1.18), we have

$$\mathbb{E}\Big[\sum_{x\in\mathbb{Z}^d}\prod_{j=1}^{2}\Big(\mathcal{F}_k^x(S_j)-(h*\mathcal{F})_k^x(S_j)\Big)\Big]^m$$
$$=\frac{1}{(2\pi)^{md}}\int_{(\mathbb{R}^d)^m}\nu(d\lambda_1)\cdots\nu(d\lambda_m)\Big|\prod_{k=1}^m\sum_{x\in\mathbb{Z}^d}e^{i\lambda_k\cdot x}\mathcal{F}_k^x(S)\Big|^2$$

where the measure $\nu(d\lambda)$ is given by

$$\nu(d\lambda)=|1-\widehat{h}(\lambda)|^2d\lambda.$$

The rest follows essentially the same argument. □

The additivity assumption is crucial for the proof of Theorem 6.1.2. In particular, it cannot be replaced by sub-additivity in our treatment. Among the special cases discussed earlier, therefore, Theorem 6.1.2 applies to the case when $\mathcal{F}_k^x(S)=l(n_k,x)$, but not to $\mathcal{F}_k^x(S)=1_{\{x\in S[1,n_k]\}}$. The absence of additivity in the latter model creates some difficulties in approximating J_n by its smoothed versions. That is the major reason why we introduce its additive version defined in (6.1.6).

6.2. Integrability of I_n and J_n

In this section, we apply Theorem 6.1.1 to establish the maximal integrability for the mutual intersection local time

$$I_n=\sum_{k_1,\cdots,k_p=1}^{n}1_{\{S_1(k_1)=\cdots=S_p(k_p)\}}$$

and the intersection of independent ranges

$$J_n=\#\big\{S_1[1,n]\times\cdots\times S_p[1,n]\big\},$$

where $S_1(k),\cdots,S_p(k)$ are independent, identically distributed symmetric random walks. We assume that $S_1(k),\cdots,S_p(k)$ are square integrable with covariance matrix Γ and the smallest sub-group of $\mathbb{Z}^d$ that supports the random walks is $\mathbb{Z}^d$ itself.

THEOREM 6.2.1. *Under $p(d-2)<d$, there is a constant $C>0$ such that*

$$\mathbb{E}I_n^m\le(m!)^{\frac{d(p-1)}{2}}C^mn^{\frac{2p-d(p-1)}{2}m}\qquad m,n=1,2,\cdots. \tag{6.2.1}$$

Consequently, there is a $\theta>0$ such that

$$\sup_{n\ge1}\mathbb{E}\exp\Big\{\theta n^{-\frac{2p-d(p-1)}{d(p-1)}}I_n^{\frac{2}{d(p-1)}}\Big\}<\infty. \tag{6.2.2}$$

Under $p(d-2)<d$, the integrability of J_n is given in the following separate statements.

(1) *When $d=1$ and $p\ge1$ (that's right, we include the case $p=1$ here)*

$$\mathbb{E}J_n^m\le(m!)^{1/2}C^mn^{m/2}\qquad m,n=1,2,\cdots. \tag{6.2.3}$$

Consequently, there is $\theta > 0$ such that

$$\sup_{n\geq 1}\mathbb{E}\exp\Big\{\theta n^{-1}J_n^2\Big\} < \infty. \tag{6.2.4}$$

(2) *When $d = 2$ and $p \geq 2$,*

$$\mathbb{E}J_n^m \leq (m!)^{p-1}C^m\Big(\frac{n}{(\log n)^p}\Big)^m \qquad m,n = 1,2,\cdots. \tag{6.2.5}$$

Consequently, there is $\theta > 0$ such that

$$\sup_{n\geq 1}\mathbb{E}\exp\Big\{\theta n^{-\frac{1}{p-1}}(\log n)^{\frac{p}{p-1}}J_n^{\frac{1}{p-1}}\Big\} < \infty. \tag{6.2.6}$$

(3) *When $d = 3$ and $p = 2$,*

$$\mathbb{E}J_n^m \leq (m!)^{3/2}C^m n^{m/2} \qquad m,n = 1,2,\cdots. \tag{6.2.7}$$

Consequently, there is $\theta > 0$ such that

$$\sup_{n\geq 1}\mathbb{E}\exp\Big\{\theta n^{-1/3}J_n^{2/3}\Big\} < \infty. \tag{6.2.8}$$

As the first step of the proof, we establish the following weak version of Theorem 6.1.1.

LEMMA 6.2.2. *Under $p(d-2) < d$, we get*

$$\mathbb{E}I_n^m \leq (m!)^p C^m n^{\frac{2p-d(p-1)}{2}} \qquad m,n = 1,2,\cdots. \tag{6.2.9}$$

For the combinations of d and p with constraint $p(d-2) < d$, the moments of J_n yield the following bounds:

(1) *When $d = 1$ and $p \geq 1$,*

$$\mathbb{E}J_n^m \leq (m!)^p C^m n^{m/2} \qquad m,n = 1,2,\cdots. \tag{6.2.10}$$

(2) *When $d = 2$ and $p \geq 2$,*

$$\mathbb{E}J_n^m \leq (m!)^p C^m\Big(\frac{n}{(\log n)^p}\Big)^m \qquad m,n = 1,2,\cdots. \tag{6.2.11}$$

(3) *When $d = 3$ and $p = 2$,*

$$\mathbb{E}J_n^m \leq (m!)^2 C^m n^{m/2} \qquad m,n = 1,2,\cdots. \tag{6.2.12}$$

Proof. For similarity we estimate the moments of J_n. It is sufficient to prove that

$$\mathbb{E}J_n^m \leq (m!)^p C^m\big(p + \mathbb{E}J_n\big)^m \qquad m,n = 1,2,\cdots. \tag{6.2.13}$$

By the representation

$$J_n = \sum_{x\in\mathbb{Z}^d}\prod_{j=1}^p 1_{\{T_x^{(j)}\leq n\}}$$

where $T_x^{(j)} = \inf\{k \ge 1;\ \ S_j(k) = x\}$, and by the independence, we get

$$\begin{aligned}
\mathbb{E}J_n^m &= \sum_{x_1,\cdots,x_m\in\mathbb{Z}^d} \Big[\mathbb{P}\{T_{x_1} \le n, \cdots, T_{x_m} \le n\}\Big]^p \\
&= \sum_{x_1,\cdots,x_m\in\mathbb{Z}^d} \Big[\sum_{\sigma\in\Sigma_m} \mathbb{P}\{T_{x_{\sigma(1)}} \le \cdots \le T_{x_{\sigma(m)}} \le n\}\Big]^p.
\end{aligned}$$

Notice that

$$\begin{aligned}
&\mathbb{P}\{T_{x_{\sigma(1)}} \le \cdots \le T_{x_{\sigma(m)}} \le n\} \\
&= \sum_{1\le j\le k\le n} \mathbb{P}\{T_{x_{\sigma(1)}} \le \cdots \le T_{x_{\sigma(m-1)}},\ T_{x_{\sigma(m-1)}} = j,\ T_{x_{\sigma(m)}} = k\} \\
&\le \sum_{1\le j\le k\le n} \mathbb{P}\{T_{x_{\sigma(1)}} \le \cdots \le T_{x_{\sigma(m-1)}},\ T_{x_{\sigma(m-1)}} = j, \\
&\qquad S(j+1) \ne x_{\sigma(m)}, \cdots, S(k-1) \ne x_{\sigma(m)},\ S(k) = x_{\sigma(m)}\} \\
&= \sum_{1\le j\le k\le n} \mathbb{P}\{T_{x_{\sigma(1)}} \le \cdots \le T_{x_{\sigma(m-1)}},\ T_{x_{\sigma(m-1)}} = j\} \\
&\qquad \times \mathbb{P}\{\widetilde{T}_{x_{\sigma(m)}-x_{\sigma(m-1)}} = k - j\}
\end{aligned}$$

where (note that the time starts at 0)

$$\widetilde{T}_x = \inf\{k \ge 0;\ \ S(k) = x\} \qquad x \in \mathbb{Z}^d.$$

Thus,

$$\begin{aligned}
&\mathbb{P}\{T_{x_{\sigma(1)}} \le \cdots \le T_{x_{\sigma(m)}} \le n\} \\
&\le \Big\{\sum_{j=1}^n \mathbb{P}\{T_{x_{\sigma(1)}} \le \cdots \le T_{x_{\sigma(m-1)}},\ T_{x_{\sigma(m-1)}} = j\}\Big\} \\
&\qquad \times \Big\{\sum_{k=0}^n \mathbb{P}\{\widetilde{T}_{x_{\sigma(m)}-x_{\sigma(m-1)}} = k\}\Big\} \\
&= \mathbb{P}\{T_{x_{\sigma(1)}} \le \cdots \le T_{x_{\sigma(m-1)}} \le n\}\mathbb{P}\{\widetilde{T}_{x_{\sigma(m)}-x_{\sigma(m-1)}} \le n\}.
\end{aligned}$$

Repeating the above procedure, we get

$$\mathbb{P}\{T_{x_{\sigma(1)}} \le \cdots \le T_{x_{\sigma(m)}} \le n\} \le \prod_{k=1}^m \mathbb{P}\{\widetilde{T}_{x_{\sigma(k)}-x_{\sigma(k-1)}} \le n\}.$$

Here we follow the convention $x_{\sigma(0)} = 0$.

Therefore, by Jensen's inequality, we get

$$\begin{aligned}
\mathbb{E}J_n^m &\le \sum_{x_1,\cdots,x_m\in\mathbb{Z}^d} \Big[\sum_{\sigma\in\Sigma_m}\prod_{k=1}^m \mathbb{P}\{\widetilde{T}_{x_{\sigma(k)}-x_{\sigma(k-1)}} \le n\}\Big]^p \\
&\le (m!)^{p-1} \sum_{x_1,\cdots,x_m\in\mathbb{Z}^d}\sum_{\sigma\in\Sigma_m}\prod_{k=1}^m \Big(\mathbb{P}\{\widetilde{T}_{x_{\sigma(k)}-x_{\sigma(k-1)}} \le n\}\Big)^p \\
&= (m!)^p\Big\{\sum_{x\in\mathbb{Z}^d}\Big(\mathbb{P}\{\widetilde{T}_x \le n\}\Big)^p\Big\}^m.
\end{aligned}$$

Finally, (6.2.13) follows from the fact that

$$\sum_{x\in\mathbb{Z}^d}\Big(\mathbb{P}\{\widetilde{T}_x\le n\}\Big)^p=\mathbb{E}\Big[\#\{S_1[0,n]\cap\cdots\cap S_p[0,n]\}\Big\}\le p+\mathbb{E}J_n.$$

□

Proof of Theorem 6.2.1. By Taylor expansion, the exponential integrabilities stated in (6.2.2), (6.2.4), (6.2.6) and (6.2.8) are the consequences of the bounds given in (6.2.1), (6.2.3), (6.2.5) and (6.2.7), respectively. Due to similarity we only prove (6.2.1). A striking feature in our argument is that we improve the bound (6.2.9) based on (6.2.9) itself.

We first deal with the case $m\le n$. By (6.2.9) and by the fact that $m\le n$, there is a constant $C>0$ (which is allowed to be different from place to place in our proof) such that

$$\mathbb{E}I_{[n/m]+1}^{k_i}\le C^{k_i}(k_i!)^p\Big(\frac{n}{m}\Big)^{\frac{2p-d(p-1)}{2}k_i}\qquad i=1,\cdots,m.$$

Taking $a=m$ and $n_k-n_{k-1}=[n/m]+1$ in the moment inequality (6.1.7), we get

$$\begin{aligned}\Big(\mathbb{E}I_n^m\Big)^{1/p}&\le\sum_{\substack{k_1+\cdots+k_m=m\\ k_1,\cdots,k_m\ge 0}}\frac{m!}{k_1!\cdots k_m!}C^{m/p}k_1!\cdots k_m!\Big(\frac{n}{m}\Big)^{\frac{2p-d(p-1)}{2p}m}\\&=m!C^{m/p}(m^m)^{-\frac{2p-d(p-1)}{2p}}n^{\frac{2p-d(p-1)}{2p}m}\binom{2m-1}{m}.\end{aligned}$$

Therefore, (6.2.1) follows from the facts that $m^m\ge m!$ and that

$$\binom{2m-1}{m}\le\sum_{k=0}^{2m-1}\binom{2m-1}{k}=2^{2m-1}.$$

As for the case $m>n$, by the fact that $I_n\le n^p$ we have the trivial bound

$$\mathbb{E}I_n^m\le n^{pm}\le m^{\frac{d(p-1)}{2}m}n^{\frac{2p-d(p-1)}{2}m}\le C^m(m!)^{\frac{d(p-1)}{2}}n^{\frac{2p-d(p-1)}{2}m}$$

where the last step follows from the Stirling formula. □

REMARK 6.2.3. *Some bounds established in Theorem 6.2.1 can be extended to higher dimensions. For example, (6.2.7) and (6.2.8) hold in the case $d\ge 4$ and $p=2$. Indeed, $\mathbb{E}J_n=O(\log n)$ when $d=4$ and $\mathbb{E}J_n=O(1)$ when $d\ge 5$. In view of (6.2.13), therefore, (6.2.12) remains to be true for $d\ge 4$. Hence, our claim follows from the same argument. Similarly, we have*

(6.2.14) $$\mathbb{E}J_n^m\le C^m(m!)^{3/2}n^{m/2}\qquad m,n=1,2,\cdots$$

(6.2.15) $$\sup_{n\ge1}\mathbb{E}\exp\Big\{\theta n^{-1/3}J_n^{2/3}\Big\}<\infty\qquad \textit{for some } \theta>0.$$

REMARK 6.2.4. *Under $p(d-2)<d$, the integrability established for I_n and J_n are best possible in the sense that there is a $\theta_0>0$ such that the left hand sides of (6.2.2), (6.2.4), (6.2.6) and (6.2.8) become infinity if $\theta>\theta_0$. For example, the* *weak convergence stated in Theorem 5.3.2 and the critical integrability stated in Remark 3.3.5 imply that*

$$\lim_{n\to\infty} \mathbb{E}\exp\left\{\theta n^{-\frac{2p-d(p-1)}{d(p-1)}} I_n^{\frac{2}{d(p-1)}}\right\} = \infty$$

for any $\theta > \frac{p}{2}\det(\Gamma)^{\frac{p-1}{2}}\kappa(d,p)^{-\frac{4p}{d(p-1)}}$.

6.3. Integrability of Q_n and R_n in low dimensions

We mainly consider the case $d = 2$ (with a few exceptions where $d = 3$ or higher), in which Q_n and R_n concentrate around their means. Our first result is on the exponential integrability of R_n.

THEOREM 6.3.1. *When* $d = 2$,

$$\sup_{n\geq 1}\mathbb{E}\exp\left\{\theta\frac{\log n}{n}R_n\right\} < \infty \qquad \forall\theta > 0. \tag{6.3.1}$$

When $d \geq 3$,

$$\sup_{n\geq 1}\mathbb{E}\exp\left\{\frac{\theta}{n}R_n\right\} < \infty \qquad \forall\theta > 0. \tag{6.3.2}$$

Proof. We apply Theorem 1.3.3 here. First, the process R_n is sub-additive in the sense that for any integers $m, n \geq 1$,

$$R_{n+m} \leq R_n + \widetilde{R}_m \tag{6.3.3}$$

where

$$\widetilde{R}_m = \#\{S_{n+1}, \cdots, S_{n+m}\} = \#\{S_{n+1} - S(n), \cdots, S_{n+m} - S(n)\}$$

is independent of $\{S(1), \cdots, S(n)\}$ and has a distribution the same as R_m.

Second, by the fact ((5.3.39)) that $\mathbb{E}R_n = O\Big(\frac{n}{\log n}\Big)$ when $d = 2$ and $\mathbb{E}R_n = O(n)$ when $d \geq 3$, and by the fact that the sub-additive sequence R_n is non-decreasing, so the assumption on stochastic boundedness posted in Theorem 1.3.3 holds with $c_n = \dfrac{n}{\log n}$ when $d = 2$ and $c_n = n$ when $d \geq 3$. □

Due to the concentration phenomena in the case $d \geq 2$, we are more interested in the integrability of $Q_n - \mathbb{E}Q_n$ and $R_n - \mathbb{E}R_n$. Here is the main result of this section.

THEOREM 6.3.2. *There is a* $\theta > 0$ *such that*

(1) *As* $d = 2$,

$$\sup_{n\geq 1}\mathbb{E}\exp\left\{\theta\frac{(\log n)^2}{n}\big|R_n - \mathbb{E}R_n\big|\right\} < \infty, \tag{6.3.4}$$

$$\sup_{n\geq 1}\mathbb{E}\exp\left\{\theta\frac{1}{n}\big|Q_n - \mathbb{E}Q_n\big|\right\} < \infty. \tag{6.3.5}$$

(2) *As* $d \geq 3$,

$$\sup_{n\geq 1} \mathbb{E}\exp\Big\{\theta\frac{1}{n^{2/3}}\big|R_n - \mathbb{E}R_n\big|\Big\} < \infty. \tag{6.3.6}$$

Proof. We first consider the case $d = 2$ and prove (6.3.4) with n being replaced by 2^n. To this end, all we need is to show that there is a $\theta > 0$ such that

$$\sup_{n\geq 1} \mathbb{E}\exp\Big\{\theta\frac{(\log 2^n)^2}{2^n}\big(R_n - \mathbb{E}R_n\big)\Big\} < \infty, \tag{6.3.7}$$

$$\sup_{n\geq 1} \mathbb{E}\exp\Big\{-\theta\frac{(\log 2^n)^2}{2^n}\big(R_n - \mathbb{E}R_n\big)\Big\} < \infty. \tag{6.3.8}$$

We only prove (6.3.7) for similarity.

The proof is based on the procedure of triangular approximation used for (2.4.7) in Theorem 2.4.2 and the reader may compare the decomposition (6.3.9) with Figure 1 in Section 2.4 for some geometric insight.

Let

$$N = [2(\log 2)^{-1}\log n].$$

It is easy to see that $2^N \sim n^2$. Set

$$\beta_k = \#\big\{S((k-1)2^{n-N}, k2^{n-N}]\big\},$$

$$\alpha_{j,k} = \#\Big\{S\big((2k-2)2^{n-j}, (2k-1)2^{n-j}\big] \cap S\big((2k-1)2^{n-j}, (2k)2^{n-j}\big]\Big\}.$$

By the inclusion-exclusion rule we have the decomposition

$$R_{2^n} - \mathbb{E}R_{2^n} = \sum_{k=1}^{2^N}\overline{\beta}_k - \sum_{j=1}^{N}\sum_{k=1}^{2^{j-1}}\overline{\alpha}_{j,k} \tag{6.3.9}$$

where $\overline{\beta}_k = \beta_k - \mathbb{E}\beta_k$ and $\overline{\alpha}_{j,k} = \alpha_{j,k} - \mathbb{E}\alpha_{j,k}$. By Theorem 6.3.1, we get

$$\sup_n \mathbb{E}\exp\Big\{\lambda\frac{\log 2^{n-N}}{2^{n-N}}|\overline{\beta}_1|\Big\} < \infty \qquad \forall\lambda > 0.$$

By Taylor expansion, there is a $C > 0$ and $\lambda_0 > 0$ independent of n such that

$$\mathbb{E}\exp\Big\{\lambda\frac{\log 2^{n-N}}{2^{n-N}}\overline{\beta}_1\Big\} \leq \exp\Big\{C\lambda^2\Big\} \qquad \forall 0 \leq \lambda \leq \lambda_0.$$

Notice that $\overline{\beta}_1, \cdots, \overline{\beta}_{2^N}$ is an i.i.d. sequence with $\mathbb{E}\overline{\beta}_1 = 0$. Thus,

$$\begin{aligned}
&\mathbb{E}\exp\Big\{\theta 2^{-N/2}\frac{\log 2^{n-N}}{2^{n-N}}\sum_{k=1}^{2^N}\overline{\beta}_k\Big\}\\
&= \bigg(\mathbb{E}\exp\Big\{\theta 2^{-N/2}\frac{\log 2^{n-N}}{2^{n-N}}\overline{\beta}_k\Big\}\bigg)^{2^N} \leq \exp\Big\{C\theta^2\Big\}.
\end{aligned}$$

Consequently, there is $\theta > 0$ such that

$$\sup_n \mathbb{E}\exp\Big\{\theta 2^{-N/2}\frac{\log 2^{n-N}}{2^{n-N}}\sum_{k=1}^{2^N}\overline{\beta}_k\Big\} < \infty.$$

By the choice of N one can see that there is $c>0$ independent of n such that

$$2^{-N/2}\frac{\log 2^{n-N}}{2^{n-N}}\geq c\frac{(\log 2^n)^2}{2^n}.$$

Therefore, there are some $\theta>0$ such that

$$\sup_n \mathbb{E}\exp\Big\{\theta\frac{(\log 2^n)^2}{2^n}\sum_{k=1}^{2^N}\overline{\beta}_k\Big\}<\infty. \tag{6.3.10}$$

We need to show that for some $\theta>0$,

$$\sup_n \mathbb{E}\exp\Big\{-\theta\frac{(\log 2^n)^2}{2^n}\sum_{j=1}^{N}\sum_{k=1}^{2^{j-1}}\overline{\alpha}_{j,k}\Big\}<\infty. \tag{6.3.11}$$

Set

$$\tilde{J}_n=\#\{S[1,n]\cap\widetilde{S}[1,n]\}\qquad n=1,2,\cdots,$$

where $\widetilde{S}$ is an independent copy of the random walk S. In our notation, for each $1\leq j\leq N$, $\{\overline{\alpha}_{j,1},\cdots,\overline{\alpha}_{j,2^{j-1}}\}$ is an i.i.d. sequence with the same distribution as $\tilde{J}_{2^{n-j}}$. By (6.2.2) in Theorem 6.2.1 (with $d=p=2$), there is a $\delta>0$ such that

$$\sup_n\sup_{j\leq N}\mathbb{E}\exp\Big\{\delta\frac{(\log 2^{n-j})^2}{2^{n-j}}\big|\overline{\alpha}_{j,1}\big|\Big\}<\infty.$$

By the argument used for (6.3.10), there is a $\bar{\theta}>0$ such that

$$\sup_n\sup_{j\leq N}\mathbb{E}\exp\Big\{-\bar{\theta}2^{-j/2}\frac{(\log 2^n)^2}{2^{n-j}}\sum_{k=1}^{2^{j-1}}\overline{\alpha}_{j,k}\Big\}<\infty.$$

Hence for some $\theta>0$,

$$C(\theta)\equiv\sup_n\sup_{j\leq N}\mathbb{E}\exp\Big\{-\theta 2^{j/2}\frac{(\log 2^n)^2}{2^n}\sum_{k=1}^{2^{j-1}}\overline{\alpha}_{j,k}\Big\}<\infty.$$

Write

$$\lambda_N=\prod_{j=1}^{N}(1-2^{-j/2})\quad\text{and}\quad\lambda_\infty=\prod_{j=1}^{\infty}(1-2^{-j/2}).$$

Using Hölder inequality with $1/p=1-2^{-N/2}$, $1/q=2^{-N/2}$ we have

$$\begin{aligned}
&\mathbb{E}\exp\Big\{-\lambda_N\theta\frac{(\log 2^n)^2}{2^n}\sum_{j=1}^{N}\sum_{k=1}^{2^{j-1}}\overline{\alpha}_{j,k}\Big\}\\
&\leq\bigg(\mathbb{E}\exp\Big\{-\lambda_{N-1}\theta\frac{(\log 2^n)^2}{2^n}\sum_{j=1}^{N-1}\sum_{k=1}^{2^{j-1}}\overline{\alpha}_{j,k}\Big\}\bigg)^{1-2^{-N/2}}\\
&\quad\times\bigg(\mathbb{E}\exp\Big\{-\lambda_N 2^{N/2}\frac{(\log 2^n)^2}{2^n}\sum_{k=1}^{2^{N-1}}\overline{\alpha}_{N,k}\Big\}\bigg)^{2^{-N/2}}\\
&\leq\mathbb{E}\exp\Big\{-\lambda_{N-1}\theta\frac{(\log 2^n)^2}{2^n}\sum_{j=1}^{N-1}\sum_{k=1}^{2^{j-1}}\overline{\alpha}_{j,k}\Big\}\cdot C(\theta)^{2^{-N/2}}
\end{aligned}$$

where the last step partially follows from the fact (by Jensen inequality) that

$$\mathbb{E}\exp\left\{-\lambda_{N-1}\theta\frac{(\log 2^n)^2}{2^n}\sum_{j=1}^{N-1}\sum_{k=1}^{2^{j-1}}\overline{\alpha}_{j,k}\right\}\geq 1.$$

Repeating this procedure, we get

$$\begin{aligned}&\mathbb{E}\exp\left\{\lambda_N\theta\frac{(\log 2^n)^2}{2^n}\Big|\sum_{j=1}^{N}\sum_{k=1}^{2^{j-1}}\overline{\alpha}_{j,k}\Big|\right\}\\ &\leq C(\theta)^{2^{-1/2}+\cdots+2^{-N/2}}\leq C(\theta)^{2^{-1/2}(1-2^{-1/2})^{-1}}.\end{aligned}$$

Thus,

$$\sup_n\mathbb{E}\exp\left\{\lambda_\infty\theta\frac{(\log 2^n)^2}{2^n}\Big|\sum_{j=1}^{N}\sum_{k=1}^{2^{j-1}}\overline{\alpha}_{j,k}\Big|\right\}\leq C(\theta)^{2^{-1/2}(1-2^{-1/2})^{-1}};$$

so we have (6.3.11).

Hence, (6.3.7) follows from (6.3.9), (6.3.10) and (6.3.11). By (6.3.7) and (6.3.8), there is $\theta>0$ such that

$$\sup_{n\geq 1}\mathbb{E}\exp\left\{\theta\frac{(\log 2^n)^2}{2^n}\big|R_{2^n}-\mathbb{E}R_{2^n}\big|\right\}<\infty. \tag{6.3.12}$$

We now extend (6.3.12) to (6.3.4). Given an integer $n\geq 2$, we have the following unique binary representation:

$$n=2^{m_1}+2^{m_2}+\cdots+2^{m_l}$$

where $m_1>m_2>\cdots m_l\geq 0$ are integers. Write

$$n_0=0 \quad\text{and}\quad n_i=2^{m_1}+\cdots+2^{m_i}\qquad i=1,\cdots,l.$$

By inclusion-exclusion, we get

$$\begin{aligned}R_n&=\sum_{i=1}^{l}\#\{S(n_{i-1},n_i]\}-\sum_{i=1}^{l-1}\#\{S(n_{i-1},n_i]\cap S(n_i,n]\}\\ &=\sum_{i=1}^{l}B_i-\sum_{i=1}^{l-1}A_i\qquad\text{(say)}.\end{aligned} \tag{6.3.13}$$

Write

$$\sum_{i=1}^{l}B_i={\sum_i}'B_i+{\sum_i}''B_i$$

where ${\sum_i}'$ is the summation over i with $2^{m_i}\geq\sqrt{n}$ and ${\sum_i}''$ is the summation over i with $2^{m_i}<\sqrt{n}$. We also define the products ${\prod_i}'$ and ${\prod_i}''$ in a similar manner. Notice that

$$B_i\overset{d}{=}R_{2^{m_i}}\qquad i=1,\cdots,l.$$

By Hölder inequality, we get

$$\mathbb{E}\exp\Big\{\theta\frac{(\log n)^2}{n}\Big|\sum_i{}'(B_i-\mathbb{E}B_i)\Big|\Big\}$$

$$\le\prod_i{}'\Big(\mathbb{E}\exp\Big\{\theta\frac{(\log n)^2}{n}2^{-m_i}\Big(\sum_j{}'2^{m_j}\Big)\big|R_{2^{m_i}}-\mathbb{E}R_{2^{m_i}}\big|\Big\}\Big)^{2^{m_i}\left(\sum_j{}'2^{m_j}\right)^{-1}}$$

$$\le\prod_i{}'\Big(\mathbb{E}\exp\Big\{4\theta\frac{(\log 2^{m_i})^2}{2^{m_i}}\big|R_{2^{m_i}}-\mathbb{E}R_{2^{m_i}}\big|\Big\}\Big)^{2^{m_i}\left(\sum_j{}'2^{m_j}\right)^{-1}}$$

$$\le\sup_m\mathbb{E}\exp\Big\{4\theta\frac{(\log 2^m)^2}{2^m}\big|R_n-\mathbb{E}R_n\big|\Big\}.$$

By (6.3.12), the right hand side is finite for some $\theta>0$.

Assume that the set $\{1\le i\le l;\ 2^{m_i}<\sqrt{n}\}$ is non-empty. We have

$$\sum_i{}''2^{m_i}\le 2\sqrt{n}.$$

So we have

$$\frac{(\log n)^2}{n}\le\frac{1}{\sqrt{n}}\le 2\Big(\sum_i{}''2^{m_i}\Big)^{-1}.$$

Hence

$$\mathbb{E}\exp\Big\{\theta\frac{(\log n)^2}{n}\Big|\sum_i{}''(B_i-\mathbb{E}B_i)\Big|\Big\}$$

$$\le\prod_i{}''\Big(\mathbb{E}\exp\Big\{\theta\frac{(\log n)^2}{n}2^{-m_i}\Big(\sum_j{}''2^{m_j}\Big)\big|R_{2^{m_i}}-\mathbb{E}R_{2^{m_i}}\big|\Big\}\Big)^{2^{m_i}\left(\sum_j{}''2^{m_j}\right)^{-1}}$$

$$\le\prod_i{}''\Big(\mathbb{E}\exp\Big\{2\theta\frac{1}{2^{m_i}}\big|R_{2^{m_i}}-\mathbb{E}R_{2^{m_i}}\big|\Big\}\Big)^{2^{m_i}\left(\sum_j{}''2^{m_j}\right)^{-1}}$$

$$\le\sup_m\mathbb{E}\exp\Big\{2\theta\frac{1}{2^m}\big|R_{2^m}-\mathbb{E}R_{2^m}\big|\Big\}.$$

By Cauchy-Schwarz inequality, therefore, there exists $\theta>0$ such that

$$\sup_{n\ge1}\mathbb{E}\exp\Big\{\theta\frac{(\log n)^2}{n}\Big|\sum_{i=1}^{l}(B_i-\mathbb{E}B_i)\Big|\Big\}<\infty. \tag{6.3.14}$$

By the fact that

$$n-n_i=2^{m_{i+1}}+\cdots+2^{m_l}\le 2^{m_i}$$

we have

$$A_i\overset{d}{=}\#\{S[1,2^{m_i}]\cap S'[1,n-n_i]\}\le J_{2^{m_i}}. \tag{6.3.15}$$

Notice that ((6.2.11) in Lemma 6.2.2, $m=1$, $p=2$)

$$\mathbb{E}J_n=O\Big(\frac{n}{(\log n)^2}\Big)\quad(n\to\infty).$$

There is a constant $C > 0$ independent of n such that

$$\begin{aligned}\sum_{i=1}^{l-1}\mathbb{E}A_i &\le \sum_{i=1}^{l-1}\mathbb{E}J_{2^{m_i}} \le C\sum_{i=1}^{l}\frac{2^{m_i}}{m_i^2}\\ &\le C\sum_{m_i<l/2}\frac{2^{m_i}}{m_i^2} + C\sum_{m_i\ge l/2}^{l}\frac{2^{m_i}}{m_i^2}\\ &\le C2^{l/2} + C\frac{n}{(\log n)^2} \le C\frac{n}{(\log n)^2}.\end{aligned}$$

By (6.3.13) and (6.3.14), therefore, we will have (6.3.4) if we can prove

$$\sup_n \mathbb{E}\exp\Big\{\theta\frac{(\log n)^2}{n}\sum_{i=1}^{l-1}A_i\Big\} < \infty.$$

Thanks to (6.2.6) in Theorem 6.2.1 (with $p = 2$) and (6.3.15), this requested conclusion follows from an argument similar to the one used for (6.3.14).

The proof of (6.3.5) is similar and easier (for having no "$\log n$" part). Indeed, applying the decomposition

(6.3.16) $$Q_{2^n} = 2^n + \sum_{j=1}^{n}\sum_{k=1}^{2^{j-1}}\sum_{i=(2k-2)2^{n-j}+1}^{(2k-1)2^{n-j}}\sum_{l=(2k-1)2^{n-j}+1}^{(2k)2^{n-j}} 1_{\{S(i)=S(l)\}}$$

and (6.2.2) (with $d = p = 2$) in a similar way gives that for some $\theta > 0$,

$$\sup_{n\ge 1}\mathbb{E}\exp\Big\{2^{-n}\theta\big|Q_{2^n} - \mathbb{E}Q_{2^n}\big|\Big\} < \infty.$$

The argument of binary expansion used above extends the above conclusion into (6.3.5).

To prove (6.3.6) we first claim that as $d \ge 3$ and $p = 2$, we get

(6.3.17) $$\sup_{n\ge 1}\mathbb{E}\exp\Big\{\theta\frac{1}{n^{2/3}}J_n\Big\} < \infty$$

for some $\theta > 0$.

Indeed, by the fact that $J_n \le n$, we get

$$\mathbb{E}\exp\Big\{\theta\frac{1}{n^{2/3}}J_n\Big\} \le \mathbb{E}\exp\Big\{\theta\frac{1}{n^{1/3}}J_n^{2/3}\Big\}.$$

Hence, (6.3.17) follows from (6.2.8) and Remark 6.2.3.

Therefore, (6.3.6) can be proved in almost the same way as (6.3.4) is proved, except that we use (6.3.17) instead of (6.2.6) in the proof.[2] □

By the fact (Theorem 5.4.3) that $Q_n - \mathbb{E}Q_n$ and $\mathbb{E}R_n - R_n$ converge, if properly normalized, to the renormalized self-intersection local time $\gamma([0,1]^2_<)$, and by the fact (Theorem 4.3.1 and Theorem 4.3.2) that $\gamma([0,1]^2_<)$ has asymmetric tails, we expect that $\mathbb{E}Q_n - Q_n$ and $R_n - \mathbb{E}R_n$ have tails lighter than the tails of $Q_n - \mathbb{E}Q_n$ and $\mathbb{E}R_n - R_n$. Indeed, we have

[2] A fact crucial to our argument given in the above proof is that the normalizing exponent, which is 1 in (6.3.4), (6.3.5) and is 2/3 in (6.3.6), is always greater than 1/2.

THEOREM 6.3.3. *The following statements hold for every* $\theta > 0$*:*

(1) *As* $d = 2$,

$$\sup_{n\geq 1} \mathbb{E}\exp\Big\{\theta\frac{(\log n)^2}{n}\big(R_n - \mathbb{E}R_n\big)\Big\} < \infty, \tag{6.3.18}$$

$$\sup_{n\geq 1} \mathbb{E}\exp\Big\{-\theta\frac{1}{n}\big(Q_n - \mathbb{E}Q_n\big)\Big\} < \infty. \tag{6.3.19}$$

(2) *As* $d \geq 3$,

$$\sup_{n\geq 1} \mathbb{E}\exp\Big\{\theta\frac{1}{n^{2/3}}\big(R_n - \mathbb{E}R_n\big)\Big\} < \infty. \tag{6.3.20}$$

Proof. Due to similarity, we only consider (6.3.18). By (6.3.4) in Theorem 6.3.2, there is $\theta_0 > 0$ such that(6.3.18) holds for $\theta \leq \theta_0$.

Assume that $\theta > \theta_0$. Take $a > 0$ such that $\theta < a\theta_0$. Write $n = n_1 + \cdots + n_a$ such that $n_j \sim a^{-1}n$ for $j = 1, \cdots a$. By sub-additivity, we get

$$R_n \leq \sum_{k=1}^{a} \widetilde{R}_{n_k}$$

where $\widetilde{R}_{n_k}$ $(k = 1, \cdots, a)$ are independent and $\widetilde{R}_{n_k} \overset{d}{=} R_{n_k}$. In addition, write $m_0 = 1$ and $m_k = n_1 + \cdots + n_k$. We have

$$\mathbb{E}R_n \geq \sum_{k=1}^{a} \mathbb{E}R_{n_k} - \sum_{k=1}^{a-1} \mathbb{E}\#\big\{S[m_k, m_{k+1}] \cap S[m_k, n]\big\}.$$

It is easy to see that the second summation on the right hand side is bounded by a (universal) constant multiple of $n/(\log n)^2$. Consequently,

$$\mathbb{E}\exp\Big\{\theta\frac{(\log n)^2}{n}\big(R_n - \mathbb{E}R_n\big)\Big\} \leq C\prod_{k=1}^{a} \mathbb{E}\exp\Big\{\theta\frac{(\log n)^2}{n}\big(R_{n_k} - \mathbb{E}R_{n_k}\big)\Big\}$$

where $C > 0$ is independent of n. By the choice of n_k, the right hand side is bounded uniformly in n. $\square$

We now comment on the integrabilities established so far. By the weak laws stated in Theorem 5.4.2 and Theorem 5.4.3, and by the critical integrability given in (4.3.3), the integrabilities stated in the case $d = 2$ ((6.3.4) and (6.3.5)) are best possible in the sense that they no longer hold if θ is too large. More precisely, we can see that

$$\lim_{n\to\infty} \mathbb{E}\exp\Big\{-\theta\frac{(\log n)^2}{n}\big(R_n - \mathbb{E}R_n\big)\Big\} = \infty \tag{6.3.21}$$

if $\theta > (2\pi)^{-2}\det(\Gamma)^{-1/2}\kappa(2,2)^{-4}$, and that

$$\lim_{n\to\infty} \mathbb{E}\exp\Big\{\theta\frac{1}{n}\big(Q_n - \mathbb{E}Q_n\big)\Big\} = \infty \tag{6.3.22}$$

if $\theta > \sqrt{\det(\Gamma)}\kappa(2,2)^{-4}$.

A natural question is to find the critical exponents θ_R and θ_Q, such that

$$\sup_{n\geq 1}\mathbb{E}\exp\Big\{-\theta\frac{(\log n)^2}{n}(R_n-\mathbb{E}R_n)\Big\}\begin{cases} <\infty & \theta<\theta_R, \\ =\infty & \theta>\theta_R, \end{cases} \tag{6.3.23}$$

$$\sup_{n\geq 1}\mathbb{E}\exp\Big\{\theta\frac{1}{n}(Q_n-\mathbb{E}Q_n)\Big\}\begin{cases} <\infty & \theta<\theta_Q, \\ =\infty & \theta>\theta_Q. \end{cases} \tag{6.3.24}$$

We may ask, in view of the weak laws stated in Theorem 5.4.2, Theorem 5.4.3, and of the critical exponent given by Theorem 4.3.1, whether or not

$$\theta_R=(2\pi)^{-2}\det(\Gamma)^{-1/2}\kappa(2,2)^{-4} \quad\text{and}\quad \theta_Q=\sqrt{\det(\Gamma)}\kappa(2,2)^{-4}.$$

The following example gives a negative answer to our question on θ_Q.

EXAMPLE 6.3.4. *Let N be an arbitrarily large integer. Let the random walk $\{S(n)\}$ be the partial sum of the i.i.d sequence of random vectors in $\mathbb{Z}^2$ that take the values $(N,0),(-N,0),(0,N),(0,-N)$ with probability $(2N^2)^{-1}$ for each, and take the value $(0,0)$ with probability $1-(2/N^2)$. Then Γ is an identity matrix. The event that $S(k)=0$ for all $k\leq n$ has probability $\big(1-(2/N^2)\big)^n$. On this event $Q_n=n(n-1)/2$. Hence,*

$$\mathbb{E}\exp\Big\{\theta\frac{1}{n}(Q_n-\mathbb{E}Q_n)\Big\}\geq\exp\Big\{-\theta\frac{\mathbb{E}Q_n}{n}\Big\}\Big(1-\frac{2}{N^2}\Big)^n\exp\Big\{\theta\frac{n-1}{2}\Big\}.$$

By (5.4.1) $\mathbb{E}Q_n=O(n\log n)$. Thus,

$$\theta_Q\leq 2\log\Big(1-\frac{2}{N^2}\Big)^{-1}.$$

Since N can be arbitrarily large, the critical exponent θ_Q can be smaller than any previously given positive number.

This example shows that the critical exponent in the continuous setting does not necessarily pass to the discrete setting through weak convergence.

The identification of θ_Q and θ_R remains open.

6.4. Integrability of Q_n and R_n in high dimensions

In an obvious way, the exponential integrabilities for $Q_n-\mathbb{E}Q_n$ and $R_n-\mathbb{E}R_n$ given in Theorem 6.3.2 and Theorem 6.3.3 are closely related to the weak laws stated in Theorem 5.4.2 and Theorem 5.4.3 in the case $d=2$. In this section, we consider the high dimensional setting $d\geq 3$ and control exponential moments for the normalized sequence appearing in Theorem 5.5.3.

THEOREM 6.4.1. *As $d=3$,*

$$\sup_n\mathbb{E}\exp\Big\{\frac{\theta}{\sqrt[3]{n\log n}}|Q_n-\mathbb{E}Q_n|^{2/3}\Big\}<\infty \quad \textit{for every } \theta>0. \tag{6.4.1}$$

As $d\geq 4$,

$$\sup_n\mathbb{E}\exp\Big\{\frac{\theta}{\sqrt[3]{n}}|Q_n-\mathbb{E}Q_n|^{2/3}\Big\}<\infty \quad \textit{for some } \theta>0. \tag{6.4.2}$$

Proof. We first consider the case $d = 3$. The argument used in the proof of Theorem 6.3.2, applicable only when the normalizing sequence is of the growth rate greater than n^b for some $b > 1/2$, is broken in our setting where the normalizing sequence (of $Q_n - \mathbb{E}Q_n$) is $\sqrt{n\log n}$.

We first prove that for any integer $m \ge 1$,

$$\mathbb{E}|Q_n - \mathbb{E}Q_n|^m = O\Big((n\log n)^{m/2}\Big) \qquad (n\to\infty). \tag{6.4.3}$$

We carry out induction on m. By (5.5.5), (6.4.3) holds as $m = 1, 2$. We let $m \ge 3$, assume that it is true for all $1 \le j \le m-1$ and prove it is true for m.

Given n, write $n_1 = [n/2]$, $n_2 = n - n_1$, and set

$$Q'_{n_2} = \sum_{n_1+1\le j<k\le n} 1_{\{S_j=S_k\}} \quad\text{and}\quad K_n = \sum_{j=1}^{n_1}\sum_{k=n_1+1}^{n} 1_{\{S_j=S_k\}}.$$

We have that

$$K_n \overset{d}{=} \sum_{j=0}^{n_1-1}\sum_{k=1}^{n-n_1} 1_{\{S_j=\widetilde{S}_k\}}$$

where $\{\widetilde{S}(k)\}$ is an independent copy of $\{S(k)\}$. Taking $d = 3$ and $p = 2$ in (6.2.1) one can see that there is $C_0 > 0$ such that

$$\mathbb{E}|K_n - \mathbb{E}K_n|^m \le C_0^m (m!)^{3/2} n^{m/2} \qquad m, n = 1, 2, \cdots. \tag{6.4.4}$$

By the independence between Q_{n_1} and Q'_{n_2}, we get

$$\begin{aligned}
&\Big(\mathbb{E}|Q_n - \mathbb{E}Q_n|^m\Big)^{1/m} \\
&\le \Big\{\mathbb{E}\Big(|Q_{n_1} - \mathbb{E}Q_{n_1}| + |Q'_{n_2} - \mathbb{E}Q'_{n_2}|\Big)^m\Big\}^{1/m} \\
&+ \Big\{\mathbb{E}|K_n - \mathbb{E}K_n|^m\Big\}^{1/m} \\
&\le \Big\{\sum_{j=0}^{m}\binom{m}{j}\mathbb{E}|Q_{n_1} - \mathbb{E}Q_{n_1}|^j \cdot \mathbb{E}|Q'_{n_2} - \mathbb{E}Q'_{n_2}|^{m-j}\Big\}^{1/m} \\
&+ C_0 (m!)^{3/(2m)}\sqrt{n}.
\end{aligned} \tag{6.4.5}$$

Combining this with the induction assumption, we get

$$\begin{aligned}
&\Big(\mathbb{E}|Q_n - \mathbb{E}Q_n|^m\Big)^{1/m} \\
&\le \Big\{O\Big((n\log n)^{m/2}\Big) + \mathbb{E}|Q_{n_1} - \mathbb{E}Q_{n_1}|^m + \mathbb{E}|Q'_{n_2} - \mathbb{E}Q'_{n_2}|^m\Big\}^{1/m} \\
&+ O\big(\sqrt{n}\big).
\end{aligned}$$

Write, for each $k \ge 1$,

$$\alpha_k = \sup\Big\{(2^k\log 2^k)^{-1/2}\Big(\mathbb{E}|Q_n - \mathbb{E}Q_n|^m\Big)^{1/m};\ \ 2^k \le n \le 2^{k+1}\Big\}.$$

Then,

$$\alpha_{k+1} \le \Big\{O(1) + 2^{-\frac{m-2}{2}}\alpha_k^m\Big\}^{1/m} + o(1) \le 2^{-\frac{m-2}{2m}}\alpha_k + O(1) \qquad (k\to\infty).$$

By the fact that $m \geq 3$ one can see that the sequence α_k is bounded. We have proved (6.4.3).

We now claim that there is $C > 0$ such that

$$\mathbb{E}|Q_n - \mathbb{E}Q_n|^m \leq C^m (m!)^{3/2} (n \log n)^{m/2} \quad m, n = 1, 2, \cdots . \tag{6.4.6}$$

Indeed, take m_0 sufficiently large so that

$$\left(1 - 2^{-\frac{m-2}{2m}}\right)^{-1} \frac{(m-1)^{1/m}}{\sqrt{2}} \leq \frac{1}{2}, \qquad 1 - 2^{-\frac{m-2}{2m}} \geq \frac{1}{4}$$

for all $m \geq m_0$. By (6.4.3), there is a constant $C > 0$ such that for all $j = 1, \cdots, m_0$,

$$\mathbb{E}|Q_n - \mathbb{E}Q_n|^j \leq C^j (j!)^{3/2} (n \log n)^{j/2} \quad n = 1, 2, \cdots .$$

We may take $C \geq 8C_0$ (recall that C_0 is given in (6.4.4)). By induction (on m), all we have to prove is that for any $m \geq m_0$, if

$$\mathbb{E}|Q_n - \mathbb{E}Q_n|^j \leq C^j (j!)^{3/2} (n \log n)^{j/2} \quad n = 1, 2, \cdots \tag{6.4.7}$$

for every $j = 1, \cdots, m-1$, then

$$\mathbb{E}|Q_n - \mathbb{E}Q_n|^m \leq C^m (m!)^{3/2} (n \log n)^{m/2} \quad n = 1, 2, \cdots . \tag{6.4.8}$$

From (6.4.5) and (6.4.7) we have

$$\begin{aligned}
&\left(\mathbb{E}|Q_n - \mathbb{E}Q_n|^m\right)^{1/m} \\
&\leq \Big\{2^{-m/2} C^m (n \log n)^{m/2} \sum_{j=1}^{m-1} \binom{m}{j} (j!)^{3/2} ((m-j)!)^{3/2} \\
&\quad + \mathbb{E}|Q_{n_1} - \mathbb{E}Q_{n_1}|^m + \mathbb{E}|Q'_{n_2} - \mathbb{E}Q'_{n_2}|^m \Big\}^{1/m} + C_0 \sqrt{n} (m!)^{3/(2m)}.
\end{aligned}$$

Notice that

$$\sum_{j=1}^{m-1} \binom{m}{j} (j!)^{3/2} ((m-j)!)^{3/2} = m! \sum_{j=1}^{m-1} (j!)^{1/2} ((m-j)!)^{1/2} \leq (m!)^{3/2} (m-1)$$

and that by (6.4.3),

$$\beta_m \equiv \sup_{n \geq 1} \left\{(n \log n)^{-m/2} \mathbb{E}|Q_n - \mathbb{E}Q_n|^m\right\} < \infty$$

for every $m \geq 1$. We have

$$\begin{aligned}
\beta_m^{1/m} &\leq \left\{\frac{m-1}{2^{m/2}} (m!)^{3/2} C^m + 2^{-\frac{m-2}{2}} \beta_m\right\}^{1/m} + C_0 (m!)^{3/(2m)} \\
&\leq C \frac{(m-1)^{1/m}}{\sqrt{2}} (m!)^{3/(2m)} + 2^{-\frac{m-2}{2m}} \beta_m^{1/m} + C_0 (m!)^{3/(2m)}.
\end{aligned}$$

Hence,

$$\begin{aligned}
\beta_m^{1/m} &\leq \left(1 - 2^{-\frac{m-2}{2m}}\right)^{-1} \left\{\frac{(m-1)^{1/m}}{\sqrt{2}} C + C_0\right\} (m!)^{3/(2m)} \\
&\leq \left(\frac{1}{2} C + 4C_0\right) (m!)^{3/(2m)} \leq C (m!)^{3/(2m)}.
\end{aligned}$$

Therefore, (6.4.3) holds.

By (6.4.3) and Taylor expansion there is $\theta_0 > 0$ such that

$$\sup_n \mathbb{E}\exp\left\{\frac{\theta_0}{\sqrt[3]{n\log n}}|Q_n - \mathbb{E}Q_n|^{2/3}\right\} < \infty. \tag{6.4.9}$$

It remains to extend (6.4.9) to all $\theta > 0$. Indeed, for any $\theta > \theta_0$, one can find an integer l such that for any n there is an integer partition $0 = n_0 < n_1, \cdots, < n_l = n$ such that $n_i - n_{i-1} < n(\theta_0/\theta)^3$ $(i = 1, \cdots, l)$. Write

$$Q_n = \sum_{i=1}^{l}\sum_{n_{i-1}<j,k\le n_i} 1_{\{S_j=S_k\}} + \sum_{i=2}^{l}\sum_{j=1}^{n_{i-1}}\sum_{k=n_{i-1}+1}^{n_i} 1_{\{S_j=S_k\}}. \tag{6.4.10}$$

Observe that for each $2 \le i \le l$,

$$\sum_{j=1}^{n_{i-1}}\sum_{k=n_{i-1}+1}^{n_i} 1_{\{S_j=S_k\}} \overset{d}{=} \sum_{j=1}^{n_{i-1}}\sum_{k=1}^{n_i-n_{i-1}} 1_{\{S_j=S'_k\}} \le I_n.$$

By (6.2.1) (with $d = 3$ and $p = 2$) and Taylor expansion there is $\lambda > 0$ such that

$$\sup_{n\ge 1}\mathbb{E}\exp\left\{\frac{\lambda}{\sqrt[3]{n}}\left(\sum_{j=1}^{n_{i-1}}\sum_{k=n_{i-1}+1}^{n_i} 1_{\{S_j=S_k\}}\right)^{2/3}\right\} < \infty.$$

By triangular inequality and Hölder inequality, therefore, we get

$$\sup_{n\ge 1}\mathbb{E}\exp\left\{\frac{\lambda}{\sqrt[3]{n\log n}}\left(\sum_{i=2}^{l}\sum_{j=1}^{n_{i-1}}\sum_{k=n_{i-1}+1}^{n_i} 1_{\{S_j=S_k\}}\right)^{2/3}\right\} < \infty$$

for every $\lambda > 0$.

By independence and by triangular inequality, we get

$$\begin{aligned}
&\mathbb{E}\exp\left\{\frac{\theta}{\sqrt[3]{n\log n}}\left(\sum_{i=1}^{l}\left|\sum_{n_{i-1}<j,k\le n_i} 1_{\{S_j=S_k\}} - \mathbb{E}\sum_{n_{i-1}<j,k\le n_i} 1_{\{S_j=S_k\}}\right|\right)^{2/3}\right\}\\
&\le \prod_{i=1}^{l}\mathbb{E}\exp\left\{\frac{\theta}{\sqrt[3]{n\log n}}|Q_{n_i-n_{i-1}} - \mathbb{E}Q_{n_i-n_{i-1}}|^{2/3}\right\}\\
&\le \prod_{i=1}^{l}\mathbb{E}\exp\left\{\frac{\theta_0}{\sqrt[3]{(n_i-n_{i-1})\log(n_i-n_{i-1})}}|Q_{n_i-n_{i-1}} - \mathbb{E}Q_{n_i-n_{i-1}}|^{2/3}\right\}.
\end{aligned}$$

Hence, the desired (6.4.1) follows from (6.4.9) and (6.4.10).

The proof of (6.4.2) can be carried out in a similar way, as (6.2.14) leads to (6.4.4) for $d \ge 4$. $\square$

By slightly modifying our argument, Theorem 6.4.1 can be extended to the renormalized range $R_n - \mathbb{E}R_n$. We give the following theorem without proof.

THEOREM 6.4.2. *As $d = 3$,*

$$\sup_n \mathbb{E}\exp\left\{\frac{\theta}{\sqrt[3]{n\log n}}|R_n - \mathbb{E}R_n|^{2/3}\right\} < \infty \quad \textit{for every } \theta > 0. \tag{6.4.11}$$

As $d \geq 4$,

$$\sup_n \mathbb{E}\exp\Big\{\frac{\theta}{\sqrt[3]{n}}|R_n - \mathbb{E}R_n|^{2/3}\Big\} < \infty \quad \textit{for some } \theta > 0. \tag{6.4.12}$$

We now address integrabilities of $\max_{1\leq k\leq n}|Q_k - \mathbb{E}Q_k|$ and $\max_{1\leq k\leq n}|R_k - \mathbb{E}R_k|$. On the one hand, we are not able to provide any sharp maximal inequalities which bound $\max_{1\leq k\leq n}|Q_k - \mathbb{E}Q_k|$ and $\max_{1\leq k\leq n}|R_k - \mathbb{E}R_k|$ by their marginals $|Q_n - \mathbb{E}Q_n|$ and $|R_n - \mathbb{E}R_n|$. On the other hand, by using the entropy method in Section D of the Appendix, we are able to prove that $\max_{1\leq k\leq n}|Q_k - \mathbb{E}Q_k|$ and $\max_{1\leq k\leq n}|R_k - \mathbb{E}R_k|$ have essentially the same integrabilities as their marginals.

THEOREM 6.4.3. *As* $d = 3$,

$$\sup_n \mathbb{E}\exp\Big\{\frac{\theta}{\sqrt[3]{n\log n}}\max_{1\leq k\leq n}|Q_k - \mathbb{E}Q_k|^{2/3}\Big\} < \infty \quad \forall\theta > 0, \tag{6.4.13}$$

$$\sup_n \mathbb{E}\exp\Big\{\frac{\theta}{\sqrt[3]{n\log n}}\max_{1\leq k\leq n}|R_k - \mathbb{E}R_k|^{2/3}\Big\} < \infty \quad \forall\theta > 0. \tag{6.4.14}$$

As $d \geq 4$,

$$\sup_n \mathbb{E}\exp\Big\{\frac{\theta}{\sqrt[3]{n}}\max_{1\leq k\leq n}|Q_k - \mathbb{E}Q_k|^{2/3}\Big\} < \infty \quad \textit{for some } \theta > 0, \tag{6.4.15}$$

$$\sup_n \mathbb{E}\exp\Big\{\frac{\theta}{\sqrt[3]{n}}\max_{1\leq k\leq n}|R_k - \mathbb{E}R_k|^{2/3}\Big\} < \infty \quad \textit{for some } \theta > 0. \tag{6.4.16}$$

Proof. For similarity we only prove (6.4.13). We first claim that there is $C > 0$ such that for any n_1, n_2 and any $m \geq 1$, we have

$$\begin{aligned}&\mathbb{E}|(Q_{n_2} - \mathbb{E}Q_{n_2}) - (Q_{n_1} - \mathbb{E}Q_{n_1})|^m \\ &\leq (m!)^{3/2}C^m \max\{n_1, n_2\}^{m/4}(\log\max\{n_1, n_2\})^{m/2}|n_2 - n_1|^{m/4}.\end{aligned} \tag{6.4.17}$$

When $n_1 < n_2$,

$$Q_{n_2} - Q_{n_1} = \sum_{n_1 < j < k \leq n_2} 1_{\{S(j)=S(k)\}} + \sum_{j=1}^{n_1}\sum_{k=n_1+1}^{n_2} 1_{\{S(j)=S(k)\}}.$$

Observe that

$$\begin{aligned}\sum_{j=1}^{n_1}\sum_{k=n_1+1}^{n_2} 1_{\{S(j)=S(k)\}} &\stackrel{d}{=} \sum_{j=0}^{n_1-1}\sum_{k=1}^{n_2-n_1} 1_{\{S(j)=\widetilde{S}(k)\}} \\ &= \sum_{x\in\mathbb{Z}^3} l_0(n_1 - 1, x)\widetilde{l}(n_2 - n_1, x)\end{aligned}$$

where $\{\widetilde{S}(k)\}$ is an independent copy of $\{S(k)\}$ and

$$l_0(n, x) = \sum_{k=0}^{n} 1_{\{S(k)=x\}}, \qquad \widetilde{l}(n, x) = \sum_{k=1}^{n} 1_{\{\widetilde{S}(k)=x\}}.$$

By Cauchy-Schwarz inequality, we get

$$\begin{aligned}
&\mathbb{E}\Big[\sum_{x\in\mathbb{Z}^3} l_0(n_1-1,x)\widetilde{l}(n_2-n_1,x)\Big]^m\\
&=\sum_{x_1,\cdots,x_m\in\mathbb{Z}^3}\Big[\mathbb{E}\prod_{k=1}^m l_0(n_1-1,x_k)\Big]\Big[\mathbb{E}\prod_{k=1}^m l(n_2-n_1,x_k)\Big]\\
&\le\Big\{\sum_{x_1,\cdots,x_m\in\mathbb{Z}^3}\Big[\mathbb{E}\prod_{k=1}^m l_0(n_1-1,x_k)\Big]^2\Big\}^{1/2}\\
&\quad\times\Big\{\sum_{x_1,\cdots,x_m\in\mathbb{Z}^3}\Big[\mathbb{E}\prod_{k=1}^m l(n_2-n_1,x_k)\Big]^2\Big\}^{1/2}\\
&=\Big\{\mathbb{E}\bar{I}_{n_1-1}^m\Big\}^{1/2}\Big\{\mathbb{E}I_{n_2-n_1}^m\Big\}^{1/2}
\end{aligned}$$

where

$$\bar{I}_n=\sum_{j=0}^n\sum_{k=1}^n 1_{\{S(j)=\widetilde{S}(k)\}},\qquad I_n=\sum_{j=1}^n\sum_{k=1}^n 1_{\{S(j)=\widetilde{S}(k)\}}.$$

By the moment bound given in (6.2.1) of Theorem 6.2.1 (with $d=3$ and $p=2$), therefore, there is a constant $c_1>0$ independent of n_1,n_2, such that for any $m\ge 1$,

$$\mathbb{E}\Big[\sum_{x\in\mathbb{Z}^3} l_0(n_1-1,x)\widetilde{l}(n_2-n_1,x)\Big]^m\le (m!)^{3/2}c_1^m n_1^{m/4}(n_2-n_1)^{m/4}.$$

Notice that

$$\sum_{n_1<j<k\le n_2} 1_{\{S(j)=S(k)\}}\overset{d}{=}Q_{n_2-n_1}.$$

By Theorem 6.4.1, there is a constant $c_2>0$ independent of n_1,n_2 and $m\ge 1$ such that

$$\begin{aligned}
&\mathbb{E}\Big|\sum_{n_1<j<k\le n_2} 1_{\{S(j)=S(k)\}}-\mathbb{E}\sum_{n_1<j<k\le n_2} 1_{\{S(j)=S(k)\}}\Big|^m\\
&\le (m!)^{3/2}c_2^m(n_2-n_1)^{m/2}\Big(\log(n_2-n_1)\Big)^{m/2}.
\end{aligned}$$

Summarizing our argument, we have proved (6.4.17).

Define the random variable $\mathcal{Z}_n(\cdot)$ taking values in $C[0,1]$ by

$$\mathcal{Z}_n(t)=\frac{Q_k-\mathbb{E}Q_k}{\sqrt{n\log n}}+(nt-k)\frac{(Q_{k+1}-\mathbb{E}Q_{k+1})-(Q_k-\mathbb{E}Q_k)}{\sqrt{n\log n}}$$

if $t\in\Big[\frac{k}{n},\frac{k+1}{n}\Big]$ for some $0\le k\le n-1$. We claim that there is a constant $C>0$ such that

$$\mathbb{E}|\mathcal{Z}_n(s)-\mathcal{Z}_n(t)|^m\le C^m(m!)^{3/2}|s-t|^{m/4}\tag{6.4.18}$$

for every $m,n\ge 1$ and for any $s,t\in[0,1]$.

Assume that $|s-t|\ge 1/n$. By (6.4.17), we get

$$\mathbb{E}|\mathcal{Z}_n(s)-\mathcal{Z}_n(t)|^m\le C^m(m!)^{3/2}n^{-m/4}\Big|[ns]-[nt]\Big|^{m/4}.$$

Hence, (6.4.18) follows from the fact that as $|s-t|\ge 1/n$,
$$\big|[ns]-[nt]\big|\le 3n|s-t|.$$

We now consider the case $|s-t|<1/n$. If s,t is in the same interval $\left[\frac{k}{n},\frac{k+1}{n}\right]$, then
$$\mathcal{Z}_n(s)-\mathcal{Z}_n(t)=n(s-t)\frac{(Q_{k+1}-\mathbb{E}Q_{k+1})-(Q_k-\mathbb{E}Q_k)}{\sqrt{n\log n}}.$$
By (6.4.17), we get
$$\begin{aligned}&\mathbb{E}|\mathcal{Z}_n(s)-\mathcal{Z}_n(t)|^m\\ &=|s-t|^m\Big(\frac{n}{\log n}\Big)^{m/2}\mathbb{E}\big|(Q_{k+1}-\mathbb{E}Q_{k+1})-(Q_k-\mathbb{E}Q_k)\big|^m\\ &\le|s-t|^m\Big(\frac{n}{\log n}\Big)^{m/2}(m!)^{3/2}C^m n^{m/4}(\log n)^{m/2}\\ &\le C^m(m!)^{3/2}|s-t|^{m/4}.\end{aligned}$$

Assume that $|s-t|<1/n$ and s,t are in two consecutive intervals. We may only consider the case $s<t$. Let $s\in\left[\frac{k-1}{n},\frac{k}{n}\right]$ and $t\in\left[\frac{k}{n},\frac{k+1}{n}\right]$. We have
$$\begin{aligned}\mathcal{Z}_n(t)-\mathcal{Z}_n(s)=(nt-k)&\frac{(Q_{k+1}-\mathbb{E}Q_{k+1})-(Q_k-\mathbb{E}Q_k)}{\sqrt{n\log n}}\\ +(k-ns)&\frac{(Q_k-\mathbb{E}Q_k)-(Q_{k-1}-\mathbb{E}Q_{k-1})}{\sqrt{n\log n}}.\end{aligned}$$
Observe that $0\le(nt-k),(k-ns)\le n(t-s)$. Applying (6.4.17) to each of the two parts on the right hand side gives (6.4.18).

By (6.4.18) and Taylor expansion, we get
$$\sup_{n\ge1}\sup_{\substack{s,t\in[0,1]\\ s\ne t}}\mathbb{E}\exp\left\{c\frac{|\mathcal{Z}(s)-\mathcal{Z}(t)|^{2/3}}{|s-t|^{1/6}}\right\}<\infty.$$
By Theorem D.6 in the Appendix, there is $\theta>0$ such that
$$\sup_{n\ge1}\mathbb{E}\exp\Big\{\theta\sup_{s,t\in[0,1]}|\mathcal{Z}(s)-\mathcal{Z}(t)|^{2/3}\Big\}<\infty.$$
Taking $s=0$ we conclude that (6.4.13) holds for some $\theta>0$. By the same argument used near the end of the proof for (6.4.1), we can extend (6.4.13) to all $\theta>0$. □

6.5. Notes and comments

Section 6.1.

The multinomial inequalities in Theorem 6.1.1 and Theorem 6.1.2 were essentially established in [**27**], [**28**] and [**8**].

EXERCISE 6.5.1. *Prove the inequality (6.1.3).*

EXERCISE 6.5.2. *Recall that the random measure $\alpha_\epsilon(A)$ is given in (2.2.4).*

(1) *Show that for any $t_1,\cdots,t_a>0$ and integer $m\ge 1$,*

$$\Big\{\mathbb{E}\Big[\alpha_\epsilon\big([0,\ t_1+\cdots+t_a]^p\big)^m\Big]\Big\}^{1/p}$$

$$\le \sum_{\substack{l_1+\cdots+l_a=m\\ l_1,\cdots,l_a\ge 0}} \frac{m!}{l_1!\cdots l_a!}\prod_{k=1}^a\Big\{\mathbb{E}\Big[\alpha_\epsilon\big([0,t_k]^p\big)^{l_k}\Big]\Big\}^{1/p}.$$

(2) *Derive the correspondent moment inequality for the intersection local time $\alpha(A)$.*

Section 6.2.

The idea of using the moment inequalities to establish exponential integrabilities (Theorem 6.2.1) for I_n and J_n first appeared in Bass, Chen and Rosen [**7**]. Prior to [**7**], the exponential integrability of intersection local times was investigated by Brydges and Slade [**19**] in a different approach.

EXERCISE 6.5.3. *Prove* (6.2.9) *in Lemma 6.2.2.*

EXERCISE 6.5.4. *Prove* (6.2.5) *in Theorem 6.2.1.*

Section 6.3.

The part "$d=2$" in Theorem 6.3.2 appears as a discrete version of exponential integrability described in (4.3.3) and was obtained in two papers by Bass, Chen and Rosen ([**7**], [**8**]). Prior to that, Brydges and Slade ([**19**]) had established (6.3.5) in the setting of random walks with continuous time.

Section 6.4.

The main results of this section was established in [**33**].

EXERCISE 6.5.5. *Prove an early result established by Brydges and Slade* ([**19**])*: For $d\ge 3$, there is a $\theta_d>0$ such that*

$$\sup_{n\ge 1}\mathbb{E}\exp\Big\{\frac{\theta}{n}Q_n\Big\}<\infty\qquad \forall\theta<\theta_d.$$

Hint: This is the same as

$$\sup_{n\ge 1}\mathbb{E}\exp\Big\{\frac{\theta}{n}(Q_n-\mathbb{E}Q_n)\Big\}<\infty\qquad \forall\theta<\theta_d.$$

CHAPTER 7

Independent random walks: large deviations

Our goal in this chapter is to establish the large deviations for the mutual intersection local times run by p independent random walks in $\mathbb{Z}^d$, and for the intersection of the ranges of these random walks. We consider three very different situations: sub-critical dimensions ($p(d-2) < d$), critical-dimensions ($p(d-2) = d$) and super-critical dimensions ($p(d-2) > d$).

The large deviations for I_n and J_n we shall establish in the case $p(d-2) < d$ are also called moderate deviations and are essentially a discrete version of the large deviation principle for $\alpha([0,1]^p)$ (Theorem 3.3.2).

In the case $p(d-2) = d$, I_n and J_n are attracted by Γ-distributions (Theorem 5.5.1). Therefore, the moderate deviations for I_n and J_n should yield Γ-tails. On the other hand, the Gagliardo-Nirenberg inequality still holds (and is called Sobolev inequality) in the critical dimensions. This raises the possibility that in addition to the moderate deviations of Γ-tails, I_n and J_n might obey a large deviation principle associated with the Sobolev constant $\kappa(d,p)$.

The independent trajectories intersect at most finitely many times when $p(d-2) > d$. Consequently, the random variables I_∞ and J_∞ are well defined for measuring the intersections in a whole lifetime. The problem in the super-critical dimensions is to study the large deviations for I_∞ and J_∞.

7.1. Feynman-Kac minorations

Despite their close relation to the large deviation principle stated in Theorem 3.3.2, the moderate deviations for I_n and J_n cannot be achieved by the method of high moment asymptotics developed in Chapter 3 for lack of the scaling property in the discrete setting. In our treatment of the moderate deviations given in the next section, the upper bounds rely on the moment inequalities established in (6.1.7), (6.1.8), and on the laws of weak convergence stated in Theorem 5.3.2, Theorem 5.3.4. For the lower bounds, we need the following partial extensions (Theorem 7.1.1 and Theorem 7.1.2) of Feynman-Kac large deviation given in Theorem 4.1.6.

As usual, $\{S(n)\}$ is a symmetric, square integrable random walk in $\mathbb{Z}^d$ with the smallest supporting group $\mathbb{Z}^d$.

THEOREM 7.1.1. *For any bounded, continuous function f on $\mathbb{R}^d$, and for any positive sequence $\{b_n\}$ satisfying*

$$b_n \longrightarrow \infty \quad \textit{and} \quad b_n = o(n) \qquad (n \to \infty),$$

$$\liminf_{n\to\infty}\frac{1}{b_n}\log\mathbb{E}\exp\left\{\frac{b_n}{n}\sum_{k=1}^{n}f\left(\sqrt{\frac{b_n}{n}}S(k)\right)\right\}$$
$$\geq \sup_{g\in\mathcal{F}_d}\left\{\int_{\mathbb{R}^d}f(\Gamma^{1/2}x)g^2(x)dx-\frac{1}{2}\int_{\mathbb{R}^d}|\nabla g(x)|^2dx\right\}$$

where $\mathcal{F}_d$ is the set of the functions g on $\mathbb{R}^d$ with

$$\int_{\mathbb{R}^d}|g(x)|^2dx=1 \quad and \quad \int_{\mathbb{R}^d}|\nabla g(x)|^2dx<\infty.$$

Proof. We first prove Theorem 7.1.1 under the aperiodicity assumption on the random walks. Let γ_n be a sequence of odd integers such that $\gamma_n\sim b_n$ $(n\to\infty)$ and write $t_n=[n/\gamma_n]$. Then $t_n\gamma_n\leq n\leq(t_n+1)\gamma_n$. By the boundedness of f, we get

$$\sum_{k=1}^{n}f\left(\sqrt{b_n n}S(k)\right)=\sum_{j=1}^{\gamma_n-1}\sum_{k=jt_n+1}^{(j+1)t_n-1}f\left(\sqrt{\frac{b_n}{n}}S(k)\right)+O(\gamma_n+t_n).$$

Hence, we need only to show that

$$\liminf_{n\to\infty}\frac{1}{b_n}\log\mathbb{E}\exp\left\{\frac{b_n}{n}\sum_{j=1}^{\gamma_n-1}\sum_{k=jt_n+1}^{(j+1)t_n-1}f\left(\sqrt{\frac{b_n}{n}}S(k)\right)\right\}\tag{7.1.1}$$
$$\geq \sup_{g\in\mathcal{F}_d}\left\{\int_{\mathbb{R}^d}f(\Gamma^{1/2}x)g^2(x)dx-\frac{1}{2}\int_{\mathbb{R}^d}|\nabla g(x)|^2dx\right\}.$$

For each n, define the linear operator $\mathfrak{A}_n$ on $\mathcal{L}^2(\mathbb{Z}^d)$ as

$$\mathfrak{A}_n\xi(x)=\mathbb{E}\left(\exp\left\{\frac{b_n}{n}\sum_{k=1}^{t_n-1}f\left(\sqrt{\frac{b_n}{n}}(x+S(k))\right)\right\}\xi(x+S(t_n))\right)$$

where $x\in\mathbb{Z}^d$ and $\xi\in\mathcal{L}^2(\mathbb{Z}^d)$. We claim that $\mathfrak{A}_n$ is symmetric:

$$\langle\xi,\mathfrak{A}_n\eta\rangle=\langle\mathfrak{A}_n\xi,\eta\rangle\qquad \xi,\eta\in\mathcal{L}^2(\mathbb{Z}^d).\tag{7.1.2}$$

Indeed,

$$\begin{aligned}
&\langle\xi,\mathfrak{A}_n\eta\rangle\\
&=\sum_{x\in\mathbb{Z}^d}\xi(x)\mathbb{E}\left(\exp\left\{\frac{b_n}{n}\sum_{k=1}^{t_n-1}f\left(\sqrt{\frac{b_n}{n}}(x+S(k))\right)\right\}\eta(x+S(t_n))\right)\\
&=\mathbb{E}\left(\sum_{x\in\mathbb{Z}^d}\xi(x)\exp\left\{\frac{b_n}{n}\sum_{k=1}^{t_n-1}f\left(\sqrt{\frac{b_n}{n}}(x+S(k))\right)\right\}\eta(x+S(t_n))\right)\\
&=\mathbb{E}\left(\sum_{x\in\mathbb{Z}^d}\xi(x-S(t_n))\exp\left\{\frac{b_n}{n}\sum_{k=1}^{t_n-1}f\left(\sqrt{\frac{b_n}{n}}(x+S(k)-S(t_n))\right)\right\}\eta(x)\right)\\
&=\sum_{x\in\mathbb{Z}^d}\eta(x)\mathbb{E}\left(\exp\left\{\frac{b_n}{n}\sum_{k=1}^{t_n-1}f\left(\sqrt{\frac{b_n}{n}}(x+S(k)-S(t_n))\right)\right\}\xi(x-S(t_n))\right)
\end{aligned}$$

where the third equality follows from translation invariance.

Observe that

$$\sum_{k=1}^{t_n-1} f\Big(\sqrt{\frac{b_n}{n}}(x+S(k)-S(t_n))\Big) = \sum_{k=1}^{t_n-1} f\Big(\sqrt{\frac{b_n}{n}}(x+S(t_n-k)-S(t_n))\Big).$$

If we write

$$\bar{S}(k) = S(t_n-k) - S(t_n) \qquad k = 1, \cdots, t_n - 1$$

by symmetry of the random walk, we have

$$\{\bar{S}(k)\}_{1\le k\le t_n-1} \stackrel{d}{=} \{S(k)\}_{1\le k\le t_n-1}.$$

Consequently,

$$\mathbb{E}\bigg(\exp\Big\{\frac{b_n}{n}\sum_{k=1}^{t_n-1} f\Big(\sqrt{\frac{b_n}{n}}(x+S(k)-S(t_n))\Big)\Big\}\xi\big(x-S(t_n)\big)\bigg)$$
$$= \mathbb{E}\bigg(\exp\Big\{\frac{b_n}{n}\sum_{k=1}^{t_n-1} f\Big(\sqrt{\frac{b_n}{n}}(x+S(k))\Big)\Big\}\xi\big(x+S(t_n)\big)\bigg).$$

Summarizing what we have proved gives us

$$\langle \xi, \mathfrak{A}_n\eta\rangle = \sum_{x\in\mathbb{Z}^d}\eta(x)\mathbb{E}\bigg(\exp\Big\{\frac{b_n}{n}\sum_{k=1}^{t_n-1} f\Big(\sqrt{\frac{b_n}{n}}(x+S(k))\Big)\Big\}\xi\big(x+S(t_n)\big)\bigg)$$

which is (7.1.2).

By the boundedness of f, it is easy to see that for each n, $\mathfrak{A}_n$ is bounded. Consequently, $\mathfrak{A}_n$ is self-adjoint.

By increment independence, we get

$$\mathbb{E}\exp\Big\{\frac{b_n}{n}\sum_{j=1}^{\gamma_n-1}\sum_{k=jt_n+1}^{(j+1)t_n-1} f\Big(\sqrt{\frac{b_n}{n}}S(k)\Big)\Big\}$$
$$= \sum_{x\in\mathbb{Z}^d}\mathbb{P}\{S(t_n)=x\}\mathbb{E}\exp\Big\{\frac{b_n}{n}\sum_{j=1}^{\gamma_n-1}\sum_{k=(j-1)t_n+1}^{jt_n-1} f\Big(\sqrt{\frac{b_n}{n}}(x+S(k))\Big)\Big\}.$$

Let $g\in\mathcal{S}(\mathbb{R}^d)$ be supported by a finite box $[-M,M]^d$ and assume that

$$\int_{\mathbb{R}^d}|g(x)|^2dx = 1$$

and write

$$\xi_n(x) = \Big(\sum_{y\in\mathbb{Z}^d} g^2\Big(\sqrt{\frac{b_n}{n}}y\Big)\Big)^{-1/2} g\Big(\sqrt{\frac{b_n}{n}}x\Big) \qquad x\in\mathbb{Z}^d.$$

Then

$$\mathbb{E}\exp\left\{\frac{b_n}{n}\sum_{j=1}^{\gamma_n-1}\sum_{k=jt_n+1}^{(j+1)t_n-1}f\Big(\sqrt{\frac{b_n}{n}}S(k)\Big)\right\}$$

$$\geq\frac{1}{||g||_\infty^2}\Big(\sum_{y\in\mathbb{Z}^d}g^2\Big(\sqrt{\frac{b_n}{n}}y\Big)\Big)\sum_{x\in\mathbb{Z}^d}\mathbb{P}\{S(t_n)=x\}\xi_n(x)$$

$$\times\mathbb{E}\bigg(\exp\left\{\frac{b_n}{n}\sum_{j=1}^{\gamma_n-1}\sum_{k=(j-1)t_n+1}^{jt_n-1}f\Big(\sqrt{\frac{b_n}{n}}(x+S(k))\Big)\right\}\xi_n(x+S((\gamma_n-1)t_n))\bigg)$$

$$=\frac{1}{||g||_\infty^2}\Big(\sum_{y\in\mathbb{Z}^d}g^2\Big(\sqrt{\frac{b_n}{n}}y\Big)\Big)\sum_{x\in\mathbb{Z}^d}\mathbb{P}\{S(t_n)=x\}\xi_n(x)\mathfrak{A}_n^{\gamma_n-1}(\xi_n)(x).$$

Notice that

$$\lim_{n\to\infty}\Big(\frac{b_n}{n}\Big)^{d/2}\sum_{y\in\mathbb{Z}^d}g^2\Big(\sqrt{\frac{b_n}{n}}y\Big)=\int_{\mathbb{R}^d}g^2(x)dx=1.$$

By (5.1.2) (with n being replaced by t_n) in Lemma 5.1.1[1], we have

$$\mathbb{P}\{S(t_n)=x\}\geq c\Big(\frac{b_n}{n}\Big)^{d/2}\qquad x\in\Big[-M\sqrt{\frac{n}{b_n}},\ M\sqrt{\frac{n}{b_n}}\Big]^d.$$

Since $\xi_n(x)=0$ outside the box

$$\Big[-M\sqrt{\frac{n}{b_n}},\ M\sqrt{\frac{n}{h}b_n}\Big]^d$$

there is a $\delta>0$ such that

$$\mathbb{E}\exp\left\{\frac{b_n}{n}\sum_{j=1}^{\gamma_n-1}\sum_{k=jt_n+1}^{(j+1)t_n-1}f\Big(\sqrt{\frac{b_n}{n}}S(k)\Big)\right\}$$

$$\geq\delta\sum_{x\in\mathbb{Z}^d}\xi_n(x)\mathfrak{A}_n^{\gamma_n-1}(\xi_n)(x)=\delta\langle\xi_n,\mathfrak{A}_n^{\gamma_n-1}\xi_n\rangle.$$

According to Theorem E.2 in the Appendix, the self-adjoint operator $\mathfrak{A}_n$ admits the spectral integral representation

$$\mathfrak{A}_n=\int_{-\infty}^{\infty}\lambda E(d\lambda).$$

Applying Corollary E.5 in the Appendix gives us

$$\mathfrak{A}_n^{\gamma_n-1}=\int_{-\infty}^{\infty}\lambda^{\gamma_n-1}E(d\lambda).$$

By (E.15) in the Appendix, the above representations lead to

$$\langle\xi_n,\mathfrak{A}_n\xi_n\rangle=\int_{-\infty}^{\infty}\lambda\mu_{\xi_n}(d\lambda),$$

$$\langle\xi_n,\mathfrak{A}_n^{\gamma_n-1}\xi_n\rangle=\int_{-\infty}^{\infty}\lambda^{\gamma_n-1}\mu_{\xi_n}(d\lambda).$$

[1]This is where we use the extra assumption on aperiodicity.

Notice ((E.5), Appendix) that $\mu_{\xi_n}(\mathbb{R}) = 1$. That means that the spectral measure μ_{ξ_n} is a probability measure on $\mathbb{R}$. By Jensen inequality (recall that $\gamma_n - 1$ is even), we get

$$\langle \xi_n, \mathfrak{A}_n^{\gamma_n - 1}\xi_n\rangle \geq \left(\int_{-\infty}^{\infty} \lambda \mu_{\xi_n}(d\lambda)\right)^{\gamma_n - 1} = \langle \xi_n, \mathfrak{A}_n \xi_n\rangle^{\gamma_n - 1}.$$

Thus,

$$\begin{aligned}&\liminf_{n\to\infty} \frac{1}{b_n}\log \mathbb{E}\exp\left\{\frac{b_n}{n}\sum_{j=1}^{\gamma_n-1}\sum_{k=jt_n+1}^{(j+1)t_n-1} f\left(\sqrt{\frac{b_n}{n}}S(k)\right)\right\}\\ &\geq \liminf_{n\to\infty}\log\langle \xi_n, \mathfrak{A}_n\xi_n\rangle.\end{aligned}$$

Notice that

$$\begin{aligned}&\langle \xi_n, \mathfrak{A}_n\xi_n\rangle\\ &= \sum_{x\in\mathbb{Z}^d}\xi_n(x)\mathbb{E}\left(\exp\left\{\frac{b_n}{n}\sum_{k=1}^{t_n-1} f\left(\sqrt{\frac{b_n}{n}}(x+S(k))\right)\right\}\xi_n(x+S(t_n))\right)\\ &= (1+o(1))\int_{\mathbb{R}^d} dx g(x)\\ &\times \mathbb{E}\left(\exp\left\{\frac{b_n}{n}\sum_{k=1}^{t_n-1} f\left(x+\sqrt{\frac{b_n}{n}}S(k)\right)\right\}g\left(x+\sqrt{\frac{b_n}{n}}S(t_n)\right)\right)\\ &\longrightarrow \int_{\mathbb{R}^d} g(x)\mathbb{E}\left(\exp\left\{\int_0^1 f(x+\Gamma^{1/2}W(t))dt\right\}g(x+\Gamma^{1/2}W(1))\right)dx\end{aligned}$$

as $n\to\infty$, where the last step partially follows from the invariant principle.

By the variable substitution $x \mapsto \Gamma^{1/2}x$, the integral on the right hand side is equal to

$$\begin{aligned}&\int_{\mathbb{R}^d}\widetilde{g}(x)\mathbb{E}\left(\exp\left\{\int_0^1 \bar{f}(x+W(t))dt\right\}\widetilde{g}(x+W(1))\right)dx\\ &\geq \exp\left\{\int_{\mathbb{R}^d}\bar{f}(x)\widetilde{g}^2(x)dx - \frac{1}{2}\int_{\mathbb{R}^d}|\nabla\widetilde{g}(x)|^2dx\right\}\end{aligned}$$

where

$$\bar{f}(x) = f(\Gamma^{1/2}x), \qquad \widetilde{g}(x) = \det(\Gamma)^{1/4}g(\Gamma^{1/2}x)$$

and where the inequality follows from (4.1.25) (with $t=1$, and with f and g being replaced by $\bar{f}$ and $\widetilde{g}$, respectively).

Summarizing our argument gives us

$$\begin{aligned}&\liminf_{n\to\infty} \frac{1}{b_n}\log \mathbb{E}\exp\left\{\frac{b_n}{n}\sum_{j=1}^{\gamma_n-1}\sum_{k=jt_n+1}^{(j+1)t_n-1} f\left(\sqrt{\frac{b_n}{n}}S(k)\right)\right\}\\ &\geq \int_{\mathbb{R}^d} f(\Gamma^{1/2}x)\widetilde{g}^2(x)dx - \frac{1}{2}\int_{\mathbb{R}^d}|\nabla\widetilde{g}(x)|^2dx.\end{aligned}$$

Notice that the set of $\widetilde{g}$ is dense in $\mathcal{F}_d$. Taking supremum over $\widetilde{g}$ on the right hand side leads to the desired conclusion (7.1.1) and therefore to Theorem 7.1.1.

We now drop the aperiodicity condition by performing resolvent approximation. Recall that $0 < \theta < 1$ is fixed and let $\{\delta_n\}$ be a Bernoulli sequence independent of $\{S(n)\}$ such that

$$\mathbb{P}\{\delta_n = 1\} = 1 - \mathbb{P}\{\delta_n = 0\} = \theta.$$

Write $\tau_0 = 0$ and

$$\tau_n = \min\{k > \tau_{n-1};\ \ \delta_k = 1\} \qquad n = 1, 2, \cdots .$$

Then $\{\tau_n - \tau_{n-1}\}_{n\geq 1}$ is an i.i.d. sequence with common (geometric) distribution

$$\mathbb{P}\{\tau_1 = k\} = \theta(1-\theta)^{k-1} \qquad k = 1, 2, \cdots .$$

Notice that $\{S(\tau_n)\}$ is a random walk whose increment has the distribution

$$\mathbb{P}\{S(\tau_1) = x\} = \sum_{k=1}^\infty \theta(1-\theta)^{k-1}\mathbb{P}\{S(k) = x\}.$$

In particular,

$$\text{Cov}\Big(S(\tau_1), S(\tau_1)\Big) = \mathbb{E}\tau_1 \cdot \Gamma = \theta^{-1}\Gamma$$

and, $\mathbb{P}\{S(\tau_1) = 0\} > 0$ implies that $\{S(\tau_n)\}$ is aperiodic. $\{S(\tau_n)\}$ is called the resolvent random walk of $\{S(n)\}$ in the literature.

Observe that

$$\sum_{k=1}^n f\Big(\sqrt{\frac{b_n}{n}}S(\tau_k)\Big) = \sum_{k=1}^{\tau_n} \delta_k f\Big(\sqrt{\frac{b_n}{n}}S(k)\Big).$$

Applying what we have proved to the aperiodic random walk $\{S(\tau_n)\}$, we get

$$\begin{aligned}(7.1.3)\qquad &\liminf_{n\to\infty}\frac{1}{b_n}\log\mathbb{E}\exp\Big\{\frac{b_n}{n}\sum_{k=1}^{\tau_n}\delta_k f\Big(\sqrt{\frac{b_n}{n}}S(k)\Big)\Big\}\\ &\geq \sup_{g\in\mathcal{F}_d}\Big\{\int_{\mathbb{R}^d} f(\theta^{-1/2}\Gamma^{1/2}x)g^2(x)dx - \frac{1}{2}\int_{\mathbb{R}^d}|\nabla g(x)|^2dx\Big\}.\end{aligned}$$

Consider the decomposition

$$\begin{aligned}\sum_{k=1}^n f\Big(\sqrt{\frac{b_n}{n}}S(k)\Big) &= \frac{1}{\theta}\sum_{k=1}^n(\theta - \delta_k)f\Big(\sqrt{\frac{b_n}{n}}S(k)\Big)\\ &+\frac{1}{\theta}\Big\{\sum_{k=1}^n\delta_k f\Big(\sqrt{\frac{b_n}{n}}S(k)\Big) - \sum_{k=1}^{\tau_n}\delta_k f\Big(\sqrt{\frac{b_n}{n}}S(k)\Big)\Big\}\\ &+\frac{1}{\theta}\sum_{k=1}^{\tau_n}\delta_k f\Big(\sqrt{\frac{b_n}{n}}S(k)\Big).\end{aligned}$$

The second term in the decomposition is bounded from below by

$$-\frac{1}{\theta}||f||_\infty(\tau_n - n)$$

To complete the proof, therefore, we need only to show that

$$(7.1.4)\qquad \lim_{n\to\infty}\frac{1}{b_n}\log\mathbb{E}\exp\Big\{\lambda\frac{b_n}{n}\Big|\sum_{k=1}^n(\theta-\delta_k)f\Big(\sqrt{\frac{b_n}{n}}S(k)\Big)\Big|\Big\} = 0$$

for any $\lambda > 0$, and that

$$\lim_{\theta\to 1^-}\limsup_{n\to\infty}\frac{1}{b_n}\log\mathbb{E}\exp\Big\{\lambda\frac{b_n}{n}(\tau_n - n)\Big\} = 0 \tag{7.1.5}$$

for any $\lambda > 0$.

Let $\{\tilde{\delta}_k\}$ be an independent copy of $\{\delta_k\}$. Conditioning on $\{S_k\}$, by Jensen inequality, by Taylor expansion and by the symmetry of $\{\tilde{\delta}_k - \delta_k\}$, gives us

$$\begin{aligned}
&\widehat{\mathbb{E}}\exp\Big\{\lambda\frac{b_n}{n}\Big|\sum_{k=1}^n(\theta-\delta_k)f\Big(\sqrt{\frac{b_n}{n}}S(k)\Big)\Big|\Big\}\\
&\le \widehat{\mathbb{E}}\exp\Big\{\lambda\frac{b_n}{n}\Big|\sum_{k=1}^n(\tilde{\delta}_k-\delta_k)f\Big(\sqrt{\frac{b_n}{n}}S(k)\Big)\Big|\Big\}\\
&\le 2\widehat{\mathbb{E}}\exp\Big\{\lambda\frac{b_n}{n}\sum_{k=1}^n(\tilde{\delta}_k-\delta_k)f\Big(\sqrt{\frac{b_n}{n}}S(k)\Big)\Big\}\\
&\le 2\exp\Big\{C\lambda^2\Big(\frac{b_n}{n}\Big)^2\sum_{k=1}^n f^2\Big(\sqrt{\frac{b_n}{n}}S(k)\Big)\Big\}
\end{aligned}$$

where the notation "$\widehat{\mathbb{E}}$" is for the expectation conditioning on $\{S_k\}$. Consequently, (7.1.4) follows from Fubini theorem and the boundedness of $f^2(x)$.

By independence, we get

$$\mathbb{E}\exp\Big\{\lambda\frac{b_n}{n}(\tau_n - n)\Big\} = \Big(\mathbb{E}\exp\Big\{\lambda\frac{b_n}{n}(\tau_1 - 1)\Big\}\Big)^n.$$

Hence, (7.1.5) follows from the fact that

$$\mathbb{E}\exp\Big\{\lambda\frac{b_n}{n}(\tau_1 - 1)\Big\} \sim \exp\Big\{\lambda\frac{b_n}{n}(\theta^{-1} - 1)\Big\}$$

as $n \to \infty$. □

In our next major theorem, we establish the Feynman-Kac minoration for the sum over the range.

THEOREM 7.1.2. *Let f be a bounded, continuous function on $\mathbb{R}^d$.*

(1) *When $d = 2$,*

$$\begin{aligned}
&\liminf_{n\to\infty}\frac{1}{b_n}\log\mathbb{E}\exp\Big\{\frac{b_n\log n}{n}\sum_{x\in S[1,n]} f\Big(\sqrt{\frac{b_n}{n}}x\Big)\Big\} \\
&\ge \sup_{g\in\mathcal{F}_2}\Big\{2\pi\sqrt{\det(\Gamma)}\int_{\mathbb{R}^2} f(\Gamma^{1/2}x)g^2(x)dx - \frac{1}{2}\int_{\mathbb{R}^2}|\nabla g(x)|^2dx\Big\}
\end{aligned} \tag{7.1.6}$$

for any positive sequence $\{b_n\}$ satisfying

$$b_n \longrightarrow \infty \quad \textit{and} \quad b_n = o(\log n) \qquad (n\to\infty).$$

(2) *When* $d \geq 3$,

$$\liminf_{n\to\infty} \frac{1}{b_n} \log \mathbb{E} \exp\left\{\frac{b_n}{n} \sum_{x\in S[1,n]} f\left(\sqrt{\frac{b_n}{n}}x\right)\right\} \tag{7.1.7}$$

$$\geq \sup_{g\in\mathcal{F}_d} \left\{\gamma_S \int_{\mathbb{R}^d} f(\Gamma^{1/2}x) g^2(x)dx - \frac{1}{2}\int_{\mathbb{R}^d} |\nabla g(x)|^2 dx\right\}$$

for any positive sequence $\{b_n\}$ *satisfying*

$$b_n \longrightarrow \infty \quad \text{and} \quad b_n = o\left(n^{1/3}\right) \qquad (n\to\infty).$$

Here we recall that $\gamma_S > 0$ *is the probability of no return defined in (5.1.12) (see also (5.1.13)).*

Compared to Theorem 7.1.1, the proof of Theorem 7.1.2 is much harder: The additivity property that

$$\sum_{i=1}^{j+k} f\left(\sqrt{\frac{b_n}{n}}S(i)\right) = \sum_{i=1}^{j} f\left(\sqrt{\frac{b_n}{n}}S(i)\right) + \sum_{i=j+1}^{j+k} f\left(\sqrt{\frac{b_n}{n}}S(i)\right) \qquad j,k\geq 1$$

is crucial in the proof of Theorem 7.1.1. Unfortunately, it is no longer the case for the process

$$\sum_{x\in S[1,n]} f\left(\sqrt{\frac{b_n}{n}}x\right).$$

Instead, we first work on a more trackable quantity

$$\sum_{j=1}^{\gamma_n-1} \sum_{x\in S(jt_n,(j+1)t_n]} f\left(\sqrt{\frac{b_n}{n}}x\right)$$

where γ_n is a sequence of odd integers such that $\gamma_n \sim b_n$ $(n\to\infty)$ and $t_n = [n/\gamma_n]$.

LEMMA 7.1.3. *Assume that* $\{S(k)\}$ *is aperiodic.*

(1) *When* $d = 2$,

$$\liminf_{n\to\infty} \frac{1}{b_n} \log \mathbb{E} \exp\left\{\frac{b_n \log n}{n} \sum_{j=0}^{\gamma_n-1} \sum_{x\in S(jt_n,(j+1)t_n]} f\left(\sqrt{\frac{b_n}{n}}x\right)\right\} \tag{7.1.8}$$

$$\geq \sup_{g\in\mathcal{F}_2} \left\{2\pi\sqrt{\det(\Gamma)} \int_{\mathbb{R}^2} f(\Gamma^{1/2}x) g^2(x)dx - \frac{1}{2}\int_{\mathbb{R}^2} |\nabla g(x)|^2 dx\right\}$$

if $\log b_n = o(\log n)$.

(2) *When* $d \geq 3$,

$$\liminf_{n\to\infty} \frac{1}{b_n} \log \mathbb{E} \exp\left\{\frac{b_n}{n} \sum_{j=0}^{\gamma_n-1} \sum_{x\in S(jt_n,(j+1)t_n]} f\left(\sqrt{\frac{b_n}{n}}x\right)\right\} \tag{7.1.9}$$

$$\geq \sup_{g\in\mathcal{F}_3} \left\{\gamma_S \int_{\mathbb{R}^3} f(\Gamma^{1/2}x) g^2(x)dx - \frac{1}{2}\int_{\mathbb{R}^3} |\nabla g(x)|^2 dx\right\}$$

if $b_n = o(n)$.

Proof. Due to similarity we only consider the case $d = 2$. The approach is very close to the one used in the proof of Theorem 7.1.1. Observe that

$$\Big|\sum_{j=0}^{\gamma_n-1}\sum_{x\in S(jt_n,(j+1)t_n]} f\Big(\sqrt{\frac{b_n}{n}}x\Big) - \sum_{j=0}^{\gamma_n-1}\sum_{x\in S(jt_n,(j+1)t_n)} f\Big(\sqrt{\frac{b_n}{n}}x\Big)\Big|$$
$$\le \|f\|_\infty \#\Big\{S(t_n), S(2t_n),\cdots, S(\gamma_n t_n)\Big\} \le \|f\|_\infty \gamma_n.$$

The right hand side is negligible. Thus, we need only to establish

$$\liminf_{n\to\infty}\frac{1}{b_n}\log \mathbb{E}\exp\Big\{\frac{b_n\log n}{n}\sum_{j=1}^{\gamma_n-1}\sum_{x\in S(jt_n,(j+1)t_n)} f\Big(\sqrt{\frac{b_n}{n}}x\Big)\Big\}$$
$$\ge \sup_{g\in\mathcal{F}_2}\Big\{2\pi\sqrt{\det(\Gamma)}\int_{\mathbb{R}^2} f(\Gamma^{1/2}x)g^2(x)dx - \frac{1}{2}\int_{\mathbb{R}^2}|\nabla g(x)|^2dx\Big\}. \tag{7.1.10}$$

Define the continuous linear operator $\mathfrak{B}_n$ on $\mathcal{L}^2(\mathbb{Z}^d)$ as

$$\mathfrak{B}_n\xi(x) = \mathbb{E}\Big(\exp\Big\{\frac{b_n\log n}{n}\sum_{y-x\in S[1,t_n-1]} f\Big(\sqrt{\frac{b_n}{n}}y\Big)\Big\}\xi(x+S(t_n))\Big).$$

Similar to (7.1.2), $\mathfrak{B}_n$ is self-adjoint. Let g be a bounded function on $\mathbb{R}^d$ and assume that g is infinitely differentiable, supported by a finite square $[-M, M]^2$ with

$$\int_{\mathbb{R}^2}|g(x)|^2dx = 1$$

and write

$$\xi_n(x) = \Big(\sum_{y\in\mathbb{Z}^2} g^2\Big(\sqrt{\frac{b_n}{n}}y\Big)\Big)^{-1/2} g\Big(\sqrt{\frac{b_n}{n}}x\Big) \qquad x\in\mathbb{Z}^2.$$

An estimate used in the proof of Theorem 7.1.1 leads to

$$\mathbb{E}\exp\Big\{\theta\frac{b_n\log n}{n}\sum_{j=1}^{\gamma_n-1}\sum_{x\in S(jt_n,(j+1)t_n)} f\Big(\sqrt{\frac{b_n}{n}}x\Big)\Big\}$$
$$\ge \delta\sum_{x\in\mathbb{Z}^2}\xi_n(x)\mathbb{E}\Big(\exp\Big\{\theta\frac{b_n\log n}{n}\sum_{j=1}^{\gamma_n-1}\sum_{y-x\in S((j-1)t_n,jt_n)} f\Big(\sqrt{\frac{b_n}{n}}y\Big)\Big\}$$
$$\times\,\xi_n\Big(x+S\big((\gamma_n-1)t_n\big)\Big)\Big)$$
$$= \delta\langle\xi_n, \mathfrak{B}_n^{\gamma_n-1}\xi_n\rangle \ge \delta\langle\xi_n, \mathfrak{B}_n\xi_n\rangle^{\gamma_n-1}$$

where the last step follows from the argument based on spectral representation and Jensen inequality.

In addition,

$$\begin{aligned}
&\langle\xi_n, \mathfrak{B}_n\xi_n\rangle \\
&= \big(1+o(1)\big)\int_{\mathbb{R}^2} dx g(x) \\
&\times \mathbb{E}\bigg(\exp\bigg\{\frac{b_n\log n}{n}\sum_{x\in S[1,t_n-1]} f\Big(x+\sqrt{\frac{b_n}{n}}y\Big)\bigg\}g\Big(x+\sqrt{\frac{b_n}{n}}S(t_n)\Big)\bigg) \\
&\longrightarrow \int_{\mathbb{R}^2} g(x)\mathbb{E}\bigg(\exp\bigg\{(2\pi)\sqrt{\det(\Gamma)}\int_0^1 f(x+\Gamma^{1/2}W(t))dt\bigg\} \\
&\qquad\qquad \times g\big(x+\Gamma^{1/2}W(1)\big)\bigg)dx \qquad (n\to\infty)
\end{aligned}$$

where the last step follows from (5.3.41) in Theorem 5.3.6 with n being replaced by $t_n - 1$.[2]

By (4.1.25) and by integration variable substitution, therefore, we get

$$\begin{aligned}
&\liminf_{n\to\infty}\frac{1}{b_n}\log\mathbb{E}\exp\bigg\{\frac{b_n\log n}{n}\sum_{j=1}^{\gamma_n-1}\sum_{x\in S(jt_n,(j+1)t_n)} f\Big(\sqrt{\frac{b_n}{n}}x\Big)\bigg\} \\
&\geq 2\pi\sqrt{\det(\Gamma)}\int_{\mathbb{R}^2} f(\Gamma^{1/2}x)\tilde{g}^2(x)dx - \frac{1}{2}\int_{\mathbb{R}^2}|\nabla\tilde{g}(x)|^2dx
\end{aligned}$$

where $\tilde{g}(x) = \det(\Gamma)^{1/4}g(\Gamma^{1/2}x)$. Taking supremum over g on the right hand side leads to (7.1.10). □

Proof Theorem 7.1.2. We first prove Theorem 7.1.2 under the assumption of aperiodicity. Observe that

$$\begin{aligned}
&\sum_{j=0}^{\gamma_n-1}\sum_{x\in S(jt_n,(j+1)t_n]} f\Big(\sqrt{\frac{b_n}{n}}x\Big) - \sum_{x\in S[1,\gamma_nt_n]} f\Big(\sqrt{\frac{b_n}{n}}x\Big) \\
&= \sum_{x\in\mathbb{Z}^d} f\Big(\sqrt{\frac{b_n}{n}}x\Big)\bigg\{\sum_{j=0}^{\gamma_n-1} 1_{\{x\in S(jt_n,(j+1)t_n]\}} - 1_{\{x\in S[1,\gamma_nt_n]\}}\bigg\}.
\end{aligned}$$

By the fact that

$$\sum_{j=0}^{\gamma_n-1} 1_{\{x\in S(jt_n,(j+1)t_n]\}} - 1_{\{x\in S[1,\gamma_nt_n]\}} \geq 0 \qquad x\in\mathbb{Z}^d,$$

[2]Our assumption on b_n implies that $\log n \sim \log(t_n - 1)$.

we have

$$\begin{aligned}&\Big|\sum_{j=0}^{\gamma_n-1}\sum_{x\in S(jt_n,(j+1)t_n]} f\Big(\sqrt{\frac{b_n}{n}}x\Big)-\sum_{x\in S[1,\gamma_nt_n]} f\Big(\sqrt{\frac{b_n}{n}}x\Big)\Big|\\
&\le ||f||_\infty\sum_{x\in\mathbb{Z}^d}\Big\{\sum_{j=0}^{\gamma_n-1}1_{\{x\in S(jt_n,(j+1)t_n]\}}-1_{\{x\in S[1,\gamma_nt_n]\}}\Big\}\\
&=||f||_\infty\Big\{\sum_{j=0}^{\gamma_n-1}\#\{S(jt_n,(j+1)t_n]-\#\{S[1,\gamma_nt_n]\}\Big\}.\end{aligned}$$

By Lemma 7.1.4 below, therefore, Theorem 7.1.2 holds under the aperiodicity assumption.

We now prove Theorem 7.1.2 without assuming aperiodicity. Let $\bar{S}(k)=S(\tau_k)$ $(k=1,2,\cdots)$ be the resolvent random walk introduced in the proof of Theorem 7.1.1. Applying what has been achieved to the aperiodic random walk $\{\bar{S}(k)\}$ gives us

$$\begin{aligned}(7.1.11)\qquad&\liminf_{n\to\infty}\frac{1}{b_n}\log\mathbb{E}\exp\Big\{\frac{b_n\log n}{n}\sum_{x\in\bar{S}[1,n]}f\Big(\sqrt{\frac{b_n}{n}}x\Big)\Big\}\\
&\ge\sup_{g\in\mathcal{F}_2}\Big\{2\pi\sqrt{\theta^{-d}\det(\Gamma)}\int_{\mathbb{R}^2}f\big(\theta^{-1/2}\Gamma^{1/2}x\big)g^2(x)dx\\
&\qquad-\frac{1}{2}\int_{\mathbb{R}^2}|\nabla g(x)|^2dx\Big\}\end{aligned}$$

when $d=2$, and

$$\begin{aligned}(7.1.12)\qquad&\liminf_{n\to\infty}\frac{1}{b_n}\log\mathbb{E}\exp\Big\{\frac{b_n}{n}\sum_{x\in\bar{S}[1,n]}f\Big(\sqrt{\frac{b_n}{n}}x\Big)\Big\}\\
&\ge\sup_{g\in\mathcal{F}_3}\Big\{\gamma_{\bar{S}}\int_{\mathbb{R}^3}f\big(\theta^{-1/2}\Gamma^{1/2}x\big)g^2(x)dx-\frac{1}{2}\int_{\mathbb{R}^3}|\nabla g(x)|^2dx\Big\}\end{aligned}$$

when $d=3$.

Consider the decomposition

$$\begin{aligned}\sum_{x\in S[1,n]}f\Big(\sqrt{\frac{b_n}{n}}x\Big)&=\sum_{x\in\bar{S}[1,n]}f\Big(\sqrt{\frac{b_n}{n}}x\Big)\\
&+\Big\{\sum_{x\in S[1,\tau_n]}f\Big(\sqrt{\frac{b_n}{n}}x\Big)-\sum_{x\in\bar{S}[1,n]}f\Big(\sqrt{\frac{b_n}{n}}x\Big)\Big\}\\
&-\Big\{\sum_{x\in S[1,\tau_n]}f\Big(\sqrt{\frac{b_n}{n}}x\Big)-\sum_{x\in S[1,n]}f\Big(\sqrt{\frac{b_n}{n}}x\Big)\Big\}.\end{aligned}$$

Notice that

$$\sum_{x\in S[1,\tau_n]} f\Big(\sqrt{\frac{b_n}{n}}x\Big) - \sum_{x\in \bar{S}[1,n]} f\Big(\sqrt{\frac{b_n}{n}}x\Big)$$
$$= \sum_{x\in\mathbb{Z}^d} f\Big(\sqrt{\frac{b_n}{n}}x\Big)\Big\{1_{\{x\in S[1,\tau_n]\}} - 1_{\{x\in\bar{S}[1,n]\}}\Big\}.$$

By the fact that

$$S[1,\tau_n] = \{S(1),\cdots,S(\tau_n)\} \supset \{S(\tau_1),\cdots,S(\tau_n)\} = \bar{S}[1,n],$$

$$\Big|\sum_{x\in S[1,\tau_n]} f\Big(\sqrt{\frac{b_n}{n}}x\Big) - \sum_{x\in \bar{S}[1,n]} f\Big(\sqrt{\frac{b_n}{n}}x\Big)\Big|$$
$$\le ||f||_\infty\Big[\#\{S(1),\cdots,S(\tau_n)\} - \#\{S(\tau_1),\cdots,S(\tau_n)\}\Big].$$

Similarly, by the relation $\tau_n \ge n$ we have that

$$\Big|\sum_{x\in S[1,\tau_n]} f\Big(\sqrt{\frac{b_n}{n}}x\Big) - \sum_{x\in S[1,n]} f\Big(\sqrt{\frac{b_n}{n}}x\Big)\Big|$$
$$\le ||f||_\infty\Big[\#\big\{S[1,\tau_n]\big\} - \#\big\{S[1,n]\big\}\Big].$$

Theorem 7.1.2 follows from (7.1.11), (7.1.12) and Lemma 7.1.5 below. □

To complete the proof of Theorem 7.1.2 and to close this section, we establish Lemma 7.1.4 and Lemma 7.1.5 below.

LEMMA 7.1.4. *Let b_n be a positive sequence with $b_n \to \infty$. For each n, let $1 = n_0 < n_1 < \cdots < n_{\gamma_n} = n$ be an integer partition of $[1,n]$ such that $\gamma_n \sim b_n$ and $n_j - n_{j-1} \sim n/b_n$ uniformly over $1 \le j \le \gamma_n$.*

(1) *When $d = 2$,*

$$\limsup_{n\to\infty}\frac{1}{b_n}\log\mathbb{E}\exp\Big\{\lambda\frac{b_n\log n}{n}\Big(\sum_{j=1}^{\gamma_n}\#\big\{S(n_{j-1},n_j]\big\} - \#\big\{S[1,n]\big\}\Big)\Big\} = 0$$

for any $\lambda > 0$, if $b_n = o(\log n)$ $(n\to\infty)$.

(2) *When $d \ge 3$,*

$$\limsup_{n\to\infty}\frac{1}{b_n}\log\mathbb{E}\exp\Big\{\lambda\frac{b_n}{n}\Big(\sum_{j=1}^{\gamma_n}\#\big\{S(n_{j-1},n_j]\big\} - \#\big\{S[1,n]\big\}\Big)\Big\} = 0$$

for any $\lambda > 0$, if $b_n = o(n^{1/3})$ $(n\to\infty)$.

Proof. Due to similarity, we only deal with the case $d = 2$. We start with the decomposition

$$\begin{aligned}
&\sum_{j=1}^{\gamma_n} \#\{S(n_{j-1}, n_j]\} - \#\{S[1,n]\} \\
&= \sum_{j=1}^{\gamma_n} \Big(\#\{S(n_{j-1}, n_j]\} - \mathbb{E}\#\{S(n_{j-1}, n_j]\}\Big) \\
&+ \Big(\mathbb{E}\#\{S[1,n]\} - \#\{S[1,n]\}\Big) \\
&+ \Big\{\sum_{j=1}^{\gamma_n} \mathbb{E}\#\{S(n_{j-1}, n_j]\} - \mathbb{E}\#\{S[1,n]\}\Big\}.
\end{aligned}$$

By (5.3.39)

$$\sum_{j=1}^{\gamma_n} \mathbb{E}\#\{S(n_{j-1}, n_j]\} - \mathbb{E}\#\{S[1,n]\} = o\Big(\frac{n}{\log n}\Big) \qquad (n \to \infty).$$

By (6.3.4) in Theorem 6.3.2 there is $c > 0$ such that

$$\sup_n \mathbb{E}\exp\Big\{c\frac{(\log n)^2}{n}\Big(\mathbb{E}\#\{S[1,n]\} - \#\{S[1,n]\}\Big)\Big\} < \infty.$$

It remains to show that

$$\begin{aligned}
\lim_{n\to\infty} \frac{1}{b_n} \log \mathbb{E}\exp\Big\{\lambda \frac{b_n \log n}{n}\Big[\sum_{j=1}^{\gamma_n}\Big(\#\{S(n_{j-1}, n_j]\} \\
- \mathbb{E}\#\{S(n_{j-1}, n_j]\}\Big)\Big]\Big\} = 0.
\end{aligned}$$

This follows from the fact (by independence) that

$$\begin{aligned}
&\mathbb{E}\exp\Big\{\lambda\frac{b_n \log n}{n}\Big[\sum_{j=1}^{\gamma_n}\Big(\#\{S(n_{j-1}, n_j]\} - \mathbb{E}\#\{S(n_{j-1}, n_j]\}\Big)\Big]\Big\} \\
&= \prod_{j=1}^{\gamma_n} \mathbb{E}\exp\Big\{\lambda\frac{b_n \log n}{n}\Big(\#\{S[1, n_j - n_{j-1}]\} - \mathbb{E}\#\{S[1, n_j - n_{j-1}]\}\Big)\Big\}
\end{aligned}$$

and from the fact (implied by (6.3.18) in Theorem 6.3.3) that

$$\lim_{n\to\infty} \mathbb{E}\exp\Big\{\lambda\frac{b_n \log n}{n}\Big(\#\{S[1, n_j - n_{j-1}]\} - \mathbb{E}\#\{S[1, n_j - n_{j-1}]\}\Big)\Big\} = 1$$

uniformly over $1 \le j \le \gamma_n$. □

Recall that the random sequence $\{\tau_k\}$ is introduced for the resolvent approximation and $0 < \theta < 1$ is the parameter associated with the distributions of $\{\tau_k\}$ (see the proof of Theorem 7.1.1).

LEMMA 7.1.5. *Suppose that the assumptions in Theorem 7.1.2 hold and let $\lambda > 0$ be fixed but arbitrary.*

(1) *When* $d = 2$,

$$\lim_{\theta\to 1^-}\limsup_{n\to\infty}\frac{1}{b_n}\log\mathbb{E}\exp\left\{\lambda\frac{b_n\log n}{n}\times\Big[\#\{S(1),\cdots,S(\tau_n)\}-\#\{S(\tau_1),\cdots,S(\tau_n)\}\Big]\right\}=0, \tag{7.1.13}$$

$$\lim_{\theta\to 1^-}\limsup_{n\to\infty}\frac{1}{b_n}\log\mathbb{E}\exp\left\{\lambda\frac{b_n\log n}{n}\times\Big[\#\{S[1,\tau_n]\}-\#\{S[1,n]\}\Big]\right\}=0. \tag{7.1.14}$$

(2) *When* $d \geq 3$,

$$\lim_{\theta\to 1^-}\limsup_{n\to\infty}\frac{1}{b_n}\log\mathbb{E}\exp\left\{\lambda\frac{b_n}{n}\times\Big[\#\{S(1),\cdots,S(\tau_n)\}-\#\{S(\tau_1),\cdots,S(\tau_n)\}\Big]\right\}=0, \tag{7.1.15}$$

$$\lim_{\theta\to 1^-}\limsup_{n\to\infty}\frac{1}{b_n}\log\mathbb{E}\exp\left\{\lambda\frac{b_n}{n}\times\Big[\#\{S[1,\tau_n]\}-\#\{S[1,n]\}\Big]\right\}=0. \tag{7.1.16}$$

Proof. Due to similarity we only work on the case $d = 2$. Write

$$Z_n = \#\{S(1), S(2),\cdots,S(\tau_n)\} - \#\{S(\tau_1), S(\tau_2),\cdots,S(\tau_n)\}.$$

It is easy to see that Z_n is non-decreasing and non-negative. By (5.3.39), we get

$$\lim_{\theta\to 1^-}\lim_{n\to\infty}\frac{\log n}{n}\mathbb{E}Z_n = 0$$

which leads to

$$\frac{\log n}{n}Z_n \xrightarrow{P} 0 \tag{7.1.17}$$

in the limiting order $n\to\infty$ first and $\theta\to 1^-$ second.

Let $m, n \geq 1$ be two integers. By the set relation

$$(A_1\cup A_2)\setminus(B_1\cup B_2)\subset(A_1\cup B_1)\cap(A_2\cup B_2),$$

$$\begin{aligned}
Z_{m+n} &\leq \#\Big\{\Big(\{S(1),\cdots,S(\tau_n)\}\setminus\{S(\tau_1),\cdots,S(\tau_n)\}\Big)\\
&\cup\Big(\{S(\tau_n+1),\cdots,S(\tau_{m+n})\}\setminus\{S(\tau_{n+1}),\cdots,S(\tau_{m+n})\}\Big)\Big\}\\
&\leq \#\Big(\{S(1),\cdots,S(\tau_n)\}\setminus\{S(\tau_1),\cdots,S(\tau_n)\}\Big)\\
&+\#\Big(\{S(\tau_n+1),\cdots,S(\tau_{m+n})\}\setminus\{S(\tau_{n+1}),\cdots,S(\tau_{m+n})\}\Big)\\
&= Z_n + Z'_m
\end{aligned}$$

where the random quantity

$$\begin{aligned}
Z'_m &= \#\{S(\tau_n+1),\cdots,S(\tau_{m+n})\} - \#\{S(\tau_{n+1}),\cdots,S(\tau_{m+n})\}\\
&= \#\{S(\tau_n+1)-S(\tau_n),\cdots,S(\tau_{m+n})-S(\tau_n)\}\\
&\quad - \#\{S(\tau_{n+1})-S(\tau_n),\cdots,S(\tau_{m+n})-S(\tau_n)\}
\end{aligned}$$

is independent of $\{Z_1,\cdots,Z_n\}$ and has a distribution the same as Z_m. That is to say, $\{Z_n\}$ is sub-additive. By Theorem 1.3.3, we get

$$\sup_{n,\theta}\mathbb{E}\exp\Big\{c\frac{\log n}{n}Z_n\Big\}<\infty\qquad\forall c>0.$$

In view of (7.1.17), we have

$$\lim_{\theta\to1^-}\limsup_{n\to\infty}\mathbb{E}\exp\Big\{c\frac{\log n}{n}Z_n\Big\}=1\qquad\forall c>0. \tag{7.1.18}$$

By sub-additivity and monotonicity of Z_n, we get

$$\mathbb{E}\exp\Big\{\lambda\frac{b_n\log n}{n}Z_n\Big\}\le\bigg(\mathbb{E}\exp\Big\{\lambda\frac{b_n\log n}{n}Z_{[n/b_n]}\Big\}\bigg)^{b_n+1}.$$

Thus, (7.1.13) follows from (7.1.18).

To prove (7.1.14), we begin with the following estimate:

$$\begin{aligned}
&\mathbb{E}\exp\Big\{\lambda\frac{b_n\log n}{n}\Big[\#\{S[1,\tau_n]\}-\#\{S[1,n]\}\Big]\Big\}\\
&\le\mathbb{E}\exp\Big\{\lambda\frac{b_n\log n}{n}\#\{S[n+1,(1+\epsilon)n]\}\Big\}\\
&+\mathbb{E}\bigg(\exp\Big\{\lambda\frac{b_n\log n}{n}\#\{S[1,\tau_n]\}\Big\}1_{\{\tau_n\ge(1+\epsilon)n\}}\bigg)\\
&\le\mathbb{E}\exp\Big\{\lambda\frac{b_n\log n}{n}\#\{S[1,\epsilon n]\}\Big\}\\
&+\bigg(\mathbb{E}\exp\Big\{2\lambda\frac{b_n\log n}{n}\#\{S[1,\tau_n]\}\Big\}\bigg)^{1/2}\Big(\mathbb{P}\{\tau_n\ge(1+\epsilon)n\}\Big)^{1/2}
\end{aligned}$$

where $\epsilon>0$.

The above bound leads to (7.1.14) due to the following facts:

First, by the argument for (7.1.13) we have that

$$\limsup_{\theta\to1^-}\limsup_{n\to\infty}\frac{1}{b_n}\log\mathbb{E}\exp\Big\{\lambda\frac{b_n\log n}{n}\#\{S[1,\tau_n]\}\Big\}<\infty,$$

$$\lim_{\epsilon\to0^+}\limsup_{n\to\infty}\frac{1}{b_n}\log\mathbb{E}\exp\Big\{\lambda\frac{b_n\log n}{n}\#\{S[1,\epsilon n]\}\Big\}=0.$$

Second, by Cramér's theorem (Theorem 1.1.5) there is a $u>0$ such that when θ is sufficiently close to 1 and when n is sufficiently large, we have

$$\mathbb{P}\{\tau_n\ge(1+\epsilon)n\}\le e^{-un}.$$

□

7.2. Moderate deviations in sub-critical dimensions

Recall that

$$I_n = \#\Big\{(k_1, \cdots, k_p); \quad S_1(k_1) = \cdots = S_p(k_p)\Big\},$$
$$J_n = \#\big\{S_1[1,n] \cap \cdots \cap S_p[1,n]\big\}.$$

Historically, the phrase "*moderate deviations*" is often used for the large deviation principles associated with some weak law of convergence. Recall, for example, that (Theorem 5.3.2) under $p(d-2) < d$, we get

$$n^{-\frac{2p-d(p-1)}{2}} I_n \xrightarrow{d} \det(\Gamma)^{-\frac{p-1}{2}} \alpha\big([0,1]^p\big)$$

where

$$\alpha\big([0,t]^p\big) = \int_{\mathbb{R}^d} \left[\prod_{j=1}^p \int_0^t \delta_x\big(W_j(s)\big)ds\right] dx$$

is the mutual intersection local time run by independent d-dimensional Brownian motions $W_1(t), \cdots, W_p(t)$.

Consequently, for each large but fixed $b > 0$, we have

$$\mathbb{P}\Big\{I_n \ge \lambda n^{\frac{2p-d(p-1)}{2}} b^{\frac{d(p-1)}{2}}\Big\} \overset{n\to\infty}{\longrightarrow} \mathbb{P}\Big\{\alpha\big([0,1]^p\big) \ge \lambda \det(\Gamma)^{\frac{p-1}{2}} b^{\frac{d(p-1)}{2}}\Big\},$$

or

$$\text{(7.2.1)} \qquad \begin{aligned} &\lim_{n\to\infty} \frac{1}{b} \log \mathbb{P}\Big\{I_n \ge \lambda n^{\frac{2p-d(p-1)}{2}} b^{\frac{d(p-1)}{2}}\Big\} \\ &= \frac{1}{b} \log \mathbb{P}\Big\{\alpha\big([0,1]^p\big) \ge \lambda \det(\Gamma)^{\frac{p-1}{2}} b^{\frac{d(p-1)}{2}}\Big\}. \end{aligned}$$

According to Theorem 3.3.2 (or its version in (3.3.4)), we have

$$\text{(7.2.2)} \qquad \lim_{t\to\infty} \frac{1}{t} \log \mathbb{P}\Big\{\alpha\big([0,1]^p\big) \ge t^{\frac{d(p-1)}{2}}\Big\} = -\frac{p}{2}\kappa(d,p)^{-\frac{dp}{d(p-1)}}.$$

Here we recall that $\kappa(d,p) > 0$ is the best constant of Gagliardo-Nirenberg inequality

$$||f||_{2p} \le C||\nabla f||_2^{\frac{d(p-1)}{2p}} ||f||_2^{1-\frac{d(p-1)}{2p}} \qquad f \in W^{1,2}(\mathbb{R}^d)$$

and

$$W^{1,2}(\mathbb{R}^d) = \{f \in \mathcal{L}^2(\mathbb{R}^d); \quad \nabla f \in \mathcal{L}^2(\mathbb{R}^d)\}.$$

Let $b \to \infty$ in (7.2.1). By (7.2.2), the right hand side of (7.2.1) converges to

$$-\frac{p}{2}\det(\Gamma)^{1/d}\kappa(d,p)^{\frac{4p}{d(p-1)}} \lambda^{\frac{2}{d(p-1)}}.$$

Therefore, it is not hard to believe (probably not hard to prove) that

$$\begin{aligned} &\lim_{n\to\infty} \frac{1}{b_n} \log \mathbb{P}\{I_n \ge \lambda n^{\frac{2p-d(p-1)}{2}} b_n^{\frac{d(p-1)}{2}}\} \\ &= -\frac{p}{2}\det(\Gamma)^{1/d}\kappa(d,p)^{-\frac{4p}{d(p-1)}} \lambda^{\frac{2}{d(p-1)}} \end{aligned}$$

if b_n increases to infinity at a sufficiently slow rate.

The general observation we make here does not have any practical value if we cannot say anything about the rate of b_n except the words "sufficiently slow". In the direction of moderate deviations, the concern is whether the result includes all "right" sequences b_n or at least those crucial to applications (such as $b_n = \log\log n$ to the law of the iterated logarithm). Our main results in this section are the moderate deviations for I_n and J_n associated to the weak laws given in Theorem 5.3.2 and in Theorem 5.3.4, respectively.

THEOREM 7.2.1. *Under $p(d-2) < d$, we have*

$$\lim_{n\to\infty} \frac{1}{b_n} \log \mathbb{P}\{I_n \ge \lambda n^{\frac{2p-d(p-1)}{2}} b_n^{\frac{d(p-1)}{2}}\} = -\frac{p}{2}\det(\Gamma)^{1/d}\kappa(d,p)^{-\frac{4p}{d(p-1)}}\lambda^{\frac{2}{d(p-1)}}$$

for any $\lambda > 0$ and for any positive sequence $\{b_n\}$ satisfying

$$b_n \longrightarrow \infty \quad \text{and} \quad b_n = o(n) \qquad (n \to \infty).$$

Proof. By Theorem 1.2.7, we need only to show that for any $\theta > 0$,

$$\lim_{n\to\infty} \frac{1}{b_n} \log \left\{ \sum_{m=0}^{\infty} \frac{\theta^m}{m!} \left(\frac{b_n}{n}\right)^{\frac{2p-d(p-1)}{2p}m} \left(\mathbb{E}I_n^m\right)^{1/p} \right\} = \Psi(\theta) \tag{7.2.3}$$

where

$$\Psi(\theta) = \frac{2p-d(p-1)}{2p}\left(\frac{d(p-1)}{p}\right)^{\frac{d(p-1)}{2p-d(p-1)}} \times \det(\Gamma)^{-\frac{p-1}{2p-d(p-1)}} \kappa(d,p)^{\frac{4p}{2p-d(p-1)}} \theta^{\frac{2p}{2p-d(p-1)}}. \tag{7.2.4}$$

We now work on the upper bound of (7.2.3). Let $t > 0$ be fixed and let $t_n = [tn/b_n]$ and $\gamma_n = [n/t_n]$. Then $n \le t_n(\gamma_n + 1)$. By taking $a = \gamma_n + 1$ in inequality (6.1.8) we have

$$\sum_{m=0}^{\infty} \frac{1}{m!}\theta^m \left(\frac{b_n}{n}\right)^{\frac{2p-d(p-1)}{2p}m} \left(\mathbb{E}I_n^m\right)^{1/p} \le \left(\sum_{m=0}^{\infty} \frac{1}{m!}\theta^m \left(\frac{b_n}{n}\right)^{\frac{2p-d(p-1)}{2p}m} \left(\mathbb{E}I_{t_n}^m\right)^{1/p} \right)^{\gamma_n+1}.$$

From Theorem 5.3.2, we have

$$\left(\frac{b_n}{n}\right)^{\frac{2p-d(p-1)}{2}} I_{t_n} \xrightarrow{d} t^{\frac{2p-d(p-1)}{2}} \det(\Gamma)^{-\frac{p-1}{2}} \alpha([0,1]^p)$$

as $n \to \infty$. By (6.2.1) in Theorem 6.2.1 and by dominated convergence,

$$\sum_{m=0}^{\infty} \frac{1}{m!}\theta^m \left(\frac{b_n}{n}\right)^{\frac{2p-d(p-1)}{2p}m} \left(\mathbb{E}I_{t_n}^m\right)^{1/p} \longrightarrow \sum_{m=0}^{\infty} \frac{1}{m!}\theta^m t^{\frac{2p-d(p-1)}{2p}m} \det(\Gamma)^{-\frac{p-1}{2p}m} \left(\mathbb{E}\alpha([0,1]^p)^m\right)^{1/p}.$$

Summarizing our argument, we get

$$\limsup_{n\to\infty}\frac{1}{b_n}\log\bigg(\sum_{m=0}^{\infty}\frac{1}{m!}\theta^m\Big(\frac{b_n}{n}\Big)^{\frac{2p-d(p-1)}{2p}m}\big(\mathbb{E}I_n^m\big)^{1/p}\bigg)$$
$$\le\frac{1}{t}\log\bigg(\sum_{m=0}^{\infty}\frac{1}{m!}\theta^m t^{\frac{2p-d(p-1)}{2p}m}\det(\Gamma)^{-\frac{p-1}{2p}m}\Big(\mathbb{E}\alpha\big([0,1]^p\big)^m\Big)^{1/p}\bigg).$$

By (7.2.2) and Theorem 1.2.9, we get

$$\lim_{t\to\infty}\frac{1}{t}\log\bigg(\sum_{m=0}^{\infty}\frac{1}{m!}\theta^m t^{\frac{2p-d(p-1)}{2p}m}\det(\Gamma)^{-\frac{p-1}{2p}m}\Big(\mathbb{E}\alpha\big([0,1]^p\big)^m\Big)^{1/p}\bigg)$$
$$=\sup_{\lambda>0}\Big\{\theta\det(\Gamma)^{-\frac{p-1}{2p}}\lambda-\frac{1}{2}\kappa(d,p)^{\frac{dp}{d(p-1)}}\lambda^{\frac{2}{d(p-1)}}\Big\}=\Psi(\theta).$$

Thus, we have established the upper bound for (7.2.3).

We now prove the lower bound for (7.2.3). Recall the representation

$$I_n=\sum_{x\in\mathbb{Z}^d}\prod_{j=1}^{p}l_j(n,x).$$

By independence, for any integer $m\ge 1$, we have

$$\mathbb{E}I_n^m=\sum_{x_1,\cdots,x_m\in\mathbb{Z}^d}\bigg[\mathbb{E}\prod_{k=1}^{m}l(n,x_k)\bigg]^p.$$

Let $q>1$ be the conjugate of p and let f be a bounded, continuous and locally supported function on $\mathbb{R}^d$ such that $||f||_q=1$. Write

$$C_n=\bigg(\sum_{x\in\mathbb{Z}^d}\Big|f\Big(\sqrt{\frac{b_n}{n}}x\Big)\Big|^q\bigg)^{1/q}.$$

By Hölder inequality, we get

$$\Big(\mathbb{E}I_n^m\Big)^{1/p}\ge C_n^{-m}\sum_{x_1,\cdots,x_m\in\mathbb{Z}^d}f\Big(\sqrt{\frac{b_n}{n}}x_1\Big)\cdots f\Big(\sqrt{\frac{b_n}{n}}x_m\Big)\mathbb{E}\prod_{k=1}^{m}l(n,x_k)$$
$$=C_n^{-m}\mathbb{E}\bigg[\sum_{x\in\mathbb{Z}^d}f\Big(\sqrt{\frac{b_n}{n}}x\Big)l(n,x)\bigg]^m=C_n^{-m}\mathbb{E}\bigg[\sum_{k=1}^{n}f\Big(\sqrt{\frac{b_n}{n}}S(k)\Big)\bigg]^m.$$

Consequently,

$$\sum_{m=0}^{\infty}\frac{1}{m!}\theta^m\Big(\frac{b_n}{n}\Big)^{\frac{2p-d(p-1)}{2p}m}\big(\mathbb{E}I_n^m\big)^{1/p}$$
$$\ge\mathbb{E}\exp\bigg\{\theta\Big(\frac{b_n}{n}\Big)^{\frac{2p-d(p-1)}{2p}}C_n^{-1}\sum_{k=1}^{n}f\Big(\sqrt{\frac{b_n}{n}}S(k)\Big)\bigg\}.$$

Observe that

$$C_n\sim\Big(\frac{n}{b_n}\Big)^{\frac{d(p-1)}{2p}}\bigg(\int_{\mathbb{R}^d}|f(x)|^q dx\bigg)^{1/q}=\Big(\frac{n}{b_n}\Big)^{\frac{d(p-1)}{2p}}\qquad(n\to\infty).$$

By Theorem 7.1.1, therefore,

$$\liminf_{n\to\infty}\frac{1}{b_n}\log\Big(\sum_{m=0}^{\infty}\frac{\theta^m}{m!}\Big(\frac{b_n}{n}\Big)^{\frac{2p-d(p-1)}{2p}m}\big(\mathbb{E}I_n^m\big)^{1/p}\Big)$$
$$\geq \sup_{g\in\mathcal{F}_d}\Big\{\theta\int_{\mathbb{R}^d} f(\Gamma^{1/2}x)g^2(x)dx-\frac{1}{2}\int_{\mathbb{R}^d}|\nabla g(x)|^2dx\Big\}.$$

Notice that the set of bounded, continuous and locally supported functions f is dense on the unit sphere of $\mathcal{L}^q(\mathbb{R}^d)$, and that when f runs over this set, so does the function

$$f_\Gamma(x)=\det(\Gamma)^{\frac{1}{2q}}f(\Gamma^{1/2}x).$$

Taking supremum over f on the right hand side gives us

$$\liminf_{n\to\infty}\frac{1}{b_n}\log\Big(\sum_{m=0}^{\infty}\frac{1}{m!}\theta^m\Big(\frac{b_n}{n}\Big)^{\frac{2p-d(p-1)}{2p}m}\big(\mathbb{E}I_n^m\big)^{1/p}\Big)$$
$$\geq \sup_{g\in\mathcal{F}_d}\Big\{\theta\det(\Gamma)^{-\frac{1}{2q}}\Big(\int_{\mathbb{R}^d}|g(x)|^{2p}dx\Big)^{1/p}-\frac{1}{2}\int_{\mathbb{R}^d}|\nabla g(x)|^2dx\Big\}$$
$$=\Psi(\theta)$$

where the last step follows from Theorem C.1 in the Appendix.

We have proved the lower bound for (7.2.3). □

THEOREM 7.2.2. (1) *As $d=2$ and $p\geq 2$,*

$$\lim_{n\to\infty}\frac{1}{b_n}\log\mathbb{P}\Big\{J_n\geq\lambda\frac{n}{(\log n)^p}b_n^{p-1}\Big\}$$
$$=-\frac{p}{2}(2\pi)^{-\frac{p}{p-1}}\det(\Gamma)^{-\frac{1}{2(p-1)}}\kappa(2,p)^{-\frac{2p}{p-1}}\lambda^{\frac{1}{p-1}}\tag{7.2.5}$$

for any $\lambda>0$ and for any positive sequence b_n satisfying

$$b_n\longrightarrow\infty\quad\text{and}\quad b_n=o(\log n)\qquad(n\to\infty).$$

(2) *As $d=3$ and $p=2$,*

$$\lim_{n\to\infty}\frac{1}{b_n}\log\mathbb{P}\Big\{J_n\geq\lambda\sqrt{nb_n^3}\Big\}=-\det(\Gamma)^{1/3}\gamma_S^{-4/3}\kappa(3,2)^{-8/3}\lambda^{2/3}\tag{7.2.6}$$

for any $\lambda>0$ and for any positive sequence b_n satisfying

$$b_n\longrightarrow\infty\quad\text{and}\quad b_n=o\big(n^{1/3}\big)\qquad(n\to\infty).$$

Here we recall that $\gamma_S>0$ is the probability of no return defined in (5.1.12) (see also (5.1.13)).

Proof. By Theorem 1.2.7, we need only to show that when $d=2$ and $p\geq 2$,

$$\lim_{n\to\infty}\frac{1}{b_n}\log\Big\{\sum_{m=0}^{\infty}\frac{\theta^m}{m!}\Big(\frac{b_n(\log n)^p}{n}\Big)^{m/p}\Big(\mathbb{E}J_n^m\Big)^{1/p}\Big\}$$
$$=\frac{1}{p}\Big(\frac{2(p-1)}{p}\Big)^{p-1}(2\pi\theta)^p\sqrt{\det(\Gamma)}\kappa(2,p)^{2p}\tag{7.2.7}$$

and that when $d = 3$ and $p = 2$,

$$\lim_{n\to\infty} \frac{1}{b_n} \log\left\{ \sum_{m=0}^{\infty} \frac{\theta^m}{m!} \left(\frac{b_n}{n}\right)^{m/4} \left(\mathbb{E}J_n^m\right)^{1/2} \right\} \tag{7.2.8}$$
$$= \frac{1}{2}\left(\frac{3}{4}\right)^3 (\gamma_S\theta)^4 \det(\Gamma)^{-1}\kappa(3,2)^8.$$

We leave their proof as exercise (Exercise 7.6.2). □

We end this section by the following comment on the restrictions that we put on the positive sequence $\{b_n\}$. By the fact that $I_n \le n^p$, it is impossible for the tail behavior of the probability

$$\mathbb{P}\{I_n \ge \lambda n^p\}$$

to fall into the pattern described in Theorem 7.2.1. Thus, the restriction "$b_n = o(n)$" is best possible for Theorem 7.2.1.

By the fact that $J_n \le n$, the restriction "$b_n = o(n^{1/3})$" is best possible for Theorem 7.2.2 in the case $d = 3$ and $p = 2$.

As for the case $d = 2$ and $p \ge 2$, we claim that for any $\lambda > 2\pi\sqrt{\det(\Gamma)}$,

$$\lim_{n\to\infty} \frac{1}{\log n} \log \mathbb{P}\Big\{ J_n \ge \lambda \frac{n}{\log n} \Big\} = -\infty \tag{7.2.9}$$

which shows that Theorem 7.2.2 fails if we take $b_n = \log n$ when $d = 2$.

By the fact that

$$J_n \le \#\big\{S_1[1,n]\big\} = R_n$$

we have

$$\mathbb{P}\Big\{J_n \ge \lambda\frac{n}{\log n}\Big\} \le \mathbb{P}\Big\{R_n \ge \lambda\frac{n}{\log n}\Big\} \le \mathbb{P}\Big\{R_n - \mathbb{E}R_n \ge \epsilon\frac{n}{\log n}\Big\}$$

for any $0 < \epsilon < \lambda - 2\pi\sqrt{\det(\Gamma)}$. Here we use the fact ((5.3.39)) that

$$\mathbb{E}R_n \sim 2\pi\sqrt{\det(\Gamma)}\frac{n}{\log n}.$$

According to (6.3.18) in Theorem 6.3.3, for any $\theta > 0$,

$$\sup_n \mathbb{E}\exp\left\{\theta\frac{(\log n)^2}{n}(R_n - \mathbb{E}R_n)\right\} < \infty,$$

which leads to (7.2.9) by a standard way of using Chebyshev inequality.

7.3. Laws of the iterated logarithm

In this section we apply the moderate deviations established in Section 7.2 to the following laws of the iterated logarithm.

THEOREM 7.3.1. *Under $p(d-2) < d$, we have*

$$\limsup_{n\to\infty} n^{-\frac{2p-d(p-1)}{2}} (\log\log n)^{-\frac{d(p-1)}{2}} I_n$$
$$= \left(\frac{2}{p}\right)^{-\frac{d(p-1)}{2}} \det(\Gamma)^{-\frac{p-1}{2}} \kappa(d,p)^{2p} \quad a.s.$$

THEOREM 7.3.2. (1) *As $d = 2$ and $p \geq 2$,*

$$\limsup_{n\to\infty} \frac{(\log n)^p}{n(\log\log n)^{p-1}} J_n = (2\pi)^p \Big(\frac{2}{p}\Big)^{p-1} \sqrt{\det(\Gamma)}\kappa(2,p)^{2p} \quad a.s. \tag{7.3.1}$$

(2) *As $d = 3$ and $p = 2$,*

$$\limsup_{n\to\infty} \frac{1}{\sqrt{n(\log\log n)^3}} J_n = \gamma_S^2 \det(\Gamma)^{-1/2}\kappa(3,2)^4 \quad a.s. \tag{7.3.2}$$

Given the moderate deviations in Section 7.2, the proof of the upper bounds for Theorem 7.3.1 and Theorem 7.3.2 is a standard practice of the Borel-Cantelli lemma. To establish the lower bounds, we introduce the following notation: For any $\bar{y} = (y_1, \cdots, y_p) \in (\mathbb{Z}^d)^p$, $\mathbb{P}_{\bar{y}}$ and $\mathbb{E}_{\bar{y}}$ are, respectively, the distribution and expectation with respect to the trajectories of the random walks $S_1(k), \cdots, S_p(k)$ with the starting points $S_1(0) = y_1, \cdots, S_p(0) = y_p$. In connection with the notation we usually use $\mathbb{P}\{\cdot\} = \mathbb{P}_{\bar{0}}\{\cdot\}$ and $\mathbb{E}\{\cdot\} = \mathbb{E}_{\bar{0}}\{\cdot\}$. Set $b_n = \log\log n$ and $t_n = [n/b_n]$.

Examining the proof of Theorem 3.3.7 carefully, all we need is to show that

$$\begin{aligned}\liminf_{n\to\infty} \frac{1}{b_n} \log \inf_{|\bar{y}|\leq\sqrt{t_n}} \mathbb{P}_{\bar{y}}\{I_n \geq \lambda n^{\frac{2p-d(p-1)}{2}} b_n^{\frac{d(p-1)}{2}}\} \\ = -\frac{p}{2}\det(\Gamma)^{1/d}\kappa(d,p)^{-\frac{4p}{d(p-1)}}\lambda^{\frac{2}{d(p-1)}}\end{aligned} \tag{7.3.3}$$

and that

$$\begin{aligned}\liminf_{n\to\infty} \frac{1}{b_n} \log \inf_{|\bar{y}|\leq\sqrt{t_n}} \mathbb{P}_{\bar{y}}\Big\{J_n \geq \lambda\frac{n}{(\log n)^p} b_n^{p-1}\Big\} \\ = -\frac{p}{2}(2\pi)^{-\frac{p}{p-1}}\det(\Gamma)^{-\frac{1}{2(p-1)}}\kappa(2,p)^{-\frac{2p}{p-1}}\lambda^{\frac{1}{p-1}}\end{aligned} \tag{7.3.4}$$

when $d = 2$, $p \geq 2$, and

$$\begin{aligned}\lim_{n\to\infty} \frac{1}{b_n} \log \inf_{|\bar{y}|\leq\sqrt{t_n}} \mathbb{P}_{\bar{y}}\Big\{J_n \geq \lambda\sqrt{nb_n^3}\Big\} \\ = -\det(\Gamma)^{1/3}\gamma_S^{-4/3}\kappa(3,2)^{-8/3}\lambda^{2/3}\end{aligned} \tag{7.3.5}$$

when $d = 3$, $p = 2$, where $\lambda > 0$ is arbitrary.

Due to similarity we only prove (7.3.3). Notice that for any $\bar{y} = (y_1, \cdots, y_p) \in (\mathbb{Z}^d)^p$,

$$\mathbb{E}_{\bar{y}} I_n^m = \sum_{x_1,\cdots,x_m\in\mathbb{Z}^d} \prod_{j=1}^p \mathbb{E}\prod_{k=1}^m l(n, x_k - y_j). \tag{7.3.6}$$

By Hölder inequality, $\mathbb{E}_{\bar{y}} I_n^m \leq \mathbb{E} I_n^m$ for all $\bar{y} \in (\mathbb{Z}^d)^p$. By (7.2.3), therefore,

$$\lim_{n\to\infty} \frac{1}{b_n} \log \sup_{\bar{y}\in(\mathbb{Z}^d)^p} \Big\{\sum_{m=0}^\infty \frac{\theta^m}{m!}\Big(\frac{b_n}{n}\Big)^{\frac{2p-d(p-1)}{2p}m}\Big(\mathbb{E}_{\bar{y}} I_n^m\Big)^{1/p}\Big\} = \Psi(\theta)$$

for all $\theta > 0$, where $\Psi(\theta)$ is written in (7.2.4). By a general argument based on Theorem 1.2.7, we will have (7.3.3) if we can show that for each $\theta > 0$,

$$\liminf_{n\to\infty} \frac{1}{b_n} \log \inf_{|\bar{y}|\leq\sqrt{t_n}} \Big\{\sum_{m=0}^\infty \frac{\theta^m}{m!}\Big(\frac{b_n}{n}\Big)^{\frac{2p-d(p-1)}{2p}m}\Big(\mathbb{E}_{\bar{y}} I_n^m\Big)^{1/p}\Big\} \geq \Psi(\theta). \tag{7.3.7}$$

By the method of resolvent approximation, we may assume that the random walks are aperiodic in our proof of (7.3.7). By (7.3.6), we get

$$\begin{aligned}
\mathbb{E}_{\bar{y}}I_n^m &\geq \sum_{x_1,\cdots,x_m\in\mathbb{Z}^d}\prod_{j=1}^p\mathbb{E}\prod_{k=1}^m\big[l(n,x_k-y_j)-l(t_n,x_k-y_j)\big]\\
&=\sum_{x_1,\cdots,x_m\in\mathbb{Z}^d}\prod_{j=1}^p\sum_{z\in\mathbb{Z}^d}\mathbb{P}\{S(t_n)=z-y_j\}\mathbb{E}\prod_{k=1}^m l(n-t_n,x_k-z)
\end{aligned}$$

where the last step follows from increment independence.

Write $B_n=\{x\in\mathbb{Z}^d;\ |x|\leq\epsilon\sqrt{t_n}\}$, where $\epsilon>0$ is fixed but arbitrary. By (5.1.2) in Lemma 5.1.1[3], there is a constant $c>0$ such that

$$\mathbb{P}\{S(t_n)=x\}\geq ct_n^{-d/2}\qquad\forall x\in 2B_n.$$

Consequently,

$$\begin{aligned}
&\inf_{|\bar{y}|\leq\sqrt{t_n}}\mathbb{E}_{\bar{y}}I_n^m\\
&\geq c^pt_n^{-dp/2}\sum_{x_1,\cdots,x_m\in\mathbb{Z}^d}\prod_{j=1}^p\sum_{z\in B_n}\mathbb{E}\prod_{k=1}^m l(n-t_n,x_k-z)\\
&=c^pt_n^{-dp/2}\sum_{z_1,\cdots,z_p\in B_n}\sum_{x_1,\cdots,x_m\in\mathbb{Z}^d}\prod_{j=1}^p\mathbb{E}\prod_{k=1}^m l(n-t_n,x_k-z_j)\\
&=c^pt_n^{-dp/2}\sum_{z_1,\cdots,z_p\in B_n}\mathbb{E}\Big(\sum_{x\in\mathbb{Z}^d}\prod_{j=1}^p l_j(n-t_n,x-y_j)\Big)^m\\
&\geq c^pt_n^{-dp/2}\#(B_n)^p\mathbb{E}\Big(\sum_{x\in\mathbb{Z}^d}\#(B_n)^{-1}\sum_{z_1,\cdots,z_p\in B_n}\prod_{j=1}^p l_j(n-t_n,x-y_j)\Big)^m\\
&=c^pt_n^{-dp/2}\#(B_n)^p\mathbb{E}\Big(\sum_{x\in\mathbb{Z}^d}\prod_{j=1}^p\zeta_j(n,x)\Big)^m
\end{aligned}$$

where the fourth inequality follows from Jensen inequality and where

$$\zeta_j(n,x)=\#(B_n)^{-1}\sum_{z\in B_n}l_j(n-t_n,x-z).$$

The problem has been reduced from (7.3.7) to

$$\begin{aligned}
&\liminf_{\epsilon\to0^+}\liminf_{n\to\infty}\frac{1}{b_n}\log\Bigg(\sum_{m=0}^\infty\frac{\theta^m}{m!}\Big(\frac{b_n}{n}\Big)^{\frac{2p-d(p-1)}{2p}m}\\
&\qquad\times\Big\{\mathbb{E}\Big(\sum_{x\in\mathbb{Z}^d}\prod_{j=1}^p\zeta_j(n,x)\Big)^m\Big\}^{1/p}\Bigg)\geq\Psi(\theta).
\end{aligned}\tag{7.3.8}$$

Let $f(x)$ be a bounded, uniformly continuous function locally supported on $\mathbb{R}^d$ with $||f||_q=1$. By independence and by a similar estimation used in the last section,

[3] This is where the aperiodicity is needed.

we get

$$\begin{aligned}\Big\{\mathbb{E}\Big(\sum_{x\in\mathbb{Z}^d}\prod_{j=1}^p\zeta_j(n,x)\Big)^m\Big\}^{1/p} &= \Big\{\sum_{x_1,\cdots,x_m\in\mathbb{Z}^d}\Big[\mathbb{E}\prod_{k=1}^m\zeta(n,x_k)\Big]^p\Big\}^{1/p}\\ &\ge C_n^{-m}\sum_{x_1,\cdots,x_m\in\mathbb{Z}^d} f\Big(\sqrt{\frac{b_n}{n}}x_1\Big)\cdots f\Big(\sqrt{\frac{b_n}{n}}x_m\Big)\mathbb{E}\prod_{k=1}^m\zeta(n,x_k)\\ &= C_n^{-m}\mathbb{E}\Big[\sum_{x\in\mathbb{Z}^d} f\Big(\sqrt{\frac{b_n}{n}}x\Big)\zeta(n,x_k)\Big]^m\end{aligned}$$

where

$$C_n = \Big\{\sum_{x\in\mathbb{Z}^d}\Big|f\Big(\sqrt{\frac{b_n}{n}}x\Big)\Big|^q\Big\}^{1/q}.$$

Summing up on both sides gives us

$$\begin{aligned}&\sum_{m=0}^\infty\frac{\theta^m}{m!}\Big(\frac{b_n}{n}\Big)^{\frac{2p-d(p-1)}{2p}m}\Big\{\mathbb{E}\Big(\sum_{x\in\mathbb{Z}^d}\prod_{j=1}^p\zeta_j(n,x)\Big)^m\Big\}^{1/p}\\ &\ge\mathbb{E}\exp\Big\{\theta\Big(\frac{b_n}{n}\Big)^{\frac{2p-d(p-1)}{2p}}C_n^{-1}\sum_{x\in\mathbb{Z}^d} f\Big(\sqrt{\frac{b_n}{n}}x\Big)\zeta(n,x)\Big\}.\end{aligned}$$

On the other hand, set

$$(\mathfrak{D}_\epsilon f)(x) = (K_d\epsilon^d)^{-1}\int_{\{|y|\le\epsilon\}} f(x+y)dy \qquad x\in\mathbb{R}^d$$

where K_d is the volume of a d-dimensional unit ball. By uniform continuity of f, we get

$$\begin{aligned}&\sum_{x\in\mathbb{Z}^d} f\Big(\sqrt{\frac{b_n}{n}}x\Big)\zeta(n,x)\\ &= \#(B_n)^{-1}\sum_{x\in\mathbb{Z}^d} l(n-t_n,x)\sum_{z\in B_n} f\Big(\sqrt{\frac{b_n}{n}}(x+z)\Big)\\ &= \#(B_n)^{-1}\sum_{k=1}^{n-t_n}\sum_{z\in B_n} f\Big(\sqrt{\frac{b_n}{n}}(S(k)+z)\Big)\\ &= o(n)+\sum_{k=1}^{n-t_n}(\mathfrak{D}_\epsilon f)\Big(\sqrt{\frac{b_n}{n}}S(k)\Big).\end{aligned}$$

By the fact that $C_n\sim(n/b_n)^{d/2q}$, and by Theorem 7.1.1 (with f being replaced by $\mathfrak{D}_\epsilon f$), we get

$$\begin{aligned}&\liminf_{n\to\infty}\frac{1}{b_n}\log\Big(\sum_{m=0}^\infty\frac{\theta^m}{m!}\Big(\frac{b_n}{n}\Big)^{\frac{2p-d(p-1)}{2p}m}\Big\{\mathbb{E}\Big(\sum_{x\in\mathbb{Z}^d}\prod_{j=1}^p\zeta_j(n,x)\Big)^m\Big\}^{1/p}\Big)\\ &\ge\sup_{g\in\mathcal{F}_d}\Big\{\theta\int_{\mathbb{R}^d}(\mathfrak{D}_\epsilon f)(\Gamma^{1/2}x)g^2(x)dx-\frac{1}{2}\int_{\mathbb{R}^d}|\nabla g(x)|^2dx\Big\}.\end{aligned}$$

Letting $\epsilon \to 0^+$ on both sides gives that

$$\liminf_{\epsilon\to 0^+}\liminf_{n\to\infty}\frac{1}{b_n}\log\left(\sum_{m=0}^{\infty}\frac{\theta^m}{m!}\left(\frac{b_n}{n}\right)^{\frac{2p-d(p-1)}{2p}m}\left\{\mathbb{E}\Big(\sum_{x\in\mathbb{Z}^d}\prod_{j=1}^{p}\zeta_j(n,x)\Big)^m\right\}^{1/p}\right)$$
$$\geq \sup_{g\in\mathcal{F}_d}\left\{\theta\int_{\mathbb{R}^d}f(\Gamma^{1/2}x)g^2(x)dx-\frac{1}{2}\int_{\mathbb{R}^d}|\nabla g(x)|^2dx\right\}.$$

Taking supremum over f on the right hand side leads to (7.3.8). □

7.4. What do we expect in critical dimensions?

The laws of the iterated logarithm for I_n and J_n in the critical dimensions have been obtained in Marcus and Rosen ([**136**], [**152**]).

THEOREM 7.4.1. *Assume additionally that the i.i.d. increments of the random walks have finite third moment.*

(1) *When $d = 4$ and $p = 2$,*

(7.4.1) $$\limsup_{n\to\infty}\frac{I_n}{\log n\cdot\log\log\log n}=\frac{1}{2\pi^2\sqrt{\det(\Gamma)}}\quad a.s.,$$

(7.4.2) $$\limsup_{n\to\infty}\frac{J_n}{\log n\cdot\log\log\log n}=\frac{\gamma_S^2}{2\pi^2\sqrt{\det(\Gamma)}}\quad a.s.$$

(2) *When $d = p = 3$,*

(7.4.3) $$\limsup_{n\to\infty}\frac{I_n}{\log n\cdot\log\log\log n}=\frac{1}{2\pi^2\det(\Gamma)}\quad a.s.,$$

(7.4.4) $$\limsup_{n\to\infty}\frac{J_n}{\log n\cdot\log\log\log n}=\frac{\gamma_S^3}{2\pi^2\det(\Gamma)}\quad a.s.$$

We skip the proof. The reader is referred to Marcus and Rosen [**136**] for the proof of part (1) and to Rosen [**152**] for the proof of part (2).

If we interpret the results given in Theorem 7.4.1 in terms of large deviations, or if we exam the proofs given in Marcus and Rosen [**136**] and Rosen [**152**], we would see that in the critical dimensions (defined by $p(d-2) = d$),

(7.4.5) $$\lim_{n\to\infty}\frac{1}{b_n}\log\mathbb{P}\Big\{I_n\geq\lambda b_n\log n\Big\}=-C_1(d,p)\lambda,$$

(7.4.6) $$\lim_{n\to\infty}\frac{1}{b_n}\log\mathbb{P}\Big\{J_n\geq\lambda b_n\log n\Big\}=-C_2(d,p)\lambda$$

for small scale b_n ($b_n = \log\log\log n$ in the papers [**136**] and [**152**]).

In (7.4.5) and (7.4.6), $I_n/\log n$ and $J_n/\log n$ have tails asymptotically the same as the tails of Γ-distributions. This behavior is due to the fact (Theorem 5.5.1) that $I_n/\log n$ and $J_n/\log n$ weakly converges to Γ-distributions. In this regard, (7.4.5) and (7.4.6) fall into the category of the moderate deviations.

On the other hand, it is highly unlikely that (7.4.5) and (7.4.6) hold for any b_n growing faster than $\log n$. What have not been clear are the forms of the large deviations for I_n and J_n when the deviation scale b_n is large. We notice, however, that all moderate deviations for I_n in sub-critical dimensions involve the best constant $\kappa(d,p)$ of the Gagliardo-Nirenberg inequality

$$||f||_{2p} \le C||\nabla f||_2^{\frac{d(p-1)}{2p}} ||f||_2^{1-\frac{d(p-1)}{2p}} \qquad f \in W^{1,2}(\mathbb{R}^d).$$

Further, the case $p(d-2) = d$ is critical to the inequality. Indeed, the Gagliardo-Nirenberg inequality becomes Sobolev inequality

$$||f||_{2p} \le C||\nabla f||_2 \qquad f \in W^{1,2}(\mathbb{R}^d).$$

In view of (7.4.5), (7.4.6) and Theorem 7.2.2, based on the facts that $b_n^p = b_n \log n$ as $b_n = (\log n)^{\frac{1}{p-1}}$ and that $I_n \le n^p$, $J_n \le n$, we make the following conjecture.

CONJECTURE 7.4.2. *When $p(d-2) = d$, the large deviation principle*

$$\lim_{n\to\infty} \frac{1}{b_n} \log \mathbb{P}\Big\{ I_n \ge \lambda b_n^p \Big\} = -\frac{p}{2} \det(\Gamma)^{1/d} \kappa(d,p)^{-2} \lambda^{1/p} \tag{7.4.7}$$

holds for $\lambda > 0$ and for positive sequence b_n with

$$\lim_{n\to\infty} b_n/(\log n)^{\frac{1}{p-1}} = \infty \quad \text{and} \quad b_n = o(n).$$

Similarly,

$$\lim_{n\to\infty} \frac{1}{b_n} \log \mathbb{P}\Big\{ J_n \ge \lambda b_n^p \Big\} = -\gamma_S^p \frac{p}{2} \det(\Gamma)^{1/d} \kappa(d,p)^{-2} \lambda^{1/p} \tag{7.4.8}$$

holds for $\lambda > 0$ and for positive sequence b_n with

$$\lim_{n\to\infty} b_n/(\log n)^{\frac{1}{p-1}} = \infty \quad \text{and} \quad b_n = o\big(n^{1/p}\big).$$

7.5. Large deviations in super-critical dimensions

We now move to the super-critical dimensions defined by the inequality $p(d-2) > d$. In this case,

$$I_\infty \equiv \#\big\{(k_1,\cdots,k_p);\ \ S_1(k_1) = \cdots = S_p(k_p)\big\} < \infty \qquad a.s.,$$

$$J_\infty = \#\big\{S_1[1,\infty) \cap \cdots \cap S_p[1,\infty)\big\} < \infty \qquad a.s.$$

We are interested in the tail probabilities

$$\mathbb{P}\{I_\infty \ge t\} \text{ and } \mathbb{P}\{J_\infty \ge t\}$$

as $t \to \infty$. In this section we shall investigate the tail behavior of I_∞.

THEOREM 7.5.1. *There is $-\infty < L < \infty$ such that*

$$\lim_{k\to\infty} \frac{1}{k} \log \frac{1}{(k!)^p} \mathbb{E} I_\infty^k = L. \tag{7.5.1}$$

Consequently,

$$\lim_{t\to\infty} t^{-1/p} \log \mathbb{P}\big\{I_\infty \ge t\big\} = -pe^{-L/p}. \tag{7.5.2}$$

Proof. We need only to show (7.5.1) as (7.5.2) is a consequence of (7.5.1) (Theorem 1.2.8). To this end we define the "augmented" intersection local time $\tilde{I}_\infty$ by

$$\tilde{I}_\infty = \sum_{i_1,\cdots,i_p=0}^{\infty} 1_{\{S_1(i_1)=\cdots=S(i_p)\}}.$$

That is, we include the intersection at time 0. We first claim that there is an extended number $-\infty \le L < \infty$ such that

$$\lim_{k\to\infty} \frac{1}{k} \log \frac{1}{(k!)^p} \mathbb{E}\tilde{I}_\infty^k = L. \tag{7.5.3}$$

By Lemma 1.3.1, it suffices for (7.5.3) to show that

$$\left(\mathbb{E}\tilde{I}_\infty^{k+l}\right)^{1/p} \le \binom{k+l}{k} \left(\mathbb{E}\tilde{I}_\infty^{k}\right)^{1/p} \left(\mathbb{E}\tilde{I}_\infty^{l}\right)^{1/p} \tag{7.5.4}$$

holds for any integers $k, l \ge 1$.

Write

$$l(x) = \sum_{j=0}^{\infty} 1_{\{S(k)=x\}}.$$

Then ((5.6.4) ensures that the quantities we deal with in the following calculation are finite.)

$$\begin{aligned}
\mathbb{E}\tilde{I}_\infty^{k+l} &= \sum_{x_1,\cdots,x_{k+l}\in\mathbb{Z}^d} \left[\mathbb{E}\prod_{j=1}^{k+l} l(x_j)\right]^p \\
&= \sum_{x_1,\cdots,x_{k+l}\in\mathbb{Z}^d} \left\{\sum_{i_1,\cdots,i_{k+l}=0}^{\infty} \mathbb{E}\prod_{j=1}^{k+l} 1_{\{S(i_j)=x_j\}}\right\} \left[\mathbb{E}\prod_{j=1}^{k+l} l(x_j)\right]^{p-1} \\
&= \sum_{i_1,\cdots,i_{k+l}=0}^{\infty} \sum_{x_1,\cdots,x_{k+l}\in\mathbb{Z}^d} \left\{\mathbb{E}\prod_{j=1}^{k+l} 1_{\{S(i_j)=x_j\}}\right\} \left[\mathbb{E}\prod_{j=1}^{k+l} l(x_j)\right]^{p-1}.
\end{aligned}$$

For any two vectors $(j_1,\cdots,j_k)$ and $(j_{k+1},\cdots,j_{k+l})$ we write

$$(j_1,\cdots,j_k) \prec (j_{k+1},\cdots,j_{k+l}) \Longleftrightarrow \max\{j_1,\cdots,j_k\} \le \min\{j_{k+1},\cdots,j_{k+l}\}.$$

For each vector $(i_1,\cdots,i_{k+l})$ we reorder its component so the resulting vector $(i'_1,\cdots,i'_{k+l})$ satisfies

$$(i'_1,\cdots,i'_k) \prec (i'_{k+1},\cdots,i'_{k+l}).$$

Let the permutation $\sigma \in \Sigma_{k+l}$ be such that $i'_{\sigma(j)} = i_j$ $(j = 1, \cdots, k+l)$. Then

$$\begin{aligned}
&\sum_{x_1,\cdots,x_{k+l}\in\mathbb{Z}^d} \Big\{\mathbb{E}\prod_{j=1}^{k+l} 1_{\{S(i_j)=x_j\}}\Big\}\Big[\mathbb{E}\prod_{j=1}^{k+l} l(x_j)\Big]^{p-1} \\
&= \sum_{x_1,\cdots,x_{k+l}\in\mathbb{Z}^d} \Big\{\mathbb{E}\prod_{j=1}^{k+l} 1_{\{S(i'_j)=x_{\sigma(j)}\}}\Big\}\Big[\mathbb{E}\prod_{j=1}^{k+l} l(x_j)\Big]^{p-1} \\
&= \sum_{x_1,\cdots,x_{k+l}\in\mathbb{Z}^d} \Big\{\mathbb{E}\prod_{j=1}^{k+l} 1_{\{S(i'_j)=x_j\}}\Big\}\Big[\mathbb{E}\prod_{j=1}^{k+l} l(x_{\sigma^{-1}(j)})\Big]^{p-1} \\
&= \sum_{x_1,\cdots,x_{k+l}\in\mathbb{Z}^d} \Big\{\mathbb{E}\prod_{j=1}^{k+l} 1_{\{S(i'_j)=x_j\}}\Big\}\Big[\mathbb{E}\prod_{j=1}^{k+l} l(x_j)\Big]^{p-1}
\end{aligned}$$

where σ^{-1} is the inverse of σ and the last step follows from the permutation invariance

$$\prod_{j=1}^{k+l} l(x_{\sigma^{-1}(j)}) = \prod_{j=1}^{k+l} l(x_j).$$

There are at most $\binom{k+l}{k}$ different ways to partition the (not necessarily distinct) indices $i_1, \cdots, i_{k+l}$ into two groups $\{i'_1, \cdots, i'_k\}$ and $\{i'_{k+1}, \cdots, i'_{k+l}\}$. Thus,

$$\begin{aligned}
\mathbb{E}\tilde{I}_\infty^{k+l} &\le \binom{k+l}{k} \sum_{\substack{(i_1,\cdots,i_k)\\ \prec(i_{k+1},\cdots,i_{k+l})}} \sum_{x_1,\cdots,x_{k+l}\in\mathbb{Z}^d} \Big[\mathbb{E}\prod_{j=1}^{k+l} 1_{\{S(i_j)=x_j\}}\Big]\Big[\mathbb{E}\prod_{j=1}^{k+l} l(x_j)\Big]^{p-1} \\
&= \binom{k+l}{k} \sum_{x_1,\cdots,x_{k+l}\in\mathbb{Z}^d} \Big[\sum_{\substack{(i_1,\cdots,i_k)\\ \prec(i_{k+1},\cdots,i_{k+l})}} \mathbb{E}\prod_{j=1}^{k+l} 1_{\{S(i_j)=x_j\}}\Big]\Big[\mathbb{E}\prod_{j=1}^{k+l} l(x_j)\Big]^{p-1}.
\end{aligned}$$

Applying Hölder inequality on the right hand side gives us

$$\mathbb{E}\tilde{I}_\infty^{k+l} \le \binom{k+l}{k}\Big\{\sum_{x_1,\cdots,x_{k+l}\in\mathbb{Z}^d}\Big[\sum_{\substack{(i_1,\cdots,i_k)\\ \prec(i_{k+1},\cdots,i_{k+l})}} \mathbb{E}\prod_{j=1}^{k+l} 1_{\{S(i_j)=x_j\}}\Big]^p\Big\}^{1/p}\Big(\mathbb{E}\tilde{I}_\infty^{k+l}\Big)^{\frac{p-1}{p}}.$$

Consequently,

$$\Big(\mathbb{E}\tilde{I}_\infty^{k+l}\Big)^{1/p} \le \binom{k+l}{k}\Big\{\sum_{x_1,\cdots,x_{k+l}\in\mathbb{Z}^d}\Big[\sum_{\substack{(i_1,\cdots,i_k)\\ \prec(i_{k+1},\cdots,i_{k+l})}} \mathbb{E}\prod_{j=1}^{k+l} 1_{\{S(i_j)=x_j\}}\Big]^p\Big\}^{1/p}.$$

For each $(i_1, \cdots, i_{k+l})$ with $(i_1, \cdots, i_k) \prec (i_{k+1}, \cdots, i_{k+l})$, we set

$$i_* = \max\{i_1, \cdots, i_k\}.$$

By increment independence, we get

$$\mathbb{E}\prod_{j=1}^{k+l} 1_{\{S(i_j)=x_j\}} = \Big\{\mathbb{E}\prod_{j=1}^{k} 1_{\{S(i_j)=x_j\}}\Big\}\Big\{\mathbb{E}\prod_{j=k+1}^{k+l} 1_{\{S(i_j-i_*)=x_j-x_{i_*}\}}\Big\}.$$

Hence,

$$\begin{aligned}
&\sum_{\substack{(i_1,\cdots,i_k)\\ \prec(i_{k+1},\cdots,i_{k+l})}} \mathbb{E}\prod_{j=1}^{k+l} 1_{\{S(i_j)=x_j\}}\\
&= \sum_{i_1,\cdots,i_k=0}^{\infty}\Big\{\mathbb{E}\prod_{j=1}^{k} 1_{\{S(i_j)=x_j\}}\Big\}\sum_{i_{k+1},\cdots,i_{k+l}=0}^{\infty}\mathbb{E}\prod_{j=k+1}^{k+l} 1_{\{S(i_j)=x_j-x_{i_*}\}}\\
&= \sum_{i_1,\cdots,i_k=0}^{\infty}\Big\{\mathbb{E}\prod_{j=1}^{k} 1_{\{S(i_j)=x_j\}}\Big\}\mathbb{E}\prod_{j=k+1}^{k+l} l(x_j-x_{i_*}).
\end{aligned}$$

For any fixed $(x_1,\cdots,x_k)$, we obtain that

$$\begin{aligned}
&\Big\{\sum_{x_{k+1},\cdots,x_{k+l}\in\mathbb{Z}^d}\Big[\sum_{\substack{(i_1,\cdots,i_k)\\ \prec(i_{k+1},\cdots,i_{k+l})}} \mathbb{E}\prod_{j=1}^{k+l} 1_{\{S(i_j)=x_j\}}\Big]^p\Big\}^{1/p}\\
&\le \sum_{i_1,\cdots,i_k=0}^{\infty}\Big[\mathbb{E}\prod_{j=1}^{k} 1_{\{S(i_j)=x_j\}}\Big]\Big\{\sum_{x_{k+1},\cdots,x_{k+l}\in\mathbb{Z}^d}\Big[\mathbb{E}\prod_{j=k+1}^{k+l} l(x_j-x_{i_*})\Big]^p\Big\}^{1/p}
\end{aligned}$$

using the triangular inequality.

By translation invariance, we get

$$\begin{aligned}
&\sum_{x_{k+1},\cdots,x_{k+l}\in\mathbb{Z}^d}\Big[\mathbb{E}\prod_{j=k+1}^{k+l} l(x_j-x_{i_*})\Big]^p\\
&= \sum_{x_{k+1},\cdots,x_{k+l}\in\mathbb{Z}^d}\Big[\mathbb{E}\prod_{j=k+1}^{k+l} l(x_j)\Big]^p = \mathbb{E}\tilde{I}_{\infty}^{l}.
\end{aligned}$$

Consequently,

$$\begin{aligned}
&\sum_{x_{k+1},\cdots,x_{k+l}\in\mathbb{Z}^d}\Big[\sum_{\substack{(i_1,\cdots,i_k)\\ \prec(i_{k+1},\cdots,i_{k+l})}} \mathbb{E}\prod_{j=1}^{k+l} 1_{\{S(i_j)=x_j\}}\Big]^p\\
&\le \mathbb{E}\tilde{I}_{\infty}^{l}\Big[\sum_{i_1,\cdots,i_k=0}^{\infty}\mathbb{E}\prod_{j=1}^{k} 1_{\{S(i_j)=x_j\}}\Big]^p = \mathbb{E}\tilde{I}_{\infty}^{l}\Big[\mathbb{E}\prod_{j=1}^{k} l(x_j)\Big]^p.
\end{aligned}$$

Summing over $(x_1,\cdots,x_k)$ on both sides gives us

$$\begin{aligned}
&\sum_{x_1,\cdots,x_{k+l}\in\mathbb{Z}^d}\Big[\sum_{\substack{(i_1,\cdots,i_k)\\ \prec(i_{k+1},\cdots,i_{k+l})}} \mathbb{E}\prod_{j=1}^{k+l} 1_{\{S(i_j)=x_j\}}\Big]^p\\
&\le \Big(\mathbb{E}\tilde{I}_{\infty}^{l}\Big)\sum_{x_1,\cdots,x_k\in\mathbb{Z}^d}\Big[\mathbb{E}\prod_{j=1}^{k} l(x_j)\Big]^p = \Big(\mathbb{E}\tilde{I}_{\infty}^{l}\Big)\Big(\mathbb{E}\tilde{I}_{\infty}^{k}\Big).
\end{aligned}$$

Summarizing our argument, we have proved (7.5.4).

As the second step, we show that $L > -\infty$. Indeed, for each $m \geq 1$,

$$\begin{aligned}\mathbb{E}\tilde{I}_\infty^k \geq \mathbb{E}I_\infty^k &= \sum_{x_1,\cdots,x_k\in\mathbb{Z}^d}\Big[\mathbb{E}\prod_{j=1}^k\sum_{i=1}^\infty 1_{\{S(i)=x_j\}}\Big]^p\\
&\geq \sum_{x_1,\cdots,x_k\in\mathbb{Z}^d}\Big[\sum_{\substack{i_1,\cdots,i_k\\ \text{distinct}}}\mathbb{P}\big\{S(i_1)=x_1,\cdots,S(i_k)=x_k\big\}\Big]^p\\
&= \sum_{x_1,\cdots,x_k\in\mathbb{Z}^d}\Big[\sum_{\sigma\in\Sigma_k}\prod_{j=1}^k\mathcal{G}(x_{\sigma(j)}-x_{\sigma(j-1)})\Big]^p\end{aligned}$$

where $\mathcal{G}(x)$ is the *Green's function* defined as

$$\mathcal{G}(x)=\sum_{k=1}^\infty\mathbb{P}\{S(k)=x\}\qquad x\in\mathbb{Z}^d.$$

Hence, by Theorem 3.1.1, we get

$$L \geq p\log\sup\Big\{\sum_{x,y\in\mathbb{Z}^d}\mathcal{G}(x-y)f(x)f(y);\ \sum_{x\in\mathbb{Z}^d}|f(x)|^{\frac{2p}{2p-1}}=1\Big\} > -\infty. \tag{7.5.5}$$

Finally, we come to the proof of (7.5.1). Noticing that $I_\infty \leq \tilde{I}_\infty$, by (7.5.3) we only have to prove that

$$\liminf_{k\to\infty}\frac{1}{k}\log\frac{1}{(k!)^p}\mathbb{E}I_\infty^k \geq L.$$

Notice that

$$\tilde{I}_\infty \leq I_\infty + \sum_{j=1}^p l_j(0)$$

where

$$l_j(x)=\sum_{k=0}^\infty 1_{\{S_j(k)=x\}}.$$

So we have

$$\Big(\mathbb{E}\tilde{I}_\infty^k\Big)^{1/k} \leq \Big(\mathbb{E}I_\infty^k\Big)^{1/k} + p\Big(\mathbb{E}l^k(0)\Big)^{1/k}.$$

For the second term,

$$\begin{aligned}\mathbb{E}l^k(0) &= \sum_{i_1,\cdots,i_k=0}^\infty \mathbb{P}\big\{S(i_1)=\cdots=S(i_k)=0\big\}\\
&\leq k!\sum_{i_1\leq\cdots\leq i_k}\mathbb{P}\big\{S(i_1)=\cdots=S(i_k)=0\big\}\\
&= k!\sum_{i_1\leq\cdots\leq i_k}\prod_{j=1}^k\mathbb{P}\{S(i_j-i_{j-1})=0\}\\
&\leq k!\Big(\sum_{i=0}^\infty\mathbb{P}\{S(i)=0\}\Big)^k.\end{aligned}$$

This confirms our claim. □

Our next investigation is to identify the constant appearing on the right hand side of (7.5.2) in terms of certain variation (the reader may choose to skip this part without compromising the later development of the book).

It should be pointed out that the relation posted in (7.5.5) cannot be further developed into an equality. To introduce the variational representation, recall the Green's function

$$\mathcal{G}(x) = \sum_{k=1}^{\infty} \mathbb{P}\{S(k) = x\} \qquad x \in \mathbb{Z}^d.$$

Notice that we follow the (slightly unusual) convention of not summing over the time $k = 0$, which has an influence on the value of $\mathcal{G}(0)$. A crucial fact in the super-critical dimensions is ((5.6.4)) that

$$\sum_{x\in\mathbb{Z}^d} \mathcal{G}^p(x) < \infty. \tag{7.5.6}$$

LEMMA 7.5.2. *For every non-negative $h \in \mathcal{L}^q(\mathbb{Z}^d)$, the linear map*

$$\mathcal{G}_h g(x) = \sqrt{e^{h(x)} - 1} \sum_{y\in\mathbb{Z}^d} \mathcal{G}(x-y) g(y) \sqrt{e^{h(y)} - 1} \qquad g \in \mathcal{L}^2(\mathbb{Z}^d)$$

defines a bounded, self-adjoint operator $\mathcal{G}_h$ on $\mathcal{L}^2(\mathbb{Z}^d)$.

Furthermore,

$$||\mathcal{G}_h|| = \sup_{|g|_2=1} \langle g, \mathcal{G}_h g\rangle \le ||\mathcal{G}||_p \Big(e^{||h||_q} - 1\Big). \tag{7.5.7}$$

Proof. By the symmetry of the random walk, it is easy to see that for $g_1, g_2 \in \mathcal{L}^2(\mathbb{Z}^d)$,

$$\langle g_1, \mathcal{G}_h g_2\rangle = \langle \mathcal{G}_h g_1, g_2\rangle$$

if both sides are well defined (indeed, this can be seen from the later development).

For any $g \in \mathcal{L}^2(\mathbb{Z}^d)$, write

$$f(x) = g(x)\sqrt{e^{h(x)} - 1} \qquad x \in \mathbb{Z}^d.$$

By Hölder inequality, we get

$$\sum_{x\in\mathbb{Z}^d} |f(x)|^{\frac{2p}{2p-1}} \le \Big\{\sum_{x\in\mathbb{Z}^d} g^2(x)\Big\}^{\frac{p}{2p-1}} \Big\{\sum_{x\in\mathbb{Z}^d} \Big(e^{h(x)} - 1\Big)^q\Big\}^{\frac{p-1}{2p-1}}.$$

By Taylor expansion and by triangular inequality, we get

$$\Big\{\sum_{x\in\mathbb{Z}^d} \Big(e^{h(x)} - 1\Big)^q\Big\}^{1/q} \le \sum_{k=1}^{\infty} \frac{1}{k!} ||h^k||_q \le e^{||h||_q} - 1$$

where the last step follows from the fact that

$$||h^k||_q^q = \sum_{x\in\mathbb{Z}^d} |h(x)|^{kq} \le \Big(\sum_{x\in\mathbb{Z}^d} |h(x)|^q\Big)^k = ||h||_q^{kq}.$$

Thus,

$$\sum_{x\in\mathbb{Z}^d} |f(x)|^{\frac{2p}{2p-1}} \le \Big(e^{||h||_q}-1\Big)^{\frac{p}{2p-1}} ||g||_2^{\frac{2p}{2p-1}}.$$

Consequently,

$$\begin{aligned}\big|\langle g,\mathcal{G}_h g\rangle\big| &= \sum_{x,y\in\mathbb{Z}^d}\mathcal{G}(x-y)|f(x)||f(y)| \le \Big\{\sum_{x\in\mathbb{Z}^d}\mathcal{G}^p(x)\Big\}^{1/p}||f||^2_{\frac{2p}{2p-1}}\\ &\le \Big\{\sum_{x\in\mathbb{Z}^d}\mathcal{G}^p(x)\Big\}^{1/p}\Big(e^{||h||_q}-1\Big)||g||_2^2,\end{aligned}$$

where the second step follows from the estimation used in the proof of Lemma 3.2.1. In particular, the above estimation gives the following bound to the spectral radius

$$||\mathcal{G}_h|| = \sup_{||g||_2=1}\big|\langle g,\mathcal{G}_h g\rangle\big| \le \Big\{\sum_{x\in\mathbb{Z}^d}\mathcal{G}^p(x)\Big\}^{1/p}\Big(e^{||h||_q}-1\Big).$$

Finally, the equality in (7.5.7) follows from the obvious fact that

$$\sup_{||g||_2=1}\big|\langle g,\mathcal{G}_h g\rangle\big| = \sup_{||g||_2=1}\langle g,\mathcal{G}_h g\rangle.$$

□

Our next theorem is formulated in terms of the variation problem

$$\mathcal{C}(S) = \inf\Big\{||h||_q;\quad h\ge 0 \text{ and } ||\mathcal{G}_h||\ge 1\Big\}. \tag{7.5.8}$$

We claim that

$$\log\Big(1+||\mathcal{G}||_p^{-1}\Big) \le \mathcal{C}(S) \le \varrho_S^{-1}$$

where

$$\varrho_S = \sup\Big\{\sum_{x,y\in\mathbb{Z}^d}\mathcal{G}(x-y)f(x)f(y);\quad \sum_{x\in\mathbb{Z}^d}|f(x)|^{\frac{2p}{2p-1}}=1\Big\}.$$

Indeed, (7.5.6) and (7.5.7) imply the lower bound. On the other hand, for any $f(x)\ge 0$ with

$$\sum_{x\in\mathbb{Z}^d} f^{\frac{2p}{2p-1}}(x) = 1$$

write $h(x) = \varrho_S^{-1} f^{\frac{2(p-1)}{2p-1}}(x)$ and $g(x) = f^{\frac{p}{2p-1}}(x)$. We have $||h||_q = \varrho_S^{-1}$ and $||g||_2 = 1$. Hence, by the inequality $e^\theta - 1 \ge \theta$ $(\theta\ge 0)$, we get

$$\begin{aligned}||\mathcal{G}_h|| &\ge \langle g,\mathcal{G}_h g\rangle \ge \sum_{x,y}\mathcal{G}(x-y)\sqrt{h(x)}g(x)\sqrt{h(y)}g(y)\\ &= \varrho_S^{-1}\sum_{x,y}\mathcal{G}(x-y)f(x)f(y).\end{aligned}$$

Taking supremum over f on the right hand side gives that $||\mathcal{G}_h||\ge 1$, which leads to the desired upper bound $\mathcal{C}(S)\le\varrho_S^{-1}$.

THEOREM 7.5.3. *The constant L in Theorem 7.5.1 is equal to $p\log \mathcal{C}(S)$. Consequently,*

$$\lim_{t\to\infty} t^{-1/p}\log \mathbb{P}\{I_\infty \ge t\} = -p\mathcal{C}(S).$$

We spend the rest of the section to prove this theorem. By Theorem 1.2.8, we need only to establish that

$$\lim_{k\to\infty}\frac{1}{k}\log\frac{1}{(k!)^p}\mathbb{E}I_\infty^k = -p\log\mathcal{C}(S). \tag{7.5.9}$$

To do that, we need to represent $\mathbb{E}I_\infty^k$ in a tractable form. What we get turns out to be much more complicated than Le Gall's identity given in Theorem 2.2.8. Notice that

$$\mathbb{E}I_\infty^k = \sum_{x_1,\cdots,x_k\in\mathbb{Z}^d}\Big[\mathbb{E}\sum_{i_1,\cdots,i_k=1}^\infty\prod_{l=1}^k 1_{\{S(i_l)=x_l\}}\Big]^p.$$

We now face the challenge of reordering the tuple $(i_1,\cdots,i_k)$, for some of $i_1,\cdots,i_k$ may take the same value. Indeed, we will miss the right constant by overcounting (summing over $i_1\le\cdots\le i_k$) and by undercounting (summing over $i_1<\cdots<i_k$).

We define, for any $1\le m\le k$, the family of all m-partitions of the set $\{1,\ldots,k\}$ as

$$\mathcal{E}_m := \Big\{\pi=(\pi_1,\cdots,\pi_m);\ \ \pi_j\neq\emptyset \ \text{ with } \ \pi_i\cap\pi_j=\emptyset$$
$$\text{for all } i\neq j \text{ and } \bigcup_{j=1}^m \pi_j=\{1,\cdots,k\}\Big\},$$

where we assume that the elements $\pi_1,\cdots,\pi_m$ of an m-partition are ordered by increasing order of their minimal elements. To any $(i_1,\cdots,i_k)$ we associate

- a tuple $(i_1^*,\cdots,i_m^*)$ of distinct natural numbers such that
  $$\{i_1,\cdots,i_k\}=\{i_1^*,\cdots,i_m^*\}$$
  and elements in the tuple $(i_1^*,\cdots,i_m^*)$ appear in the order in which they appear first in $(i_1,\cdots,i_k)$;
- an m-partition $(\pi_1,\cdots,\pi_m)\in\mathcal{E}_m$ with $i_j=i_\ell^*$ whenever $j\in\pi_\ell$.

Conversely, given a tuple $(j_1,\cdots,j_m)$ of distinct natural numbers and an m-partition π we can find $(i_1,\cdots,i_k)$ such that the induced m-tuple is $(j_1,\cdots,j_m)$ and the induced m-partition is π.

For any $\pi\in\mathcal{E}_m$, define

$$\mathcal{A}(\pi)=\big\{(x_1,\cdots,x_k)\in(\mathbb{Z}^d)^k\ :\ x_i=x_j \text{ for all } i,j\in\pi_\ell \text{ and } \ell\in\{1,\cdots,m\}\big\}.$$

For any $(x_1,\cdots,x_k)\in\mathcal{A}(\pi)$ and for any $1\le\ell\le m$, we use x_{π_ℓ} for the common value of $\{x_j\ :\ j\in\pi_\ell\}$. There are two facts which we shall use frequently in the later argument:

First, for any $j_1,\cdots,j_m\ge1$ with $j_1+\cdots+j_m=k$,

$$\#\Big\{\pi=(\pi_1,\cdots,\pi_m)\in\mathcal{E}_m;\ \ \#(\pi_l)=j_l\ \ \forall 1\le l\le m\Big\}=\frac{1}{m!}\frac{k!}{j_1!\cdots j_m!}.$$

Second, for any $\pi \in \mathcal{E}_m$,

$$\Big\{(x_{\pi_1},\cdots,x_{\pi_m});\quad (x_1,\cdots,x_k)\in\mathcal{A}(\pi)\Big\} = (\mathbb{Z}^d)^m$$

and the map $(x_1,\cdots,x_k)\mapsto(x_{\pi_1},\cdots,x_{\pi_m})$ is a one-to-one correspondence between $\mathcal{A}(\pi)$ and $(\mathbb{Z}^d)^m$.

Observing that $S(i_\ell)=x_\ell$ for all $\ell\in\{1,\cdots,k\}$ if and only if $(x_1,\cdots,x_k)\in\mathcal{A}(\pi)$ and $S(i_\ell^*)=x_{\pi_\ell}$ for all $\ell\in\{1,\cdots,m\}$, we get that

$$\begin{aligned}
&\mathbb{E}\sum_{i_1,\cdots,i_k=1}^\infty\prod_{l=1}^k 1_{\{S(i_l)=x_l\}}\\
&=\sum_{m=1}^k\sum_{\pi\in\mathcal{E}_m}1_{\{(x_1,\cdots,x_k)\in\mathcal{A}(\pi)\}}\sum_{\substack{j_1,\cdots,j_m\\ \text{distinct}}}\mathbb{E}\prod_{l=1}^m 1_{\{S(j_l)=x_{\pi_l}\}}\\
&=\sum_{m=1}^k\sum_{\pi\in\mathcal{E}_m}1_{\{(x_1,\cdots,x_k)\in\mathcal{A}(\pi)\}}\sum_{\sigma\in\Sigma_m}\prod_{l=1}^m\mathcal{G}(x_{\pi_{\sigma(l)}}-x_{\pi_{\sigma(l-1)}}).
\end{aligned}$$

Here, and elsewhere in the remaining of this section, we follow the convention that $x_{\pi_{\sigma(0)}}=0$.

Therefore,

$$\begin{aligned}
\mathbb{E}I_\infty^k=\sum_{x_1,\cdots,x_k\in\mathbb{Z}^d}\Bigg[&\sum_{m=1}^k\sum_{\pi\in\mathcal{E}_m}1_{\{(x_1,\cdots,x_k)\in\mathcal{A}(\pi)\}}\\
&\times\sum_{\sigma\in\Sigma_m}\prod_{l=1}^m\mathcal{G}(x_{\pi_{\sigma(l)}}-x_{\pi_{\sigma(l-1)}})\Bigg]^p.
\end{aligned}\tag{7.5.10}$$

By (7.5.1) in Theorem 7.5.1, the next lemma gives the lower bound of (7.5.9).

LEMMA 7.5.4. *If there exists $h\ge 0$ with $\|h\|_q\le\theta$ such that $\|\mathcal{G}_h\|>1$, then*

$$\sum_{k=1}^\infty\frac{\theta^k}{k!}\big[\mathbb{E}I_\infty^k\big]^{\frac1p}=\infty.$$

Proof. Let $h\ge0$ with $\|h\|_q\le\theta$. Applying Hölder's inequality to (7.5.10),

$$\begin{aligned}
&\theta^k\big[\mathbb{E}I_\infty^k\big]^{\frac1p}\\
&\ge\sum_{x_1,\cdots,x_k\in\mathbb{Z}^d}\Big(\prod_{j=1}^k h(x_j)\Big)\sum_{m=1}^k\sum_{\pi\in\mathcal{E}_m}1_{\{(x_1,\cdots,x_k)\in\mathcal{A}(\pi)\}}\\
&\qquad\times\sum_{\sigma\in\Sigma_m}\prod_{\ell=1}^m\mathcal{G}(x_{\pi_{\sigma(\ell)}}-x_{\pi_{\sigma(\ell-1)}})\\
&=\sum_{m=1}^k\sum_{\sigma\in\Sigma_m}\sum_{\pi\in\mathcal{E}_m}\sum_{x_1,\cdots,x_m\in\mathbb{Z}^d}\Big(\prod_{\ell=1}^m h(x_\ell)^{\#(\pi_\ell)}\Big)\Big(\prod_{\ell=1}^m\mathcal{G}(x_{\sigma(\ell)}-x_{\sigma(\ell-1)})\Big),
\end{aligned}$$

where the last step follows from the one-to-one correspondence between $\mathcal{A}(\pi)$ and $\mathbb{Z}^m$. Therefore,

$$\theta^k\left[\mathbb{E}I_\infty^k\right]^{\frac{1}{p}} \ge \sum_{m=1}^k m! \sum_{x_1,\cdots,x_m\in\mathbb{Z}^d}\sum_{\pi\in\mathcal{E}_m}\Big(\prod_{\ell=1}^m h(x_\ell)^{\#(\pi_\ell)}\Big)\Big(\prod_{\ell=1}^m \mathcal{G}(x_\ell - x_{\ell-1})\Big)$$

$$= \sum_{m=1}^k \sum_{x_1,\cdots,x_m\in\mathbb{Z}^d}\sum_{\substack{j_1+\cdots+j_m=k\\ j_1,\cdots,j_m\ge 1}} \frac{k!}{j_1!\cdots j_m!}\Big(\prod_{\ell=1}^m h(x_\ell)^{j_\ell}\Big)\Big(\prod_{\ell=1}^m \mathcal{G}(x_\ell - x_{\ell-1})\Big).$$

Summing up both sides, we get

$$\begin{aligned}
&\sum_{k=1}^\infty \frac{\theta^k}{k!}\left[\mathbb{E}I_\infty^k\right]^{\frac{1}{p}}\\
&\ge \sum_{m=1}^\infty\sum_{k=m}^\infty\sum_{x_1,\cdots,x_m}\Big\{\sum_{\substack{j_1+\cdots+j_m=k\\ j_1,\cdots,j_m\ge 1}}\frac{1}{j_1!\cdots j_m!}\prod_{\ell=1}^m h(x_\ell)^{j_\ell}\Big\}\prod_{\ell=1}^m \mathcal{G}(x_\ell - x_{\ell-1})\\
&= \sum_{m=1}^\infty\sum_{x_1,\cdots,x_m}\Big\{\prod_{\ell=1}^m\sum_{j=1}^\infty\frac{h(x_\ell)^j}{j!}\Big\}\prod_{\ell=1}^m \mathcal{G}(x_\ell - x_{\ell-1})\\
&= \sum_{m=1}^\infty\sum_{x_1,\cdots,x_m}\prod_{\ell=1}^m\left(e^{h(x_\ell)}-1\right)\mathcal{G}(x_\ell - x_{\ell-1}).
\end{aligned}$$

By the fact that $\|\mathcal{G}_h\| > 1$ we can pick $g \ge 0$ with $\|g\|_2 = 1$ such that $\langle g, \mathcal{G}_h g\rangle > 1$. By monotone convergence we may find a finite set $A \subset \mathbb{Z}^d$ such that

$$\Big\langle g1_A, \mathcal{G}_h\big(g1_A\big)\Big\rangle > 1$$

and additionally, but without loss of generality, $g, h > 0$ on A. Using spectral theory in the same way as in the proof of Theorem 3.1.1, we can prove that

$$\begin{aligned}
&\liminf_{m\to\infty}\frac{1}{m}\log\sum_{x_1,\cdots,x_m}\prod_{\ell=1}^m\left(e^{h(x_\ell)}-1\right)\mathcal{G}(x_\ell - x_{\ell-1})\\
&\ge p\log\Big\langle g1_A, \mathcal{G}_h\big(g1_A\big)\Big\rangle > 1.
\end{aligned}$$

This leads to the desired conclusion. □

It remains to prove the upper bound of (7.5.9). Our starting point is the formula (7.5.10) and we first project the problem from the full lattice space $\mathbb{Z}^d$ onto a finite set. We fix a large integer N and write $A_N = [-N, N)^d$ for a box of lattice points. Then every $x \in \mathbb{Z}^d$ can uniquely be written as $x = 2Nz + y$ for $z \in \mathbb{Z}^d$ and $y \in A_N$. Hence,

$$\begin{aligned}
\mathbb{E}I_\infty^k = &\sum_{y_1,\cdots,y_k\in A_N}\sum_{z_1,\cdots,z_k\in\mathbb{Z}^d}\Bigg[\sum_{m=1}^k\sum_{\pi\in\mathcal{E}_m} 1_{\{(2Nz_1+y_1,\cdots,2Nz_k+y_k)\in\mathcal{A}(\pi)\}}\\
&\times\sum_{\sigma\in\Sigma_m}\prod_{\ell=1}^m\mathcal{G}\big(2N(z_{\pi_{\sigma(\ell)}} - z_{\pi_{\sigma(\ell-1)}}) + (y_{\pi_{\sigma(\ell)}} - y_{\pi_{\sigma(\ell-1)}})\big)\Bigg]^p.
\end{aligned}$$

Fix $y_1, \cdots, y_k \in A_N$. By triangular inequality,

$$\begin{aligned}&\Big\{\sum_{z_1,\cdots,z_k\in\mathbb{Z}^d}\Big[\sum_{m=1}^k\sum_{\pi\in\mathcal{E}_m}1_{\{(2Nz_1+y_1,\cdots,2Nz_k+y_k)\in\mathcal{A}(\pi)\}}\\&\times\sum_{\sigma\in\Sigma_m}\prod_{\ell=1}^m\mathcal{G}\big(2N(z_{\pi_\sigma(\ell)}-z_{\pi_\sigma(\ell-1)})+(y_{\pi_\sigma(\ell)}-y_{\pi_\sigma(\ell-1)})\big)\Big]^p\Big\}^{1/p}\\&\le\sum_{m=1}^k\sum_{\pi\in\mathcal{E}_m}\sum_{\sigma\in\Sigma_m}\Big\{\sum_{z_1,\dots,z_k\in\mathbb{Z}^d}1_{\{(2Nz_1+y_1,\cdots,2Nz_k+y_k)\in\mathcal{A}(\pi)\}}\\&\times\prod_{\ell=1}^m\mathcal{G}^p\big(2N(z_{\pi_\sigma(\ell)}-z_{\pi_\sigma(\ell-1)})+(y_{\pi_\sigma(\ell)}-y_{\pi_\sigma(\ell-1)})\big)\Big\}^{1/p}.\end{aligned}$$

From the uniqueness of the decomposition $x = 2Nz + y$ we infer that

$$1_{\{(2Nz_1+y_1,\cdots,2Nz_k+y_k)\in\mathcal{A}(\pi)\}} = 1_{\{(y_1,\cdots,y_k)\in\mathcal{A}(\pi)\}}\,1_{\{(z_1,\cdots,z_k)\in\mathcal{A}(\pi)\}}.$$

Therefore,

$$\begin{aligned}&\sum_{z_1,\cdots,z_k\in\mathbb{Z}^d}1_{\{(2Nz_1+y_1,\cdots,2Nz_k+y_k)\in\mathcal{A}(\pi)\}}\\&\times\prod_{\ell=1}^m\mathcal{G}^p\big(2N(z_{\pi_\sigma(\ell)}-z_{\pi_\sigma(\ell-1)})+(y_{\pi_\sigma(\ell)}-y_{\pi_\sigma(\ell-1)})\big)\\&=1_{\{(y_1,\cdots,y_k)\in\mathcal{A}(\pi)\}}\sum_{z_1,\cdots,z_k\in\mathbb{Z}^d}1_{\{(z_1,\cdots,z_k)\in\mathcal{A}(\pi)\}}\\&\times\prod_{\ell=1}^m\mathcal{G}^p\big(2N(z_{\pi_\sigma(\ell)}-z_{\pi_\sigma(\ell-1)})+(y_{\pi_\sigma(\ell)}-y_{\pi_\sigma(\ell-1)})\big).\end{aligned}$$

By the one-to-one correspondence between $\mathcal{A}(\pi)$ and $\mathbb{Z}^m$, the summation on the right hight side is equal to

$$\begin{aligned}&\sum_{z_1,\cdots,z_m\in\mathbb{Z}^d}\prod_{\ell=1}^m\mathcal{G}^p\big(2N(z_{\sigma(\ell)}-z_{\sigma(\ell-1)})+(y_{\pi_\sigma(\ell)}-y_{\pi_\sigma(\ell-1)})\big)\\&=\prod_{\ell=1}^m\widetilde{\mathcal{G}}^p(y_{\pi_\sigma(\ell)}-y_{\pi_\sigma(\ell-1)})\end{aligned}$$

where

$$\widetilde{\mathcal{G}}(x)=\Big\{\sum_{z\in\mathbb{Z}^d}\mathcal{G}^p(2Nz+x)\Big\}^{1/p}.$$

Summarizing our argument, we have

$$\begin{aligned}\text{(7.5.11)}\qquad \mathbb{E}I_\infty^k\le\sum_{y_1,\cdots,y_k\in A_N}&\Big[\sum_{m=1}^k\sum_{\pi\in\mathcal{E}_m}1_{\{(y_1,\cdots,y_k)\in\mathcal{A}(\pi)\}}\\&\times\sum_{\sigma\in\Sigma_m}\prod_{\ell=1}^m\widetilde{\mathcal{G}}(y_{\pi_\sigma(\ell)}-y_{\pi_\sigma(\ell-1)})\Big]^p.\end{aligned}$$

For each $h \in l^q(A_N)$, let $\widetilde{\mathcal{G}}_h$ be the bounded, self-adjoint operator on $l^2(A_N)$ defined as

$$\widetilde{\mathcal{G}}_h g(x) = \sqrt{e^{h(x)} - 1} \sum_{y \in A_N} \widetilde{\mathcal{G}}(x-y) g(y) \sqrt{e^{h(y)} - 1} \qquad g \in l^2(A_N).$$

Write

$$\mathcal{C}_N(S) = \inf\Big\{||h||_q; \quad h \geq 0 \text{ and } ||\widetilde{\mathcal{G}}_h|| \geq 1\Big\}$$

and

$$H_k = \sum_{y_1,\cdots,y_k \in A_N} \Big[\sum_{m=1}^k \sum_{\pi \in \mathcal{E}_m} 1_{\{(y_1,\cdots,y_k) \in \mathcal{A}(\pi)\}} \sum_{\sigma \in \Sigma_m} \prod_{\ell=1}^m \widetilde{\mathcal{G}}(y_{\pi_{\sigma(\ell)}} - y_{\pi_{\sigma(\ell-1)}})\Big]^p.$$

LEMMA 7.5.5.

$$\limsup_{k \to \infty} \frac{1}{k} \log \frac{1}{(k!)^p} H_k \leq -p \log \mathcal{C}_N(S).$$

Proof. The following combinatorial estimate is partially similar to the one used in the proof of Theorem 3.1.2. Fix a vector $\mathbf{x} = (x_1, \cdots, x_k)$ of length k with entries from the set A_N and associate its empirical measure $L_k^{\mathbf{x}}$ by letting

$$L_k^{\mathbf{x}} = \frac{1}{k} \sum_{j=1}^k \delta_{x_j}.$$

For each $\tau \in \Sigma_k$ and $\pi \in \mathcal{E}_m$ we denote by $\tau(\pi) \in \mathcal{E}_m$ the partition consisting of the sets $\tau(\pi)_\ell := \tau(\pi_\ell)$ for $\ell \in \{1, \cdots, m\}$. Then, for any $\tau \in \Sigma_k$ and $\mathbf{x} = (x_1, \cdots, x_k)$, we get

$$\begin{aligned}
&\sum_{m=1}^k \sum_{\pi \in \mathcal{E}_m} 1_{\{(x_1,\cdots,x_k) \in \mathcal{A}(\pi)\}} \sum_{\sigma \in \Sigma_m} \prod_{\ell=1}^m \widetilde{\mathcal{G}}(x_{\pi_{\sigma(\ell)}} - x_{\pi_{\sigma(\ell-1)}}) \\
&= \sum_{y_1,\cdots,y_k \in A_N} 1_{\{L_k^{\mathbf{y}} = L_k^{\mathbf{x}}\}} 1_{\{\mathbf{x} = \mathbf{y} \circ \tau\}} \sum_{m=1}^k \sum_{\pi \in \mathcal{E}_m} 1_{\{(y_1,\cdots,y_k) \in \mathcal{A}(\pi)\}} \\
&\quad \times \sum_{\sigma \in \Sigma_m} \prod_{\ell=1}^m \widetilde{\mathcal{G}}(y_{\pi_{\sigma(\ell)}} - y_{\pi_{\sigma(\ell-1)}}).
\end{aligned}$$

Observe that, abbreviating $\mu = L_k^{\mathbf{x}}$ and assuming $L_k^{\mathbf{y}} = L_k^{\mathbf{x}}$, we have

$$\sum_{\tau \in \Sigma_k} 1_{\{\mathbf{x} = \mathbf{y} \circ \tau\}} = \prod_{x \in A_N} (k\mu(x))!,$$

and hence summing the previous expression over all permutations $\tau \in \Sigma_k$ gives

$$\begin{aligned}
&k! \sum_{m=1}^k \sum_{\pi \in \mathcal{E}_m} 1_{\{(x_1,\cdots,x_k) \in \mathcal{A}(\pi)\}} \sum_{\sigma \in \Sigma_m} \prod_{\ell=1}^m \widetilde{\mathcal{G}}(x_{\pi_{\sigma(\ell)}} - x_{\pi_{\sigma(\ell-1)}}) \\
&= \Big(\prod_{x \in A_N} (k\mu(x))!\Big) \sum_{y_1,\cdots,y_k \in A_N} 1_{\{L_k^{\mathbf{y}} = \mu\}} \sum_{m=1}^k \sum_{\pi \in \mathcal{E}_m} 1_{\{(y_1,\cdots,y_k) \in \mathcal{A}(\pi)\}} \\
&\quad \times \sum_{\sigma \in \Sigma_m} \prod_{\ell=1}^m \widetilde{\mathcal{G}}(y_{\pi_{\sigma(\ell)}} - y_{\pi_{\sigma(\ell-1)}}).
\end{aligned}$$

Write $\varphi_\mu(x) = \mu(x)^{1/q}$ for all $x \in A_N$. The summation over $y_1, \cdots, y_k$ on the right hand side is less than or equal to

$$\begin{aligned}
&\Big(\prod_{x\in A_N} \varphi_\mu(x)^{-k\mu(x)}\Big) \sum_{y_1,\cdots,y_k} \varphi_\mu(y_1)\cdots\varphi_\mu(y_k)\\
&\times \sum_{m=1}^k \sum_{\pi\in\mathcal{E}_m} 1_{\{(y_1,\cdots,y_k)\in\mathcal{A}(\pi)\}} \sum_{\sigma\in\Sigma_m}\prod_{\ell=1}^m \widetilde{\mathcal{G}}(y_{\pi_{\sigma(\ell)}} - y_{\pi_{\sigma(\ell-1)}})\\
&= \Big(\prod_{x\in A_N} \varphi_\mu(x)^{-k\mu(x)}\Big) \sum_{m=1}^k \sum_{\pi\in\mathcal{E}_m} \sum_{y_1,\dots,y_k} 1_{\{(y_1,\cdots,y_k)\in\mathcal{A}(\pi)\}}\\
&\times \Big\{\prod_{\ell=1}^m \varphi_\mu(y_{\pi_\ell})^{\#(\pi_\ell)}\Big\} \sum_{\sigma\in\Sigma_m}\prod_{\ell=1}^m \widetilde{\mathcal{G}}(y_{\pi_{\sigma(\ell)}} - y_{\pi_{\sigma(\ell-1)}}).
\end{aligned}$$

By the one-to-one correspondence between $\mathcal{A}(\pi)$ and $\mathbb{Z}^m$, the summation over m on the right hand side is equal to

$$\begin{aligned}
&\sum_{m=1}^k \sum_{\pi\in\mathcal{E}_m} \sum_{y_1,\cdots,y_m} \Big\{\prod_{\ell=1}^m \varphi_\mu(y_\ell)^{\#(\pi_\ell)}\Big\} \sum_{\sigma\in\Sigma_m}\prod_{\ell=1}^m \widetilde{\mathcal{G}}(y_{\sigma(\ell)} - y_{\sigma(\ell-1)})\\
&= \sum_{m=1}^k m! \sum_{y_1,\cdots,y_m} \Big\{\sum_{\pi\in\mathcal{E}_m}\prod_{\ell=1}^m \varphi_\mu(y_\ell)^{\#(\pi_\ell)}\Big\} \prod_{\ell=1}^m \widetilde{\mathcal{G}}(y_\ell - y_{\ell-1})\\
&= \sum_{m=1}^k \sum_{y_1,\cdots,y_m} \Big\{\sum_{\substack{j_1+\cdots+j_m=k\\ j_1,\cdots,j_m\ge 1}} \frac{k!}{j_1!\cdots j_m!}\prod_{\ell=1}^m \varphi_\mu(y_\ell)^{j_\ell}\Big\} \prod_{\ell=1}^m \widetilde{\mathcal{G}}(y_\ell - y_{\ell-1}).
\end{aligned}$$

Further, using Stirling's formula, there is a fixed polynomial $P(\cdot)$, depending only on the cardinality of A_N, such that

$$\Big\{\prod_{x\in A_N} (k\mu(x))!\Big\}\Big\{\prod_{x\in A_N} \varphi_\mu(x)^{-k\mu(x)}\Big\} \le P(k)\, k! \Big\{\prod_{j=1}^k \mu(x_j)^{\frac{1}{p}}\Big\}.$$

Summarizing what we have obtained so far, gives us

$$\begin{aligned}
&\sum_{m=1}^k \sum_{\pi\in\mathcal{E}_m} 1_{\{(x_1,\cdots,x_k)\in\mathcal{A}(\pi)\}} \sum_{\sigma\in\Sigma_m}\prod_{\ell=1}^m \widetilde{\mathcal{G}}(x_{\pi_{\sigma(\ell)}} - x_{\pi_{\sigma(\ell-1)}})\\
&\le P(k)\, k! \Big\{\prod_{j=1}^k \mu(x_j)^{\frac{1}{p}}\Big\} \sum_{m=1}^k \sum_{y_1,\cdots,y_m\in A_N}\\
&\times \Big\{\sum_{\substack{j_1+\cdots+j_m=k\\ j_1,\cdots,j_m\ge 1}} \frac{1}{j_1!\cdots j_m!}\prod_{\ell=1}^m \varphi_\mu^{j_\ell}(y_\ell)\Big\} \prod_{\ell=1}^m \widetilde{\mathcal{G}}(y_\ell - y_{\ell-1}).
\end{aligned}$$

Summing over all possible vectors $\mathbf{x} = (x_1, \cdots, x_k)$ gives

$$H_k \le P(k)^p\,(k!)^p\Big(\sum_{x_1,\cdots,x_k\in A_N}\prod_{j=1}^k \mu(x_j)\Big)\sup_{\nu\in\mathcal{P}_k(A_N)}\Big[\sum_{m=1}^k\sum_{y_1,\cdots,y_m\in A_N}$$
$$\times\Big(\sum_{\substack{j_1+\cdots+j_m=k\\ j_1,\cdots,j_m\ge 1}}\frac{1}{j_1!\cdots j_m!}\prod_{\ell=1}^m\varphi_\nu^{j_\ell}(y_\ell)\Big)\prod_{\ell=1}^m\widetilde{\mathcal{G}}(y_\ell-y_{\ell-1})\Big]^p,$$

where $\mathcal{P}_k(A_N)$ is the set of probability densities ν on A_N such that $\nu(x)$ is of the form i/k. Recall that $\varphi_\mu \ge 0$ satisfies $\|\varphi_\mu\|_q = 1$ and hence we may let the supremum take over all L^q-normalized non-negative functions f on A_N with $\|f\|_q = 1$.

Similar to the estimate in the proof of Theorem 3.1.2, there is a polynomial $P_1(\cdot)$ such that

$$P(k)^p\,(k!)^p\sum_{x_1,\cdots,x_k\in A_N}\prod_{j=1}^k\mu(x_j)\le \big(P_1(k)\big)^p$$

for all $k \ge 1$.

Thus, for every $\theta > 0$,

$$\frac{\theta^k}{k!}H_k^{\frac{1}{p}} \le P_1(k)\sup_{\|f\|_q=1}\sum_{n=1}^\infty\theta^n\sum_{m=1}^n\sum_{y_1,\cdots,y_m\in A_N}$$
$$\times\Big(\sum_{\substack{j_1+\cdots+j_m=k\\ j_1,\cdots,j_m\ge 1}}\frac{1}{j_1!\cdots j_m!}\prod_{\ell=1}^m f^{j_\ell}(y_\ell)\Big)\prod_{\ell=1}^m\widetilde{\mathcal{G}}(y_\ell-y_{\ell-1})$$
$$= P_1(k)\sup_{\|h\|_q=\theta}\sum_{m=1}^\infty\sum_{y_1,\cdots,y_m\in A_N}\Big\{\prod_{\ell=1}^m\sum_{j=1}^\infty\frac{h(y_\ell)^j}{j!}\Big\}\prod_{\ell=1}^m\widetilde{\mathcal{G}}(y_\ell-y_{\ell-1})$$
$$= P_1(k)\sup_{\|h\|_q=\theta}\sum_{m=1}^\infty\sum_{y_1,\cdots,y_m\in A_N}\prod_{\ell=1}^m\Big(e^{h(y_\ell)}-1\Big)\widetilde{\mathcal{G}}(y_\ell-y_{\ell-1}).$$

Similarly to (3.1.10), we have

$$\limsup_{m\to\infty}\frac{1}{m}\log\sup_{\|h\|_q=\theta}\sum_{y_1,\cdots,y_m\in A_N}\prod_{\ell=1}^m\Big(e^{h(y_\ell)}-1\Big)\widetilde{\mathcal{G}}(y_\ell-y_{\ell-1})$$
$$\le \log\sup_{\|h\|_q=\theta}\sup_{||g||_2=1}\Big|\langle g,\widetilde{\mathcal{G}}_h g\rangle\Big| = \log\sup_{\|h\|_q=\theta}||\widetilde{\mathcal{G}}_h||.$$

Suppose that $\theta < \mathcal{C}_N(S)$, then there exists $\epsilon > 0$ such that every h with $\|h\|_q = \theta$ satisfies $\|\widetilde{\mathcal{G}}_h\| < 1-\epsilon$. Consequently, there is a constant $C > 0$ such that

$$\frac{\theta^k}{k!}H_k^{\frac{1}{p}} \le CP_1(k)$$

and hence

$$\limsup_{k\to\infty}\frac{1}{k}\log\frac{1}{(k!)^p}H_k \le -p\log\theta\,,$$

as required to complete the proof. □

By (7.5.11) and Lemma 7.5.5, the next lemma leads to the upper bound of (7.5.9) and therefore completes the proof of Theorem 7.5.3.

LEMMA 7.5.6.

$$\liminf_{N\to\infty} \mathcal{C}_N(S) \geq \mathcal{C}(S).$$

Proof. Fix positive integers $M < N$. We decompose $\mathcal{G} = \mathcal{G}_+ + \mathcal{G}_-$ with $\mathcal{G}_+(x) = \mathcal{G}(x)1_{\{x\notin[M,M)^d\}}$ and $\mathcal{G}_-(x) = \mathcal{G}(x)1_{\{x\in[M,M)^d\}}$. In analogy to $\widetilde{\mathcal{G}}$ we define symmetric, periodic functions $\widetilde{\mathcal{G}}_+$ and $\widetilde{\mathcal{G}}_-$ by

$$\widetilde{\mathcal{G}}_\pm(y) = \Big\{\sum_{z\in\mathbb{Z}^d} \mathcal{G}_\pm^p(2Nz+y)\Big\}^{\frac{1}{p}} \qquad \text{for any } y\in\mathbb{Z}^d.$$

By the triangle inequality we have $\widetilde{\mathcal{G}} \leq \widetilde{\mathcal{G}}_+ + \widetilde{\mathcal{G}}_-$. The induced operators

$$\widetilde{\mathcal{G}}_{h,\pm} : l^2(A_N) \to l^2(A_N)$$

are defined by

$$\widetilde{\mathcal{G}}_{h,\pm} g(x) = \sqrt{e^{h(x)}-1}\sum_{y\in A_N} \widetilde{\mathcal{G}}_\pm(x-y)\sqrt{e^{h(y)}-1}\, g(y).$$

For the norms of these operators we thus obtain

$$\|\widetilde{\mathcal{G}}_h\| \leq \|\widetilde{\mathcal{G}}_{h,+}\| + \|\widetilde{\mathcal{G}}_{h,-}\|.$$

Similarly to (7.5.7), we have

$$\begin{aligned}\|\widetilde{\mathcal{G}}_{h,+}\| &\leq \Big(e^{||h||_q}-1\Big)\|\widetilde{\mathcal{G}}_+\|_p = \Big(e^{||h||_q}-1\Big)\Big\{\sum_{x\in\mathbb{Z}^d}\mathcal{G}_+^p(x)\Big\}^{1/p}\\ &= \Big(e^{||h||_q}-1\Big)\Big\{\sum_{x\notin[-M,M)^d}\mathcal{G}^p(x)\Big\}^{1/p}.\end{aligned}$$

Define

$$E_N = [-N,N)^d \setminus [-M,M)^d.$$

For any non-negative functions g and h on A_N, we extend them (without changing the notations) into periodic functions (with the period $2N$) on $\mathbb{Z}^d$. Similar to (3.2.14),

$$\begin{aligned}\|\widetilde{\mathcal{G}}_{h,-}\| &= \sup_{||g||_2=1}\langle g, \mathcal{G}_{h,-}g\rangle\\ &\leq \|\mathcal{G}_h\| + 2\Big(e^{||h||_q}-1\Big)\Big(\frac{|E_N|}{(2N)^d}\Big)^{\frac{p-1}{p}}\Big\{\sum_{x\in\mathbb{Z}^d}\mathcal{G}^p(x)\Big\}^{1/p}.\end{aligned}$$

Summarizing our argument, we have

$$\begin{aligned}\|\widetilde{\mathcal{G}}_h\| \leq \|\mathcal{G}_h\| + \Big(e^{||h||_q}-1\Big)\Bigg(&\Big\{\sum_{x\notin[-M,M)^d}\mathcal{G}^p(x)\Big\}^{1/p}\\ &+2\Big(\frac{|E_N|}{(2N)^d}\Big)^{\frac{p-1}{p}}\Big\{\sum_{x\in\mathbb{Z}^d}\mathcal{G}^p(x)\Big\}^{1/p}\Bigg).\end{aligned}\tag{7.5.12}$$

We now complete the proof: First, we may assume that

$$\liminf_{N\to\infty} \mathcal{C}_N(S) < \infty.$$

Hence, there is bounded sub-sequence of $\{\mathcal{C}_N(S)\}$. For simplicity of notation we may assume that the sub-sequence is $\{\mathcal{C}_N(S)\}$. For each N, we pick up a non-negative function h_N supported on A_N such that $\|\widetilde{\mathcal{G}}_{h_N}\| \ge 1$ and $||h_N||_q \le \mathcal{C}_N(S) + N^{-1}$. Take $M = \sqrt{N}$ in (7.5.12) and notice that $\{||h_N||_q\}$ is bounded. From (7.5.12) we have $\|\mathcal{G}_{h_N}\| \ge 1 - o(1)$ as $N \to \infty$.

For any $\theta > 0$, define

$$\mathcal{C}(S,\theta) = \inf\Big\{||h||_q;\quad h \ge 0 \text{ and } ||\mathcal{G}_h|| \ge \theta\Big\}$$

(so $\mathcal{C}(S,1) = \mathcal{C}(S)$). By our argument given above, we get

$$\liminf_{N\to\infty} \mathcal{C}_N(S) \ge \mathcal{C}(S, 1-\epsilon) \qquad \forall \epsilon > 0. \tag{7.5.13}$$

On the other hand, by the inequality[4]

$$\frac{1}{1-\epsilon}\Big(e^{\theta} - 1\Big) \le e^{\frac{\theta}{1-\epsilon}} - 1$$

and by the definition of the operator $\mathcal{G}_h$, one can see that

$$||\mathcal{G}_{(1-\epsilon)^{-1}h}|| \ge (1-\epsilon)^{-1}||\mathcal{G}_h||,$$

which implies that $\mathcal{C}(S, 1-\epsilon) \ge (1-\epsilon)\mathcal{C}(S)$. Bringing this back to (7.5.13) and letting $\epsilon \to 0^+$ we have proved Lemma 7.5.6. □

The story of J_∞ is quite different. It is shown in Khanin, Mazel, Shlosman and Sinai ([**108**]) that in the special case $p = 2$, $d \ge 5$, we have

$$\exp\Big\{-t^{\frac{d-2}{d}+\delta}\Big\} \le \mathbb{P}\{J_\infty \ge t\} \le \exp\Big\{-t^{\frac{d-2}{d}-\delta}\Big\} \tag{7.5.14}$$

for large $t > 0$, where $\delta > 0$ can be any previously given number.

The estimate made in (7.5.14) exhibits a behavior pattern drastically different from the one stated in Theorem 7.5.1 and Theorem 7.5.3. The discovery of Khanin *et al.* came as a surprise for there was a belief that in the high dimensions the intersection of independent ranges might behave like a constant multiple of the intersection local time.[5] Here is an explanation for that. For I_∞ to be large, the "optimal strategy" is to let the random walks stay inside a big but fixed ball, and repeatedly intersect at the same sites. While for J_∞ to be large, the random walks have to intersect at many different sites. Consequently, the random walks behave in a diffusive way.

Providing a sharp estimate for the tail probability of J_∞ is a challenging task, and the problem still remains open.

CONJECTURE 7.5.7. *When $p(d-2) > d$, there is a $\widetilde{\mathcal{C}}(d,p) > 0$ such that*

$$\lim_{t\to\infty} t^{-\frac{d-2}{d}} \log \mathbb{P}\{J_\infty \ge t\} = -\widetilde{\mathcal{C}}(d,p).$$

[4]This can be proved by comparing the Taylor expansion in θ on both sides.

[5]One may want to compare Theorem 7.2.1 and Theorem 7.2.2 in the case when $d = 3$ and $p = 2$.

See the "Notes and comments" section for the background of the conjecture. If the conjecture turns out to be right, it is of great interest to identify the constant $\widetilde{C}(d,p)$.

7.6. Notes and comments

Section 7.1

Theorem 7.1.1 and Theorem 7.1.2 were first established in [**27**] and [**28**], respectively. They are obviously linked to Theorem 4.1.6. The correspondent upper bounds are very likely.

A further problem is to study Theorem 7.1.2 in the critical cases $b_n = \log n$ (when $d=2$) and $n = n^{1/3}$ (when $d=3$). There is reason to believe that

$$\lim_{n\to\infty}\frac{1}{\log n}\log\mathbb{E}\exp\left\{\frac{(\log n)^2}{n}\sum_{x\in S[1,n]} f\left(\sqrt{\frac{\log n}{n}}x\right)\right\} = \Lambda_2(f) \tag{7.6.1}$$

when $d=2$ and that

$$\lim_{n\to\infty}\frac{1}{\sqrt[3]{n}}\log\mathbb{E}\exp\left\{\frac{1}{n^{2/3}}\sum_{x\in S[1,n]} f\left(\frac{x}{\sqrt[3]{n}}\right)\right\} = \Lambda_3(f) \tag{7.6.2}$$

when $d=3$, where f is reasonably nice and the functionals $\Lambda_2(f)$, $\Lambda_3(f)$ are non-trivial.

The investigation in this direction may lead to some interesting discoveries. Take $d=2$ and $f(x) = \theta 1_{D_1}(x)$, where $\theta>0$ and $D_r = \{x\in\mathbb{R}^2;\ \ |x|\le r\}$ is the disk with radius r. From (7.6.1) we might obtain that for $\theta>0$,

$$\lim_{n\to\infty}\frac{1}{\log n}\log\mathbb{E}\exp\left\{\theta\frac{(\log n)^2}{n}\#\left\{S[1,n]\cap D_{\sqrt{n/\log n}}\right\}\right\} = \Psi(\theta).$$

Write $R = \sqrt{n/\log n}$. In light of the Gärtner-Ellis theorem, it is expected that

$$\lim_{R\to\infty}\frac{1}{\log R}\log\mathbb{P}\left\{\#\{S[1,R^2\log R]\cap D_R\}\ge\lambda t^2\right\} = -I(\lambda).$$

Notice that the number of lattice points in D_R is roughly πR^2. So $I(\lambda)=\infty$ as $\lambda>\pi$. For any $0<\gamma\le 1$, let $Z_R(\gamma)$ be the time needed for the planar random walk $\{S(k)\}$ to cover the percentage γ of the lattice points in D_R. Our above analysis suggests a lower tail for $Z_R(\gamma)$ of the form

$$\lim_{R\to\infty}\frac{1}{\log R}\log\mathbb{P}\left\{Z_R(\gamma)\le R^2\log R\right\} = -I(\gamma\pi). \tag{7.6.3}$$

With $\gamma=1$, $Z_R = Z_R(1)$ is the cover time defined as the steps needed for the planar random walk $\{S(k)\}$ to cover D_R. The latest development on this famous model is the weak law

$$\lim_{R\to\infty}\mathbb{P}\left\{\log Z_R\le t(\log R)^2\right\} = e^{-4/t} \tag{7.6.4}$$

obtained by Dembo, Peres, Rosen and Zeitouni ([**49**]) (see the references therein for an historic account of this problem).

If it holds, (7.6.3) is sharply contrary to (7.6.4). It will be interesting to see what will happen to $I(\gamma\pi)$ when γ gets close to 1.

EXERCISE 7.6.1. *We show how Theorem 7.1.2 can be pushed further in the special case $f \equiv 1$.*

(1) *Prove that when $d = 2$,*

$$\lim_{n\to\infty} \frac{1}{b_n} \log \mathbb{E} \exp\left\{\theta \frac{b_n \log n}{n} \#\{S[1,n]\}\right\} = 2\pi\sqrt{\det(\Gamma)}\theta$$

for any $\theta > 0$ and any positive sequence b_n satisfying

$$b_n \longrightarrow \infty \quad \text{and} \quad \log b_n = o(\log n) \qquad (n \to \infty).$$

(2) *Prove that when $d \geq 3$,*

$$\lim_{n\to\infty} \frac{1}{b_n} \log \mathbb{E} \exp\left\{\theta \frac{b_n}{n} \#\{S[1,n]\}\right\} = \gamma_S \theta$$

for any $\theta > 0$ and any positive sequence b_n satisfying

$$b_n \longrightarrow \infty \quad \text{and} \quad b_n = o(n) \qquad (n \to \infty).$$

Hint: You may use (5.3.39) and sub-additivity.

Section 7.2

There does not seem to be an agreement in literature on the definition of moderate deviations. Here we view the moderate deviations as a kind of large deviations governed by some weak laws. In some other works such as [**11**], on the other hand, the words "moderate deviations" refer to the deviations away from the mean that are smaller than the scale of the mean.

Theorem 7.2.1 and Theorem 7.2.2 were obtained in [**27**] and [**28**], respectively. An improvement made here is the optimization of the condition on b_n in Theorem 7.2.1, thanks to Lemma 7.1.4.

A careful reader may notice that the case $d = 1$ is missing from Theorem 7.2.2. The reason behind it is that J_n has has a drastically different behavior in the dimension $d = 1$. In light of the weak convergence given Theorem 5.2.4, a large deviation of the Gaussian tail is expected when $d = 1$. Indeed, it was proved in [**30**] that for any $p \geq 1$ ($J_n = \#\{S[1,n]\} = R_n$ when $p = 1$), and for any $b_n \to \infty$ with $b_n = o(n)$, we get

$$\lim_{n\to\infty} \frac{1}{b_n} \log \mathbb{P}\{J_n \geq \lambda\sqrt{nb_n}\} = -\frac{p\lambda^2}{2\sigma^2} \qquad (\lambda > 0).$$

EXERCISE 7.6.2. *Complete the proof of Theorem 7.2.2.*

EXERCISE 7.6.3. *Here we propose an alternative solution to Corollary 3.3.4. The fact crucial to our approach is that*

$$\left(\int_{-\infty}^{\infty} \left[\prod_{j=1}^{p} f_j(x)\right] dx\right)^{1/p} \leq \frac{1}{p} \sum_{j=1}^{p} \left(\int_{-\infty}^{\infty} f_j^p(x) dx\right)^{1/p}$$

where $f_1, \cdots, f_p \in \mathcal{L}^p(\mathbb{R})$ *are non-negative functions.*

(1) *Recall (see (2.2.18)) that*

$$\alpha([0,t]^p) = \int_{-\infty}^{\infty} \Big[\prod_{j=1}^p L_j(t,x)\Big] dx$$

as $d = 1$. *Based on the above inequality and (4.2.2), establish an upper bound for the limit*

$$\limsup_{t\to\infty} \frac{1}{t} \log \mathbb{E} \exp\Big\{\Big(\alpha([0,t]^p)\Big)^{1/p}\Big\}$$

where $\theta > 0$. *Then transform it into an upper bound for*

$$\limsup_{t\to\infty} \frac{1}{t} \log\bigg(\sum_{m=0}^\infty \frac{\theta^m}{m!}\Big\{\mathbb{E}\Big[\alpha^m\big([0,t]^p\big)\Big]\Big\}^{1/p}\bigg).$$

(2) *Establish a lower bound for*

$$\liminf_{t\to\infty} \frac{1}{t} \log\bigg(\sum_{m=0}^\infty \frac{\theta^m}{m!}\Big\{\mathbb{E}\Big[\alpha^m\big([0,t]^p\big)\Big]\Big\}^{1/p}\bigg).$$

(3) *Combine the above results to establish Corollary 3.3.4.*

We now compare Theorem 7.2.2 to the work by van den Berg, Bolthausen and den Hollander ([**12**]) on the large deviations for the volume of *Wiener sausage* intersection. For $\epsilon > 0$, let $W^\epsilon(t)$ be the ϵ-neighborhood of the Brownian path $W(s)$ up to time t. Using Donsker-Varadhan large deviation for Wiener empirical measures ([**55**]), van den Berg, Bolthausen and den Hollander were able to establish the large deviations for the volume

$$|W^\epsilon(t) \cap \widetilde{W}^\epsilon(t)|$$

of the intersection of two independent Wiener sausages $W^\epsilon(t)$ and $\widetilde{W}^\epsilon(t)$. They proved that for any $c > 0$,

$$\lim_{t\to\infty} \frac{1}{\log t} \log \mathbb{P}\Big\{|W^\epsilon(ct) \cap \widetilde{W}^\epsilon(ct)| \ge \frac{t}{\log t}\Big\} = -I_2(c) \tag{7.6.5}$$

as $d = 2$; and

$$\lim_{t\to\infty} t^{-\frac{d-2}{d}} \log \mathbb{P}\Big\{|W^\epsilon(ct) \cap \widetilde{W}^\epsilon(ct)| \ge t\Big\} = -I_d(c) \tag{7.6.6}$$

as $d \ge 3$, where $I_d(c)$ is given in the form of variation. Further, van den Berg, Bolthausen and den Hollander pointed out that

$$\lim_{c\to\infty} c^{\frac{4-d}{d}} I_d(c) = \begin{cases} (2\pi)^{-2}\mu_d & d = 2, \\ \theta_d(\epsilon)^{-4/d}\mu_d & d = 3, 4, \end{cases} \tag{7.6.7}$$

where

$$\theta_d(\epsilon) = \epsilon^{d-2} 2\pi^{d/2} / \Gamma\Big(\frac{d-2}{2}\Big), \tag{7.6.8}$$

$$\mu_d = \begin{cases} \inf\Big\{||\nabla\psi||_2^2;\ \ \psi\in W^{1,2}(\mathbb{R}^d),\ ||\psi||_2=1,\ ||\psi||_4=1\Big\} & d=2,3,\\ \inf\Big\{||\nabla\psi||_2^2;\ \ \psi\in D(\mathbb{R}^4),\ ||\psi||_4=1\Big\} & d=4,\end{cases}$$

$$W^{1,2}(\mathbb{R}^d)=\big\{f\in\mathcal{L}^2(\mathbb{R}^d);\ \ \nabla f\in\mathcal{L}^2(\mathbb{R}^d)\big\},$$

and

$$D(\mathbb{R}^d)=\big\{f\ge 0;\ \ \nabla f\in\mathcal{L}^2(\mathbb{R}^d)\big\}.$$

It is not hard to see that $\mu_d=\kappa(d,2)^{-8/d}$ for $d\le 4$. On the other hand, the Brownian scaling implies that

$$|W^\epsilon(ct)\cap\widetilde{W}^\epsilon(ct)|\overset{d}{=}c^{d/2}|W^{\epsilon/\sqrt{c}}(t)\cap\widetilde{W}^{\epsilon/\sqrt{c}}(t)|.$$

In the case $d=2$, the above discussion leads to

$$\lim_{c\to\infty}\lim_{t\to\infty}\frac{c}{\log t}\log\mathbb{P}\Big\{|W^{\epsilon/\sqrt{c}}(t)\cap\widetilde{W}^{\epsilon/\sqrt{c}}(t)|\ge\frac{t}{c\log t}\Big\}=-(2\pi)^{-2}\kappa(2,2)^{-4}.$$

This is related to (7.2.5) in Theorem 7.2.2 with $p=2$, where b_n corresponds to $c^{-1}\log t$.

In the case $d=3$,

$$\log\mathbb{P}\Big\{|W^{\epsilon/\sqrt{c}}(t)\cap\widetilde{W}^{\epsilon/\sqrt{c}}(t)|\ge c^{-3/2}t\Big\}\sim-\Big(\frac{t^{1/3}}{c}\Big)\theta_3(\epsilon/\sqrt{c})^{-4/3}\kappa(3,2)^{-8/3}$$

as the limit is taken in the order $t\to\infty$, $c\to\infty$. Clearly, this is related to (7.2.6) in Theorem 7.2.2 with b_n corresponding to $c^{-1}t^{1/3}$.

Section 7.3.

EXERCISE 7.6.4. *Complete the proof of the law of the iterated logarithm given in (7.3.1).*

Section 7.4.

Part of the support for Conjecture 7.4.2 comes from van den Berg, Bolthausen and den Hollander's work ([**12**]). Taking $d=4$ in (7.6.6), we get

$$\log\mathbb{P}\Big\{|W^{\epsilon/\sqrt{c}}(t)\cap\widetilde{W}^{\epsilon/\sqrt{c}}(t)|\ge c^{-2}t\Big\}\sim-\Big(\frac{\sqrt{t}}{c}\Big)\theta_4(\epsilon/\sqrt{c})^{-1}\kappa(4,2)^{-2}$$

in the limiting order $t\to\infty$, $c\to\infty$. This is a version of (7.4.8) with b_n corresponding to $\sqrt{t}/c$.

In addition, we point out a very recent work by Castell [**25**] for some partial progress in this direction.

Section 7.5

The tail probabilities of I_∞ and J_∞ in the special case $d\ge 5$ and $p=2$ were first investigated by Khanin, Mazal, Shlosman and Sinai ([**108**]). Theorem 7.5.3 responded to the demand on the sharp tail asymptotics for I_∞, a problem left by Khanin, Mazel, Shlosman and Sinai.

Perhaps the most important progress since the paper [**108**] was made by van den Berg, Bolthausen and den Hollander ([**12**]). They pointed out that when $d \geq 5$, there is a $c_d > 0$ such that the rate function $I_d(\cdot)$ given in (7.6.6) satisfies

$$I_d(c) = \theta_d(\epsilon)^{-1} \inf\left\{ ||\nabla\psi||_2^2;\ \ \psi \in D(\mathbb{R}^d),\ \int_{\mathbb{R}^d} \left[1 - e^{-\psi^2(x)}\right]^2 dx = 1 \right\}$$

for $c \geq c_d$. Consequently,

$$\lim_{t\to\infty} t^{-\frac{d-2}{d}} \log \mathbb{P}\Big\{ |W^\epsilon(ct) \cap \widetilde{W}^\epsilon(ct)| \geq t \Big\} = -I_d(c_d)$$

as $c \geq c_d$. Based on this, van den Berg, Bolthausen and den Hollander ([**12**]) conjectured that

$$\lim_{t\to\infty} t^{-\frac{d-2}{d}} \log \mathbb{P}\Big\{ |W^\epsilon(\infty) \cap \widetilde{W}^\epsilon(\infty)| \geq t \Big\} = -I_d(c_d).$$

There is a noticeable similarity between this conjecture and Conjecture 7.5.7.

EXERCISE 7.6.5 (hard). *The inequality (7.5.4) reminds us of Exercise 3.4.3. Here we propose an alternative treatment which does not need Le Gall's moment identity. Let* $\epsilon > 0$ *be fixed and let the random measure* $\alpha_\epsilon(A)$ *be defined by (2.2.4). Let* $\tau_1, \cdots, \tau_p$ *be i.i.d. exponential times with parameter 1 and assume the independence between* $\{\tau_1, \cdots, \tau_p\}$ *and* $\{W_1(t), \cdots, W_p(t)\}$. *Prove that for any integers* $m, n \geq 1$,

$$\mathbb{E}\Big[\alpha_\epsilon\big([0,\tau_1] \times \cdots \times [0,\tau_p]\big)^{m+n}\Big]$$
$$\leq \binom{m+n}{n}^p \mathbb{E}\Big[\alpha_\epsilon\big([0,\tau_1] \times \cdots \times [0,\tau_p]\big)^m\Big] \mathbb{E}\Big[\alpha_\epsilon\big([0,\tau_1] \times \cdots \times [0,\tau_p]\big)^n\Big].$$

Now extend it to the intersection local time $\alpha(A)$ *when* $p(d-2) < d$.

EXERCISE 7.6.6 (hard). *Recall that a Lévy process* $X(t)$ *is an* $\mathbb{R}^d$*-valued stochastic process with independent and stationary increments (see Bertoin* [**13**] *for a systematic account of the Lévy process). Lévy processes appear as a continuous-time version of random walks. Here we consider the case where* $X(t)$ *takes lattice values. Assume that* $X(t)$ *is non-degenerate, symmetric and square integrable; and that the smallest sub-group of* $\mathbb{Z}^d$ *that supports* $X(t)$ *is* $\mathbb{Z}^d$ *itself. Let* $p \geq 2$ *be an integer satisfying* $p(d-2) > d$ *and write*

$$G_\infty(x) = \int_0^\infty \mathbb{P}\{X(t) = x\} dt \qquad x \in \mathbb{Z}^d.$$

Similarly to what we concluded in Exercise 5.6.6, we have

$$\sum_{x\in\mathbb{Z}^d} G_\infty^p(x) < \infty.$$

Let $X_1(t), \cdots, X_p(t)$ *be independent copies of* $X(t)$. *In this exercise we establish a continuous-time version of Theorem 7.5.3.*

(1) *Prove that*

$$\chi \equiv \int_0^\infty \cdots \int_0^\infty 1_{\{X_1(s_1)=\cdots=X_p(s_p)\}} ds_1 \cdots ds_p < \infty \qquad a.s.$$

(2) *Prove that*

$$\lim_{t\to\infty} t^{-1/p}\log\mathbb{P}\{\chi\ge t\} = -\frac{p}{\varrho}$$

where

$$\varrho = \sup\Big\{\sum_{x,y\in\mathbb{Z}^d} G_\infty(x-y)f(x)f(y);\ \sum_{x\in\mathbb{Z}^d}|f(x)|^{\frac{2p}{2p-1}} = 1\Big\}.$$

EXERCISE 7.6.7 (hard). *Assume that $p(d-2)>d$.*

Prove that for any $k\ge 1$,

$$\mathbb{E}J_\infty^k = \sum_{m=1}^k \frac{1}{m!}\Big\{\sum_{\substack{j_1+\cdots+j_m=k\\ j_1,\cdots,j_m\ge 1}} \frac{k!}{j_1!\cdots j_m!}\Big\} \sum_{\substack{x_1,\cdots,x_m\in\mathbb{Z}^d\\ \text{distinct}}} \Big[\mathbb{P}\{\{x_1,\cdots,x_m\}\subset S[1,\infty)\}\Big]^p.$$

Hint: Read the proof of (7.5.10).

CHAPTER 8

Single random walk: large deviations

In this chapter we establish the large deviations for the self-intersection local time and the range of a single d-dimensional random walk. Comparing with the previous chapter, our main theorems in this chapter are more dimension-dependent. Indeed, the results in the dimensions $d = 1$ and $d = 2$ appear as the moderate deviations in connection to the large deviations for Brownian intersection local times, which are stated in Theorem 4.2.1, Theorem 4.3.1 and Theorem 4.3.2. When $d = 3, 4$, the Brownian self-intersection local times cannot be defined even under renormalization. Further, the central limit theorems given in Theorem 5.5.3 suggest the moderate deviations with Gaussian tails for $Q_n - \mathbb{E}Q_n$ and $R_n - \mathbb{E}R_n$, at least when the deviation scale b_n is sufficiently small. On the other hand, the validity of Galiardo-Nirenberg inequality (when $d = 3$) and Sobolev inequality (when $d = 4$) indicates that some large deviations with non-Gaussian tails may hold for a larger deviation scale. Finally, the lack of Galiardo-Nirenberg inequality or Sobolev inequality makes the dimensions $d \geq 5$ very different from the case $d \leq 4$.

8.1. Self-intersection in one dimension

Let $S(n)$ be a symmetric random walk on $\mathbb{Z}$ with variance $\sigma^2 > 0$. Assume that the smallest sub-group of $\mathbb{Z}$ that supports $S(n)$ is $\mathbb{Z}$. Recall that $l(n, x)$ is the local time of $S(n)$ defined in (5.2.1). By (5.2.2) and (5.2.3), the study of the large deviations for the p-multiple self-intersection local time is equivalent to the study of the large deviations for the p-square of the local time given by

$$\sum_{x\in\mathbb{Z}} l^p(n,x) \qquad n = 1, 2, \cdots .$$

The following is the main theorem of this section.

THEOREM 8.1.1. *Let $p > 1$ be fixed, then*

$$\lim_{n\to\infty} \frac{1}{b_n} \log \mathbb{P}\Big\{ \sum_{x\in\mathbb{Z}} l^p(n,x) \geq n^{\frac{p+1}{2}} b_n^{\frac{p-1}{2}} \Big\}$$
$$= -\frac{\sigma^2}{4(p-1)} \Big(\frac{p+1}{2}\Big)^{\frac{3-p}{p-1}} B\Big(\frac{1}{p-1}, \frac{1}{2}\Big)$$

for any positive sequence b_n satisfying

$$b_n \longrightarrow \infty \text{ and } b_n = o(n) \qquad (n \to \infty)$$

where $B(\cdot, \cdot)$ is the beta function.

Proof. By Corollary 1.2.5 and by Theorem C.1 in the Appendix, it suffices to show that for any $\theta > 0$,

(8.1.1)
$$\lim_{n\to\infty}\frac{1}{b_n}\log\mathbb{E}\exp\left\{\theta\Big(\frac{b_n}{n}\Big)^{\frac{p+1}{2p}}\Big(\sum_{x\in\mathbb{Z}}l^p(n,x)\Big)^{1/p}\right\}$$
$$=\sup_{g\in\mathcal{F}_1}\left\{\sigma^{-\frac{p-1}{p}}\theta\Big(\int_{-\infty}^{\infty}|g(x)|^{2p}dx\Big)^{1/p}-\frac{1}{2}\int_{-\infty}^{\infty}|g'(x)|^2dx\right\}.$$

Here we recall that

$$\mathcal{F}_1=\left\{g\in\mathcal{L}^2(\mathbb{R});\ \int_{-\infty}^{\infty}g^2(x)dx=1\text{ and }\int_{-\infty}^{\infty}|g'(x)|^2dx<\infty\right\}.$$

Let $q > 1$ be the conjugate of p and let $f(x)$ be a bounded, locally supported continuous function on $(-\infty,\infty)$ such that $||f||_q = 1$. By Hölder inequality, we get

$$\Big(\sum_{x\in\mathbb{Z}}l^p(n,x)\Big)^{1/p}\ge C_n^{-1/q}\sum_{x\in\mathbb{Z}}f\Big(\sqrt{\frac{b_n}{n}}x\Big)l(n,x)=C_n^{-1/q}\sum_{k=1}^{n}f\Big(\sqrt{\frac{b_n}{n}}S(k)\Big)$$

where

$$C_n=\sum_{x\in\mathbb{Z}}\Big|f\Big(\sqrt{\frac{b_n}{n}}x\Big)\Big|^q\sim\sqrt{\frac{n}{b_n}}\int_{-\infty}^{\infty}|f(x)|^qdx=\sqrt{\frac{n}{b_n}}\qquad(n\to\infty).$$

By Feynman-Kac minoration given in Theorem 7.1.1, we get

$$\liminf_{n\to\infty}\frac{1}{b_n}\log\mathbb{E}\exp\left\{\theta\Big(\frac{b_n}{n}\Big)^{\frac{p+1}{2p}}\Big(\sum_{x\in\mathbb{Z}}l^p(n,x)\Big)^{1/p}\right\}$$
$$\ge\sup_{g\in\mathcal{F}_1}\left\{\theta\int_{-\infty}^{\infty}f(\sigma x)g^2(x)dx-\frac{1}{2}\int_{-\infty}^{\infty}|g'(x)|^2dx\right\}.$$

Taking supremum over f leads to the lower bound

(8.1.2)
$$\liminf_{n\to\infty}\frac{1}{b_n}\log\mathbb{E}\exp\left\{\theta\Big(\frac{b_n}{n}\Big)^{\frac{p+1}{2p}}\Big(\sum_{x\in\mathbb{Z}}l^p(n,x)\Big)^{1/p}\right\}$$
$$\ge\sup_{g\in\mathcal{F}_1}\left\{\sigma^{-\frac{p-1}{p}}\theta\Big(\int_{-\infty}^{\infty}|g(x)|^{2p}dx\Big)^{1/p}-\frac{1}{2}\int_{-\infty}^{\infty}|g'(x)|^2dx\right\}.$$

On the other hand, by the triangular inequality, we get

$$\Big(\sum_{x\in\mathbb{Z}}l^p(m+n,x)\Big)^{1/p}$$
$$\le\Big(\sum_{x\in\mathbb{Z}}l^p(m,x)\Big)^{1/p}+\Big(\sum_{x\in\mathbb{Z}}\big[l(m+n,x)-l(m,x)\big]^p\Big)^{1/p}$$

for any integers $m, n \ge 1$. Observe that

$$\sum_{x\in\mathbb{Z}}\big[l(m+n,x)-l(m,x)\big]^p\overset{d}{=}\sum_{x\in\mathbb{Z}}l^p(n,x)$$

and that the left hand side is independent of $\{S(1), \cdots, S(m)\}$. Hence, the random sequence

$$\Big(\sum_{x\in\mathbb{Z}} l^p(n,x)\Big)^{1/p} \qquad n = 1, 2, \cdots$$

is sub-additive.

Let $t > 0$ be fixed for a moment and write $t_n = [tn/b_n]$, $\gamma_n = [n/b_n]$. By sub-additivity, we get

$$\begin{aligned}
&\mathbb{E}\exp\Big\{\theta\Big(\frac{b_n}{n}\Big)^{\frac{p+1}{2p}}\Big(\sum_{x\in\mathbb{Z}} l^p(n,x)\Big)^{1/p}\Big\}\\
&\le \Big(\mathbb{E}\exp\Big\{\theta\Big(\frac{b_n}{n}\Big)^{\frac{p+1}{2p}}\Big(\sum_{x\in\mathbb{Z}} l^p(t_n,x)\Big)^{1/p}\Big\}\Big)^{\gamma_n+1}.
\end{aligned}$$

Replacing n by t_n in Theorem 5.2.3, gives us

$$\Big(\frac{b_n}{n}\Big)^{\frac{p+1}{2p}}\Big(\sum_{x\in\mathbb{Z}} l^p(t_n,x)\Big)^{1/p} \xrightarrow{d} \sigma^{-\frac{p-1}{p}}\Big(\int_{-\infty}^{\infty} L^p(t,x)dx\Big)^{1/p}$$

where $L(t,x)$ is the local time of a 1-dimensional Brownian motion. By Theorem 1.3.3 (with n being replaced by t_n), the sub-additivity also implies that

$$\sup_{n\ge 1}\mathbb{E}\exp\Big\{\lambda\Big(\frac{b_n}{n}\Big)^{\frac{p+1}{2p}}\Big(\sum_{x\in\mathbb{Z}} l^p(t_n,x)\Big)^{1/p}\Big\} < \infty \qquad \lambda > 0.$$

Therefore,

$$\begin{aligned}
&\mathbb{E}\exp\Big\{\theta\Big(\frac{b_n}{n}\Big)^{\frac{p+1}{2p}}\Big(\sum_{x\in\mathbb{Z}} l^p(t_n,x)\Big)^{1/p}\Big\}\\
&\longrightarrow \mathbb{E}\exp\Big\{\theta\sigma^{-\frac{p-1}{p}}\Big(\int_{-\infty}^{\infty} L^p(t,x)dx\Big)^{1/p}\Big\} \qquad (n\to\infty).
\end{aligned}$$

Summarizing our argument, we have

$$\begin{aligned}
&\limsup_{n\to\infty}\frac{1}{b_n}\log\mathbb{E}\exp\Big\{\theta\Big(\frac{b_n}{n}\Big)^{\frac{p+1}{2p}}\Big(\sum_{x\in\mathbb{Z}} l^p(n,x)\Big)^{1/p}\Big\}\\
&\le \frac{1}{t}\log\mathbb{E}\exp\Big\{\theta\sigma^{-\frac{p-1}{p}}\Big(\int_{-\infty}^{\infty} L^p(t,x)dx\Big)^{1/p}\Big\}.
\end{aligned}$$

By (4.2.3), letting $t\to\infty$ on the right hand side leads to the upper bound

$$\begin{aligned}
&\limsup_{n\to\infty}\frac{1}{b_n}\log\mathbb{E}\exp\Big\{\theta\Big(\frac{b_n}{n}\Big)^{\frac{p+1}{2p}}\Big(\sum_{x\in\mathbb{Z}} l^p(n,x)\Big)^{1/p}\Big\}\\
&\le \sup_{g\in\mathcal{F}_1}\Big\{\sigma^{-\frac{p-1}{p}}\theta\Big(\int_{-\infty}^{\infty}|g(x)|^{2p}dx\Big)^{1/p} - \frac{1}{2}\int_{-\infty}^{\infty}|g'(x)|^2dx\Big\}.
\end{aligned}$$

Combining this with the lower bound given in (8.1.2) gives us (8.1.1). □

Some ideas used in the proof of (8.1.1) can be partially applied to the multi-dimensional setting. In the rest of the section $S(n)$ is a symmetric, square integrable random walk supported by $\mathbb{Z}^d$ with $d \ge 1$. Let $\epsilon > 0$ be fixed and write

$$l(n,x,\epsilon) = \Big(\frac{b_n}{n}\Big)^{d/2} \det(\Gamma)^{-1/2} \sum_{k=1}^{n} p_\epsilon\Big(\sqrt{\frac{b_n}{n}}\Gamma^{-1/2}\big(S(k)-x\big)\Big), \tag{8.1.3}$$

$$R(n,x,\epsilon) = \Big(\frac{b_n}{n}\Big)^{d/2} \det(\Gamma)^{-1/2} \sum_{y\in S[1,n]} p_\epsilon\Big(\sqrt{\frac{b_n}{n}}\Gamma^{-1/2}(y-x)\Big). \tag{8.1.4}$$

THEOREM 8.1.2. *For any $\theta > 0$ and $p > 1$,*

$$\lim_{n\to\infty} \frac{1}{b_n} \log \mathbb{E} \exp\Big\{\theta\Big(\frac{b_n}{n}\Big)^{\frac{2p-d(p-1)}{2p}} \Big(\sum_{x\in\mathbb{Z}^d} l^p(n,x,\epsilon)\Big)^{1/p}\Big\} \tag{8.1.5}$$
$$= \sup_{g\in\mathcal{F}_d} \Big\{\theta \det(\Gamma)^{-\frac{p-1}{2p}} \Big(\int_{\mathbb{R}^d} |g(x,\epsilon)|^{2p} dx\Big)^{1/p} - \frac{1}{2}\int_{\mathbb{R}^d} |\nabla g(x)|^2 dx\Big\}$$

for any positive sequence b_n satisfying

$$b_n \longrightarrow \infty \text{ and } b_n = o(n) \quad (n\to\infty)$$

where

$$g(x,\epsilon) = \Big\{\int_{\mathbb{R}^d} g^2(y) p_\epsilon(y-x)dy\Big\}^{1/2}.$$

THEOREM 8.1.3. *Let $\theta > 0$ and $p \ge 1$.*

(1) *When $d = 2$,*

$$\lim_{n\to\infty} \frac{1}{b_n} \log \mathbb{E} \exp\Big\{\theta\Big(\frac{b_n}{n}\Big)^{\frac{1}{p}} \log n \Big(\sum_{x\in\mathbb{Z}^2} R^p(n,x,\epsilon)\Big)^{1/p}\Big\}$$
$$= \sup_{g\in\mathcal{F}_2} \Big\{2\pi\theta \det(\Gamma)^{\frac{1}{2p}} \Big(\int_{\mathbb{R}^2} |g(x,\epsilon)|^{2p} dx\Big)^{1/p} - \frac{1}{2}\int_{\mathbb{R}^2} |\nabla g(x)|^2 dx\Big\}$$

for any positive sequence b_n satisfying

$$b_n \longrightarrow \infty \text{ and } b_n = o\big(\log n\big) \quad (n\to\infty).$$

(2) *When $d \ge 3$,*

$$\lim_{n\to\infty} \frac{1}{b_n} \log \mathbb{E} \exp\Big\{\theta\Big(\frac{b_n}{n}\Big)^{\frac{2p-d(p-1)}{2p}} \Big(\sum_{x\in\mathbb{Z}^d} R^p(n,x,\epsilon)\Big)^{1/p}\Big\}$$
$$= \sup_{g\in\mathcal{F}_d} \Big\{\gamma_S\theta \det(\Gamma)^{-\frac{p-1}{2p}} \Big(\int_{\mathbb{R}^d} |g(x,\epsilon)|^{2p} dx\Big)^{1/p} - \frac{1}{2}\int_{\mathbb{R}^d} |\nabla g(x)|^2 dx\Big\}$$

for any positive sequence b_n satisfying

$$b_n \longrightarrow \infty \text{ and } b_n = o\big(n^{1/3}\big) \quad (n\to\infty).$$

Here we recall that $\gamma_S = \mathbb{P}\{S(k) \ne 0 \ \ \forall n \ge 1\}$.

The approach for the lower bounds in Theorem 8.1.2 and Theorem 8.1.3 are the same as the correspondent argument given in the proof of Theorem 8.1.1. For the upper bounds, the only difference is that here we use Theorem 4.2.3 instead of (4.2.3). We omit the details of the proof here.

8.2. Self-intersection in $d = 2$

In multi-dimensions, we work with the double self-intersection local times (i.e., $p = 2$). Recall that

$$Q_n = \sum_{1\le j<k\le n} 1_{\{S(j)=S(k)\}} \qquad n = 1, 2, \cdots$$

where $\{S(k)\}_{k\ge 1}$ is a symmetric random walk in $\mathbb{Z}^d$ with covariance matrix Γ, and the smallest sub-group of $\mathbb{Z}^d$ that supports the random walk $\{S(k)\}_{k\ge 1}$ is $\mathbb{Z}^d$ itself. Due to the concentration phenomena when $d \ge 2$, we are interested in the renormalized self-intersection local time $Q_n - \mathbb{E}Q_n$.

In this section we let $d = 2$. By Theorem 5.4.2, $n^{-1}(Q_n - \mathbb{E}Q_n)$ weakly converges to $\det(\Gamma)^{-1/2}\gamma([0,1]^2_<)$. Here we recall that

$$\begin{aligned}\gamma([0,t]^2_<) = &\iint_{\{0\le r<s\le t\}} \delta_0\big(W(r) - W(s)\big)drds \\ &- \mathbb{E}\iint_{\{0\le r<s\le t\}} \delta_0\big(W(r) - W(s)\big)drds\end{aligned}$$

is the renormalized self-intersection local time run by the planar Brownian motion $W(t)$. Hence, the tail behavior of the type given in Theorem 4.3.1 and Theorem 4.3.2 is naturally expected for $n^{-1}(Q_n - \mathbb{E}Q_n)$. The following theorem is the main result in this section.

THEOREM 8.2.1. *For any $\lambda > 0$ and for any positive sequence with*

$$b_n \longrightarrow \infty \quad \textit{and} \quad b_n = o(n) \quad (n\to\infty), \tag{8.2.1}$$

we have

$$\lim_{n\to\infty} \frac{1}{b_n} \log \mathbb{P}\Big\{Q_n - \mathbb{E}Q_n \ge \lambda n b_n\Big\} = -\lambda\sqrt{\det(\Gamma)}\kappa(2,2)^{-4} \tag{8.2.2}$$

where $\kappa(2,2) > 0$ is the best constant of the Gagliardo-Nirenberg inequality defined in (4.3.4).

Proof. By (6.3.19) in Theorem 6.3.3 and by Chebyshev inequality, one can see that

$$\lim_{n\to\infty} \frac{1}{b_n} \log \mathbb{P}\Big\{ - (Q_n - \mathbb{E}Q_n) \ge \lambda n b_n\Big\} = -\infty$$

for every $\lambda > 0$. Consequently, (8.2.2) is equivalent to

$$\lim_{n\to\infty} \frac{1}{b_n} \log \mathbb{P}\Big\{|Q_n - \mathbb{E}Q_n| \ge \lambda n b_n\Big\} = -\lambda\sqrt{\det(\Gamma)}\kappa(2,2)^{-4}.$$

By Corollary 1.2.5 (with $p = 2$), it suffices to show that

$$\lim_{n\to\infty} \frac{1}{b_n} \log \mathbb{E} \exp\left\{\theta\sqrt{\frac{b_n}{n}}|Q_n - \mathbb{E}Q_n|^{1/2}\right\} = \frac{1}{4}\theta^2 \det(\Gamma)^{-1/2}\kappa(2,2)^4 \qquad \forall\theta > 0. \tag{8.2.3}$$

According to Theorem 6.3.2, there is a $c > 0$ such that

$$\sup_{n\geq 1} \mathbb{E} \exp\left\{\frac{c}{n}|Q_n - \mathbb{E}Q_n|\right\} < \infty.$$

A standard argument of Chebyshev type shows that there is a constant $C_Q > 0$ such that

$$\limsup_{n\to\infty} \frac{1}{b_n} \log \mathbb{E} \exp\left\{\theta\sqrt{\frac{b_n}{n}}|Q_n - \mathbb{E}Q_n|^{1/2}\right\} \leq C_Q\theta^2 \tag{8.2.4}$$

for every sequence b_n satisfying (8.2.1) and for every $\theta > 0$. In the following we strengthen (8.2.4) into (8.2.3).

One of the key ideas is the following decomposition:

$$Q_n = \sum_{j=1}^{2^N} \eta_j + \sum_{j=1}^{N}\sum_{k=1}^{2^{j-1}} \xi_{j,k} \tag{8.2.5}$$

where N is a large but fixed number,

$$\eta_j = \sum_{(j-1)n2^{-N}<i<i'\leq jn2^{-N}} 1_{\{S(i)=S(i')\}},$$

$$\xi_{j,k} = \sum_{\substack{(2k-2)n2^{-j}<i\leq(2k-1)n2^{-j}\\ (2k-1)n2^{-j}<i'\leq(2k)n2^{-j}}} 1_{\{S(i)=S(i')\}}.$$

The decomposition follows from a geometric idea similar to the one exploited in Figure 1 of Section 2.4. Essentially, our approach here is the argument of triangular approximation.

The sequence $\{\eta_j\}_{1\leq j\leq 2^N}$ is independent with

$$\eta_j \stackrel{d}{=} Q_{n_j} \qquad j = 1,\cdots 2^N \tag{8.2.6}$$

where

$$n_j = \#((j-1)n2^{-N}, jn2^{-N}] = [2^{-N}n] \text{ or } [2^{-N}n] \pm 1.$$

Here and elsewhere, we sometimes use the notation $(a, b]$ for, without notification when it becomes obvious, the set of the integers in that interval.

For each $1 \leq j \leq N$, $\{\xi_{j,k}\}_{1\leq k\leq 2^{j-1}}$ is an independent sequence with

$$\xi_{j,k} \stackrel{d}{=} \sum_{i=0}^{n_{j,k}-1}\sum_{i'=1}^{n'_{j,k}} 1_{S(i)=\widetilde{S}(i')\}} \qquad k = 1,\cdots, 2^{j-1} \tag{8.2.7}$$

where $\{\widetilde{S}(k)\}$ is an independent copy of the random walk $\{S(k)\}$ and

$$n_{j,k} = \#((2k-1)n2^{-j},\ (2k)n2^{-j}] = 2^{-j}n \text{ or } 2^{-j}n \pm 1,$$

$$n'_{j,k} = \#((2k)n2^{-j}\ (2k+1)n2^{-j}] = 2^{-j}n, \text{ or } 2^{-j}n \pm 1.$$

Recall that the smoothed local time $l(n, x, \epsilon)$ is defined by

$$l(n,x,\epsilon) = \det(\Gamma)^{-1/2}\Big(\frac{b_n}{n}\Big)\sum_{k=1}^n p_\epsilon\Big(\sqrt{\frac{b_n}{n}}\Gamma^{-1/2}\big(S(k)-x\big)\Big).$$

According to Theorem 8.1.2 (with $d = p = 2$), for each $\epsilon > 0$,

$$\lim_{n\to\infty}\frac{1}{b_n}\log\mathbb{E}\exp\Big\{\theta\sqrt{\frac{b_n}{n}}\Big(\sum_{x\in\mathbb{Z}^2} l^2(n,x,\epsilon)\Big)^{1/2}\Big\} \tag{8.2.8}$$
$$= \sup_{g\in\mathcal{F}_2}\Big\{\theta\det(\Gamma)^{-1/4}\Big(\int_{\mathbb{R}^2}|g(x,\epsilon)|^4dx\Big)^{1/2} - \frac{1}{2}\int_{\mathbb{R}^2}|\nabla g(x)|^2dx\Big\}$$

where

$$\mathcal{F}_2 = \Big\{g\in\mathcal{L}^2(\mathbb{R}^2);\ \ ||g||_2 = 1 \ \text{ and } \ \nabla g\in\mathcal{L}^2(\mathbb{R}^2)\Big\}.$$

In the following we approximate (8.2.3) by (8.1.5). Define

$$\xi_{j,k}(\epsilon) = \det(\Gamma)^{-1}\Big(\frac{b_n}{n}\Big)^2\sum_{x\in\mathbb{Z}^2}\sum_{\substack{(2k-2)n2^{-j}<i\le(2k-1)n2^{-j}\\ (2k-1)n2^{-j}<i'\le(2k)n2^{-j}}}$$
$$\times p_\epsilon\Big(\sqrt{\frac{b_n}{n}}\Gamma^{-1/2}\big(S(i)-x\big)\Big)p_\epsilon\Big(\sqrt{\frac{b_n}{n}}\Gamma^{-1/2}\big(S(i')-x\big)\Big)$$

where $k = 1,\cdots,2^{j-1}$ and $j = 1,\cdots,N$.

Define

$$\eta_j(\epsilon) = \frac{1}{\sqrt{\det(\Gamma)}}\frac{b_n}{n}\Big\{\sum_{x\in\mathbb{Z}^2}\Big[\sum_{(j-1)\frac{n}{2^N}<i\le j\frac{n}{2^N}} p_\epsilon\Big(\sqrt{\frac{b_n}{n}}\Gamma^{-1/2}\big(S(i)-x\big)\Big)\Big]^2\Big\}^{1/2}$$

where $j = 1,\cdots,2^N$.

Correspondent to (8.2.5) we have that

$$\sum_{x\in\mathbb{Z}^2} l^2(n,x,\epsilon) = \sum_{j=1}^{2^N}\eta_j^2(\epsilon) + 2\sum_{j=1}^N\sum_{k=1}^{2^{j-1}}\xi_{j,k}(\epsilon). \tag{8.2.9}$$

In addition, $\eta_1(\epsilon),\cdots,\eta_{2^N}(\epsilon)$ are independent with

$$\eta_j(\epsilon) \overset{d}{=} \frac{1}{\sqrt{\det(\Gamma)}}\frac{b_n}{n}\Big\{\sum_{x\in\mathbb{Z}^2}\Big[\sum_{i=1}^{n_j}p_\epsilon\Big(\sqrt{\frac{b_n}{n}}\Gamma^{-1/2}\big(S(i)-x\big)\Big)\Big]^2\Big\}^{1/2}. \tag{8.2.10}$$

For each $1 \le j \le N$, $1 \le k \le 2^{j-1}$,

$$\xi_{j,k} - \xi_{j,k}(\epsilon) \stackrel{d}{=} \sum_{x\in\mathbb{Z}^2} \sum_{\substack{0\le i\le n_{j,k}-1 \\ 1\le i'\le n'_{j,k}}} 1_{\{S(i)=\widetilde{S}(i')=x\}} \tag{8.2.11}$$
$$- \det(\Gamma)^{-1}\Big(\frac{b_n}{n}\Big)^2 \sum_{x\in\mathbb{Z}^2} \sum_{\substack{0\le i\le n_{j,k}-1 \\ 1\le i'\le n'_{j,k}}} p_\epsilon\Big(\sqrt{\frac{b_n}{n}}\Gamma^{-1/2}\big(S(i)-x\big)\Big)$$
$$\times\, p_\epsilon\Big(\sqrt{\frac{b_n}{n}}\Gamma^{-1/2}\big(\widetilde{S}(i')-x\big)\Big)$$

where $n_j, n_j, n_{j,k}, n'_{j,k}$ are defined the same as above.

By (8.2.4) and (8.2.6) for each $1 \le j \le 2^N$, we get

$$\limsup_{n\to\infty} \frac{1}{b_n} \log \mathbb{E}\exp\Big\{\theta\sqrt{\frac{b_n}{n}}|\eta_j - \mathbb{E}\eta_j|^{1/2}\Big\} \le 2^{-N}C_Q\theta^2.$$

According to part (b), Theorem 1.2.2, for any $\theta > 0$,

$$\limsup_{n\to\infty} \frac{1}{b_n} \log \mathbb{E}\exp\Big\{\theta\sqrt{\frac{b_n}{n}}\Big(\sum_{j=1}^{2^N}|\eta_j - \mathbb{E}\eta_j|\Big)^{1/2}\Big\} \le 2^{-N}C_Q\theta^2. \tag{8.2.12}$$

By (8.2.8) and (8.2.10), we get

$$\limsup_{n\to\infty} \frac{1}{b_n} \log \mathbb{E}\exp\Big\{\theta\sqrt{\frac{b_n}{n}}\eta_j(\epsilon)\Big\}$$
$$\le 2^{-N}\sup_{g\in\mathcal{F}_2}\Big\{\theta\det(\Gamma)^{-1/4}\Big(\int_{\mathbb{R}^2}|g(x,\epsilon)|^4dx\Big)^{1/2} - \frac{1}{2}\int_{\mathbb{R}^2}|\nabla g(x)|^2dx\Big\}$$
$$\le 2^{-N}\sup_{g\in\mathcal{F}_2}\Big\{\theta\det(\Gamma)^{-1/4}\Big(\int_{\mathbb{R}^2}|g(x)|^4dx\Big)^{1/2} - \frac{1}{2}\int_{\mathbb{R}^2}|\nabla g(x)|^2dx\Big\}$$

where the last step follows from Jensen inequality.

Taking $d = p = 2$ in Theorem C.1 in the Appendix, for each $\theta > 0$,

$$\sup_{g\in\mathcal{F}_2}\Big\{\theta\Big(\int_{\mathbb{R}^2}|g(x)|^4dx\Big)^{1/2} - \frac{1}{2}\int_{\mathbb{R}^2}|\nabla g(x)|^2dx\Big\} = 2\kappa(2,2)^4\theta^2. \tag{8.2.13}$$

Hence,

$$\limsup_{n\to\infty} \frac{1}{b_n} \log \mathbb{E}\exp\Big\{\theta\sqrt{\frac{b_n}{n}}\eta_j(\epsilon)\Big\} \le 2^{-(N-1)}\det(\Gamma)^{-1/2}\kappa(2,2)^4\theta^2.$$

By part (b), Theorem 1.2.2 again, for any $\theta > 0$, we get

$$\limsup_{n\to\infty} \frac{1}{b_n} \log \mathbb{E}\exp\Big\{\theta\sqrt{\frac{b_n}{n}}\Big(\sum_{j=1}^{2^N}\eta_j(\epsilon)\Big)\Big\} \tag{8.2.14}$$
$$\le 2^{-(N-1)}\det(\Gamma)^{-1/2}\kappa(2,2)^4\theta^2.$$

The bounds in (8.2.12) and (8.2.14) indicate that for large N, it is the second summations in the decompositions (8.2.5) and (8.2.9) that play a dominating role. By (8.2.11) and by the exponential approximation established in Lemma 8.2.2 below, we get

$$\lim_{\epsilon\to 0^+}\limsup_{n\to\infty}\frac{1}{b_n}\log\mathbb{E}\exp\left\{\theta\sqrt{\frac{b_n}{n}}|\xi_{j,k}-\xi_{j,k}(\epsilon)|^{1/2}\right\}=0.$$

For any fixed but arbitrary N and for any $\theta > 0$,

$$\lim_{\epsilon\to 0^+}\limsup_{n\to\infty}\frac{1}{b_n}\log\mathbb{E}\exp\left\{\theta\sqrt{\frac{b_n}{n}}\Big(\sum_{j=1}^{N}\sum_{k=1}^{2^{j-1}}|\xi_{j,k}-\xi_{j,k}(\epsilon)|\Big)^{1/2}\right\}=0. \tag{8.2.15}$$

In addition, from (8.2.7) one can see that

$$\sum_{j=1}^{N}\sum_{k=1}^{2^{j-1}}\mathbb{E}\xi_{j,k}=O(n) \qquad (n\to\infty).$$

Combining this with (8.2.12), (8.2.14) and (8.2.15), a standard argument by Hölder inequality gives

$$\lim_{\epsilon\to 0^+}\limsup_{n\to\infty}\frac{1}{b_n}\log\mathbb{E}\exp\left\{\theta\sqrt{\frac{b_n}{n}}\Big|(Q_n-\mathbb{E}Q_n)-\frac{1}{2}\sum_{x\in\mathbb{Z}^2}l^2(n,x,\epsilon)\Big|^{1/2}\right\}=0.$$

This, together with (8.2.6) (with θ being replaced by $\theta/\sqrt{2}$), gives us

$$\begin{aligned}&\lim_{n\to\infty}\frac{1}{b_n}\log\mathbb{E}\exp\left\{\theta\sqrt{\frac{b_n}{n}}|Q_n-\mathbb{E}Q_n|^{1/2}\right\}\\&=\lim_{\epsilon\to 0^+}\sup_{g\in\mathcal{F}_2}\left\{\frac{\theta}{\sqrt{2}}\det(\Gamma)^{-1/4}\Big(\int_{\mathbb{R}^2}|g(x,\epsilon)|^4dx\Big)^{1/2}-\frac{1}{2}\int_{\mathbb{R}^2}|\nabla g(x)|^2dx\right\}.\end{aligned}$$

Here the existence of the limit (as $\epsilon\to 0^+$) on the right hand side is briefly demonstrated as follows: By the Jensen inequality,

$$\begin{aligned}&\sup_{g\in\mathcal{F}_2}\left\{\frac{\theta}{\sqrt{2}}\det(\Gamma)^{-1/4}\Big(\int_{\mathbb{R}^2}|g(x,\epsilon)|^4dx\Big)^{1/2}-\frac{1}{2}\int_{\mathbb{R}^2}|\nabla g(x)|^2dx\right\}\\&\le\sup_{g\in\mathcal{F}_2}\left\{\frac{\theta}{\sqrt{2}}\det(\Gamma)^{-1/4}\Big(\int_{\mathbb{R}^2}|g(x)|^4dx\Big)^{1/2}-\frac{1}{2}\int_{\mathbb{R}^2}|\nabla g(x)|^2dx\right\}.\end{aligned}$$

On the other hand, for any $f\in\mathcal{F}_2$,

$$\begin{aligned}&\sup_{g\in\mathcal{F}_2}\left\{\frac{\theta}{\sqrt{2}}\det(\Gamma)^{-1/4}\Big(\int_{\mathbb{R}^2}|g(x,\epsilon)|^4dx\Big)^{1/2}-\frac{1}{2}\int_{\mathbb{R}^2}|\nabla g(x)|^2dx\right\}\\&\ge\frac{\theta}{\sqrt{2}}\det(\Gamma)^{-1/4}\Big(\int_{\mathbb{R}^2}|f(x,\epsilon)|^4dx\Big)^{1/2}-\frac{1}{2}\int_{\mathbb{R}^2}|\nabla f(x)|^2dx.\end{aligned}$$

By Gagliardo-Nirenberg inequality defined in (4.3.4), $\mathcal{F}_2\subset\mathcal{L}^4(\mathbb{R}^2)$. Applying Lemma 2.2.2 to f^2 we have that

$$\lim_{\epsilon\to 0^+}\int_{\mathbb{R}^2}|f(x,\epsilon)|^4dx=\int_{\mathbb{R}^d}|f(x)|^4dx.$$

Letting $\epsilon \to 0^+$ on both sides and then taking supremum over f on the right hand side of the previous display we have

$$\begin{aligned}&\liminf_{\epsilon\to 0^+}\sup_{g\in\mathcal{F}_2}\Big\{\frac{\theta}{\sqrt{2}}\det(\Gamma)^{-1/4}\Big(\int_{\mathbb{R}^2}|g(x,\epsilon)|^4dx\Big)^{1/2}-\frac{1}{2}\int_{\mathbb{R}^2}|\nabla g(x)|^2dx\Big\}\\ &\ge \sup_{f\in\mathcal{F}_2}\Big\{\frac{\theta}{\sqrt{2}}\det(\Gamma)^{-1/4}\Big(\int_{\mathbb{R}^2}|f(x)|^4dx\Big)^{1/2}-\frac{1}{2}\int_{\mathbb{R}^2}|\nabla f(x)|^2dx\Big\}.\end{aligned}$$

Finally, the desired (8.2.3) follows from (8.2.13). □

As we have seen in the above proof, our argument substantially relies on approximating the intersection local time of two independent random walks by its smoothed version. Here we provide support to such an approach.

Write

$$I_n = \sum_{x\in\mathbb{Z}^2} l(n,x)\widetilde{l}(n,x)$$

and for any $\epsilon > 0$, write

$$I_n(\epsilon) = \sum_{x\in\mathbb{Z}^2} l(n,x,\epsilon)\widetilde{l}(n,x,\epsilon)$$

where $l(n,x)$ and $\widetilde{l}(n,x)$ are the local times of the independent 2-dimensional random walks $\{S(k)\}$ and $\{\widetilde{S}(k)\}$, respectively, and $\widetilde{l}(n,x,\epsilon)$ is generated by $\{\widetilde{S}(k)\}$ in the same way as $l(n,x,\epsilon)$ is generated by $\{S(k)\}$.

Lemma 8.2.2. *For any $\theta > 0$,*

$$\lim_{\epsilon\to 0^+}\limsup_{n\to\infty}\frac{1}{b_n}\log\mathbb{E}\exp\Big\{\theta\sqrt{\frac{b_n}{n}}|I_n - I_n(\epsilon)|^{1/2}\Big\} = 0.$$

Proof. Define

$$\begin{aligned}\bar{I}_n(\epsilon) = K_{n,\epsilon}^{-2}\sum_{x\in\mathbb{Z}^2}&\Big[\sum_{k=1}^n p_\epsilon\Big(\sqrt{\frac{b_n}{n}}\Gamma^{-1/2}(S(k)-x)\Big)\Big]\\ &\times\Big[\sum_{k=1}^n p_\epsilon\Big(\sqrt{\frac{b_n}{n}}\Gamma^{-1/2}(\widetilde{S}(k)-x)\Big)\Big]\end{aligned}$$

where

$$K_{n,\epsilon} = \sum_{y\in\mathbb{Z}^2} p_\epsilon\Big(\sqrt{\frac{b_n}{n}}\Gamma^{-1/2}y\Big) \sim \sqrt{\det(\Gamma)}\frac{n}{b_n} \qquad (n\to\infty).$$

One can see that for each $\epsilon > 0$ there is a constant sequence $C_n \to 1$ such that $I_n(\epsilon) = C_n\bar{I}_n(\epsilon)$.

From (7.2.3) (with $d = p = 2$), one can see that

$$\limsup_{n\to\infty}\frac{1}{b_n}\log\mathbb{E}\exp\Big\{\theta\sqrt{\frac{b_n}{n}}I_n^{1/2}\Big\} < \infty \qquad \forall\theta > 0.$$

Consequently, it suffices to establish that

$$\lim_{\epsilon\to 0^+}\limsup_{n\to\infty}\frac{1}{b_n}\log\mathbb{E}\exp\left\{\theta\sqrt{\frac{b_n}{n}}|I_n-\bar{I}_n(\epsilon)|^{1/2}\right\}=0. \tag{8.2.16}$$

By Taylor expansion and by Cauchy-Schwarz inequality, we get

$$\begin{aligned}&\mathbb{E}\exp\left\{\theta\sqrt{\frac{b_n}{n}}|I_n-\bar{I}_n(\epsilon)|^{1/2}\right\}\\&\le\sum_{m=0}^{\infty}\frac{\theta^m}{m!}\Big(\frac{b_n}{n}\Big)^{m/2}\Big\{\mathbb{E}|I_n-\bar{I}_n(\epsilon)|^m\Big\}^{1/2}.\end{aligned}$$

By the fact (see the representation (6.1.18)) that $\mathbb{E}\big[I_n-\bar{I}_n(\epsilon)\big]^m\ge 0$ for all integers $m\ge 0$, and the fact that the terms with odd power in the summation on the right hand side can be dominated properly by the terms with even power through Cauchy-Schwarz inequality, we need only to show that

$$\lim_{\epsilon\to 0^+}\limsup_{n\to\infty}\frac{1}{b_n}\log\left(\sum_{m=0}^{\infty}\frac{\theta^m}{m!}\Big(\frac{b_n}{n}\Big)^{m/2}\Big\{\mathbb{E}\big[I_n-\bar{I}_n(\epsilon)\big]^m\Big\}^{1/2}\right)=0. \tag{8.2.17}$$

For sufficiently large n, let $0=n_0<n_1<\cdots<n_{[b_n]}=n$ be an integer partition of $[0,n]$ with $n_k-n_{k-1}=[n/b_n]$ or $[n/b_n]+1$ for each $1\le k\le[b_n]$. Taking the additive functional $\mathcal{F}_k^x(S)=l(n_k,x)$ $(k=1,\cdots,[b_n])$ on the increasing sequence $n_1<\cdots<n_{[b_n]}$ and the probability density function

$$h(x)=K_{n,\epsilon}^{-1}p_\epsilon\Big(\sqrt{\frac{b_n}{n}}\Gamma^{-1/2}x\Big)$$

in (6.1.15), Theorem 6.1.2, we get

$$\begin{aligned}&\sum_{m=0}^{\infty}\frac{\theta^m}{m!}\Big(\frac{b_n}{n}\Big)^{m/2}\Big\{\mathbb{E}\big[I_n-\bar{I}_n(\epsilon)\big]^m\Big\}^{1/2}\\&\le\prod_{k=1}^{[b_n]}\sum_{m=0}^{\infty}\frac{\theta^m}{m!}\Big(\frac{b_n}{n}\Big)^{m/2}\Big\{\mathbb{E}\big|I_{n_k-n_{k-1}}-\bar{I}_{n_k-n_{k-1},\epsilon}\big|^m\Big\}^{1/2},\end{aligned} \tag{8.2.18}$$

where, in connection to (5.3.1) (with $d=p=2$),

$$\begin{aligned}\bar{I}_{n_k-n_{k-1},\epsilon}&\equiv K_{n,\epsilon}^{-2}\sum_{x\in\mathbb{Z}^2}\Big[\sum_{j=1}^{n_k-n_{k-1}}p_\epsilon\Big(\sqrt{\frac{b_n}{n}}\Gamma^{-1/2}\big(S(j)-x\big)\Big)\Big]\\&\quad\times\Big[\sum_{j=1}^{n_k-n_{k-1}}p_\epsilon\Big(\sqrt{\frac{b_n}{n}}\Gamma^{-1/2}\big(\widetilde{S}(j)-x\big)\Big)\Big]\\&=\big(1+o(1)\big)I_{n_k-n_{k-1},\epsilon}\qquad(n\to\infty).\end{aligned}$$

Replacing n by n_k-n_{k-1} in Theorem 5.3.1, we have

$$\lim_{\epsilon\to 0^+}\limsup_{n\to\infty}\Big(\frac{b_n}{n}\Big)^2\mathbb{E}\Big[I_{n_k-n_{k-1}}-\bar{I}_{n_k-n_{k-1},\epsilon}\Big]^2=0.$$

Further, (6.2.1) in Theorem 6.2.1 (with $d = p = 2$) ensures that the dominated convergence theorem applies to our setting so that

$$\lim_{\epsilon\to 0^+}\limsup_{n\to\infty}\sum_{m=0}^\infty \frac{\theta^m}{m!}\Big(\frac{b_n}{n}\Big)^{m/2}\Big\{\mathbb{E}\big|I_{n_k-n_{k-1}}-\bar{I}_{n_k-n_{k-1},\epsilon}\big|^m\Big\}^{1/2}=1.$$

Obviously, the limit holds uniformly for $1 \le k \le [b_n]$. Consequently, the requested (8.2.17) follows from (8.2.18). $\square$

The lower tail asymptotics established in Theorem 4.3.2 can be partially extended to $Q_n - \mathbb{E}Q_n$.

THEOREM 8.2.3. *Suppose that* $\mathbb{E}|S(1)|^{2+\delta} < \infty$ *for some* $\delta > 0$. *Then there exist* $C_1, C_2 > 0$ *such that for any positive sequence* b_n *satisfying*

$$b_n \longrightarrow \infty \quad \text{and} \quad \log b_n = o(\log n) \qquad (n \to \infty)$$

and for any $\lambda > 0$,

$$\limsup_{n\to\infty}\frac{1}{b_n^\lambda}\log\mathbb{P}\Big\{Q_n-\mathbb{E}Q_n\le -\lambda\frac{n\log b_n}{2\pi\sqrt{\det(\Gamma)}}\Big\}\le -C_1,$$

$$\liminf_{n\to\infty}\frac{1}{b_n^\lambda}\log\mathbb{P}\Big\{Q_n-\mathbb{E}Q_n\le -\lambda\frac{n\log b_n}{2\pi\sqrt{\det(\Gamma)}}\Big\}\ge -C_2.$$

This result was established in [**7**], Theorem 1.2. We omit its proof.

8.3. LDP of Gaussian tail in $d = 3$

According to Theorem 5.5.3, $(Q_n - \mathbb{E}Q_n)/\sqrt{n\log n}$ weakly converges to a normal distribution when $d = 3$. Naturally, a large deviation principle (also called moderate deviation principle for its connection to the weak law) is expected at least for small deviation scale b_n. Throughout this section we let $d = 3$. The following is the main result.

THEOREM 8.3.1. *For any positive sequence*

$$b_n \longrightarrow \infty \quad \text{and} \quad b_n = o\big(\sqrt{\log n}\big) \qquad (n \to \infty)$$

and for any $\lambda > 0$,

$$\lim_{n\to\infty}\frac{1}{b_n}\log\mathbb{P}\Big\{\pm(Q_n-\mathbb{E}Q_n)\ge\lambda\sqrt{nb_n\log n}\Big\}=-\lambda^2\pi^2\det(\Gamma). \tag{8.3.1}$$

Proof. Our starting point is again the triangular decomposition (8.2.5). Here we take $N = [\gamma^{-1}\log_2 b_n]$ for large n, where $\log_2(\cdot)$ is the logarithmic function with the base 2, and $0 < \gamma < 1/2$ is a constant which will be specified later. Unlike the argument in the proof of Theorem 8.2.1, we shall show that it is the first sum in the decomposition (the diagonal part) that makes a major contribution.

Write

$$\bar{\eta}_j = \eta_j - \mathbb{E}\eta_j \quad \text{and} \quad \bar{\xi}_{j,k} = \xi_{j,k} - \mathbb{E}\xi_{j,k}.$$

We first prove that for any $L > 0$ there is a constant $a > 0$ such that

$$\limsup_{n\to\infty} \frac{1}{b_n} \log \mathbb{P}\Big\{\Big|\sum_{j=1}^N \sum_{k=1}^{2^{j-1}} \bar{\xi}_{j,k}\Big| \geq a\sqrt{nb_n^3}\Big\} \leq -L. \tag{8.3.2}$$

By the fact that $2^{-\gamma}+\cdots+2^{-\gamma N} \leq 2^{-\gamma}(1-2^{-\gamma})^{-1}$ and by triangular inequality, for any $a > 0$ we get

$$\begin{aligned} &\mathbb{P}\Big\{\Big|\sum_{j=1}^N \sum_{k=1}^{2^{j-1}} \bar{\xi}_{j,k}\Big| \geq a\sqrt{nb_n^3}\Big\} \\ &\leq \sum_{j=1}^N \mathbb{P}\Big\{\Big|\sum_{k=1}^{2^{j-1}} \bar{\xi}_{j,k}\Big| \geq a2^\gamma(1-2^{-\gamma})2^{-\gamma j}\sqrt{nb_n^3}\Big\}. \end{aligned} \tag{8.3.3}$$

Let $C > 0$ be a fixed but arbitrary number and set

$$\xi'_{j,k} = \bar{\xi}_{j,k} 1_{\{|\bar{\xi}_{j,k}| \leq C\sqrt{n2^{-j}b_n^3}\}} - \mathbb{E}\bar{\xi}_{j,k} 1_{\{|\bar{\xi}_{j,k}| \leq C\sqrt{n2^{-j}b_n^3}\}}$$

and $\xi''_{j,k} = \bar{\xi}_{j,k} - \xi'_{j,k}$. Notice that uniformly on $1 \leq j \leq N$, we have

$$\sum_{k=1}^{2^{j-1}} \mathbb{E}\xi_{j,k} = o\big(n\sqrt{2^{-j}b_n^3}\big).$$

Thus, for any $\delta > 0$,

$$\mathbb{P}\Big\{\Big|\sum_{k=1}^{2^{j-1}} \xi''_{j,k}\Big| \geq \delta\sqrt{n2^{-j}b_n^3}\Big\} \leq \sum_{1\leq k\leq 2^{j-1}} \mathbb{P}\Big\{|\bar{\xi}_{j,k}| \geq C\sqrt{n2^{-j}b_n^3}\Big\},$$

as n is sufficiently large.

Taking $d = 3$ and $p = 2$ in (6.2.1), Theorem 6.2.1, by the in-law-identity (8.2.7) we have

$$\sup_{n,j,k} \mathbb{E}\exp\Big\{\frac{\theta}{(n2^{-j})^{1/3}}|\bar{\xi}_{j,k}|^{2/3}\Big\} < \infty \tag{8.3.4}$$

for some $\theta > 0$. Using Chebyshev inequality in a standard way, one can show that for any $L > 0$, there is $C > 0$ such that

$$\limsup_{n\to\infty} \frac{1}{b_n} \log \sup_{j,k} \mathbb{P}\Big\{|\bar{\xi}_{j,k}| \geq C\sqrt{n2^{-j}b_n^3}\Big\} \leq -L.$$

Consequently, by the fact that $2^{j-1} \leq 2^N \leq b_n^{\gamma^{-1}}$, we get

$$\limsup_{n\to\infty} \frac{1}{b_n} \log \max_{1\leq j\leq N} \mathbb{P}\Big\{\Big|\sum_{k=1}^{2^{j-1}} \xi''_{j,k}\Big| \geq \delta\sqrt{n2^{-j}b_n^3}\Big\} \leq -L \tag{8.3.5}$$

for any $\delta > 0$.

By independence, we get

$$\mathbb{E}\exp\Big\{\frac{2^{\gamma j}\theta}{\sqrt{nb_n}} \sum_{k=1}^{2^{j-1}} \xi'_{j,k}\Big\} = \prod_{k=1}^{2^{j-1}} \mathbb{E}\exp\Big\{\frac{2^{\gamma j}\theta}{\sqrt{nb_n}} \xi'_{j,k}\Big\}.$$

For each $1 \le k \le 2^{j-1}$,

$$|\xi'_{j,k}| \le 2C\sqrt{n2^{-j}b_n^3}.$$

Using (8.3.4), there is a $u > 0$ such that

$$\sup_{j,k,n} \mathbb{E}\exp\left\{\frac{u}{\sqrt{n2^{-j}b_n}}|\xi'_{j,k}|\right\} \le \sup_{j,k,n} \mathbb{E}\exp\left\{\frac{u(2C)^{1/3}}{(n2^{-j})^{1/3}}|\xi'_{j,k}|^{2/3}\right\} < \infty.$$

By Taylor expansion, there is a $M > 0$ such that when $|\theta|$ is sufficiently small,

$$\mathbb{E}\exp\left\{\frac{2^{\gamma j}\theta}{\sqrt{nb_n}}\xi'_{j,k}\right\} \le \exp\left\{\frac{M2^{2\gamma j}\theta^2}{nb_n}\mathbb{E}\xi_{j,k}^2\right\}.$$

In view of (8.2.7) and (5.3.14) (with $d = 3$ and $p = 2$ and n being replaced by $2^{-j}n$), $\mathbb{E}\xi_{j,k}^2 = O(n2^{-j})$. Consequently, there is a constant $M' > 0$ such that

$$\max_{1\le j\le N} \mathbb{E}\exp\left\{\frac{\theta}{\sqrt{nb_n}}\sum_{k=1}^{2^{j-1}}\xi'_{j,k}\right\} \le \exp\left\{\frac{M'\theta^2 2^{2\gamma j}}{b_n}\right\} \le \exp\left\{\frac{M'\theta^2 2^{2\gamma N}}{b_n}\right\}. \tag{8.3.6}$$

By the fact that $2^{2\gamma N} \sim b_n^2$, and by Chebyshev inequality, for any $L > 0$ there is a constant $a > 0$ such that

$$\limsup_{n\to\infty}\frac{1}{b_n}\log\max_{1\le j\le N}\mathbb{P}\left\{\left|\sum_{k=1}^{2^{j-1}}\xi'_{j,k}\right| \ge a2^{-\gamma j}\sqrt{nb_n^3}\right\} \le -L.$$

Combining this with (8.3.5) gives us

$$\limsup_{n\to\infty}\frac{1}{b_n}\log\max_{1\le j\le N}\mathbb{P}\left\{\left|\sum_{k=1}^{2^{j-1}}\bar{\xi}_{j,k}\right| \ge (a+\delta)2^{-\gamma j}\sqrt{nb_n^3}\right\} \le -L. \tag{8.3.7}$$

Thus, (8.3.2) follows from (8.3.3).

Notice that $\sqrt{nb_n^3} = o\big(\sqrt{nb_n\log n}\big)$.[1] Consequently, for any $\epsilon > 0$,

$$\lim_{n\to\infty}\frac{1}{b_n}\log\mathbb{P}\left\{\left|\sum_{j=1}^{N}\sum_{k=1}^{2^{j-1}}\bar{\xi}_{j,k}\right| \ge \epsilon\sqrt{nb_n\log n}\right\} = -\infty.$$

To prove (8.3.1), therefore, it remains to show that

$$\lim_{n\to\infty}\frac{1}{b_n}\log\mathbb{P}\left\{\pm\sum_{j=1}^{2^N}\bar{\eta}_j \ge \lambda\sqrt{nb_n\log n}\right\} = -\lambda^2\pi^2\det(\Gamma). \tag{8.3.8}$$

Write

$$\eta'_j = \bar{\eta}_j 1\left\{|\bar{\eta}_j| \le \sqrt{2^{-N}nb_n^3\log n}\right\} - \mathbb{E}\bar{\eta}_j 1\left\{|\bar{\eta}_j| \le \sqrt{2^{-N}nb_n^3\log n}\right\},$$

$$\eta''_j = \bar{\eta}_j 1\left\{|\bar{\eta}_j| > \sqrt{2^{-N}nb_n^3\log n}\right\} - \mathbb{E}\bar{\eta}_j 1\left\{|\bar{\eta}_j| > \sqrt{2^{-N}nb_n^3\log n}\right\}.$$

By the in-law identity (8.2.6) and by Theorem 6.4.1, we get

$$\sup_{j,n}\mathbb{E}\exp\left\{\frac{\theta}{(2^{-N}n\log n)^{1/3}}|\eta_j|^{2/3}\right\} < \infty \qquad \forall\theta > 0. \tag{8.3.9}$$

---

[1] This is the only place in the whole proof where the assumption $b_n = o(\sqrt{\log n})$ is needed.

Replacing $\left\{\bar{\xi}_{N,k};\ 1 \le k \le 2^{N-1}\right\}$ by the sequence $\left\{\bar{\eta}_j/\sqrt{\log n};\ 1 \le j \le 2^N\right\}$ in the argument used for (8.3.5) gives that for any $L > 0$ and $\delta > 0$, we have

$$\limsup_{n\to\infty} \frac{1}{b_n} \log \mathbb{P}\left\{\left|\sum_{j=1}^{2^N} \eta_j''\right| \ge \delta\sqrt{2^{-N} n b_n^3 \log n}\right\} \le -L.$$

We now make $\gamma < 1/4$. Then

$$\sqrt{2^{-N} n b_n^3 \log n} = o\left(\sqrt{\frac{n\log n}{b_n}}\right) \qquad (n \to \infty). \tag{8.3.10}$$

In particular,

$$\lim_{n\to\infty} \frac{1}{b_n} \log \mathbb{P}\left\{\left|\sum_{j=1}^{2^N} \eta_j''\right| \ge \epsilon\sqrt{n b_n \log n}\right\} = -\infty.$$

Therefore, it remains to show that

$$\lim_{n\to\infty} \frac{1}{b_n} \log \mathbb{P}\left\{ \pm \sum_{j=1}^{2^N} \eta_j' \ge \lambda\sqrt{n b_n \log n}\right\} = -\lambda^2\pi^2 \det(\Gamma). \tag{8.3.11}$$

By Theorem 1.1.4, all we need is to establish

$$\lim_{n\to\infty} \frac{1}{b_n} \log \mathbb{E}\exp\left\{\theta\sqrt{\frac{b_n}{n\log n}} \sum_{j=1}^{2^N} \eta_j'\right\} = \frac{\theta^2}{(2\pi)^2\det(\Gamma)} \tag{8.3.12}$$

for every real number θ.

By independence, Taylor expansion and (8.3.10), we get

$$\begin{aligned}\mathbb{E}\exp\left\{\theta\sqrt{\frac{b_n}{n\log n}} \sum_{j=1}^{2^N} \eta_j'\right\} &= \prod_{j=1}^{2^N} \mathbb{E}\exp\left\{\theta\sqrt{\frac{b_n}{n\log n}}\eta_j'\right\} \\ &= \prod_{j=1}^{2^N} \exp\left\{\big(1+o(1)\big)\frac{\theta^2 b_n}{2n\log n}\mathbb{E}(\eta_j')^2\right\} \qquad (n\to\infty).\end{aligned}$$

Observe from (8.2.6) that uniformly on $1 \le j \le 2^N$, we have

$$\begin{aligned}\mathbb{E}|\bar{\eta}_j| &\sim \mathbb{E}|Q_{[2^{-N}n]} - \mathbb{E}Q_{[2^{-N}n]}| \\ &= O\big(\sqrt{2^{-N}n\log n}\big) = o\big(\sqrt{2^{-N}nb_n^3\log n}\big).\end{aligned}$$

Hence, the uniform integrability of the family $|\bar{\eta}_j|^2$ $(1 \le j \le 2^N)$ leads to

$$\mathbb{E}|\eta_j''|^2 = o(2^{-N}n\log n).$$

Consequently, by the weak law given in Theorem 5.5.3 we get

$$\mathbb{E}(\eta_j')^2 \sim \mathbb{E}(\bar{\eta}_j)^2 \sim \frac{1}{(2\pi)^2\det(\Gamma)}2^{-N}n\log n$$

uniformly over $1 \le j \le 2^N$ as $n \to \infty$. This leads to (8.3.12). □

If we focus only on the lower tail of $Q_n - \mathbb{E}Q_n$ in exchange for relaxing the condition on b_n, we establish the following sub-Gaussian tail.

THEOREM 8.3.2. *For any positive sequence*

$$b_n \longrightarrow \infty \quad \text{and} \quad b_n = o(n) \qquad (n \to \infty)$$

and for every $\lambda > 0$,

$$\limsup_{n\to\infty} \frac{1}{b_n} \log \mathbb{P}\Big\{Q_n - \mathbb{E}Q_n \le -\lambda\sqrt{nb_n \log(n/b_n)}\Big\} \le -\lambda^2\pi^2 \det(\Gamma). \tag{8.3.13}$$

Proof. The key idea is argument by sub-additivity. Introduce the sequence,

$$q(n) = \sum_{j=1}^{n}\sum_{k=j}^{\infty} \mathbb{P}\{S(k) = 0\} \qquad n = 1, 2, \cdots.$$

By an argument used in the proof of (5.1.6), we get

$$q(n) \sim \frac{1}{(2\pi)^{3/2}\sqrt{\det(\Gamma)}} \sum_{j=1}^{n}\sum_{k=j}^{\infty} \frac{1}{k^{3/2}} \sim \frac{4}{(2\pi)^{3/2}\sqrt{\det(\Gamma)}}\sqrt{n}. \tag{8.3.14}$$

We claim that for any integers $m, n \ge 1$,

$$\mathbb{E}\bigg(\sum_{j=1}^{n}\sum_{k=n+1}^{m+n} 1_{\{S(j)=S(k)\}}\bigg) = q(n) + q(m) - q(m+n). \tag{8.3.15}$$

Indeed,

$$\begin{aligned}
&\mathbb{E}\bigg(\sum_{j=1}^{n}\sum_{k=n+1}^{m+n} 1_{\{S(j)=S(k)\}}\bigg)\\
&= \sum_{j=1}^{n}\sum_{k=0}^{m-1} \mathbb{P}\{S(j+k) = 0\} = \sum_{j=1}^{n}\sum_{k=j}^{m+j-1} \mathbb{P}\{S(k) = 0\}\\
&= \sum_{j=1}^{n}\bigg(\sum_{k=j}^{\infty}\mathbb{P}\{S(k) = 0\} - \sum_{k=m+j}^{\infty} \mathbb{P}\{S(k) = 0\}\bigg)\\
&= \sum_{j=1}^{n}\sum_{k=j}^{\infty}\mathbb{P}\{S(k) = 0\} - \sum_{j=m+1}^{m+n}\sum_{k=j}^{\infty}\mathbb{P}\{S(k) = 0\}\\
&= q(n) + q(m) - q(m+n).
\end{aligned}$$

Write

$$\xi(n) = q(n) + (\mathbb{E}Q_n - Q_n) \qquad n = 1, 2, \cdots.$$

Then

$$\begin{aligned}
\xi(m+n) &= q(m+n) + (\mathbb{E}Q_{m+n} - Q_{m+n})\\
&= q(m+n) + (\mathbb{E}Q_n - Q_n) + (\mathbb{E}\widetilde{Q}_m - \widetilde{Q}_m)\\
&\quad + \mathbb{E}\bigg(\sum_{j=1}^{n}\sum_{k=n+1}^{m+n} 1_{\{S(j)=S(k)\}}\bigg) - \sum_{j=1}^{n}\sum_{k=n+1}^{m+n} 1_{\{S(j)=S(k)\}}\\
&\le \Big\{q(n) + (\mathbb{E}Q_n - Q_n)\Big\} + \Big\{q(m) + (\mathbb{E}\widetilde{Q}_m - \widetilde{Q}_m)\Big\}
\end{aligned}$$

where

$$\widetilde{Q}_m = \sum_{n+1\le j<k\le n+m} 1_{\{S(j)=S(k)\}}$$

is independent of $\{S_1, \cdots, S_n\}$ and is equal in law to Q_m.

Therefore, $\xi(n)$ is a sub-additive sequence:

$$\xi(n+m) \le \xi(n) + \widetilde{\xi}(m) \qquad m, n \ge 1 \tag{8.3.16}$$

with $\widetilde{\xi}(m) \overset{d}{=} \xi(m)$ and with independence between $\widetilde{\xi}(m)$ and $\{\xi(1), \cdots, \xi(n)\}$.

By a straightforward calculation we have that for each $k \ge 1$, we get

$$\xi(k) - \xi(k-1) = \sum_{j=1}^{\infty} \mathbb{P}\{S(j) = 0\} - \sum_{j=1}^{k-1} 1_{\{S(j)=S(k)\}}.$$

Consequently,

$$\sup_{k\ge 1} \{\xi(k) - \xi(k-1)\} \le \sum_{k=1}^{\infty} \mathbb{P}\{S(k) = 0\}.$$

The transience in $d = 3$ implies that the right hand side is a finite constant. By Theorem 6.4.3 and (8.3.14), the normalized sequence

$$\max_{1\le k\le n} \xi(k)/\sqrt{n\log n} \qquad n = 2, 3, \cdots$$

is bounded in probability.

By Theorem 1.3.3, therefore, we get

$$\sup_n \mathbb{E}\exp\left\{\frac{\theta}{\sqrt{n\log n}}\xi(n)\right\} < \infty \qquad \forall\theta > 0. \tag{8.3.17}$$

Let $\theta > 0$. By the sub-additivity given in (8.3.16), we get

$$\mathbb{E}\exp\left\{\theta\sqrt{\frac{b_n}{n\log(n/b_n)}}\xi(n)\right\} \le \left(\mathbb{E}\exp\left\{\theta\sqrt{\frac{b_n}{n\log(n/b_n)}}\xi([n/b_n])\right\}\right)^{1+b_n}.$$

By the central limit theorem given in Theorem 5.5.3, by (8.3.17) (with n being replaced by $[n/b_n]$) and by dominated convergence theorem, we get

$$\mathbb{E}\exp\left\{\theta\sqrt{\frac{b_n}{n\log(n/b_n)}}\xi([n/b_n])\right\} \longrightarrow \exp\left\{\frac{\theta^2}{(2\pi)^2\det(\Gamma)}\right\} \qquad (n\to\infty).$$

Consequently, for any $\theta > 0$,

$$\limsup_{n\to\infty} \frac{1}{b_n}\log\mathbb{E}\exp\left\{\theta\sqrt{\frac{b_n}{n\log(n/b_n)}}\xi(n)\right\} \le \frac{\theta^2}{(2\pi)^2\det(\Gamma)}.$$

In view of (8.3.14), this is equivalent to

$$\limsup_{n\to\infty} \frac{1}{b_n}\log\mathbb{E}\exp\left\{\theta\sqrt{\frac{b_n}{n\log(n/b_n)}}(\mathbb{E}Q_n - Q_n)\right\} \le \frac{\theta^2}{(2\pi)^2\det(\Gamma)}$$

which leads to the conclusion in Theorem 8.3.2 by a standard application of Chebyshev inequality. □

We conjecture that (8.3.13) can be strengthened into

$$\lim_{n\to\infty}\frac{1}{b_n}\log\mathbb{P}\Big\{Q_n-\mathbb{E}Q_n\le -\lambda\sqrt{nb_n\log(n/b_n)}\Big\}=-\lambda^2\pi^2\det(\Gamma)$$

under the assumptions given in Theorem 8.3.2.

8.4. LDP of non-Gaussian tail in $d=3$

The decomposition (8.2.5) shows that the self-intersection of a random walk consists of two parts: short and long range intersections. When $b_n=o(\sqrt{\log n})$, it is the short range intersection that dominates. Consequently, the large deviation principle for $Q_n-\mathbb{E}Q_n$ has a Gaussian tail in this case. In this section, we shall see that increasing the scale b_n may change the dynamics.

Throughout this section, $d=3$.

LEMMA 8.4.1. *For any $L>0$, there is a $\lambda>0$ such that*

$$\limsup_{n\to\infty}\frac{1}{b_n}\log\mathbb{P}\Big\{|Q_n-\mathbb{E}Q_n|\ge\lambda\sqrt{nb_n^3}\Big\}\le -L$$

for any $b_n=o(n)$ with $\lim_{n\to\infty}b_n/\sqrt{\log n}=\infty$.

Proof. Again, our starting point is the decomposition (8.2.5). Here we let $N=[\log_2(\log n)]$ (Recall that $\log_2(\cdot)$ is the logarithmic function with base 2). Then (8.3.2) remains true. Here we emphasize two facts: First, the constant $\lambda>0$ exists uniformly for b_n satisfying the condition posted in the lemma.[2] Second, to establish (8.3.2) in the current setting, we take $\gamma=1/2$ in (8.3.3), and we use our assumption $\lim_{n\to\infty}b_n/\sqrt{\log n}=\infty$ in (8.3.6), where the right hand side (note $\gamma=1/2$) is bounded by

$$\exp\left\{\frac{M'\theta^2\log n}{b_n}\right\}=\exp\{o(b_n)\}\qquad(n\to\infty).$$

It remains to show that for any $L>0$, there is a $\lambda>0$ such that

$$\limsup_{n\to\infty}\frac{1}{b_n}\log\mathbb{P}\Big\{\Big|\sum_{j=1}^{2^N}\overline{\eta}_j\Big|\ge\lambda\sqrt{nb_n^3}\Big\}\le -L. \tag{8.4.1}$$

Write

$$\zeta_j=\eta_j/\sqrt{\log n}\qquad j=1,\cdots,2^N.$$

From (8.3.9),

$$\sup_{n,j}\mathbb{E}\exp\left\{\frac{\theta}{(2^{-N}n)^{1/3}}|\zeta_j|^{2/3}\right\}<\infty.$$

---

[2]When the assumption on b_n is weakened as $\liminf_{n\to\infty}b_n/\sqrt{\log n}>0$, (8.3.2) remains true, but we are unable to establish its uniformity over b_n.

The argument used for (8.3.5) (the assumption on b_n is needed here) leads to the fact that for any $L > 0$, there is a $\lambda > 0$ such that

$$\limsup_{n\to\infty} \frac{1}{b_n} \log \mathbb{P}\Big\{\Big|\sum_{j=1}^{2^N} \eta_j\Big| \geq \lambda\sqrt{2^{-N}nb_n^3}\Big\} \leq -L.$$

Equivalently, we have proved (8.4.1). □

The following is the main result of this section.

THEOREM 8.4.2. *For any b_n satisfying*

$$\lim_{n\to\infty} b_n/\sqrt{\log n} = \infty \quad \text{and} \quad b_n = o(n) \qquad (n \to \infty)$$

and for any $\lambda > 0$,

$$\lim_{n\to\infty} \frac{1}{b_n} \log \mathbb{P}\Big\{Q_n - \mathbb{E}Q_n \geq \lambda\sqrt{nb_n^3}\Big\} = -\sqrt[3]{\frac{\det(\Gamma)}{2}}\kappa(3,2)^{-8/3}\lambda^{2/3}. \tag{8.4.2}$$

Here we recall that under the definition given in Remark 3.3.3, $\kappa(3,2) > 0$ is the best constant of Gagliardo-Nirenberg inequality in $\mathbb{R}^3$:

$$||f||_4 \leq C||\nabla f||_2^{3/4}||f||_2^{1/4} \qquad f \in W^{1,2}(\mathbb{R}^3).$$

Proof. Under our assumption on b_n, $\sqrt{nb_n\log(n/b_n)} = o\big(\sqrt{nb_n^3}\big)$ as $n \to \infty$. By the sub-Gaussian tail given in Theorem 8.3.2, we get

$$\lim_{n\to\infty} \frac{1}{b_n} \log \mathbb{P}\Big\{Q_n - \mathbb{E}Q_n \leq -\lambda\sqrt{nb_n^3}\Big\} = -\infty \qquad \forall\lambda > 0.$$

Consequently, (8.4.2) is equivalent to

$$\lim_{n\to\infty} \frac{1}{b_n} \log \mathbb{P}\Big\{|Q_n - \mathbb{E}Q_n| \geq \lambda\sqrt{nb_n^3}\Big\} = -\sqrt[3]{\frac{\det(\Gamma)}{2}}\kappa(3,2)^{-8/3}\lambda^{2/3}. \tag{8.4.3}$$

We now claim that the statement in Lemma 8.4.1 can be slightly strengthened into: There is a constant $C_Q > 0$ such that

$$\limsup_{n\to\infty} \frac{1}{b_n} \log \mathbb{P}\Big\{|Q_n - \mathbb{E}Q_n| \geq \lambda\sqrt{nb_n^3}\Big\} \leq -C_Q\lambda^{2/3} \qquad \forall\lambda > 0 \tag{8.4.4}$$

for any b_n with $b_n/\sqrt{\log n} \to \infty$.

Indeed, by Lemma 8.4.1 there is a constant $a > 0$ such that

$$\limsup_{n\to\infty} \frac{1}{b_n} \log \mathbb{P}\Big\{|Q_n - \mathbb{E}Q_n| \geq a\sqrt{nb_n^3}\Big\} \leq -1$$

for every $b_n = o(n)$ with $b_n/\sqrt{\log n} \to \infty$. Replacing b_n by $(a^{-1}\lambda)^{2/3}b_n$ in the above bound and letting $C_Q = a^{-2/3}$ we have (8.4.4).

By Theorem 1.1.4, a possible approach is to compute the logarithmic generating function

$$\lim_{n\to\infty} \frac{1}{b_n} \log \mathbb{E}\exp\Big\{\theta\Big(\frac{b_n}{n}\Big)^{1/4}|Q_n - \mathbb{E}Q_n|^{1/2}\Big\}.$$

Unfortunately, we are not able to do this. As a matter of fact, we don't even know whether or not the limsup,

$$\limsup_{n\to\infty}\frac{1}{b_n}\log\mathbb{E}\exp\left\{\theta\Big(\frac{b_n}{n}\Big)^{1/4}|Q_n-\mathbb{E}Q_n|^{1/2}\right\}$$

is finite for any $\theta>0$, despite (8.4.4).

Instead, we condition on that $|Q_n-\mathbb{E}Q_n|$ is suitably bounded. More precisely, let $M>0$ be large but fixed and define the event

$$\Omega_n=\left\{|Q_n-\mathbb{E}Q_n|\le M\sqrt{nb_n^3}\right\}.$$

We prove that

$$\begin{aligned}(8.4.5)\qquad &\lim_{n\to\infty}\frac{1}{b_n}\log\mathbb{E}\left(\exp\left\{\theta\Big(\frac{b_n}{n}\Big)^{1/4}|Q_n-\mathbb{E}Q_n|^{1/2}\right\}\Big|\Omega_n\right)\\ &=\frac{1}{16}\Big(\frac{3}{2}\Big)^3\det(\Gamma)^{-1}\kappa(3,2)^8\theta^4\qquad 0<\theta<cM^{1/6}\end{aligned}$$

where $c>0$ is a constant independent of M.

The triangular decomposition (8.2.5) plays an important role in our proof of (8.4.5). As in section 8.2, we let N be large but fixed. By (8.2.6) and (8.4.4), we get

$$\limsup_{n\to\infty}\frac{1}{b_n}\log\mathbb{P}\left\{|\overline{\eta}_j|\ge\lambda\sqrt{nb_n^3}\right\}\le -C_Q2^{N/3}\lambda^{2/3}$$

for every $\lambda>0$ (recall that $\overline{\eta}_j=\eta_j-\mathbb{E}\eta_j$). By Theorem 1.2.2, we get

$$\limsup_{n\to\infty}\frac{1}{b_n}\log\mathbb{P}\left\{\Big|\sum_{j=1}^{2^N}\overline{\eta}_j\Big|\ge\lambda\sqrt{nb_n^3}\right\}\le -C_Q2^{N/3}\lambda^{2/3}$$

for every $\lambda>0$. In view of (8.4.4), we have

$$(8.4.6)\qquad \limsup_{n\to\infty}\frac{1}{b_n}\log\mathbb{P}\left\{\Big|\sum_{j=1}^{2^N}\overline{\eta}_j\Big|\ge\lambda\sqrt{nb_n^3}\Big|\Omega_n\right\}\le -C_Q2^{N/3}\lambda^{2/3}.$$

We now claim that

$$(8.4.7)\quad \limsup_{n\to\infty}\frac{1}{b_n}\log\mathbb{E}\left(\exp\left\{\theta\Big(\frac{b_n}{n}\Big)^{1/4}\Big|\sum_{j=1}^{2^N}\overline{\eta}_j\Big|^{1/2}\right\}\Big|\Omega_n\right)\le\frac{27}{256}2^{-N}C_Q^{-3}\theta^4.$$

By (8.2.7) and by (6.2.1) in Theorem 6.2.1, there is a $\theta>0$ (possibly depending on N) such that

$$\sup_{j,k,n}\mathbb{E}\exp\left\{\frac{\theta}{\sqrt[3]{n}}\xi_{j,k}^{2/3}\right\}<\infty$$

for some $\theta>0$. Notice that N is fixed. We obtain that

$$\sup_n\mathbb{E}\exp\left\{\frac{\theta}{\sqrt[3]{n}}\left||Q_n-\mathbb{E}Q_n|-\Big|\sum_{j=1}^{2^N}\overline{\eta}_j\Big|\right|^{2/3}\right\}<\infty$$

for some $\theta>0$, using decomposition (8.2.5).

This leads to, by a Chebyshev type estimate, that

$$\limsup_{n\to\infty} \frac{1}{b_n} \log \mathbb{E} \exp\left\{\theta\Big(\frac{b_n}{n}\Big)^{1/4} \left| |Q_n - \mathbb{E}Q_n| - \Big|\sum_{j=1}^{2^N} \overline{\eta}_j\Big| \right|^{1/2}\right\} < \infty$$

for every $\theta > 0$. Consequently,

$$\limsup_{n\to\infty} \frac{1}{b_n} \log \mathbb{E}\left(\exp\left\{\theta\Big(\frac{b_n}{n}\Big)^{1/4} \Big|\sum_{j=1}^{2^N} \overline{\eta}_j\Big|^{1/2}\right\} \Big| \Omega_n\right) < \infty \qquad \forall \theta > 0.$$

Combined with the observation (8.4.6), this legitimizes the use of Varadhan's integral lemma (Theorem 1.1.6). Thus, for any $\theta > 0$,

$$\begin{aligned}&\limsup_{n\to\infty} \frac{1}{b_n} \log \mathbb{E}\left(\exp\left\{\theta\Big(\frac{b_n}{n}\Big)^{1/4} \Big|\sum_{j=1}^{2^N} \overline{\eta}_j\Big|^{1/2}\right\} \Big| \Omega_n\right)\\ &\le \sup_{\lambda>0}\left\{\lambda\theta - C_Q 2^{N/3}\lambda^{4/3}\right\} = \frac{27}{256} 2^{-N} C_Q^{-3}\theta^4.\end{aligned}$$

We have confirmed (8.4.7).

Recall (Theorem 8.1.2, with $d = 3$ and $p = 2$) that

$$\begin{aligned}(8.4.8)\qquad &\lim_{n\to\infty} \frac{1}{b_n} \log \mathbb{E} \exp\left\{\theta\Big(\frac{b_n}{n}\Big)^{1/4} \Big(\sum_{x\in\mathbb{Z}^3} l^2(n,x,\epsilon)\Big)^{1/2}\right\}\\ &= \sup_{g\in\mathcal{F}_3}\left\{\theta \det(\Gamma)^{-1/4}\Big(\int_{\mathbb{R}^3} |g(x,\epsilon)|^4 dx\Big)^{1/2} - \frac{1}{2}\int_{\mathbb{R}^3} |\nabla g(x)|^2 dx\right\}\\ &= M_\epsilon(\theta) \qquad \text{(say)}\end{aligned}$$

where

$$l(n,x,\epsilon) = \det(\Gamma)^{-1/2}\Big(\frac{b_n}{n}\Big)^{3/2} \sum_{k=1}^n p_\epsilon\Big(\sqrt{\frac{b_n}{n}}\Gamma^{-1/2}(S(k) - x)\Big)$$

$$\mathcal{F}_3 = \left\{g \in \mathcal{L}^2(\mathbb{R}^3);\ \ ||g||_2 = 1 \text{ and } \nabla g \in \mathcal{L}^2(\mathbb{R}^3)\right\}$$

$$g(x,\epsilon) = \left\{\int_{\mathbb{R}^3} p_\epsilon(x-y)g^2(y)dy\right\}^{1/2}.$$

Clearly, (8.4.8) implies that

$$(8.4.9)\qquad \limsup_{n\to\infty} \frac{1}{b_n} \log \mathbb{E}\left[\exp\left\{\theta\Big(\frac{b_n}{n}\Big)^{1/4} \Big(\sum_{x\in\mathbb{Z}^3} l^2(n,x,\epsilon)\Big)^{1/2}\right\} \Big| \Omega_n\right] \le M_\epsilon(\theta).$$

On the other hand,

$$\begin{aligned}
&\mathbb{E}\exp\left\{\theta\Big(\frac{b_n}{n}\Big)^{1/4}\Big(\sum_{x\in\mathbb{Z}^3} l^2(n,x,\epsilon)\Big)^{1/2}\right\}\\
&=\mathbb{E}\left[\exp\left\{\theta\Big(\frac{b_n}{n}\Big)^{1/4}\Big(\sum_{x\in\mathbb{Z}^3} l^2(n,x,\epsilon)\Big)^{1/2}\right\}1_{\Omega_n}\right]\\
&\quad+\mathbb{E}\left[\exp\left\{\theta\Big(\frac{b_n}{n}\Big)^{1/4}\Big(\sum_{x\in\mathbb{Z}^3} l^2(n,x,\epsilon)\Big)^{1/2}\right\}1_{\Omega_n^c}\right].
\end{aligned}$$

Consequently,

$$\begin{aligned}
&\max\left\{\liminf_{n\to\infty}\frac{1}{b_n}\log\mathbb{E}\left[\exp\left\{\theta\Big(\frac{b_n}{n}\Big)^{1/4}\Big(\sum_{x\in\mathbb{Z}^3} l^2(n,x,\epsilon)\Big)^{1/2}\right\}\Big|\Omega_n\right],\right.\\
&\left.\limsup_{n\to\infty}\frac{1}{b_n}\log\mathbb{E}\left[\exp\left\{\theta\Big(\frac{b_n}{n}\Big)^{1/4}\Big(\sum_{x\in\mathbb{Z}^3} l^2(n,x,\epsilon)\Big)^{1/2}\right\}1_{\Omega_n^c}\right]\right\}\\
&\geq M_\epsilon(\theta).
\end{aligned}$$

By Cauchy-Schwarz inequality,

$$\begin{aligned}
&\mathbb{E}\left[\exp\left\{\theta\Big(\frac{b_n}{n}\Big)^{1/4}\Big(\sum_{x\in\mathbb{Z}^3} l^2(n,x,\epsilon)\Big)^{1/2}\right\}1_{\Omega_n^c}\right]\\
&\leq\left[\mathbb{E}\exp\left\{2\theta\Big(\frac{b_n}{n}\Big)^{1/4}\Big(\sum_{x\in\mathbb{Z}^3} l^2(n,x,\epsilon)\Big)^{1/2}\right\}\right]^{1/2}\Big\{\mathbb{P}(\Omega_n^c)\Big\}^{1/2}.
\end{aligned}$$

Combining this with (8.4.8) (with θ replaced by 2θ) and (8.4.4) (with $\lambda = M$),

$$\begin{aligned}
&\limsup_{n\to\infty}\frac{1}{b_n}\log\mathbb{E}\left[\exp\left\{\theta\Big(\frac{b_n}{n}\Big)^{1/4}\Big(\sum_{x\in\mathbb{Z}^3} l^2(n,x,\epsilon)\Big)^{1/2}\right\}1_{\Omega_n^c}\right]\\
&\leq\frac{1}{2}M_\epsilon(2\theta)-\frac{1}{2}C_Q M^{2/3}.
\end{aligned}$$

By Jensen inequality, we get

$$M_\epsilon(2\theta)\leq\sup_{g\in\mathcal{F}_3}\left\{2\theta\det(\Gamma)^{-1/4}\Big(\int_{\mathbb{R}^3}|g(x)|^4dx\Big)^{1/2}-\frac{1}{2}\int_{\mathbb{R}^3}|\nabla g(x)|^2dx\right\}.$$

According to Theorem C.1 in the Appendix, for any $\theta > 0$,

$$\sup_{g\in\mathcal{F}_3}\left\{\theta\Big(\int_{\mathbb{R}^3}|g(x)|^4dx\Big)^{1/2}-\frac{1}{2}\int_{\mathbb{R}^3}|\nabla g(x)|^2dx\right\}=\frac{1}{4}\Big(\frac{3}{2}\Big)^3\kappa(3,2)^8\theta^4. \tag{8.4.10}$$

Hence, we have proved that

$$\begin{aligned}
&\limsup_{n\to\infty}\frac{1}{b_n}\log\mathbb{E}\left[\exp\left\{\theta\Big(\frac{b_n}{n}\Big)^{1/4}\Big(\sum_{x\in\mathbb{Z}^3} \xi^2(n,x,\epsilon)\Big)^{1/2}\right\}1_{\Omega_n^c}\right]\\
&\leq\frac{27}{4}\det(\Gamma)^{-1}\kappa(3,2)^8\theta^4-\frac{1}{2}C_Q M^{2/3}.
\end{aligned}$$

Notice that there exists a constant $c > 0$ independent of M, such that the right hand side is negative for all $0 < \theta < cM^{1/6}$. Summarizing our argument after (8.4.9), therefore, we get

$$\liminf_{n\to\infty} \frac{1}{b_n} \log \mathbb{E}\left[\exp\left\{\theta\Big(\frac{b_n}{n}\Big)^{1/4}\Big(\sum_{x\in\mathbb{Z}^3} l^2(n,x,\epsilon)\Big)^{1/2}\right\}\Big|\Omega_n\right] \ge M_\epsilon(\theta) \tag{8.4.11}$$

for $0 < \theta < cM^{1/6}$.

Combining (8.4.9) and (8.4.11), for any $0 < \theta < cM^{1/6}$,

$$\lim_{n\to\infty} \frac{1}{b_n} \log \mathbb{E}\left[\exp\left\{\theta\Big(\frac{b_n}{n}\Big)^{1/4}\Big(\sum_{x\in\mathbb{Z}^3} l^2(n,x,\epsilon)\Big)^{1/2}\right\}\Big|\Omega_n\right] = M_\epsilon(\theta). \tag{8.4.12}$$

Like the proof of Theorem 8.2.1, a key idea is the exponential approximation. For this we need the decomposition given in (8.2.9). Similarly to (8.2.14), for any $\theta > 0$, we have

$$\begin{aligned}&\limsup_{n\to\infty} \frac{1}{b_n} \log \mathbb{E}\exp\left\{\theta\Big(\frac{b_n}{n}\Big)^{1/4}\Big(\sum_{j=1}^{2^N} \eta_j^2(\epsilon)\Big)^{1/2}\right\}\\ &\le \frac{1}{2^{N+2}}\Big(\frac{3}{2}\Big)^3 \det(\Gamma)^{-1}\kappa(3,2)^8\theta^4.\end{aligned} \tag{8.4.13}$$

Similarly to (8.2.15), we have

$$\lim_{\epsilon\to 0^+}\limsup_{n\to\infty} \frac{1}{b_n} \log \mathbb{E}\exp\left\{\theta\Big(\frac{b_n}{n}\Big)^{1/4}\Big(\sum_{j=1}^{N}\sum_{k=1}^{2^{j-1}} \big|\xi_{j,k} - \xi_{j,k}(\epsilon)\big|\Big)^{1/2}\right\} = 0. \tag{8.4.14}$$

In addition, by (8.2.7) one can see that

$$\sum_{j=1}^{N}\sum_{k=1}^{2^{j-1}} \mathbb{E}\xi_{j,k} = O(\sqrt{n}) \qquad (n\to\infty).$$

Combining this with (8.4.7), (8.4.13) and (8.4.14), and applying Hölder inequality we have that for any $\theta > 0$,

$$\begin{aligned}&\lim_{\epsilon\to 0^+}\limsup_{n\to\infty} \frac{1}{b_n} \log \mathbb{E}\Bigg[\exp\left\{\theta\Big(\frac{b_n}{n}\Big)^{1/4}\right.\\ &\qquad\left.\times\left|(Q_n - \mathbb{E}Q_n) - \frac{1}{2}\sum_{x\in\mathbb{Z}^3} \xi^2(n,x,\epsilon)\right|^{1/2}\right\}\Bigg|\Omega_n\Bigg] = 0.\end{aligned}$$

In view of (8.4.12) (with θ replaced by $\theta/\sqrt{2}$), the exponential approximation established above gives that for $0 < \theta < cM^{1/6}$,

$$\begin{aligned}
&\lim_{n\to\infty}\frac{1}{b_n}\log\mathbb{E}\bigg(\exp\bigg\{\theta\Big(\frac{b_n}{n}\Big)^{1/4}|Q_n-\mathbb{E}Q_n|^{1/2}\bigg\}\bigg|\Omega_n\bigg)\\
&=\sup_{g\in\mathcal{F}_3}\bigg\{\frac{\theta}{\sqrt{2}}\det(\Gamma)^{-1/4}\bigg(\int_{\mathbb{R}^3}|g(x)|^4dx\bigg)^{1/2}-\frac{1}{2}\int_{\mathbb{R}^3}|\nabla g(x)|^2dx\bigg\}\\
&=\frac{1}{16}\Big(\frac{3}{2}\Big)^3\det(\Gamma)^{-1}\kappa(3,2)^8\theta^4
\end{aligned}$$

where the last step follows from (8.4.10).

We have completed the proof of (8.4.5).

Finally, we come to the proof of (8.4.3). Due to the limitation on θ posted in (8.4.5), the Gärtner-Ellis theorem does not directly apply to our setting. However, by Chebyshev inequality and (8.4.5) we have the upper bound

$$\begin{aligned}
&\limsup_{n\to\infty}\frac{1}{b_n}\log\mathbb{P}\Big\{|Q_n-\mathbb{E}Q_n|\ge\lambda\sqrt{nb_n^3}\Big|\Omega_n\Big\}\\
&\le-\sup_{0<\theta<cM^{1/6}}\Big\{\theta\lambda^{1/2}-\frac{1}{16}\Big(\frac{3}{2}\Big)^3\det(\Gamma)^{-1}\kappa(3,2)^8\theta^4\Big\}\\
&=-\sqrt[3]{\frac{\det(\Gamma)}{2}}\kappa(3,2)^{-8/3}\lambda^{2/3}
\end{aligned}$$

for

$$0<\lambda<\frac{1}{16}\Big(\frac{3}{2}\Big)^6\det(\Gamma)^{-2}\kappa(3,2)^{16}c^6M.$$

Since $M > 0$ can be arbitrarily large, by (8.4.4) (with $\lambda = M$) and a standard argument of exponential approximation, we get

(8.4.15)
$$\begin{aligned}
&\limsup_{n\to\infty}\frac{1}{b_n}\log\mathbb{P}\Big\{|Q_n-\mathbb{E}Q_n|\ge\lambda\sqrt{nb_n^3}\Big\}\\
&\le-\sqrt[3]{\frac{\det(\Gamma)}{2}}\kappa(3,2)^{-8/3}\lambda^{2/3}\qquad\forall\lambda>0.
\end{aligned}$$

To establish the lower bound, we introduce an i.i.d. sequence $\{\tau_k\}$ of random variables with the common probability density

$$f(x)=\frac{1}{2}e^{-|x|}\qquad x\in\mathbb{R}.$$

Let ξ be a Bernoulli random variable taking values -1 and 1 with probability $1/2$.

We assume independence among ξ, $\{\tau_k\}$ and $\{S(k)\}$. Write

$$Y_n=\frac{1}{b_n}\sum_{k=1}^{[b_n]}\tau_k\qquad n=1,2,\cdots.$$

By (8.4.5), we get

$$\lim_{n\to\infty}\frac{1}{b_n}\log\mathbb{E}\bigg[\exp\bigg\{\theta b_n\Big(\frac{\xi|Q_n-\mathbb{E}Q_n|^{1/2}}{\sqrt[4]{nb_n^3}}+\frac{Y_n}{cM^{1/6}}\Big)\bigg\}\bigg|\Omega_n\bigg]=\Lambda(\theta)$$

for every $\theta \in \mathbb{R}$, where the logarithmic moment generating function

$$\Lambda(\theta) = \begin{cases} \frac{1}{16}\Big(\frac{3}{2}\Big)^3 \det(\Gamma)^{-1}\kappa(3,2)^8\theta^4 + \log\Big\{1 - \Big(\frac{\theta}{cM^{1/6}}\Big)^2\Big\}^{-1} & |\theta| < cM^{1/6}, \\ \infty & |\theta| \geq cM^{1/6} \end{cases}$$

is essentially smooth on $\mathbb{R}$ according to Definition 1.1.3. Applying Theorem 1.1.4 to the random sequence

$$\frac{\xi|Q_n - \mathbb{E}Q_n|^{1/2}}{\sqrt[4]{nb_n^3}} + \frac{Y_n}{cM^{1/6}} \qquad n = 1, 2, \cdots$$

gives that for any $\lambda > 0$,

$$\lim_{n\to\infty} \frac{1}{b_n} \log \mathbb{P}\Big\{\Big|\frac{\xi|Q_n - \mathbb{E}Q_n|^{1/2}}{\sqrt[4]{nb_n^3}} + \frac{Y_n}{cM^{1/6}}\Big| \geq \lambda \Big| \Omega_n\Big\} = -\inf_{|x|>\lambda} \Lambda^*(x)$$

where

$$\begin{aligned} \Lambda^*(x) &= \sup_{\theta\in\mathbb{R}} \Big\{\theta x - \Lambda(\theta)\Big\} = \sup_{|\theta|<cM^{1/6}} \Big\{\theta x - \Lambda(\theta)\Big\} \\ &\leq \sup_{|\theta|<cM^{1/6}} \Big\{\theta x - \frac{1}{16}\Big(\frac{3}{2}\Big)^3 \det(\Gamma)^{-1}\kappa(3,2)^8\theta^4\Big\} \qquad x \in \mathbb{R}. \end{aligned}$$

It is straightforward to check that the right hand side is equal to

$$\sqrt[3]{\frac{\det(\Gamma)}{2}}\kappa(3,2)^{-8/3}|x|^{4/3}$$

when

$$|x| < \frac{1}{4}\Big(\frac{3}{2}\Big)^3 \det(\Gamma)^{-1}\kappa(3,2)^8 c^3 M^{1/2}.$$

Therefore,

$$\begin{aligned} \text{(8.4.16)} \qquad &\liminf_{n\to\infty} \frac{1}{b_n} \log \mathbb{P}\Big\{\Big|\frac{\xi|Q_n - \mathbb{E}Q_n|^{1/2}}{\sqrt[4]{nb_n^3}} + \frac{Y_n}{cM^{1/6}}\Big| \geq \lambda \Big| \Omega_n\Big\} \\ &\geq -\sqrt[3]{\frac{\det(\Gamma)}{2}}\kappa(3,2)^{-8/3}\lambda^{4/3} \end{aligned}$$

whenever

$$0 < \lambda < \frac{1}{4}\Big(\frac{3}{2}\Big)^3 \det(\Gamma)^{-1}\kappa(3,2)^8 c^3 M^{1/2}.$$

In addition, by Cramér's large deviation principle (Theorem 1.1.5) one concludes that for any $\epsilon > 0$,

$$\lim_{M\to\infty} \limsup_{n\to\infty} \frac{1}{b_n} \log \mathbb{P}\Big\{|Y_n| \geq \epsilon c M^{1/6}\Big\} = -\infty.$$

Combining this with Lemma 8.4.1 and (8.4.16), a standard argument by exponential approximation gives us

$$\liminf_{n\to\infty} \frac{1}{b_n} \log \mathbb{P}\Big\{|Q_n - \mathbb{E}Q_n|^{1/2} \geq \lambda\sqrt[4]{nb_n^3}\Big\} \geq -\sqrt[3]{\frac{\det(\Gamma)}{2}}\kappa(3,2)^{-8/3}\lambda^{4/3}.$$

Replacing λ by $\sqrt{\lambda}$, we have

$$\liminf_{n\to\infty}\frac{1}{b_n}\log\mathbb{P}\Big\{|Q_n-\mathbb{E}Q_n|\ge\lambda\sqrt{nb_n^3}\Big\}\ge-\sqrt[3]{\frac{\det(\Gamma)}{2}}\kappa(3,2)^{-8/3}\lambda^{2/3}. \tag{8.4.17}$$

Finally, the desired (8.4.3) follows from (8.4.15) and (8.4.17). □

8.5. LDP for renormalized range in $d = 2, 3$

Recall that by our notation, the range of a lattice valued random walk $\{S(k)\}$ is given by

$$R_n = \#\{S[1,n]\} = \#\{S(1),\cdots,S(n)\}, \qquad n = 1,2,\cdots.$$

In this section we study the large deviations for the renormalized range $R_n - \mathbb{E}R_n$.

It is intuitively clear that Q_n and R_n are negatively correlated. Indeed, a spread-out trajectory rarely intersects. As a consequence, the LDP behavior of $R_n - \mathbb{E}R_n$ is naturally linked to that of $-(Q_n - \mathbb{E}Q_n)$.

Similarly to (8.2.5), given an integer $N \ge 1$, we get

$$R_n = \sum_{j=1}^{2^N}\beta_j - \sum_{j=1}^{N}\sum_{k=1}^{2^{j-1}}\alpha_{j,k} \tag{8.5.1}$$

where

$$\beta_j = \#\Big\{S\big((j-1)n2^{-N}, jn2^{-N}\big]\Big\},$$

$$\alpha_{j,k} = \#\Big\{S\big((2k-2)n2^{-j},\ (2k-1)n2^{-j}\big]\cap S\big((2k-1)n2^{-j},\ (2k)n2^{-j}\big]\Big\}.$$

The sequence $\{\beta_j\}_{1\le j\le 2^N}$ is an independent sequence with

$$\beta_j \overset{d}{=} R_{n_j} \qquad j = 1,\cdots,2^N \tag{8.5.2}$$

where

$$n_j = \#\big((j-1)n2^{-N}, jn2^{-N}\big], \qquad j = 1,\cdots,2^N.$$

For each $1 \le j \le N$, $\{\alpha_{j,k}\}_{1\le k\le 2^{j-1}}$ is an independent sequence with

$$\alpha_{j,k} \overset{d}{=} \#\Big\{S\big[0, n_{j,k}-1\big]\cap\widetilde{S}\big[1, n'_{j,k}\big]\Big\} \qquad k = 1,\cdots 2^{j-1} \tag{8.5.3}$$

where $\{\widetilde{S}(k)\}$ is an independent copy of $\{S(k)\}$ and

$$n_{j,k} = \#\big\{(2k-1)n2^{-j}, (2k)n2^{-j}\big],$$

$$n'_{j,k} = \#\big\{(2k)n2^{-j}, (2k+1)n2^{-j}\big].$$

Corespondent to Theorem 8.2.3, we have

THEOREM 8.5.1. *Let $d = 2$. Suppose that $\mathbb{E}|S(1)|^{2+\delta} < \infty$ for some $\delta > 0$. Then there exist $K_1, K_2 > 0$ such that for any positive sequence b_n satisfying*

$$b_n \longrightarrow \infty \quad \textit{and} \quad \log b_n = o(\sqrt{\log n}) \qquad (n\to\infty)$$

and for any $\lambda > 0$,

$$\limsup_{n\to\infty} \frac{1}{b_n^\lambda} \log \mathbb{P}\Big\{ R_n - \mathbb{E}R_n \ge 2\lambda\pi\sqrt{\det(\Gamma)}\frac{n\log b_n}{(\log n)^2}\Big\} \le -K_1,$$

$$\liminf_{n\to\infty} \frac{1}{b_n^\lambda} \log \mathbb{P}\Big\{ R_n - \mathbb{E}R_n \ge 2\lambda\pi\sqrt{\det(\Gamma)}\frac{n\log b_n}{(\log n)^2}\Big\} \ge -K_2.$$

The proof is omitted here and the reader is referred to [**8**] (Corollary 1.3) for a proof.

The following theorem is an analogue to the results given in Theorem 8.3.1 and Theorem 8.3.2.

THEOREM 8.5.2. *Let* $d = 3$.

(1) *For any positive sequence* b_n *satisfying*

$$b_n \longrightarrow \infty \quad \text{and} \quad b_n = o\big(\sqrt{\log n}\big) \qquad (n \to \infty)$$

and for any $\lambda > 0$,

$$\lim_{n\to\infty} \frac{1}{b_n} \log \mathbb{P}\Big\{ \pm (R_n - \mathbb{E}R_n) \ge \lambda\sqrt{nb_n \log n}\Big\} = -\gamma_S^{-4}\lambda^2\pi^2 \det(\Gamma). \tag{8.5.4}$$

(2) *If the condition on* b_n *is relaxed to*

$$b_n \longrightarrow \infty \quad \text{and} \quad b_n = o\big(n\big) \qquad (n \to \infty),$$

we have

$$\limsup_{n\to\infty} \frac{1}{b_n} \log \mathbb{P}\Big\{ R_n - \mathbb{E}R_n \ge \lambda\sqrt{nb_n \log(n/b_n)}\Big\} \le -\gamma_S^{-4}\lambda^2\pi^2 \det(\Gamma) \tag{8.5.5}$$

for every $\lambda > 0$.

Here we recall our notation

$$\gamma_S = \mathbb{P}\{S(k) \neq 0; \ \ \forall k \ge 1\}.$$

Proof. Replacing the decomposition (8.2.5) by (8.5.1), the proof of (8.5.4) is an obvious modification of the argument used in the argument for (8.3.1) of Theorem 8.3.1.

We now prove (8.5.5). Recall that

$$q(n) = \sum_{j=1}^\infty \sum_{k=j}^\infty \mathbb{P}\{S(k) = 0\}.$$

First notice that

$$\begin{aligned}\sum_{j=1}^n \sum_{k=n+1}^{m+n} 1_{\{S(j)=S(k)\}} &= \sum_x \sum_{j=1}^n \sum_{k=n+1}^{m+n} 1_{\{S(j)=S(k)=x\}} \\ &\ge \sum_x 1_{\{x\in S[1,n]\}} 1_{\{x \in S[n+1,n+m]\}} \\ &= \#\big\{S[1,n] \cap S[n+1, n+m]\big\}.\end{aligned}$$

By (8.3.14), for any integers $m, n \geq 1$, we have

$$\begin{aligned}&\mathbb{E}\#\big\{S[1,n]\cap S[n+1,n+m]\big\}\\&\leq \mathbb{E}\sum_{j=1}^{n}\sum_{k=n+1}^{m+n} 1_{\{S(j)=S(k)\}} = q(n)+q(m)-q(m+n).\end{aligned}$$

Notice that

$$R_{m+n} = R_n + \#\big\{S[n+1,n+m]\big\} - \#\big\{S[1,n]\cap S[n+1,n+m]\big\}.$$

Similarly to (8.3.16), therefore, the sequence $\{\eta(n)\}$ defined by

$$\eta(n) = q(n) + R_n - \mathbb{E}R_n \qquad n = 1, 2, \cdots$$

is sub-additive:

$$\eta(m+n) \leq \eta(n) + \widetilde{\eta}(m) \qquad m, n \geq 1 \tag{8.5.6}$$

where

$$\widetilde{\eta}(m) = q(m) + \#\big\{S[n+1,n+m]\big\} - \mathbb{E}\#\big\{S[n+1,n+m]\big\}$$

is independent of $\{\eta(1), \cdots, \eta(n)\}$ and has the same distribution as $\eta(m)$.

The rest of the proof can be carried out in the same way as the proof of Theorem 8.3.2. □

Perhaps the following theorem is the most interesting result of this section.

THEOREM 8.5.3. *Let $\lambda > 0$ be fixed but arbitrary.*

When $d = 2$,

$$\begin{aligned}&\lim_{n\to\infty}\frac{1}{b_n}\log\mathbb{P}\Big\{R_n - \mathbb{E}R_n \leq -\lambda\frac{nb_n}{(\log n)^2}\Big\}\\&= -(2\pi)^{-2}\det(\Gamma)^{-1/2}\kappa(2,2)^{-4}\lambda\end{aligned} \tag{8.5.7}$$

for any positive sequence b_n satisfying

$$b_n \longrightarrow \infty \quad \text{and} \quad b_n = o\big(\log n\big) \qquad (n \to \infty).$$

When $d = 3$,

$$\begin{aligned}&\lim_{n\to\infty}\frac{1}{b_n}\log\mathbb{P}\Big\{R_n - \mathbb{E}R_n \leq -\lambda\sqrt{nb_n^3}\Big\}\\&= -\sqrt[3]{\frac{\det(\Gamma)}{2}}\kappa(3,2)^{-8/3}\gamma_S^{-4/3}\lambda^{2/3}\end{aligned} \tag{8.5.8}$$

for any positive sequence b_n satisfying

$$\lim_{n\to\infty} b_n/\sqrt{\log n} = \infty \quad \text{and} \quad b_n = o\big(n^{1/3}\big) \qquad (n \to \infty).$$

We now comment on the assumptions on b_n. By the fact that $R_n \leq n$, the condition $b_n = o(n^{1/3})$ is best possible for (8.5.8) in the case $d = 3$. When $d = 2$, by the fact that

$$\mathbb{E}R_n - R_n \leq \mathbb{E}R_n = O\Big(\frac{n}{\log n}\Big) \qquad (n \to \infty)$$

the condition $b_n = o(\log n)$ is best possible for (8.5.7).

Most of the proof of Theorem 8.5.3 is essentially the same as those for Theorem 8.2.1 and Theorem 8.3.1: In view of Theorem 8.1.3 (with $p = 2$), we use the random quantity

$$\frac{1}{2} \sum_{x \in \mathbb{Z}^d} R^2(n, x, \epsilon)$$

to approximate $|\mathbb{E} R_n - R_n|$ exponentially. Every step appears as an obvious modification of its correspondent counterpart used in the proof of Theorem 8.2.1 and Theorem 8.3.1.

The only exception is the part correspondent to Lemma 8.2.2. To complete the argument for Theorem 8.5.3, we prove Lemma 8.5.4 below in the rest of this section.

To begin, recall that

$$J_n = \#\{S[1, n] \cap \widetilde{S}[1, n]\}$$

and write, for $\epsilon > 0$,

$$J_n(\epsilon) = \sum_{x \in \mathbb{Z}^d} R(n, x, \epsilon) \widetilde{R}(n, x, \epsilon)$$

where $R(n, x, \epsilon)$ is defined in (8.1.4) and $\widetilde{R}(n, x, \epsilon)$ is generated by $\{\widetilde{S}(k)\}$ in the same way as $R(n, x, \epsilon)$ is generated by $\{S(k)\}$.

LEMMA 8.5.4. (1) *When $d = 2$,*

$$\lim_{\epsilon \to 0^+} \limsup_{n \to \infty} \frac{1}{b_n} \log \mathbb{E} \exp\left\{ \theta \sqrt{\frac{b_n}{n}} \log n |J_n - J_n(\epsilon)|^{1/2} \right\} = 0 \tag{8.5.9}$$

for any $\theta > 0$ and for any positive sequence b_n satisfying

$$b_n \longrightarrow \infty \quad \text{and} \quad b_n = o(\log n) \qquad (n \to \infty).$$

(2) *When $d = 3$,*

$$\lim_{\epsilon \to 0^+} \limsup_{n \to \infty} \frac{1}{b_n} \log \mathbb{E} \exp\left\{ \theta \left(\frac{b_n}{n}\right)^{1/4} |J_n - J_n(\epsilon)|^{1/2} \right\} = 0 \tag{8.5.10}$$

for any $\theta > 0$ and for any positive sequence b_n satisfying

$$b_n \longrightarrow \infty \quad \text{and} \quad b_n = o(n^{1/3}) \qquad (n \to \infty).$$

Proof. For similarity we only deal with the case $d = 2$. The challenge we face here is much more substantial than the proof of Lemma 8.2.2. Contrary to (6.1.13) (which is used in the proof of Lemma 8.2.2), there is no moment inequality providing bounds directly for $J_n - J_n(\epsilon)$.

Let $0 = n_0 < n_1 < \cdots < n_{[b_n]} = n$ be an integer partition of the interval $[0, n]$ such that

$$n_k - n_{k-1} = [n/b_n] \text{ or } [n/b_n] \pm 1 \qquad k = 1, \cdots [b_n].$$

Recall that

$$R(n, x, \epsilon) = \left(\frac{b_n}{n}\right) \det(\Gamma)^{-1/2} \sum_{y \in S[1,n]} p_\epsilon\left(\sqrt{\frac{b_n}{n}} \Gamma^{-1/2}(y - x)\right).$$

Write

$$\xi(n,x,\epsilon)=\Big(\frac{b_n}{n}\Big)\det(\Gamma)^{-1/2}\sum_{j=1}^{[b_n]}\sum_{y\in S(n_{j-1},n_j]}p_\epsilon\Big(\sqrt{\frac{b_n}{n}}\Gamma^{-1/2}(y-x)\Big).$$

Let $h_{n,\epsilon}$ be a function on $\mathbb{Z}^2$ defined by

$$h_{n,\epsilon}(x)=\Big(\frac{b_n}{n}\Big)\det(\Gamma)^{-1}\sum_{y\in\mathbb{Z}^2}p_\epsilon\Big(x-\sqrt{\frac{b_n}{n}}\Gamma^{-1/2}y\Big)p_\epsilon\Big(\sqrt{\frac{b_n}{n}}\Gamma^{-1/2}y\Big).$$

At the risk of over-simplification of the notation, we define

$$R(x)=1_{\{x\in S[1,n]\}},$$

$$\xi(x)=\sum_{j=1}^{[b_n]}1_{\{x\in S_j(n_{j-1},n_j]\}}$$

and

$$R_\epsilon(x)=\sum_{y\in S[1,n]}h_{n,\epsilon}\Big(\sqrt{\frac{b_n}{n}}\Gamma^{-1/2}(y-x)\Big),$$

$$\xi_\epsilon(x)=\sum_{j=1}^{[b_n]}\sum_{y\in S(n_{j-1},n_j]}h_{n,\epsilon}\Big(\sqrt{\frac{b_n}{n}}\Gamma^{-1/2}(y-x)\Big).$$

We define $\widetilde{R}(n,x,\epsilon)$, $\widetilde{\xi}(n,x,\epsilon)$, $\widetilde{R}(x)$, $\widetilde{\xi}(x)$, $\widetilde{R}_\epsilon(x)$, $\widetilde{\xi}_\epsilon(x)$ in an obvious way.

Observe that

$$J_n(\epsilon)=\sum_{x\in\mathbb{Z}^2}R(n,x,\epsilon)\widetilde{R}(n,x,\epsilon)=\sum_{x\in Z^2}1_{\{x\in S[1,n]\}}\widetilde{R}_\epsilon(x).$$

Therefore, with the notation used in the Hilbert space $l^2(\mathbb{Z}^2)$, we get

$$J_n-J_n(\epsilon)=\langle R,\widetilde{R}\rangle-\langle R,\widetilde{R}_\epsilon\rangle=\langle R,\widetilde{R}-\widetilde{R}_\epsilon\rangle.$$

Hence,

$$\big|J_n-J_n(\epsilon)\big|\le\big|\langle\xi,\widetilde{R}-\widetilde{R}_\epsilon\rangle\big|+\big|\langle\xi-R,\widetilde{R}-\widetilde{R}_\epsilon\rangle\big|.$$

For the first term on the right hand side,

$$\langle\xi,\widetilde{R}-\widetilde{R}_\epsilon\rangle=\langle\xi,\widetilde{R}\rangle-\langle\xi,\widetilde{R}_\epsilon\rangle=\langle\xi,\widetilde{R}\rangle-\langle\xi_\epsilon,\widetilde{R}\rangle=\langle\xi-\xi_\epsilon,\widetilde{R}\rangle.$$

As for the second term, observe that $\xi(x)-R(x)\ge 0$. Therefore,

$$\begin{aligned}\big|\langle\xi-R,\widetilde{R}-\widetilde{R}_\epsilon\rangle\big|&\le\langle\xi-R,\widetilde{\xi}\rangle+\langle\xi-R,\widetilde{\xi}_\epsilon\rangle\\&=\langle\xi-R,\widetilde{\xi}-\widetilde{\xi}_\epsilon\rangle+2\langle\xi-R,\widetilde{\xi}_\epsilon\rangle\\&=\langle\xi-R,\widetilde{\xi}-\widetilde{\xi}_\epsilon\rangle+2\langle\xi_\epsilon-R_\epsilon,\widetilde{\xi}\rangle.\end{aligned}$$

In summary, we have

$$\big|J_n-J_n(\epsilon)\big|\le\big|\langle\xi-\xi_\epsilon,\widetilde{R}\rangle\big|+\langle\xi-R,\widetilde{\xi}-\widetilde{\xi}_\epsilon\rangle+2\langle\xi_\epsilon-R_\epsilon,\widetilde{\xi}\rangle.$$

Notice that

$$\langle\xi-R,\widetilde{\xi}-\widetilde{\xi}_\epsilon\rangle\overset{d}{=}\langle\xi-\xi_\epsilon,\widetilde{\xi}-\widetilde{R}\rangle\le\big|\langle\xi-\xi_\epsilon,\widetilde{\xi}\rangle\big|+\big|\langle\xi-\xi_\epsilon,\widetilde{R}\rangle\big|.$$

For (8.5.9) to hold, therefore, we need only to show that for any $\theta > 0$,

$$\lim_{\epsilon\to 0^+}\limsup_{n\to\infty}\frac{1}{b_n}\log\mathbb{E}\exp\left\{\theta\sqrt{\frac{b_n}{n}}\log n\big|\langle\xi-\xi_\epsilon,\widetilde{R}\rangle\big|^{1/2}\right\}=0, \tag{8.5.11}$$

$$\lim_{\epsilon\to 0^+}\limsup_{n\to\infty}\frac{1}{b_n}\log\mathbb{E}\exp\left\{\theta\sqrt{\frac{b_n}{n}}\log n\big|\langle\xi-\xi_\epsilon,\widetilde{\xi}\rangle\big|^{1/2}\right\}=0, \tag{8.5.12}$$

and

$$\lim_{n\to\infty}\frac{1}{b_n}\log\mathbb{E}\exp\left\{\theta\sqrt{\frac{b_n}{n}}\log n\langle\xi_\epsilon-R_\epsilon,\widetilde{\xi}\rangle^{1/2}\right\}=0 \tag{8.5.13}$$

for any fixed $\epsilon > 0$.

Observe that for any integer $m \ge 1$,

$$\begin{aligned}
&\mathbb{E}\langle\xi-\xi_\epsilon,\widetilde{R}\rangle^m\\
&=\sum_{x_1,\cdots,x_m\in\mathbb{Z}^2}\left[\mathbb{E}\prod_{k=1}^m\Big(\xi(x_k)-\xi_\epsilon(x_k)\Big)\right]\left[\mathbb{E}\prod_{k=1}^m 1_{\{x_k\in S[1,n]\}}\right]\\
&\le\left\{\sum_{x_1,\cdots,x_m\in\mathbb{Z}^2}\left[\mathbb{E}\prod_{k=1}^m\Big(\xi(x_k)-\xi_\epsilon(x_k)\Big)\right]^2\right\}^{1/2}\\
&\quad\times\left\{\sum_{x_1,\cdots,x_m\in\mathbb{Z}^2}\left[\mathbb{E}\prod_{k=1}^m 1_{\{x_k\in S[1,n]\}}\right]^2\right\}^{1/2}\\
&=\Big\{\mathbb{E}\langle\xi-\xi_\epsilon,\widetilde{\xi}-\widetilde{\xi}_\epsilon\rangle^m\Big\}^{1/2}\Big\{\mathbb{E}J_n^m\Big\}^{1/2}.
\end{aligned}$$

For (8.5.11) to hold, it is sufficient to show that

$$\begin{aligned}
&\lim_{\epsilon\to 0^+}\limsup_{n\to\infty}\frac{1}{b_n}\log\sum_{m=0}^\infty\frac{\theta^m}{m!}\Big(\sqrt{\frac{b_n}{n}}\log n\Big)^m\\
&\qquad\times\left[\Big\{\mathbb{E}\langle\xi-\xi_\epsilon,\widetilde{\xi}-\widetilde{\xi}_\epsilon\rangle^m\Big\}\Big\{\mathbb{E}J_n^m\Big\}\right]^{1/4}=0.
\end{aligned} \tag{8.5.14}$$

By Cauchy-Schwarz inequality, for any $\lambda > 0$,

$$\begin{aligned}
&\sum_{m=0}^\infty\frac{\theta^m}{m!}\Big(\sqrt{\frac{b_n}{n}}\log n\Big)^m\left[\Big\{\mathbb{E}\langle\xi-\xi_\epsilon,\widetilde{\xi}-\widetilde{\xi}_\epsilon\rangle^m\Big\}\Big\{\mathbb{E}J_n^m\Big\}\right]^{1/4}\\
&\le\left\{\sum_{m=0}^\infty\frac{(\lambda\theta)^m}{m!}\Big(\sqrt{\frac{b_n}{n}}\log n\Big)^m\Big\{\mathbb{E}\langle\xi-\xi_\epsilon,\widetilde{\xi}-\widetilde{\xi}_\epsilon\rangle^m\Big\}^{1/2}\right\}^{1/2}\\
&\quad\times\left\{\sum_{m=0}^\infty\frac{(\lambda^{-1}\theta)^m}{m!}\Big(\sqrt{\frac{b_n}{n}}\log n\Big)^m\Big\{\mathbb{E}J_n^m\Big\}^{1/2}\right\}^{1/2}.
\end{aligned}$$

Taking $p = 2$ in (7.2.7), we have the bound

$$\lim_{n\to\infty}\frac{1}{b_n}\log\sum_{m=0}^\infty\frac{(\lambda^{-1}\theta)^m}{m!}\Big(\sqrt{\frac{b_n}{n}}\log n\Big)^m\Big\{\mathbb{E}J_n^m\Big\}^{1/2}=C(\lambda^{-1}\theta)^2 \tag{8.5.15}$$

for a universal constant $C > 0$. One can make the right hand side arbitrarily small by making λ sufficiently large. To establish (8.5.14), therefore, we only need show that

$$\lim_{\epsilon\to 0^+}\limsup_{n\to\infty}\frac{1}{b_n}\log\sum_{m=0}^{\infty}\frac{\theta^m}{m!}\Big(\sqrt{\frac{b_n}{n}}\log n\Big)^m\Big\{\mathbb{E}\langle\xi-\xi_\epsilon,\widetilde{\xi}-\widetilde{\xi}_\epsilon\rangle^m\Big\}^{1/2}=0$$

for every $\theta > 0$.

Indeed, this follows from an argument similar to the one used in the proof of Lemma 8.2.2. We describe it here briefly.

Write

$$\xi'_\epsilon = \lambda_{n,\epsilon}^{-2}\xi_\epsilon$$

where we choose $\lambda_{n,\epsilon}$ to make the function

$$\bar{h}_{n,\epsilon}(x) = \lambda_{n,\epsilon}^{-1}h_{n,\epsilon}(x) \qquad x\in\mathbb{Z}^2$$

a probability density on $\mathbb{Z}^2$. It is easy to see that for each $\epsilon > 0$, $\lambda_{n,\epsilon}\to 1$ as $n\to\infty$. Therefore, our claim is equivalent to

$$\lim_{\epsilon\to 0^+}\limsup_{n\to\infty}\frac{1}{b_n}\log\bigg(\sum_{m=0}^{\infty}\frac{\theta^m}{m!}\Big(\sqrt{\frac{b_n}{n}}\log n\Big)^m \times\Big\{\mathbb{E}\langle\xi-\xi'_\epsilon,\widetilde{\xi}-\widetilde{\xi'}_\epsilon\rangle^m\Big\}^{1/2}\bigg)=0. \tag{8.5.16}$$

Here we point out that $\mathbb{E}\langle\xi-\xi'_\epsilon,\widetilde{\xi}-\widetilde{\xi'}_\epsilon\rangle^m\ge 0$ for every integer $m\ge 0$.

Take the additive functional

$$\mathcal{F}_k^x(S)=\sum_{j=1}^k 1_{\{x\in S(n_{j-1},n_j]\}}\qquad k=1,\cdots,[b_n]$$

on the increasing sequence $n_1<\cdots<n_{[b_n]}$ and take the probability density $h(x)=\bar{h}_{n,\epsilon}(x)$ in the inequality (6.1.16) of Theorem 6.1.2. We have

$$\begin{aligned}&\sum_{m=0}^{\infty}\frac{\theta^m}{m!}\Big(\sqrt{\frac{b_n}{n}}\log n\Big)^m\Big\{\mathbb{E}\langle\xi-\xi'_\epsilon,\widetilde{\xi}-\widetilde{\xi'}_\epsilon\rangle^m\Big\}^{1/2}\\ &\le\prod_{k=1}^{[b_n]}\sum_{m=0}^{\infty}\frac{\theta^m}{m!}\Big(\sqrt{\frac{b_n}{n}}\log n\Big)^m\Big\{\mathbb{E}\big[\bar{D}^{(2)}_{n_k-n_{k-1},\epsilon}\big]^m\Big\}^{1/2},\end{aligned}$$

where, in connection to (5.3.32) with $d=p=2$,

$$\begin{aligned}\bar{D}^{(2)}_{n_k-n_{k-1},\epsilon}&\equiv\sum_{x\in\mathbb{Z}^2}\bigg[1_{\{x\in S[1,n_k-n_{k-1}]\}}-\sum_{z\in S[1,n_k-n_{k-1}]}\bar{h}_{n,\epsilon}(z-x)\bigg]\\ &\qquad\times\bigg[1_{\{x\in\widetilde{S}[1,n_k-n_{k-1}]\}}-\sum_{z\in\widetilde{S}[1,n_k-n_{k-1}]}\bar{h}_{n,\epsilon}(z-x)\bigg]\\ &=\big(1+o(1)\big)D_{n_k-n_{k-1},2\epsilon}\qquad(n\to\infty).\end{aligned}$$

By (5.3.33) in Theorem 5.3.5 (with $p=2$ and n being replaced by n_k-n_{k-1}),

$$\lim_{\epsilon\to 0^+}\limsup_{n\to\infty}\Big(\frac{b_n}{n}\Big)^2(\log n)^4\mathbb{E}\big[\bar{D}^{(2)}_{n_k-n_{k-1},\epsilon}\big]^2=0$$

uniformly over $1 \le k \le [b_n]$. Consequently,

$$\lim_{\epsilon\to 0^+}\limsup_{n\to\infty}\sum_{m=0}^{\infty}\frac{\theta^m}{m!}\Big(\sqrt{\frac{b_n}{n}}\log n\Big)^m\Big\{\mathbb{E}\big[\bar{D}^{(2)}_{n_k-n_{k-1},\epsilon}\big]^m\Big\}^{1/2}=1$$

uniformly over $1 \le k \le [b_n]$. This leads to the requested (8.5.16) and therefore to (8.5.11).

Observe that by the inequality (6.1.12) with $a = [b_n]$ and by (5.3.26) in Theorem 5.3.4 (with n being replaced by $[n/b_n]$), the bound given in (8.5.15) remains true even when J_n is replaced by $\mathcal{J}_n \equiv \langle \xi, \widetilde{\xi}\rangle$. Consequently, (8.5.12) follows from the same argument as (8.5.11).

It remains to prove (8.5.13). Notice that

$$\langle \xi_\epsilon - R_\epsilon, \widetilde{\xi}\rangle = \langle \xi_\epsilon - R_\epsilon, \widetilde{R}\rangle + \langle \xi_\epsilon - R_\epsilon, \widetilde{\xi}-\widetilde{R}\rangle$$

and that

$$\langle \xi_\epsilon - R_\epsilon, \widetilde{\xi}-\widetilde{R}\rangle = \sum_{x\in\mathbb{Z}^2}\big[\xi(n,x,\epsilon)-R(n,x,\epsilon)\big]\big[\widetilde{\xi}(n,x,\epsilon)-\widetilde{R}(n,x,\epsilon)\big].$$

From the second equation

$$\begin{aligned}\langle \xi_\epsilon - R_\epsilon, \widetilde{\xi}-\widetilde{R}\rangle^{1/2} \le &\frac{1}{2}\Big\{\sum_{x\in\mathbb{Z}^2}\big[\xi(n,x,\epsilon)-R(n,x,\epsilon)\big]^2\Big\}^{1/2}\\ &+\frac{1}{2}\Big\{\sum_{x\in\mathbb{Z}^2}\big[\widetilde{\xi}(n,x,\epsilon)-\widetilde{R}(n,x,\epsilon)\big]^2\Big\}^{1/2}.\end{aligned}$$

In addition,

$$\begin{aligned}\langle \xi_\epsilon - R_\epsilon, \widetilde{R}\rangle &= \sum_{x\in\mathbb{Z}^2}\big[\xi(n,x,\epsilon)-R(n,x,\epsilon)\big]\widetilde{R}(n,x,\epsilon)\\ &\le \Big\{\sum_{x\in\mathbb{Z}^2}\big[\xi(n,x,\epsilon)-R(n,x,\epsilon)\big]^2\Big\}^{1/2}\Big\{\sum_{x\in\mathbb{Z}^2}\widetilde{R}^2(n,x,\epsilon)\Big\}^{1/2}.\end{aligned}$$

Hence, for any fixed but arbitrary $\lambda > 0$,

$$\begin{aligned}&\langle \xi_\epsilon - R_\epsilon, \widetilde{R}\rangle^{1/2}\\ &\le \frac{\lambda}{2}\sum_{x\in\mathbb{Z}^2}\big[\xi(n,x,\epsilon)-R(n,x,\epsilon)\big]^2\Big\}^{1/2}+\frac{\lambda^{-1}}{2}\Big\{\sum_{x\in\mathbb{Z}^2}\widetilde{R}^2(n,x,\epsilon)\Big\}^{1/2}.\end{aligned}$$

Furthermore, notice that

$$\sum_{x\in\mathbb{Z}^2}\big[\xi(n,x,\epsilon)-R(n,x,\epsilon)\big]^2 \overset{d}{=} \sum_{x\in\mathbb{Z}^2}\big[\widetilde{\xi}(n,x,\epsilon)-\widetilde{R}(n,x,\epsilon)\big]^2,$$

$$\sum_{x\in\mathbb{Z}^2}\widetilde{R}^2(n,x,\epsilon) \overset{d}{=} \sum_{x\in\mathbb{Z}^2}R^2(n,x,\epsilon)$$

and that by Theorem 8.1.3 (with $d = p = 2$),

$$\begin{aligned}
&\lim_{n\to\infty}\frac{1}{b_n}\log\mathbb{E}\exp\Big\{\theta\sqrt{\frac{b_n}{n}}\log n\Big(\sum_{x\in\mathbb{Z}^2}R^2(n,x,\epsilon)\Big)^{1/2}\Big\}\\
&=\sup_{g\in\mathcal{F}_2}\Big\{2\pi\theta\det(\Gamma)^{1/4}\Big(\int_{\mathbb{R}^2}|g(x,\epsilon)|^4dx\Big)^{1/2}-\frac{1}{2}\int_{\mathbb{R}^2}|\nabla g(x)|^2dx\Big\}\\
&\le\sup_{g\in\mathcal{F}_2}\Big\{2\pi\theta\det(\Gamma)^{1/4}\Big(\int_{\mathbb{R}^2}|g(x)|^4dx\Big)^{1/2}-\frac{1}{2}\int_{\mathbb{R}^2}|\nabla g(x)|^2dx\Big\}\\
&=2\pi^2\sqrt{\det(\Gamma)}\kappa(2,2)^4\theta^2
\end{aligned}$$

for every $\theta > 0$ and $\epsilon > 0$, where the second step follows from Jensen inequality and the last step from Theorem C.1 in the Appendix.

For (8.5.13) to hold, therefore, it is sufficient to establish that

$$\begin{aligned}
&\lim_{n\to\infty}\frac{1}{b_n}\log\mathbb{E}\exp\Big\{\theta\sqrt{\frac{b_n}{n}}\log n\\
&\quad\times\Big(\sum_{x\in\mathbb{Z}^2}\big[\xi(n,x,\epsilon)-R(n,x,\epsilon)\big]^2\Big)^{1/2}\Big\}=0
\end{aligned}\tag{8.5.17}$$

for every fixed $\theta > 0$, $\epsilon > 0$.

Let $f \in l^2(\mathbb{Z}^2)$ satisfy $||f||_2 = 1$. We have

$$\begin{aligned}
&\sum_{x\in\mathbb{Z}^2}f(x)\big[\xi(n,x,\epsilon)-R(n,x,\epsilon)\big]\\
&=\Big(\frac{b_n}{n}\Big)\det(\Gamma)^{-1/2}\sum_{x\in\mathbb{Z}^2}T_{n,\epsilon}f(x)\Big\{\sum_{j=1}^{[b_n]}1_{\{x\in S(n_{j-1},n_j]\}}-1_{\{x\in S[1,n]\}}\Big\}
\end{aligned}$$

where, by Cauchy-Schwarz inequality, we have

$$\begin{aligned}
T_{n,\epsilon}f(x)&\equiv\sum_{y\in\mathbb{Z}^2}f(y)p_\epsilon\Big(\sqrt{\frac{b_n}{n}}\Gamma^{-1/2}(x-y)\Big)\\
&\le\Big\{\sum_{y\in\mathbb{Z}^2}p_\epsilon^2\Big(\sqrt{\frac{b_n}{n}}\Gamma^{-1/2}(x-y)\Big)\Big\}^{1/2}\\
&=\Big\{\sum_{y\in\mathbb{Z}^2}p_\epsilon^2\Big(\sqrt{\frac{b_n}{n}}\Gamma^{-1/2}y\Big)\Big\}^{1/2}\equiv K_{n,\epsilon}.
\end{aligned}$$

By the fact that

$$\sum_{j=1}^{[b_n]}1_{\{x\in S(n_{j-1},n_j]\}}-1_{\{x\in S[1,n]\}}\ge 0\qquad x\in\mathbb{Z}^2,$$

we obtain

$$\begin{aligned}
&\sum_{x\in\mathbb{Z}^2} f(x)\big[\xi(n,x,\epsilon)-R(n,x,\epsilon)\big]\\
&\le\Big(\frac{b_n}{n}\Big)\det(\Gamma)^{-1/2}\sum_{x\in\mathbb{Z}^2}K_{n,\epsilon}\bigg\{\sum_{j=1}^{[b_n]}1_{\{x\in S(n_{j-1},n_j]\}}-1_{\{x\in S[1,n]\}}\bigg\}\\
&=\Big(\frac{b_n}{n}\Big)\det(\Gamma)^{-1/2}K_{n,\epsilon}\bigg\{\sum_{j=1}^{[b_n]}\#\big\{S(n_{j-1},n_j]\big\}-\#\big\{S[1,n]\big\}\bigg\}.
\end{aligned}$$

Taking supremum over f on the left hand side gives us

$$\begin{aligned}
&\bigg(\sum_{x\in\mathbb{Z}^2}\big[\xi(n,x,\epsilon)-R(n,x,\epsilon)\big]^2\bigg)^{1/2}\\
&\le\Big(\frac{b_n}{n}\Big)\det(\Gamma)^{-1/2}K_{n,\epsilon}\bigg\{\sum_{j=1}^{[b_n]}\#\big\{S(n_{j-1},n_j]\big\}-\#\big\{S[1,n]\big\}\bigg\}.
\end{aligned}$$

In addition,

$$\lim_{n\to\infty}\Big(\frac{b_n}{n}\Big)\det(\Gamma)^{-1/2}K_{n,\epsilon}^2=\int_{\mathbb{R}^2}p_\epsilon^2(x)dx=p_{2\epsilon}(0)=\frac{1}{4\pi\epsilon}.$$

Therefore, the requested (8.5.17) follows from part (1) of Lemma 7.1.4. □

8.6. Laws of the iterated logarithm

Among the large deviations established in this chapter, those including $b_n = \log\log n$ (and $b_n = \log\log\log n$ sometimes) apply to the law of the iterated logarithm. The following theorem appears as a consequence of Theorem 8.1.1.

THEOREM 8.6.1. *For $d=1$ and $p>1$,*

$$\begin{aligned}
&\limsup_{n\to\infty} n^{-\frac{p+1}{2}}(\log\log n)^{-\frac{p-1}{2}}\sum_{x\in\mathbb{Z}}l^p(n,x)\\
&=2^{\frac{p+1}{2}}\sigma^{-(p-1)}(p-1)^{\frac{p-1}{2}}(p+1)^{\frac{p-3}{2}}B\Big(\frac{1}{p-1},\frac{1}{2}\Big)^{-(p-1)}\qquad a.s.
\end{aligned}$$

Proof. Exercise 8.8.7. □

Corresponding to Theorem 8.2.1, Theorem 8.2.3, Theorem 8.5.1 and (8.5.7) in Theorem 8.5.3, we have[3]

[3]Without the assumption on $(2+\delta)$-moment in Theorem 8.2.3 and in Theorem 8.5.1, one still can have some weaker versions of large deviations leading to (8.6.2) and (8.6.4), we refer the interested reader to Theorem 1.1 in [**8**].

THEOREM 8.6.2. *For $d = 2$,*

$$\limsup_{n\to\infty} \frac{1}{n\log\log n}(Q_n - \mathbb{E}Q_n) = \det(\Gamma)^{-1/2}\kappa(2,2)^4 \quad a.s., \tag{8.6.1}$$

$$\liminf_{n\to\infty} \frac{1}{n\log\log\log n}(Q_n - \mathbb{E}Q_n) = -(2\pi)^{-1}\det(\Gamma)^{-1/2} \quad a.s., \tag{8.6.2}$$

$$\liminf_{n\to\infty} \frac{(\log n)^2}{n\log\log n}(R_n - \mathbb{E}R_n) = -(2\pi)^2\sqrt{\det(\Gamma)}\kappa(2,2)^4 \quad a.s., \tag{8.6.3}$$

$$\limsup_{n\to\infty} \frac{(\log n)^2}{n\log\log\log n}(R_n - \mathbb{E}R_n) = 2\pi\sqrt{\det(\Gamma)} \quad a.s. \tag{8.6.4}$$

Proof. Due to similarity we only take (8.6.1) as an example. By examining the proof of Theorem 4.4.1 carefully, one can see that the only thing that needs to be proved is that

$$\sup_{n\geq 1}\mathbb{E}\exp\left\{\frac{\theta}{n}\max_{1\leq k\leq n}|Q_k - \mathbb{E}Q_k|\right\} < \infty \tag{8.6.5}$$

for some $\theta > 0$. This can be done by an argument used in the proof of Theorem 6.4.3 □

In the case $d = 3$, the large deviations given in Theorem 8.3.1 and in Theorem 8.5.2 lead to the following theorem.

THEOREM 8.6.3. *For $d = 3$,*

$$\limsup_{n\to\infty} \frac{\pm(Q_n - \mathbb{E}Q_n)}{\sqrt{n\log n\log\log n}} = \frac{1}{\pi\sqrt{\det(\Gamma)}} \quad a.s., \tag{8.6.6}$$

$$\limsup_{n\to\infty} \frac{\pm(R_n - \mathbb{E}R_n)}{\sqrt{n\log n\log\log n}} = \frac{\gamma_S^2}{\pi\sqrt{\det(\Gamma)}} \quad a.s. \tag{8.6.7}$$

Proof. Due to similarity we only prove (8.6.6). Using the integrability established in (6.4.13), Theorem 6.4.3 instead of (6.4.1) in Theorem 6.4.1, a slight modification of the estimate conducted in the proof of Theorem 8.3.1 gives that

$$\limsup_{n\to\infty}\frac{1}{b_n}\log\mathbb{P}\Big\{\max_{1\leq k\leq n}|Q_k - \mathbb{E}Q_k| \geq \lambda\sqrt{nb_n\log n}\Big\} \leq -2$$

for sufficiently large $\lambda > 0$. Taking $b_n = \log\log n$, by Theorem 8.3.1 and a Borel-Cantelli-type argument (see the proof of of Theorem 4.4.1) we have (8.6.6). □

In addition to the results established in the dimensions $d \leq 3$, the laws of the iterated logarithm for $Q_n - \mathbb{E}Q_n$ and for $R_n - \mathbb{E}R_n$ were studied in the early days in the case $d \geq 4$. The next theorem essentially belongs to Jain and Pruitt ([**103**]).[4]

[4]Jain and Pruitt only proved (8.6.9), but their argument can be easily applied to the proof of (8.6.8).

THEOREM 8.6.4. *For $d \geq 4$,*

$$\limsup_{n\to\infty} \frac{\pm(Q_n - \mathbb{E}Q_n)}{\sqrt{n\log\log n}} = \sqrt{2}\mathfrak{D} \quad a.s., \tag{8.6.8}$$

$$\limsup_{n\to\infty} \frac{\pm(R_n - \mathbb{E}R_n)}{\sqrt{n\log\log n}} = \sqrt{2}\widetilde{\mathfrak{D}} \quad a.s. \tag{8.6.9}$$

where the constants $\mathfrak{D} > 0$ and $\widetilde{\mathfrak{D}} > 0$ are given in Lemma 5.5.2.

We omit the proof for which we refer an interested reader to [**103**]. Instead we point out the fact the main part of the proof relies on some large deviation tail estimates in the special case $b_n = \log\log n$.[5] The forms of these large deviations will be discussed in the next section.

8.7. What do we expect in $d \geq 4$?

In light of the central limit theorem given in Theorem 5.5.3, it is natural to have a large deviation principle with Gaussian tail when $d \geq 4$, at least for the small deviation scale b_n. As shown at the end of the previous section, this includes the case when $b_n = \log\log n$. As for the full spectrum of b_n responding to the large deviation with Gaussian tail, we have the following conjecture.

CONJECTURE 8.7.1. *In the case $d \geq 4$,*

$$\lim_{n\to\infty} \frac{1}{b_n} \log \mathbb{P}\big\{ \pm (Q_n - \mathbb{E}Q_n) \geq \lambda\sqrt{nb_n}\big\} = -\frac{\lambda^2}{2\mathfrak{D}^2} \tag{8.7.1}$$

for any $\lambda > 0$ and for any positive sequence b_n satisfying

$$b_n \longrightarrow \infty \quad and \quad b_n = o(n^{1/3}) \qquad (n \to \infty)$$

and

$$\lim_{n\to\infty} \frac{1}{b_n} \log \mathbb{P}\big\{ \pm (R_n - \mathbb{E}R_n) \geq \lambda\sqrt{nb_n}\big\} = -\frac{\lambda^2}{2\widetilde{\mathfrak{D}}^2} \tag{8.7.2}$$

for any $\lambda > 0$ and for any positive sequence b_n satisfying

$$b_n \longrightarrow \infty \quad and \quad b_n = o\Big(n^{\frac{d-2}{d+2}}\Big) \qquad (n \to \infty)$$

where the constants $\mathfrak{D}^2$ and $\widetilde{\mathfrak{D}}^2$ are given in Lemma 5.5.2.

Here is what we can do in connection to the large deviations with Gaussian tails: By an argument the same as the one used for Theorem 8.3.2 and for (8.5.5) in Theorem 8.5.2 we can claim the following bounds of sub-Gaussian tails.

THEOREM 8.7.2. *When $d \geq 4$,*

$$\limsup_{n\to\infty} \frac{1}{b_n} \log \mathbb{P}\big\{Q_n - \mathbb{E}Q_n \leq -\lambda\sqrt{nb_n}\big\} \leq -\frac{\lambda^2}{2\mathfrak{D}^2}, \tag{8.7.3}$$

$$\limsup_{n\to\infty} \frac{1}{b_n} \log \mathbb{P}\big\{R_n - \mathbb{E}R_n \geq \lambda\sqrt{nb_n}\big\} \leq -\frac{\lambda^2}{2\widetilde{\mathfrak{D}}^2} \tag{8.7.4}$$

[5] The paper [**103**] did not formulate them explicitly into the form of large deviations.

for $\lambda > 0$ and for any positive sequence b_n satisfying

$$b_n \longrightarrow \infty \quad \text{and} \quad b_n = o(n) \qquad (n \to \infty).$$

Again, we conjecture that the sub-Gaussian tails can be strengthened into Gaussian tails.

Our major concern is on the possibility of the large deviations with non-Gaussian tails in the case $d \geq 4$. When $d \geq 5$, Asselah ([**3**]) obtained a non-Gaussian large deviation principle for $Q_n - \mathbb{E}Q_n$ which states that

$$\lim_{n\to\infty} \frac{1}{\sqrt{n}} \log \mathbb{P}\{Q_n - \mathbb{E}Q_n \geq \lambda n\} = -\sqrt{2\lambda}\mathcal{C}(S) \tag{8.7.5}$$

where $\mathcal{C}(S) > 0$ is defined defined in (7.5.8) (with $p = 2$). One can see that (8.7.5) is related to Theorem 7.5.3 with $b_n = \sqrt{n}$. A natural question is to find a full spectrum of b_n such that (8.7.5) holds.

Another open problem is on the non-Gaussian large deviations for $R_n - \mathbb{E}R_n$ in the case $d \geq 5$.

CONJECTURE 8.7.3. *When $d \geq 5$,*

$$\lim_{n\to\infty} \frac{1}{b_n} \log \mathbb{P}\{Q_n - \mathbb{E}Q_n \geq \lambda b_n^2\} = -\sqrt{2\lambda}\mathcal{C}(S) \tag{8.7.6}$$

for any $\lambda > 0$ and for any b_n satisfying

$$b_n/\sqrt[3]{n} \longrightarrow \infty \quad \text{and} \quad b_n = o(n) \qquad (n \to \infty)$$

and

$$\lim_{n\to\infty} \frac{1}{b_n} \log \mathbb{P}\Big\{R_n - \mathbb{E}R_n \leq -\lambda b_n^{\frac{d}{d-2}}\Big\} = \widetilde{\mathcal{C}}(S)\lambda^{\frac{d-2}{d}} \tag{8.7.7}$$

for any $\lambda > 0$ and for any b_n satisfying

$$b_n/n^{\frac{d-2}{d+2}} \longrightarrow \infty \quad \text{and} \quad b_n = o\Big(n^{\frac{d-2}{d}}\Big) \qquad (n \to \infty).$$

The tail behaviors of $Q_n - \mathbb{E}Q_n$ and $R_n - \mathbb{E}R_n$ are closely related to those of, respectively, I_∞ and J_∞. The reason why $Q_n - \mathbb{E}Q_n$ and $R_n - \mathbb{E}R_n$ should behave differently in the super-critical dimension $d \geq 5$ is caused by the difference in LDP pattern between I_∞ and J_∞ (see Theorem 7.5.3 and Conjecture 8.7.3). If Conjecture 8.7.3 holds, a natural problem is to identify the constant $\widetilde{\mathcal{C}}(S)$. We believe that the value of $\widetilde{\mathcal{C}}(S)$ is related to that of $\widetilde{\mathcal{C}}(d,p)$ (with $p = 2$) given in Conjecture 7.5.7

The extension from (8.7.5) to (8.7.6) can be very delicate. Indeed, by the fact that $\mathbb{E}Q_n = O(n)$, the large deviation in (8.7.5) is essentially a story of the tail behavior of Q_n. When $b_n = o(\sqrt{n})$, on the other hand, the renormalization makes the model much less tractable. In addition, $Q_n - \mathbb{E}Q_n$ is no longer approximatable by its smoothed version in the super-critical dimension $d \geq 5$.

Perhaps the dimension $d = 4$ is the most interesting and most challenging case. Much less regarding to the large deviations for $Q_n - \mathbb{E}Q_n$ and $R_n - \mathbb{E}R_n$ has been known in the case $d = 4$, especially the one with non-Gaussian tail. On the analytic side, the Gagliardo-Nirenberg inequality given in (3.3.2) still holds in the form of the Sobolev inequality

$$||f||_4 \le C||\nabla f||_2 \qquad f \in W^{1,2}(\mathbb{R}^4) \tag{8.7.8}$$

when $d = 4$ and $p = 2$. On the other hand, the L_2-norm $||f||_2$ is no longer relevant. All of these indicate that in the sense of the non-Gaussian tail large deviations for $Q_n - \mathbb{E}Q_n$ and $R_n - \mathbb{E}R_n$, the case $d = 4$ is critical between a diffusive behavior pattern (when $d \le 3$) and a non-diffusive behavior pattern (when $d \ge 5$) of the random walks.

CONJECTURE 8.7.4. *When $d = 4$,*

$$\lim_{n\to\infty} \frac{1}{b_n} \log \mathbb{P}\{Q_n - \mathbb{E}Q_n \ge \lambda b_n^2\} = -\frac{\sqrt{\lambda}}{4\pi} \det(\Gamma)^{-1/4}\kappa(4,2)^{-2} \tag{8.7.9}$$

for any $\lambda > 0$ and for any positive sequence b_n satisfying

$$b_n/n^{1/3} \longrightarrow \infty \quad \text{and} \quad b_n = o(n) \qquad (n \to \infty)$$

and

$$\lim_{n\to\infty} \frac{1}{b_n} \log \mathbb{P}\{R_n - \mathbb{E}R_n \le -\lambda b_n^2\} = -\frac{\sqrt{\lambda}}{4\pi\gamma_S} \det(\Gamma)^{-1/4}\kappa(4,2)^{-2} \tag{8.7.10}$$

for any $\lambda > 0$ and for any positive sequence b_n satisfying

$$b_n/n^{1/3} \longrightarrow \infty \quad \text{and} \quad b_n = o(\sqrt{n}) \qquad (n \to \infty).$$

In addition to the difficulty we face in the super-critical dimensions, the extra challenge here is the lack of sharp tail estimate for I_n and J_n in the case $d = 4$ and $p = 2$. In this sense, the final solution of Conjecture 8.7.4 might depend on the progress on Conjecture 7.4.2. On the positive side, the conjecture made in (8.7.10) is partially supported by van den Berg, Bolthausen and den Hollander's large deviation for the Wiener sausage defined in (8.8.6) ($d = 4$) in the "Notes and comments" section below. Finally, we refer the interested reader to the recent work by Castell ([**25**]) for partial progress on this problem.

8.8. Notes and comments

Section 8.1

Theorem 8.1.1 is a discrete version of Theorem 4.2.1 and was established by Chen and Li ([**35**]). We refer the reader to [**27**] (Theorem 3.1, [**27**]) for a detailed proof of Theorem 8.1.2.

EXERCISE 8.8.1. *Prove Theorem 8.1.3.*

EXERCISE 8.8.2. *Establish the following LDP in the case $d = 1$:*

$$\lim_{n\to\infty} \frac{1}{b_n} \log \mathbb{P}\Big\{ \sup_{x\in\mathbb{Z}} l(n,x) \ge \lambda\sqrt{nb_n}\Big\} = -\frac{\sigma^2\lambda^2}{2} \qquad \lambda > 0 \tag{8.8.1}$$

for any positive sequence satisfying

$$b_n \longrightarrow \infty \quad \text{and} \quad b_n = o(n) \qquad (n \to \infty).$$

The following two exercises originate from Khoshnevisan and Lewis ([**111**]), Chen ([**33**]).

EXERCISE 8.8.3 (hard). *Let ζ_n be the random walk in random scenery defined in (5.6.7). In addition, assume that for some $\theta > 0$,*

$$\mathbb{E}\exp\big\{\theta\xi(0)\big\} < \infty.$$

Prove that as $d = 1$,

$$\lim_{n\to\infty}\frac{1}{b_n}\log\mathbb{P}\big\{\pm\zeta_n \geq \lambda(nb_n)^{3/4}\big\} = -\frac{1}{4}\sigma^{2/3}(3\lambda)^{4/3} \qquad \lambda > 0$$

for any positive sequence b_n satisfying $b_n \to \infty$ with $b_n = o\big(n^{1/3}\big)$.

EXERCISE 8.8.4 (hard). *Let H_n be the energy of the charged polymer defined in (5.6.11). In addition, assume that ω_1 is symmetric and that for some $\theta > 0$,*

$$\mathbb{E}\exp\big\{\theta\omega_1^2\big\} < \infty.$$

Prove that as $d = 1$,

$$\lim_{n\to\infty}\frac{1}{b_n}\log\mathbb{P}\big\{\pm H_n \geq \lambda(nb_n)^{3/4}\big\} = -\frac{1}{2}\sigma^{2/3}(3\lambda)^{4/3} \qquad \lambda > 0$$

for any positive sequence b_n satisfying $b_n \to \infty$ with $b_n = o\big(n^{1/3}\big)$.

Section 8.2

Theorem 8.2.2 was proved in [**7**] and appears as the discrete version of Theorem 4.3.1.

EXERCISE 8.8.5 (hard). *Let $X(t)$ be a symmetric and square integrable Lévy process taking values in $\mathbb{Z}^2$ and define the local time of $X(t)$:*

$$\bar{l}(t,x) = \int_0^t 1_{\{X(s)=x\}}ds.$$

The process $X(t)$ is also known as a compound Poisson process due to the representation

$$X(t) = S(N_t) \qquad t \geq 0$$

where $S(k)$ is a symmetric random walk $\mathbb{Z}^2$ and N_t is a Poisson process independent of $S(k)$, and

$$\mathbb{P}\{N_t = n\} = e^{-t}\frac{t^n}{n!} \qquad n = 0,1,2,\cdots.$$

Prove that

$$\lim_{n\to\infty}\frac{1}{b_n}\log\mathbb{E}\exp\left\{\theta\sqrt{\frac{b_n}{n}}\Big(\sum_{x\in\mathbb{Z}^d}\big[l(n,x) - \bar{l}(n,x)\big]^2\Big)^{1/2}\right\} = 0$$

for any $\theta > 0$ and for any positive sequence b_n satisfying

$$b_n \longrightarrow \infty \quad \text{and} \quad b_n = o(n) \qquad (n\to\infty).$$

Hint: Write $T_k = \inf\{t \geq 0;\ \ N_t \geq k+1\}$ $(k = 0, 1, \cdots)$ *and* $\tau_k = T_k - T_{k-1}$ $(k = 1, 2, \cdots)$. *Then* $\{\tau_k\}$ *is an i.i.d. sequence of exponential random variable with parameter 1, and*

$$\bar{l}(T_n, x) = 1_0(x) + \sum_{k=1}^{n} \tau_k 1_{\{S(k)=x\}} \qquad x \in \mathbb{Z}^d.$$

Sections 8.3 and 8.4

In light of the central limit theorem given in Theorem 5.5.3, the large deviation (also called moderate deviation) stated in Theorem 8.3.1 and Theorem 8.3.2, is naturally expected.

Prior to [**3**], Asselah [**2**] published a paper which sets rough bounds for the tail probability of Q_n in the case $b_n = n^{1/3}$. Further, Fleischmann, Mörters and Wachtel pointed out (Proposition 11, [**79**]) that

$$\mathbb{P}\{|Q_n - \mathbb{E}Q_n| \geq \sqrt{nb_n^3}\} \leq \exp\{-cb_n\}$$

for $b_n \geq (\log n)^3$.

Section 8.5

An early result by Donsker and Varadhan ([**59**]) shows that

$$\lim_{n\to\infty} n^{-\frac{d}{d+2}} \log \mathbb{E}\exp\{-\theta R_n\} = -\theta^{\frac{2}{2+d}} \Big(\frac{d+2}{2}\Big)\Big(\frac{2\lambda_d}{d}\Big)^{\frac{d}{d+2}}$$

where $\lambda_d > 0$ is the lowest eigenvalue of the generator $-(\Delta/2)$ on the ball of unit volume with zero boundary values. See Donsker and Varadhan [**57**], Bolthausen [**16**] and Sznitman [**157**] for some versions of this result in the setting of Wiener sausages. For the large deviations of different scales, Hamana ([**89**]), Hamana and Kesten ([**90**], [**91**]) proved that for any $\lambda > 0$,

$$\lim_{n\to\infty} \frac{1}{n} \log \mathbb{P}\{R_n \geq \lambda n\}$$

yields a non-trivial limit. This is complementary to (8.5.5) and to (8.7.4) with $b_n = n$.

Historically, the limit laws for the renormalized range $R_n - \mathbb{E}R_n$ were first investigated in high dimensions ($d \geq 3$). The central limit theorems given in Theorem 5.5.3 and the related Gaussian tails were realized in the early days. The study of the large deviations were often formulated in the forms of the law of the iterated logarithm, where b_n increases in the iterated logarithmic rates — too slow for non-Gaussian tails. Here we mention the papers by Jain and Pruitt ([**103**], Hamana ([**88**]) for $d \geq 4$ and the paper by Bass and Kumagai ([**10**]) for $d = 3$. Theorem 8.5.2 summarizes the Gaussian-type large deviations for $R_n - \mathbb{E}R_n$ in $d = 3$.

We point out the paper by van den Berg, Bolthausen and den Hollander ([**11**]), for a possible link to our non-Gaussian-type large deviations stated in Theorem 8.5.3. Recall that

$$W^\epsilon(t) = \bigcup_{0\leq s\leq t} \{x \in \mathbb{R}^d;\ |W(s) - x| \leq \epsilon\}$$

is the Wiener sausage. In [**11**], van den Berg, Bolthausen and den Hollander proved that for any $b > 0$,

$$\lim_{t\to\infty} \frac{1}{\log t} \log \mathbb{P}\Big\{|W^\epsilon(t)| \le \frac{bt}{\log t}\Big\} = -\Psi_2(b) \tag{8.8.2}$$

when $d = 2$, and that

$$\lim_{t\to\infty} t^{-\frac{d-2}{d}} \log \mathbb{P}\big\{|W^\epsilon(t)| \le bt\big\} = -\Psi_d(b) \tag{8.8.3}$$

when $d \ge 3$, where $\Psi_d(b)$ is given in terms of variation. Furthermore,

$$\lim_{b\to 2\pi^-} (2\pi - b)^{-1}\Psi_2(b) = (2\pi)^{-2}\kappa(2,2)^{-4}$$

and

$$\lim_{b\to\theta_d(\epsilon)^-} \big(\theta_d(\epsilon) - b\big)^{-2/d}\Psi_d(b) = (2\pi)^{-\frac{d-2}{d}}\theta_d(\epsilon)^{-4/d}\kappa(d,2)^{-8/d}$$

for $d = 3, 4$, where $\theta_d(\epsilon)$ is defined in (7.6.8).

The "$d = 2$" part of Theorem 8.5.3 was established by Bass, Chen and Rosen ([**8**]), where it was required that $b_n = o\big((\log n)^{1/5}\big)$. Here we improve this result by optimizing the conditions on b_n and therefore close the gap between the van den Berg, Bolthausen and den Hollander's large deviations and the large deviations in Theorem 8.5.3.

By comparing Theorem 8.5.3 and Theorems 8.2.1 and 8.4.2, one can see how $Q_n - \mathbb{E}Q_n$ and $R_n - \mathbb{E}R_n$ are related as far as the large deviations are concerned.

Let $p \ge 1$ be an integer. Recall that $\widehat{R}_n^{(p)}$ is the number of distinct sites visited exactly p ($p \ge 1$) times by $S(k)$ in the first n steps, and that $R_n^{(p)}$ is the number of distinct sites visited at least p ($p \ge 1$) times by $S(k)$ in the first n steps. In view of Hamana's weak laws given in (5.6.5) and (5.6.6), and of the LDP for $\gamma\big([0,1]^2_<\big)$ given in Theorem 4.3.1, it is natural to believe the following:

CONJECTURE 8.8.6. *For any positive sequence* $\{b_n\}$ *satisfying*

$$b_n \longrightarrow \infty \quad \text{and} \quad b_n = o(\log n) \qquad (n \to \infty)$$

and for any $\lambda > 0$,

$$\lim_{n\to\infty} \frac{1}{b_n} \log \mathbb{P}\Big\{R_n^{(p)} - \mathbb{E}R_n^{(p)} \le -\lambda\frac{nb_n}{(\log n)^2}\Big\} = -\frac{\lambda}{(2\pi)^2\sqrt{\det(\Gamma)}}\kappa(2,2)^{-4}, \tag{8.8.4}$$

$$\lim_{n\to\infty} \frac{1}{b_n} \log \mathbb{P}\Big\{\widehat{R}_n^{(p)} - \mathbb{E}\widehat{R}_n^{(p)} \le -\lambda\frac{nb_n}{(\log n)^3}\Big\} = -\frac{\lambda}{16\pi^3\det(\Gamma)}\kappa(2,2)^{-4}. \tag{8.8.5}$$

EXERCISE 8.8.7. *Prove (8.5.6).*

Section 8.6

Theorem 8.6.1 was obtained in [**35**]. The laws of the iterated logarithm (8.6.1) and (8.6.2) were obtained in [**7**]; (8.6.2) and (8.6.4) were obtained in [**8**]. Prior to [**8**], Bass and Kumagai ([**10**]) had established (8.6.4) without identifying the constant on the right hand side. As mentioned before, (8.6.9) is due to Jain and Pruitt ([**103**]).

EXERCISE 8.8.8. *Prove Theorem 8.6.1.*

EXERCISE 8.8.9. *Establish the following law of the iterated logarithm: When* $d = 1$,

$$\limsup_{n\to\infty}(n\log\log n)^{-1/2}\sup_{x\in\mathbb{Z}} l(n,x) = \sqrt{2}\sigma^{-1} \quad a.s. \tag{8.8.6}$$

Section 8.7

EXERCISE 8.8.10. *Prove Theorem 8.7.2.*

Appendix

A. Green's function

Recall that the Green's function $G(x)$ of a d-dimensional Brownian motion is defined as

$$G(x) = \int_0^\infty e^{-t} p_t(x) dx = \frac{1}{(2\pi)^{d/2}} \int_0^\infty t^{-d/2} \exp\left\{ -\frac{1}{2}\Big(2t + \frac{|x|^2}{t}\Big)\right\} dt.$$

The major concern is the p-square integrability of $G(\cdot)$ on $\mathbb{R}^d$. It is easy to see that this is totally determined by the behavior of $G(\cdot)$ near $x = 0$.

THEOREM A.1. *When* $|x| \to 0$,

$$G(x) \sim \begin{cases} \dfrac{1}{\sqrt{2}} & d = 1, \\[2ex] -\dfrac{1}{\pi} \log |x| & d = 2, \\[2ex] \dfrac{1}{2\pi^{d/2}} \Gamma\Big(\dfrac{d-2}{2}\Big) |x|^{-(d-2)} & d \geq 3. \end{cases}$$

In particular,

$$\int_{\mathbb{R}^d} G^p(x) dx < \infty$$

if and only if $p(d-2) < d$.

Proof. We first consider the case $d \geq 2$. For any $v \geq 0$, set

$$\mathcal{K}_v(\theta) = \frac{1}{2} \int_0^\infty \frac{1}{t^{v+1}} \exp\left\{ -t - \frac{\theta}{t}\right\} dt \qquad \theta > 0.$$

The function $\mathcal{K}_v(\theta)$ is called the *Bessel-Clifford function of the second kind*, which was first introduced by Clifford ([**39**]). We refer to the book by Watson ([**163**]) for some facts about $\mathcal{K}_v(\theta)$ used in this book. The function $K_v(\theta)$ solves the *modified Bessel equation* (see [**163**] for details)

$$\theta^2 \frac{d^2 y}{d\theta^2} + \theta \frac{dy}{d\theta} - (\theta^2 + v^2) y = 0.$$

It is known that

$$\mathcal{K}_v(\theta) = \theta^{-v/2} K_v(2\sqrt{\theta}) \sim \begin{cases} -\dfrac{1}{2} \log \theta & v = 0, \\[2ex] \dfrac{\Gamma(v)}{2} \theta^{-v} & v > 0, \end{cases}$$

as $\theta \to 0^+$. Thus, the desired conclusion follows from the fact that

$$G(x) = \frac{2}{(2\pi)^{d/2}} \mathcal{K}_v\Big(\frac{|x|^2}{2}\Big)$$

with $v = \dfrac{d-2}{2}$.

The case $d = 1$ follows from (A.1) below. □

The closed form of $G(x)$ can be reached for some d. We start with the case $d = 1$ and we show that

$$\text{(A.1)} \qquad G(x) = \frac{1}{\sqrt{2}} e^{-\sqrt{2}|x|} \qquad x \in \mathbb{R}.$$

Indeed,

$$\begin{aligned} G(x) &= \frac{1}{\sqrt{2\pi}} \int_0^\infty t^{-1/2} \exp\Big\{-\frac{1}{2}\Big(2t + \frac{|x|^2}{t}\Big)\Big\} dt \\ &\overset{u=\sqrt{2t}}{=} \frac{1}{\sqrt{\pi}} \int_0^\infty \exp\Big\{-\frac{1}{2}\Big(u^2 + \frac{2|x|^2}{u^2}\Big)\Big\} du \\ &= \frac{1}{\sqrt{\pi}} e^{-\sqrt{2}|x|} \int_0^\infty \exp\Big\{-\frac{1}{2}\Big(u - \frac{\sqrt{2}|x|}{u}\Big)^2\Big\} du \\ &= \frac{1}{\sqrt{\pi}} e^{-\sqrt{2}|x|} I(|x|) \end{aligned}$$

where

$$\begin{aligned} I(|x|) &\equiv \int_0^\infty \exp\Big\{-\frac{1}{2}\Big(u - \frac{\sqrt{2}|x|}{u}\Big)^2\Big\} du \\ &= \int_0^\infty \Big(1 + \frac{\sqrt{2}|x|}{u^2}\Big) \exp\Big\{-\frac{1}{2}\Big(u - \frac{\sqrt{2}|x|}{u}\Big)^2\Big\} du \\ &\quad - \int_0^\infty \frac{\sqrt{2}|x|}{u^2} \exp\Big\{-\frac{1}{2}\Big(u - \frac{\sqrt{2}|x|}{u}\Big)^2\Big\} du \\ &\overset{v=\frac{\sqrt{2}|x|}{u}}{=} \int_{-\infty}^\infty \exp\Big\{-\frac{1}{2}u^2\Big\} du - \int_0^\infty \exp\Big\{-\frac{1}{2}\Big(v - \frac{\sqrt{2}|x|}{v}\Big)^2\Big\} dv \\ &= \sqrt{2\pi} - I(|x|). \end{aligned}$$

Consequently, $I(|x|) = \sqrt{\pi/2}$, which leads to (A.1).

In addition, write

$$G(x) = (2\pi)^{-d/2} F_d(|x|^2)$$

where

$$F_d(\theta) = \int_0^\infty t^{-d/2} \exp\Big\{-\frac{1}{2}\Big(2t + \frac{\theta}{t}\Big)\Big\} dt \qquad \theta > 0.$$

We now claim that for any $d \geq 3$,

$$\text{(A.2)} \qquad F_d(\theta) = -2F'_{d-2}(\theta).$$

Indeed, for any $\theta > 0$,

$$\int_\theta^\infty F_d(u)du = \int_0^\infty t^{-d/2}\Big[\int_\theta^\infty \exp\Big\{-\frac{1}{2}\Big(2t+\frac{u}{t}\Big)\Big\}du\Big]dt$$
$$= 2\int_0^\infty t^{-(d-2)/2}\exp\Big\{-\frac{1}{2}\Big(2t+\frac{\theta}{t}\Big)\Big\}dt = 2F_{d-2}(\theta).$$

Taking derivatives on both sides proves (A.2).

In principle, by (A.1) and (A.2) one can find the closed form for all Green's functions $G(x)$ when the dimension d is odd. In particular, it is straightforward to check that when $d = 3$,

$$G(x) = \frac{1}{2\pi}|x|^{-1}e^{-\sqrt{2}|x|} \qquad x \in \mathbb{R}^3. \tag{A.3}$$

B. Fourier transformation

We refer the reader to the books by Edwards ([**71**], [**72**]) and by Donoghue ([**54**]) for the general theory of Fourier transformation.

For any real function f on $\mathbb{Z}^d$ such that

$$\sum_{x\in\mathbb{Z}^d}|f(x)|^2 < \infty,$$

its *Fourier transform* is defined as

$$\widehat{f}(\lambda) = \sum_{x\in\mathbb{Z}^d} f(x)e^{i\lambda\cdot x} \qquad \lambda \in \mathbb{R}^d. \tag{B.1}$$

Here we point out that the summation on the right converges in $\mathcal{L}^2\{[-\pi,\pi]^d\}$ when λ is limited to $[-\pi,\pi]^d$ and that $\widehat{f}(\lambda)$ is periodic with period 2π.

One can view the above equality as the Fourier expansion of the periodic function $\widehat{f}(\lambda)$. So the classic theory of Fourier series applies here. In particular, taking the norm square and then integrating on both sides of (B.1), by orthogonality we obtain the *Parseval identity* (also known as the *Bessel identity*):

$$\sum_{x\in\mathbb{Z}^d}|f(x)|^2 = \frac{1}{(2\pi)^d}\int_{[-\pi,\pi]^d}|\widehat{f}(\lambda)|^2d\lambda. \tag{B.2}$$

Fourier transform defines an *isometric linear operator* from $l^2(\mathbb{Z}^d)$ to $\mathcal{L}^2[-\pi,\pi]$. By linearity, Parseval identity leads to the following:

$$\sum_{x\in\mathbb{Z}^d} f(x)g(x) = \frac{1}{(2\pi)^d}\int_{[-\pi,\pi]^d}\widehat{f}(\lambda)\overline{\widehat{g}(\lambda)}d\lambda \qquad f,g\in l^2(\mathbb{Z}^d). \tag{B.3}$$

In addition, multiplying $e^{-i\lambda\cdot x}$ and then integrating on both sides of (B.1), by orthogonality again we obtain the *Fourier inversion*

$$f(x) = \frac{1}{(2\pi)^d}\int_{[-\pi,\pi]^d} e^{-i\lambda\cdot x}\widehat{f}(\lambda)d\lambda \qquad x\in\mathbb{Z}^d. \tag{B.4}$$

Further, given $f \in l^2(\mathbb{Z}^d)$ and a probability density on $\mathbb{Z}^d$, define *convolution* $f * h$ as

$$f * h(x) = \sum_{y \in \mathbb{Z}^d} f(y)h(x-y) = \sum_{y \in \mathbb{Z}^d} f(x-y)h(y) \qquad x \in \mathbb{Z}^d.$$

By Jensen inequality, we get

$$\sum_{x \in \mathbb{Z}^d} |f * h(x)|^2 \le \sum_{x \in \mathbb{Z}^d} |f(x)|^2 < \infty.$$

It is straightforward to see that

$$\widehat{f * h}(\lambda) = \widehat{f}(\lambda)\widehat{h}(\lambda). \tag{B.5}$$

For a real function $f(x)$ on $\mathbb{R}^d$, we intend to define its Fourier transform as

$$\widehat{f}(\lambda) = \int_{\mathbb{R}^d} f(x) e^{i\lambda \cdot x} dx \qquad \lambda \in \mathbb{R}^d \tag{B.6}$$

and to establish the *Fourier inversion*

$$f(x) = \frac{1}{(2\pi)^d} \int_{\mathbb{R}^d} e^{-i\lambda \cdot x} \widehat{f}(\lambda) d\lambda \qquad x \in \mathbb{R}^d. \tag{B.7}$$

The issue of convergence has to be addressed when we talk about Fourier transform and its inversion in the continuous setting. Even under the assumption of integrability on f, the Fourier transform $\widehat{f}$ is not necessarily integrable on $\mathbb{R}^d$. In the following theorem we establish the Fourier inversion for rapidly decreasing functions. Recall that an infinitely smooth function f is said to be rapidly decreasing, if its derivatives of all orders decay at infinity faster than any negative power of $|x|$. The set of all rapidly decreasing functions is denoted by $\mathcal{S}(\mathbb{R}^d)$, which is called Schwartz space.

THEOREM B.1. *For any $f \in \mathcal{S}(\mathbb{R}^d)$, the Fourier inversion (B.7) holds. More generally, for any integers $k_1, \cdots, k_d \ge 0$,*

$$\Big(\frac{\partial^{k_1 + \cdots + k_d}}{\partial x_1^{k_1} \cdots \partial x_d^{k_d}} f\Big)(x) = \frac{1}{(2\pi)^d} \int_{\mathbb{R}^d} \Big(\prod_{j=1}^d (-i\lambda_j)^{k_j}\Big) e^{-i\lambda \cdot x} \widehat{f}(\lambda) d\lambda \tag{B.8}$$

for all $x \in \mathbb{R}^d$. Here we adopt the notation $x = (x_1, \cdots, x_d)$ and $\lambda = (\lambda_1, \cdots, \lambda_d)$.

Proof. The argument we use here is essentially provided by Robert ([**143**]). By the obvious fact that $\widehat{f}$ is also rapidly decreasing, the right hand sides of (B.7) and (B.8) are well defined, and (B.7) clearly implies (B.8).

We need only to show (B.7) in the case $x = 0$, for otherwise we replace $f(\cdot)$ by $f(x + \cdot)$ in the argument below.

Given $k > 0$, notice that the function

$$F_k(z) \equiv \sum_{y \in \mathbb{Z}^d} f(z + 2ky) \qquad z \in \mathbb{R}^d$$

is well defined and periodic with the period $2k$. For each $x \in \mathbb{Z}^d$,

$$\begin{aligned}
&\frac{1}{(2k)^d}\int_{[-k,k]^d}\exp\Big\{-iz\cdot\Big(\frac{\pi x}{k}\Big)\Big\}F(z)dz\\
&=\frac{1}{(2k)^d}\int_{\mathbb{R}^d}\exp\Big\{-iz\cdot\Big(\frac{\pi x}{k}\Big)\Big\}f(z)dz\\
&=\frac{1}{(2k)^d}\widehat{f}\Big(-\frac{\pi x}{k}\Big).
\end{aligned}$$

By Fourier expansion, we get

$$F_k(z)=\frac{1}{(2k)^d}\sum_{x\in\mathbb{Z}^d}\widehat{f}\Big(-\frac{\pi x}{k}\Big)e^{iz\cdot x}\qquad z\in\mathbb{R}^d.$$

Taking $z=0$, we obtain

$$\sum_{y\in\mathbb{Z}^d}f(2ky)=\frac{1}{(2k)^d}\sum_{x\in\mathbb{Z}^d}\widehat{f}\Big(\frac{\pi x}{k}\Big).$$

Letting $k\to\infty$ on both sides leads to the desired conclusion:

$$f(0)=\frac{1}{(2\pi)^d}\int_{\mathbb{R}^d}\widehat{f}(\lambda)d\lambda.$$

□

Given two real rapidly decreasing functions f, g, multiplying $g(x)$ and then integrating on both sides of (B.7), we have the *Parseval identity*

$$\int_{\mathbb{R}^d}f(x)g(x)dx=\frac{1}{(2\pi)^d}\int_{\mathbb{R}^d}\widehat{f}(\lambda)\overline{\widehat{g}(\lambda)}d\lambda. \tag{B.9}$$

The convolution $f*h$ between two functions f and h is defined as

$$f*h(x)=\int_{\mathbb{R}^d}f(y)h(x-y)dy=\int_{\mathbb{R}^d}f(x-y)h(y)dy\qquad x\in\mathbb{R}^d$$

whenever the involved integrals are well defined. When $f\in\mathcal{S}(\mathbb{R}^d)$ and h is a probability density on $\mathbb{R}^d$, then

$$\widehat{f*h}(\lambda)=\widehat{f}(\lambda)\widehat{h}(\lambda). \tag{B.10}$$

In addition, by integration by parts and by (B.8), we get

$$\begin{aligned}
\int_{\mathbb{R}^d}\nabla f(x)\nabla g(x)dx&=-\int_{\mathbb{R}^d}\Delta f(x)g(x)dx\\
&=\frac{1}{(2\pi)^d}\int_{\mathbb{R}^d}|\lambda|^2\widehat{f}(\lambda)\overline{\widehat{g}(\lambda)}d\lambda\qquad f,g\in\mathcal{S}(\mathbb{R}^d).
\end{aligned} \tag{B.11}$$

Since $\mathcal{S}(\mathbb{R}^d)$ is dense in $\mathcal{L}^2(\mathbb{R}^d)$, a standard procedure of approximation based on (B.9) extends uniquely the Fourier transform into a linear isometric operator from $\mathcal{L}^2(\mathbb{R}^d)$ to the complex L^2-space on $\mathbb{R}^d$, such that (B.9) holds for any $f,g\in\mathcal{L}^2(\mathbb{R}^d)$, and that (B.10) holds for every $f\in\mathcal{L}^2(\mathbb{R}^d)$ and every probability density h on $\mathbb{R}^d$.

Recall our notation

$$W^{1,2}(\mathbb{R}^d)=\Big\{g\in\mathcal{L}^2(\mathbb{R}^d);\;\;\nabla g\in\mathcal{L}^2(\mathbb{R}^d)\Big\}.$$

It is a well known fact that $W^{1,2}(\mathbb{R}^d)$ becomes a Hilbert space under the Sobolev norm

$$\sqrt{||g||_2^2 + ||\nabla g||_2^2}.$$

The isometric relation (B.11) can also be extended to $W^{1,2}(\mathbb{R}^d)$. More precisely, we have the following theorem.

THEOREM B.2. *Given that $f \in \mathcal{L}^2(\mathbb{R}^d)$, $f \in W^{1,2}(\mathbb{R}^d)$ if and only if*

$$\int_{\mathbb{R}^d} |\lambda|^2 |\widehat{f}(\lambda)|^2 d\lambda < \infty.$$

Further, for any $f, g \in W^{1,2}(\mathbb{R}^d)$.

$$\int_{\mathbb{R}^d} \nabla f(x) \nabla g(x) dx = \frac{1}{(2\pi)^d} \int_{\mathbb{R}^d} |\lambda|^2 \widehat{f}(\lambda) \overline{\widehat{g}(\lambda)} d\lambda.$$

Given a measure μ on $\mathbb{R}^d$, the *Fourier transform* $\widehat{\mu}(\cdot)$ of μ is defined as the function on $\mathbb{R}^d$ given as

$$\widehat{\mu}(\lambda) = \int_{\mathbb{R}^d} e^{i\lambda \cdot x} \mu(dx) \qquad \lambda \in \mathbb{R}^d \tag{B.12}$$

whenever the integral on the right hand side is well defined.

In particular, there is an one-to-one correspondence between μ and its Fourier transform $\widehat{\mu}$. In this book we are particularly interested in two opposite special cases; when μ is distributed on the lattice $\mathbb{Z}^d$ and when μ is absolutely continuous with respect to the Lebesgue measure on $\mathbb{R}^d$. It is a classic fact that μ is supported by the lattice $\mathbb{Z}^d$ if and only if $\widehat{\mu}(\lambda)$ is a *periodic function* with the period 2π:

$$\widehat{\mu}(\lambda + 2\pi y) = \widehat{\mu}(\lambda) \qquad \lambda \in \mathbb{R}^d, \quad y \in \mathbb{Z}^d.$$

In this case $\widehat{\mu}(\lambda) = \widehat{f}(\lambda)$, where $f(x)$ is the density of μ with respect to the counting measure on $\mathbb{Z}^d$ and $\widehat{f}(\lambda)$ is given by (B.1).

As for the test of absolute continuity, we put the correspondent results into the following theorem known as the *Plancherel-Parseval theorem*.

THEOREM B.3. *A finite measure μ on $\mathbb{R}^d$ is absolutely continuous with respect to the Lebesgue measure on $\mathbb{R}^d$ if*

$$\int_{\mathbb{R}^d} |\widehat{\mu}(\lambda)|^2 d\lambda < \infty.$$

In this case we have the Parseval identity,

$$\int_{\mathbb{R}^d} f^2(x) dx = \frac{1}{(2\pi)^d} \int_{\mathbb{R}^d} |\widehat{\mu}(\lambda)|^2 d\lambda$$

where $f(x)$ is the related density of μ.

Proof. For each $\epsilon > 0$, define

$$f_\epsilon(x) = \int_{\mathbb{R}^d} p_\epsilon(y - x) \mu(dy) \qquad x \in \mathbb{R}^d$$

where $p_\epsilon(x)$ is the normal density with mean zero and variance ϵ. We have

$$\widehat{f}_\epsilon(\lambda)=\exp\Big\{-\frac{\epsilon}{2}|\lambda|^2\Big\}\widehat{\mu}(\lambda).$$

In particular, $\widehat{f}_\epsilon(\cdot)$ ($\epsilon>0$) is a Cauchy sequence in $\mathcal{L}^2(\mathbb{R}^d)$ as $\epsilon\to 0^+$; so is the sequence $f_\epsilon(\cdot)$ ($\epsilon>0$) by Parseval identity. Let $f(\cdot)$ be the L^2-limit of $f_\epsilon(\cdot)$. One can clearly see that $f(x)$ is a density of μ and satisfies the Parseval identity. □

C. Constant $\kappa(d,p)$ and related variations

Assume that $p(d-2)\le d$. The central topic is the best constant $\kappa(d,p)$ of the Gagliardo-Nirenberg inequality

$$\text{(C.1)}\qquad ||f||_{2p}\le C||\nabla f||_2^{\frac{d(p-1)}{2p}}||f||_2^{1-\frac{d(p-1)}{2p}}\qquad f\in W^{1,2}(\mathbb{R}^d)$$

where for any $r\ge 1$, $W^{1,r}(\mathbb{R}^d)$ denotes the *Sobolev space*

$$W^{1,r}(\mathbb{R}^d)=\big\{f\in\mathcal{L}^2(\mathbb{R}^d);\ \ \nabla f\in\mathcal{L}^r(\mathbb{R}^d)\big\}.$$

That is,

$$\kappa(d,p)=\inf\Big\{C>0;\ \ ||f||_{2p}\le C||\nabla f||_2^{\frac{d(p-1)}{2p}}||f||_2^{1-\frac{d(p-1)}{2p}}\ \text{ for all } f\in W^{1,2}(\mathbb{R}^d)\Big\}.$$

Gagliardo-Nirenberg inequality is closed related to *Sobolev inequality*, which claims (Theorem 2.4.1, p. 56, [**176**]) that for any $1\le r<d$, there is a constant $K>0$ such that

$$\text{(C.2)}\qquad ||f||_{r*}\le K||\nabla f||_r\qquad f\in W^{1,r}(\mathbb{R}^d)$$

where $r*=dr(d-r)^{-1}$ and $W^{1,r}(\mathbb{R}^d)$ is the Sobolev space defined as

$$W^{1,r}(\mathbb{R}^d)=\big\{f\in\mathcal{L}^r(\mathbb{R}^d);\ \ \nabla f\in\mathcal{L}^r(\mathbb{R}^d)\big\}.$$

In the special case $p(d-2)=d$, Gagliardo-Nirenberg inequality becomes the Sobolev inequality with $r=2$. In the remaining case $p(d-2)<d$, Gagliardo-Nirenberg inequality can be derived from Sobolev inequality. Indeed, taking $r=d(p-1)p^{-1}$ in (C.2) we have

$$||f||_{d(p-1)}\le K||\nabla f||_{d(p-1)p^{-1}}.$$

Replacing f by $|f|^{\frac{2p}{d(p-1)}}$ gives

$$\begin{aligned}||f||_{2p}&\le\Big(\frac{2pK}{d(p-1)}\Big)^{\frac{d(p-1)}{2p}}\Big(\int_{\mathbb{R}^d}|\nabla f(x)|^{\frac{d(p-1)}{p}}|f(x)|^{\frac{2p-d(p-1)}{p}}dx\Big)^{1/2}\\&\le\Big(\frac{2pK}{d(p-1)}\Big)^{\frac{d(p-1)}{2p}}||\nabla f||_2^{\frac{d(p-1)}{2p}}||f||_2^{1-\frac{d(p-1)}{2p}}\end{aligned}$$

where the second step follows from Cauchy-Schwarz inequality.

Other variations associated with $\kappa(d,p)$ are $M(\theta)$ and ρ defined below:

$$\text{(C.3)}\qquad M(\theta)=\sup_{g\in\mathcal{F}_d}\Big\{\theta\Big(\int_{\mathbb{R}^d}|g(x)|^{2p}dx\Big)^{1/p}-\int_{\mathbb{R}^d}|\nabla g(x)|^2dx\Big\}$$

where

$$\mathcal{F}_d = \Big\{ g \in \mathcal{L}^2(\mathbb{R}^d);\ \int_{\mathbb{R}^d} g^2(x)dx = 1 \text{ and } \int_{\mathbb{R}^d} |\nabla g(x)|^2 dx < \infty \Big\},$$

(C.4) $$\rho = \sup\Big\{ \iint_{\mathbb{R}^d\times\mathbb{R}^d} G(x-y)f(x)f(y)dxdy;\ \int_{\mathbb{R}^d} |f(x)|^{\frac{2p}{2p-1}}dx = 1 \Big\}$$

where

$$G(x) = \int_0^\infty e^{-t}p_t(x)dt \qquad x \in \mathbb{R}^d$$

is Green's function of a d-dimensional Brownian motion.

THEOREM C.1. *Assume that $p(d-2) < d$. For any $\theta > 0$,*

$$M(\theta) = \frac{2p - d(p-1)}{2p}\Big(\frac{d(p-1)}{p}\Big)^{\frac{d(p-1)}{2p-d(p-1)}} \kappa(d,p)^{\frac{4p}{2p-d(p-1)}}\theta^{\frac{2p}{2p-d(p-1)}}.$$

Proof. For any $f \in \mathcal{F}_d$,

$$\begin{aligned}
&\theta\Big(\int_{\mathbb{R}^d} |f(x)|^{2p}dx\Big)^{1/p} - \int_{\mathbb{R}^d} |\nabla f(x)|^2 dx\\
&\le \theta\kappa(d,p)^2 ||\nabla f||_2^{\frac{d(p-1)}{p}} - \frac{1}{2}||\nabla f||_2^2\\
&\le \sup_{\lambda\ge 0}\Big\{\theta\kappa(d,p)^2\lambda^{\frac{d(p-1)}{p}} - \frac{1}{2}\lambda^2\Big\}\\
&= \frac{2p - d(p-1)}{2p}\Big(\frac{d(p-1)}{p}\Big)^{\frac{d(p-1)}{2p-d(p-1)}} \kappa(d,p)^{\frac{4p}{2p-d(p-1)}}\theta^{\frac{2p}{2p-d(p-1)}}.
\end{aligned}$$

Taking supremum over $f \in \mathcal{F}_d$ on the left, we have

$$M(\theta) \le \frac{2p - d(p-1)}{2p}\Big(\frac{d(p-1)}{p}\Big)^{\frac{d(p-1)}{2p-d(p-1)}} \kappa(d,p)^{\frac{4p}{2p-d(p-1)}}\theta^{\frac{2p}{2p-d(p-1)}}.$$

On the other hand, for any $C < \kappa(d,p)$ there is a g such that

$$||g||_{2p} > C||\nabla g||_2^{\frac{d(p-1)}{2p}}||g||_2^{1-\frac{d(p-1)}{2p}}.$$

By homogeneity we may assume that $||g||_2 = 1$. Given $\lambda > 0$, let $f(x) = \lambda^{d/2}g(\lambda x)$. Then $f \in \mathcal{F}_d$,

$$||\nabla f||_2 = \lambda||\nabla g||_2$$

and

$$||f||_{2p} = \lambda^{\frac{d(p-1)}{2p}}||g||_{2p} > C(\lambda||\nabla g||_2)^{\frac{d(p-1)}{2p}}.$$

Hence,

$$M(\theta) \ge \theta||f||_{2p}^2 - \frac{1}{2}||\nabla f||_2^2 > \theta C^2(\lambda||\nabla g||_2)^{\frac{d(p-1)}{p}} - \frac{1}{2}(\lambda||\nabla g||_2)^2.$$

Since $\lambda > 0$ is arbitrary, we have

$$\begin{aligned}
M(\theta) &\ge \sup_{x>0}\Big\{\theta C^2 x^{\frac{d(p-1)}{p}} - \frac{1}{2}x^2\Big\}\\
&= \frac{2p - d(p-1)}{2p}\Big(\frac{d(p-1)}{p}\Big)^{\frac{d(p-1)}{2p-d(p-1)}} C^{\frac{4p}{2p-d(p-1)}}\theta^{\frac{2p}{2p-d(p-1)}}.
\end{aligned}$$

Letting $C \to \kappa(d,p)$ on the right hand side gives

$$M(\theta) \geq \frac{2p - d(p-1)}{2p}\Big(\frac{d(p-1)}{p}\Big)^{\frac{d(p-1)}{2p-d(p-1)}} \kappa(d,p)^{\frac{4p}{2p-d(p-1)}} \theta^{\frac{2p}{2p-d(p-1)}}.$$

□

THEOREM C.2. *Assume that $p(d-2) < d$.*

$$\rho = \Big(\frac{2p - d(p-1)}{2p}\Big)^{\frac{2p-d(p-1)}{2p}} \Big(\frac{d(p-1)}{p}\Big)^{\frac{d(p-1)}{2p}} \kappa(d,p)^2.$$

Proof. By Theorem C.1, we need only to show that

$$M\Big(\frac{1}{\rho}\Big) = 1. \tag{C.5}$$

First, for any function f on $\mathbb{R}^d$ we introduce the notation

$$Gf(x) = \int_{\mathbb{R}^d} G(y-x)f(y)dx \qquad x \in \mathbb{R}^d$$

whenever the integral on the right hand side is well defined.

According to Theorem B in [**164**], the best constant $\kappa(d,p)$ can be attained at an infinitely smooth positive function. By examining the argument for Theorem C.1, one concludes that there is a $f \in \mathcal{F}_d$ such that f is positive and infinitely smooth, and

$$M\Big(\frac{1}{\rho}\Big) = \frac{1}{\rho}\Big(\int_{\mathbb{R}^d} f^{2p}(x)dx\Big)^{1/p} - \frac{1}{2}\int_{\mathbb{R}^d} |\nabla f(x)|^2 dx.$$

By the *Lagrange multiplier*,

$$\frac{1}{\rho}\Big(\int_{\mathbb{R}^d} f^{2p}(y)dy\Big)^{-\frac{p-1}{p}} f^{2p-1}(x) + \frac{1}{2}\triangle f(x) = M\Big(\frac{1}{\rho}\Big)f(x).$$

Write $\bar{f}(x) = f(x)/||f||_{2p}$. Then

$$\frac{1}{\rho}\bar{f}^{2p-1}(x) + \frac{1}{2}\triangle \bar{f}(x) = M\Big(\frac{1}{\rho}\Big)\bar{f}(x).$$

Multiplying $G\bar{f}^{2p-1}(x)$ and integrating on both sides, we get

$$\frac{1}{\rho}\int_{\mathbb{R}^d} \bar{f}^{2p-1}(x)G\bar{f}^{2p-1}(x)dx + \frac{1}{2}\int_{\mathbb{R}^d} \triangle \bar{f}(x)G\bar{f}^{2p-1}(x)dx$$
$$= M\Big(\frac{1}{\rho}\Big)\int_{\mathbb{R}^d} \bar{f}(x)G\bar{f}^{2p-1}(x)dx.$$

So we have

$$1 + \frac{1}{2}\int_{\mathbb{R}^d} \triangle \bar{f}(x)G\bar{f}^{2p-1}(x)dx \geq M\Big(\frac{1}{\rho}\Big)\int_{\mathbb{R}^d} \bar{f}(x)G\bar{f}^{2p-1}(x)dx.$$

Notice that

$$\frac{1}{2}\int_{\mathbb{R}^d}\Delta\bar{f}(x)G\bar{f}^{2p-1}(x)dx=\frac{1}{2}\int_{\mathbb{R}^d}\bar{f}(x)\Delta\circ G\bar{f}^{2p-1}(x)dx$$

$$=\int_{\mathbb{R}^d}\bar{f}(x)(G-I)\bar{f}^{2p-1}(x)dx=\int_{\mathbb{R}^d}\bar{f}(x)G\bar{f}^{2p-1}(x)dx-\int_{\mathbb{R}^d}\bar{f}^{2p}(x)dx$$

$$=\int_{\mathbb{R}^d}\bar{f}(x)G\bar{f}^{2p-1}(x)dx-1,$$

where I is the identity operator on the function space, and the second equality follows from the *resolvent equation* (see (4.1.18) with $f\equiv 0$ and $\lambda=1$)

$$G=I+2^{-1}\Delta\circ G. \tag{C.6}$$

Hence,

$$\int_{\mathbb{R}^d}\bar{f}(x)G\bar{f}^{2p-1}(x)dx\geq M\Big(\frac{1}{\rho}\Big)\int_{\mathbb{R}^d}\bar{f}(x)G\bar{f}^{2p-1}(x)dx,$$

which leads to

$$M\Big(\frac{1}{\rho}\Big)\leq 1.$$

On the other hand, let

$$c_0=\inf\Big\{\int_{\mathbb{R}^d}|f(x)|^2dx+\frac{1}{2}\int_{\mathbb{R}^d}|\nabla f(x)|^2dx;\quad\int_{\mathbb{R}^d}|f(x)|^{2p}dx=1\Big\}.$$

We first claim that

$$\rho\leq c_0^{-1}. \tag{C.7}$$

To this end, we show that for any function f on $\mathbb{R}^d$ with

$$\int_{\mathbb{R}^d}|f(x)|^{\frac{2p}{2p-1}}dx=1,$$

$$A\equiv\iint_{\mathbb{R}^d\times\mathbb{R}^d}G(x-y)f(x)f(y)dxdy\leq c_0^{-1}.$$

Without loss of generality, we may assume that $A>0$. Notice that

$$\iint_{\mathbb{R}^d\times\mathbb{R}^d}G(x-y)f(x)f(y)dxdy=\int_{\mathbb{R}^d}f(x)Gf(x)dx\leq\Big\{\int_{\mathbb{R}^d}|Gf(x)|^{2p}dx\Big\}^{\frac{1}{2p}}.$$

Consequently,

$$\Big\{\int_{\mathbb{R}^d}|Gf(x)|^{2p}dx\Big\}^{\frac{1}{2p}}\geq A.$$

By the resolvent equation in (C.6), we get

$$A=\langle Gf,f\rangle=\langle Gf,\ (I-2^{-1}\Delta)\circ Gf\rangle$$

$$=A^2\Big\langle\frac{Gf}{A},\ (I-2^{-1}\Delta)\frac{Gf}{A}\Big\rangle\geq A^2c_0.$$

This leads to (C.7).

By (C.7), for any $0<\epsilon<\rho$ there is an f on $\mathbb{R}^d$ such that

$$\int_{\mathbb{R}^d}|f(x)|^{2p}dx=1$$

and that

$$\frac{1}{\rho-\epsilon} > \int_{\mathbb{R}^d} |f(x)|^2 dx + \frac{1}{2}\int_{\mathbb{R}^d} |\nabla f(x)|^2 dx.$$

Let $g(x) = f(x)/||f||_2$. Then $g \in \mathcal{F}_d$ and

$$\begin{aligned}
&\frac{1}{\rho-\epsilon}\left\{\int_{\mathbb{R}^d} |g(x)|^{2p} dx\right\}^{1/p} - \frac{1}{2}\int_{\mathbb{R}^d} |\nabla g(x)|^2 dx \\
&\geq \left\{||f||_2 + \frac{1}{2}||\nabla f||^2\right\}||f||_2^{-2} - \frac{1}{2}||f||_2^{-2}||\nabla f||^2 \\
&= 1.
\end{aligned}$$

Letting $\epsilon \to 0^+$ on the left hand side leads to

$$M\Big(\frac{1}{\rho}\Big) \geq 1.$$

□

With some obvious modification, Theorem C.1 and Theorem C.2 can be extended to the critical dimensions $p(d-2) = d$ (the case $d = p = 3$ and the case $d = 4$, $p = 2$).

THEOREM C.3. *In the critical dimensions defined by $p(d-2) = d$,*

$$M(\theta) = \begin{cases} 0 & \theta < 2\kappa(d,p)^2, \\ \\ \infty & \theta > 2\kappa(d,p)^2, \end{cases}$$

$$\rho = 2\kappa(d,p)^2.$$

Finally, we compute the best constant of Gagliardo-Nirenberg inequality in the case $d = 1$.

THEOREM C.4. *Let $d = 1$. For any $p > 1$,*

$$\text{(C.8)} \qquad M(\theta) = p^{-\frac{2p}{p+1}}\left(\frac{\sqrt{2}}{(p-1)(p+1)}B\Big(\frac{1}{p-1},\frac{1}{2}\Big)\right)^{-\frac{2(p-1)}{p+1}}\theta^{\frac{2p}{p+1}} \qquad (\theta > 0),$$

$$\text{(C.9)} \qquad \kappa(1,p) = 2^{\frac{1}{2p}}(p-1)^{\frac{p-1}{4p}}(p+1)^{\frac{p-3}{4p}}B\Big(\frac{1}{p-1},\frac{1}{2}\Big)^{-\frac{p-1}{2p}}$$

where $B(\cdot,\cdot)$ is beta function defined by

$$B(a,b) = \int_0^1 x^{a-1}(1-x)^{b-1}dx \qquad a,b > 0.$$

Proof. By Theorem C.1, we need only to show (C.8) in the case $\theta = 1$. Our approach relies on Lagrange multiplier. By Theorem B in [**164**], there is an infinitely smooth, positive function $f(x)$ such that $f(-x) = f(x)$ for all $x \in \mathbb{R}$, that $f(x) \geq f(y)$ for $|x| \leq |y|$, and that

$$M(1) = \left(\int_{-\infty}^{\infty} f^{2p}(x)dx\right)^{1/p} - \frac{1}{2}\int_{-\infty}^{\infty} |f'(x)|^2 dx.$$

By Lagrange multipliers, there is a $\lambda \in \mathbb{R}$ such that

$$2\Big(\int_{-\infty}^{\infty} f^{2p}(x)dx\Big)^{-\frac{p-1}{p}} \int_{-\infty}^{\infty} f^{2p-1}(x)g(x)dx - \int_{-\infty}^{\infty} f'(x)g'(x)dx$$
$$= 2\lambda \int_{-\infty}^{\infty} f(x)g(x)dx$$

for any $g \in W^{1,2}(\mathbb{R})$. Applying integration by parts, we get

$$-2\Big(\int_{-\infty}^{\infty} f^{2p}(x)dx\Big)^{-\frac{p-1}{p}} \int_{-\infty}^{\infty} g'(x) \int_0^x f^{2p-1}(y)dydx - \int_{-\infty}^{\infty} f'(x)g'(x)dx$$
$$= -2\lambda \int_{\infty}^{\infty} g'(x) \int_0^x f(y)dydx \qquad (x \in \mathbb{R}).$$

Consequently,

$$\Big(\int_{-\infty}^{\infty} f^{2p}(x)dx\Big)^{-\frac{p-1}{p}} \int_0^x f^{2p-1}(y)dy + \frac{1}{2}f'(x) = \lambda \int_0^x f(y)dy.$$

Hence, $f'(0) = 0$ and

(C.10) $$\Big(\int_{-\infty}^{\infty} f^{2p}(z)dz\Big)^{-\frac{p-1}{p}} f^{2p-1}(x) + \frac{1}{2}f''(x) = \lambda f(x) \qquad x \in \mathbb{R}.$$

Multiplying both sides by $f(x)$ and integrating, gives us

(C.11) $$\lambda = \Big(\int_{-\infty}^{\infty} f^{2p}(x)dx\Big)^{1/p} - \frac{1}{2}\int_{-\infty}^{\infty} |f'(x)|^2 dx = M(1).$$

Multiplying by $f'(x)$ and then integrating both sides of (C.10) we have, after simplification, that

$$\big(f'(x)\big)^2 = 2\Big\{\lambda f^2(x) - \frac{1}{2p}\Big(\int_{-\infty}^{\infty} f^{2p}(z)dz\Big)^{-\frac{p-1}{p}} f^{2p}(x) + C\Big\}$$

where, using the fact that $f'(0) = 0$,

(C.12) $$C = \frac{1}{2p}\Big(\int_{-\infty}^{\infty} f^{2p}(z)dz\Big)^{-\frac{p-1}{p}} f^{2p}(0) - \lambda f^2(0).$$

Thus, by the fact that $f'(x) \le 0$ for $x \ge 0$, we get

(C.13) $$dx = -\frac{1}{\sqrt{2}}\Big\{\lambda f^2(x) - \frac{1}{2p}\Big(\int_{-\infty}^{\infty} f^{2p}(z)dz\Big)^{-\frac{p-1}{p}} f^{2p}(x) + C\Big\}^{-1/2} df(x)$$

for all $x \ge 0$. Consequently,

$$x = \frac{1}{\sqrt{2}}\int_{f(x)}^{f(0)} \Big\{\lambda y^2 - \frac{1}{2p}\Big(\int_{-\infty}^{\infty} f^{2p}(z)dz\Big)^{-\frac{p-1}{p}} y^{2p} + C\Big\}^{-1/2} dy \qquad (x \ge 0).$$

This is impossible (can be seen when $x \to \infty$) unless $C = 0$. So by (C.12), we get

(C.14) $$f(0) = \Big(\int_{-\infty}^{\infty} f^{2p}(x)dx\Big)^{\frac{1}{2p}} (2p)^{\frac{1}{2p-2}} \lambda^{\frac{1}{2p-2}}.$$

Combining (C.13) and (C.14), by the fact that $f(\infty) = 0$, gives us

$$\begin{aligned}
\int_{-\infty}^{\infty} f^{2p}(x)dx &= 2\int_0^{\infty} f^{2p}(x)dx \\
&= \sqrt{2}\int_0^{f(0)} y^{2p}\left\{\lambda y^2 - \frac{1}{2p}\left(\int_{-\infty}^{\infty} f^{2p}(z)dz\right)^{-\frac{p-1}{p}} y^{2p}\right\}^{-1/2} dy \\
&= f^{2p}(0)\sqrt{\frac{2}{\lambda}}\int_0^1 \frac{u^{2p}du}{\sqrt{u^2 - u^{2p}}} \\
&= \sqrt{2}p^{\frac{p}{p-1}}\lambda^{\frac{p+1}{2(p-1)}}\left(\int_{-\infty}^{\infty} f^{2p}(x)dx\right)\int_0^1 \frac{u^{2p}du}{\sqrt{u^2 - u^{2p}}}.
\end{aligned}$$

Therefore

$$1 = \sqrt{2}p^{\frac{p}{p-1}}\lambda^{\frac{p+1}{2(p-1)}}\int_0^1 \frac{u^{2p}du}{\sqrt{u^2 - u^{2p}}}.$$

Hence,

$$\lambda = p^{-\frac{2p}{p+1}}\left(\frac{\sqrt{2}}{(p-1)(p+1)}B\left(\frac{1}{p-1}, \frac{1}{2}\right)\right)^{-\frac{2(p-1)}{p+1}}.$$

Finally, the desired conclusion follows from (C.11). □

D. Regularity of stochastic processes

Let $X = (X_t)_{t\in T}$ be a real stochastic process, where the index set T is compact under the metrics $d(s,t)$ $(s,t \in T)$. Our concern is the regularities of the process which include continuity of its path and the integrability of its supremum norm.

A milestone paper in the study of regularity of the stochastic processes is due to Dudley ([**62**]). Dudley's idea can be roughly stated in the following way: One can install sample path continuity and boundedness for $X = (X_t)_{t\in T}$ by putting average Lipschitz conditions (with respect to the distance $d(s,t)$) on the process and by controlling the size of the index set T. In Dudley's work, the size of T is measured by counting the small balls (under the metrics $d(s,t)$) necessary for covering T.

This powerful approach, known as the *entropy method* , has been developed into various versions. The main part of this section is chosen from the book by Ledoux and Talagrand (Section 11.1, [**121**]) to fit the needs of our book.

A function Ψ: $\mathbb{R}^+ \longrightarrow \mathbb{R}^+$ is called a *Young function,* if it is increasing and convex with $\Psi(0) = 0$. Given a probability space $(\Omega, \mathcal{A}, \mathbb{P})$ and a Young function Ψ, the *Orlicz space* $\mathcal{L}_\Psi(\Omega, \mathcal{A}, \mathbb{P})$ is defined as the linear space of all random variables ξ on $(\Omega, \mathcal{A}, \mathbb{P})$ such that

$$||\xi||_\Psi \equiv \inf\left\{c > 0;\ \ \mathbb{E}\Psi(c^{-1}|\xi|) \le 1\right\} < \infty.$$

It is classic knowledge that $||\cdot||_\Psi$ defines a norm on $\mathcal{L}_\Psi(\Omega, \mathcal{A}, \mathbb{P})$ known as the *Orlicz norm,* and that $\mathcal{L}_\Psi(\Omega, \mathcal{A}, \mathbb{P})$ becomes a Banach space under the norm $||\cdot||_\Psi$.

For each $\epsilon > 0$, let $N(T, d, \epsilon)$ be the minimal number of open balls of radius ϵ which are necessary to cover T. A basic assumption is the following *entropy condition:*

$$\int_0^{D(T)} \Psi^{-1}\big(N(T,d,\epsilon)\big)d\epsilon < \infty \tag{D.1}$$

where $\Psi^{-1}(\cdot)$ is the inverse of $\Psi(\cdot)$, and

$$D(T) = \max\{d(s,t);\quad s,t \in T\}$$

is the diameter of T.

THEOREM D.1. *Let $X = (X_t)_{t\in T}$ be a stochastic process in $\mathcal{L}(\Omega,\mathcal{A},\mathbb{P})$ such that for any event $A \in \mathcal{A}$ and $s,t \in T$,*

$$\int_A |X_s - X_t| d\mathbb{P} \le d(s,t)\mathbb{P}(A)\Psi^{-1}\Big(\frac{1}{\mathbb{P}(A)}\Big). \tag{D.2}$$

Then, for any $A \in \mathcal{A}$,

$$\int_A \sup_{s,t\in T} |X_s - X_t| d\mathbb{P} \le 8d(s,t)\mathbb{P}(A)\int_0^{D(T)} \Psi^{-1}\Big(\frac{N(T,d,\epsilon)}{\mathbb{P}(A)}\Big)d\epsilon. \tag{D.3}$$

This result was established in Theorem 11.2, [**121**]. By slightly sharpening their estimate, Ledoux and Talagrand (Theorem 11.6, [**121**]) derived a version of what is known as *Kolmogorov's continuity theorem.*

THEOREM D.2. *Let $X = (X_t)_{t\in T}$ be a stochastic process in $\mathcal{L}(\Omega,\mathcal{A},\mathbb{P})$ and assume (D.1) and (D.2). Then $X = (X_t)_{t\in T}$ admits a modification $\widetilde{X} = (\widetilde{X}_t)_{t\in T}$ with all sample paths bounded and (uniformly) continuous on (T,d). Moreover, for each $\epsilon > 0$, there is a $\delta > 0$ such that*

$$\mathbb{E} \sup_{d(s,t)<\delta} |\widetilde{X}_s - \widetilde{X}_t| < \epsilon. \tag{D.4}$$

For a stochastic process $X = (X_t)_{t\in T}$ with continuous sample path on T, X can be viewed as a random variable taking values in the Banach space $C(T)$, the space of continuous functions on T equipped with supremum norm. Given a family $\mathcal{X}$ of stochastic process $X = (X_t)_{t\in T}$ on T with continuous sample path on T, an important issue is the relative compactness of the family $\mathcal{X}$ with respect to the topology of weak convergence. For this purpose we recall some of the classic material on this subject.

Given a separable Banach space B, let $\mathcal{P}(B)$ be the space of all probability measures on B. Under the topology of weak convergence, $\mathcal{P}(B)$ becomes a separable complete metric space. A family $\{\mu_i\}_{i\in I}$ in $\mathcal{P}(B)$ is said to be *uniformly tight*, if for any $\epsilon > 0$ there is a compact set $K \subset B$ such that

$$\mu_i(K^c) < \epsilon \qquad i \in I. \tag{D.5}$$

The following well-known *Prokhorov criterion* characterizes the relatively compact sets in $\mathcal{P}(B)$ in terms of the uniform tightness (see, e.g., Theorem 2.1 in [**121**]).

THEOREM D.3. *The family $\mathcal{P}(B)$ is relatively compact in $\mathcal{P}(B)$ if and only if it is uniformly tight.*

To exam uniform tightness, one needs to look for the compact set $K \subset B$ such that (D.5) holds. When $B = C(T)$, the following classic *Arzelá-Ascoli theorem* provides criterion for relatively compact sets in $C(T)$.

THEOREM D.4. *A family $\mathcal{F}$ in $C(T)$ is relatively compact if and only if for any (or equivalently, for some) $t \in T$, the set $\{x(t);\ \ x \in \mathcal{F}\}$ of real numbers is bounded and $\mathcal{F}$ is equicontinuous: For each $\epsilon > 0$, there is $\delta > 0$ such that*

$$\sup_{x\in\mathcal{F}} |x(s) - x(t)| < \epsilon$$

as soon as $d(s,t) < \delta$.

Given a family $\mathcal{X}$ of the stochastic processes on T such that every $X \in \mathcal{X}$ has a continuous sample path and that (D.1) and (D.2) holds for every $X \in \mathcal{X}$, Ledoux and Talagrand's estimate allows the uniformity of (D.4) over $\mathcal{X}$, which, by continuity of X, states that for each $\epsilon > 0$ there is $\delta > 0$ such that

$$\sup_{X\in\mathcal{X}} \mathbb{E} \sup_{d(s,t)<\delta} |X_s - X_t| < \epsilon. \tag{D.6}$$

By this observation, Ledoux and Talagrand were able to establish the following theorem.

THEOREM D.5. *Let $\mathcal{X}$ be a family of the continuous stochastic processes on T which are in $\mathcal{L}(\Omega, \mathcal{A}, \mathbb{P})$ and satisfy (D.1) and (D.2). Assume that for some $t_0 \in T$, the family $\{X_{t_0};\ \ X \in \mathcal{X}\}$ of the real random variables is uniformly tight in $\mathbb{R}$. Then $\mathcal{X}$ is uniformly tight in $C(T)$ and is therefore relatively compact in $\mathcal{P}\big(C(T)\big)$.*

Proof. Let $\epsilon > 0$ be fixed. By uniform tightness of $\{X_{t_0};\ \ X \in \mathcal{X}\}$ there is a $M > 0$ such that

$$\mathbb{P}\{|X_{t_0}| > M\} < \frac{\epsilon}{2} \qquad X \in \mathcal{X}.$$

For each $k \geq 1$ by (D.6) and Chebyshev inequality there is a $\delta_k > 0$ such that

$$\mathbb{P}\Big\{\sup_{d(s,t)<\delta_k} |X_s - X_t| \geq \frac{1}{k}\Big\} \leq \frac{\epsilon}{2^{k+1}}.$$

By the Arzelá-Ascoli theorem, the set

$$A = \bigcap_{k=1}^{\infty} \Big\{x \in C(T);\ \ |x(t_0)| \leq M \text{ and } \sup_{d(s,t)<\delta_k} |x(s) - x(t)| \leq \frac{1}{k}\Big\}$$

is relative compact in $C(T)$. By our construction, $\mathbb{P}\{X \notin A\} < \epsilon$. Taking the compact set K as the closure of A in $C(T)$ leads to the uniform tightness of $\mathcal{X}$. □

To make them more practical to application, we now derive some consequences out of Theorem D.1, Theorem D.2 and Theorem D.5. The elegant argument we shall present comes from the book by Ledoux and Talagrand ([**121**]).

First, the condition (D.2) holds if

$$||X_s - X_t||_\Psi \leq d(s,t) \qquad s, t \in T. \tag{D.7}$$

Indeed, by Jensen's inequality, we get

$$\begin{aligned}\int_A |X_s - X_t| d\mathbb{P} &= d(s,t)\mathbb{P}(A) \int_A \Psi^{-1} \circ \Psi\Big(\frac{|X_s - X_t|}{d(s,t)}\Big) \frac{d\mathbb{P}}{\mathbb{P}(A)} \\ &\le d(s,t)\mathbb{P}(A)\Psi^{-1}\Big(\frac{1}{\mathbb{P}(A)}\mathbb{E}\Psi\Big(\frac{|X_s - X_t|}{d(s,t)}\Big)\Big) \\ &\le d(s,t)\mathbb{P}(A)\Psi^{-1}\Big(\frac{1}{\mathbb{P}(A)}\Big).\end{aligned}$$

Second, if we assume that

$$\Psi^{-1}(xy) \le C\Psi^{-1}(x)\Psi^{-1}(y) \qquad x, y \ge 1 \tag{D.8}$$

then (D.3) leads to

$$\mathbb{P}\Big\{\sup_{s,t\in T} |X_s - X_t| \ge u\Big\} \le \Big(\Psi\Big(\frac{u}{8EC}\Big)\Big)^{-1} \qquad u > 0 \tag{D.9}$$

where

$$E = \int_0^{D(T)} \Psi^{-1}\big(N(T, d, \epsilon)\big) d\epsilon.$$

Indeed, by (D.3) and (D.8)

$$\int_A |X_s - X_t| d\mathbb{P} \le 8CE\mathbb{P}(A)\Psi^{-1}\Big(\frac{1}{\mathbb{P}(A)}\Big).$$

Take $A = \Big\{\sup_{s,t\in T} |X_s - X_t| \ge u\Big\}$. By Chebyshev inequality, we get

$$\mathbb{P}(A) \le \frac{1}{u} 8CE\mathbb{P}(A)\Psi^{-1}\Big(\frac{1}{\mathbb{P}(A)}\Big),$$

which proves to (D.9).

Roughly speaking, (D.8) means that the increasing rate of $\Psi(x)$ is no less than a linear function, an assumption that agrees at large with the monotonicity and convexity of Ψ.

For the applications to this book, we take T as a compact domain in $\mathbb{R}^N$ with $N \ge 1$, and $d(s,t) = c|s - t|^\delta$, where $0 < \delta \le 1$ and $|\cdot|$ is the Euclidean norm. One can see that in this case

$$N(T, d, \epsilon) = O\Big(\epsilon^{-N\delta^{-1}}\Big) \qquad (\epsilon \to 0^+).$$

Consequently, (D.1) holds if

$$\int_0^1 \Psi^{-1}\big(\epsilon^{-N\delta^{-1}}\big) d\epsilon < \infty. \tag{D.10}$$

As for the Young functions, we only consider two special cases $\Psi(\theta) = \exp\{\theta^r\} - 1$ for some $r > 0$ and $\Psi(\theta) = \theta^m$ for some $m > N$.

In the first case, $\Psi(\theta) = \exp\{\theta^r\} - 1$ is not a Young function when $r < 1$, as the convexity fails for small θ. To fix this problem, one can easily construct a Young function $\widetilde{\Psi}$ satisfying (D.8) such that $\Psi(\theta) = \widetilde{\Psi}(\theta)$ for large θ and apply what has been achieved to $\widetilde{\Psi}$. Notice that in this case (D.10) automatically holds. Here is what we get.

THEOREM D.6. *Let $X = (X_t)_{t\in T}$ be a stochastic process in $\mathcal{L}(\Omega, \mathcal{A}, \mathbb{P})$, where T is a compact domain in $\mathbb{R}^N$. Let $0 < \delta \leq 1$ and $r > 0$ be fixed and assume that*

$$\sup_{\substack{s,t\in T\\ s\neq t}} \mathbb{E}\exp\left\{c\frac{|X_s - X_t|^r}{|s-t|^\delta}\right\} < \infty \tag{D.11}$$

for some $c > 0$. Then $X = (X_t)_{t\in T}$ admits a modification $\widetilde{X} = (\widetilde{X}_t)_{t\in T}$ with all sample paths bounded and (uniformly) continuous on (T, d).

In addition, there is $c' > 0$ such that

$$\mathbb{E}\exp\left\{c' \sup_{s,t\in T} |X_s - X_t|^r\right\} < \infty. \tag{D.12}$$

Given a family $\mathcal{X}$ of stochastic processes on T such that (D.11) holds uniformly for all $X \in \mathcal{X}$,

$$\sup_{X\in\mathcal{X}} \mathbb{E}\exp\left\{c'' \sup_{s,t\in T} |X_s - X_t|^r\right\} < \infty \tag{D.13}$$

for some $c'' > 0$.

Proof. All we need to say is that (D.9) leads to (D.12) (and (D.13) in the latter case) for our choice of Ψ. □

As for the second choice $\Psi(\theta) = \theta^m$, notice that (D.10) holds if $n \equiv m\delta > N$. Therefore, we have the following version of Kolmogorov's continuity theorem.

THEOREM D.7. *Let $X = (X_t)_{t\in T}$ be a stochastic process in $\mathcal{L}(\Omega, \mathcal{A}, \mathbb{P})$, where T is a compact domain in $\mathbb{R}^N$. Assume that there are constants $m \geq 1$ and $n > N$ such that $m\delta > N$ and that*

$$\mathbb{E}|X_s - X_t|^m \leq M|s-t|^n \qquad s,t\in T \tag{D.14}$$

for some constant $M > 0$. Then $X = (X_t)_{t\in T}$ admits a modification $\widetilde{X} = (\widetilde{X}_t)_{t\in T}$ with all sample paths bounded and (uniformly) continuous on (T, d).

Further, a family $\mathcal{X}$ of $C(T)$-valued random variables is uniformly tight in $C(T)$ and is therefore relatively compact in $\mathcal{P}(C(T))$, if for some $t \in T$, the family $\{X_{t_0};\ X \in \mathcal{X}\}$ of the real random variables is uniformly tight in $\mathbb{R}$, and if (D.14) holds uniformly for all $X \in \mathcal{X}$.

E. Self-adjoint operators

Given an $N \times N$ symmetric matrix with real elements, there is an $N \times N$ orthogonal matrix T such that $A = TDT^*$, where D is the diagonal matrix with the eigenvalues $\lambda_1 \leq \cdots \leq \lambda_m$ of A as its diagonal elements (in increasing order), and T^* is the transpose of T. This representation can be rewritten as

$$A = \sum_{k=1}^N \lambda_k\big(E(k) - E(k-1)\big) \tag{E.1}$$

where $E(0)$ is an $N \times N$ 0-matrix and $E(k) = TD_kT^*$, where D_k is the diagonal matrix whose first k diagonal elements are 1's and the remaining are 0's. For each $0 \leq k \leq N$, $E(k)$ projects $\mathbb{R}^N$ into a k-dimensional subspace of $\mathbb{R}^N$ and for $1 \leq j \leq k \leq N$, $E(j)E(k) = E(k)E(j) = E(j)$. For any $x \in \mathbb{R}^N$,

$$\mu_x(k) \equiv \Big\langle x, \big(E(k) - E(k-1)\big)x\Big\rangle \geq 0 \qquad 1 \leq k \leq N$$

and

$$\sum_{k=1}^{N} \mu_x(k) = |x|^2. \tag{E.2}$$

The representation in (E.1) is the most classic example of *spectral decomposition.* For any integer $m \geq 0$, A^m is symmetric and the equality $A^m = TD^mT^*$ holds for the same T and D. Therefore,

$$A^m = \sum_{k=1}^{N} \lambda_k^m \big(E(k) - E(k-1)\big). \tag{E.3}$$

Given a real polynomial

$$\xi(\lambda) = \sum_{i=0}^{m} c_i \lambda^i,$$

one can define the matrix $\xi(A)$ in the following natural way:

$$\xi(A) = \sum_{i=0}^{m} c_i A^i.$$

By linearity and (E.3), we have

$$\xi(A) = \sum_{k=1}^{N} \xi(\lambda_k)\big(E(k) - E(k-1)\big). \tag{E.4}$$

For a more general function $\xi(\lambda)$, one can define the matrix $\xi(A)$ by (E.4).

This elementary observation can be extended to a class of linear operators on Hilbert spaces known as self-adjoint operators, and the topic is called *spectral representation.* The case of bounded self-adjoint operators has become standard material for a graduate course on functional analysis. In the following we consider the general case in which A is not necessarily bounded. Most of the material on the general theory of self-adjoint operators in this section is taken from the book by Yosida ([**175**]).

In the following discussion, H is a separable real Hilbert space. A linear operator A on H is said to be *densely defined* if its domain, denoted by $D(A)$, is dense in H. The *adjoint operator* of A is the operator A^*: $y \mapsto y^*$ defined by the relation

$$\langle Ax, y\rangle = \langle x, y^*\rangle \qquad x \in D(A).$$

A densely defined operator is said to be *symmetric*, if

$$\langle Ax, y\rangle = \langle x, Ay\rangle \qquad x, y \in D(A).$$

Clearly, $D(A) \subset D(A^*)$ and $Ay = A^*y$ for each $y \in D(A)$ when A is symmetric. A symmetric operator is said to be *self-adjoint* , if $D(A) = D(A^*)$. In other words, A and A^* are the same operator when A is self-adjoint. In the special case when A is bounded, there is no difference between being symmetric and being self-adjoint. This is no longer the case when it comes to unbounded operators. See Example 4,

Section VII in [**175**] for an example of non-self-adjoint but symmetric operators. In many aspects, it is the self-adjoint operators that appear as a natural extension of symmetric matrices. In practice, however, it can be very difficult to tell whether or not a symmetric but unbounded operator is self-adjoint. A way of resolving this problem is to extend a symmetric operator into a self-adjoint operator.

A linear operator A is said to be *upper semi-bounded*, (or *lower semi-bounded*), if

$$\sup_{\substack{x\in\mathcal{D}(A)\\ ||x||=1}} \langle x, Ax\rangle < \infty \qquad (\text{or} \quad \inf_{\substack{x\in\mathcal{D}(A)\\ ||x||=1}} \langle x, Ax\rangle > -\infty).$$

It is easy to see that a linear operator A is bounded if and only if it is upper and lower semi-bounded. For any bounded self-adjoint operator A,

$$||A|| = \sup_{||x||=1} \big|\langle x, Ax\rangle\big|.$$

THEOREM E.1. *A semi-bounded symmetric operator A can be extended into a self-adjoint operator. More precisely, there is a self-adjoint operator $\widetilde{A}$ on H such that $\mathcal{D}(A) \subset \mathcal{D}(\widetilde{A})$ and that $Ax = \widetilde{A}x$ for every $x \in \mathcal{D}(A)$.*

The above theorem is known as *Friedrichs' extension theorem* (see, e.g., Theorem 2, Section 7, Chapter XI, [**175**]).

A *projection operator* P is a bounded linear operator on H with $P^2 = P$. It is easy to see that $||P|| \le 1$ (the equality holds when P is not a zero operator) and that P is symmetric. Consequently, a projection operator is self-adjoint.

A family $\{E(\lambda);\ -\infty < \lambda < \infty\}$ of projection operators on H is called a *resolution of the identity*, if

(1) $E(\lambda) \circ E(\gamma) = E(\lambda \wedge \gamma)$ for any $-\infty < \lambda, \gamma < \infty$;

(2) $E(-\infty)$ is a zero operator, $E(\infty)$ is an identity operator, and $E(\lambda+0) = E(\lambda)$ for every $\lambda \in \mathbb{R}$, where $E(-\infty)$, $E(\infty)$ and $E(\lambda+0)$ are the linear operators defined as

$$E(\pm\infty)(x) \overset{||\cdot||}{=} \lim_{\lambda\to\pm\infty} E(\lambda)(x), \qquad E(\lambda+0)(x) \overset{||\cdot||}{=} \lim_{\gamma\to\lambda+} E(\gamma)(x) \qquad \forall x \in H.$$

By definition, for any $x \in H$, the function

$$F_x(\lambda) = \langle E(\lambda)(x), x\rangle = ||E(\lambda)(x)||^2$$

is a distribution function on $\mathbb{R}$. We write μ_x for the measure generated by $F_x(\cdot)$ and call μ_x a *spectral measure*. Clearly, μ_x is finite and, correspondent to (E.2), so

$$\mu_x(\mathbb{R}) = ||x||^2 \qquad x \in H. \tag{E.5}$$

There is an obvious similarity between the definitions of resolution of the identity and the probability distribution function. Recall that a probability distribution function is defined by four properties: monotonicity, right continuity, vanishing at $-\infty$, and increasing to 1 at ∞. To see the similarity, all we need is to define the order $P_1 \le P_2$ between two projection operators P_1 and P_2 on H by the relation $P_1(H) \subset P_2(H)$.

Given a suitable function $\xi(\lambda)$ on $\mathbb{R}$ and a distribution function $F(\lambda)$, the integral

$$\int_{-\infty}^{\infty} \xi(\lambda)F(d\lambda)$$

is extensively discussed in every graduate text book in probability. Similarly, we intend to define the linear (possibly unbounded) operator in the form

$$\int_{-\infty}^{\infty} \xi(\lambda)E(d\lambda)$$

and call it the *spectral integral.*

In the special case when ξ is a step function supported on a finite interval $[a, b]$ and is piecewisely defined with respect to a partition $a = \lambda_0 < \lambda_1 \cdots < \lambda_n = b$, we define

$$\int_{-\infty}^{\infty} \xi(\lambda)E(d\lambda) = \sum_{k=1}^{n} c_k \big(E(\lambda_k) - E(\lambda_{k-1})\big)$$

where c_k is the value of the function $\xi(\cdot)$ on the kth sub-interval $(\lambda_{k-1}, \lambda_k)$.

By this definition one can immediately see that for each $x \in H$, we get

$$\left\|\left(\int_{-\infty}^{\infty} \xi(\lambda)E(d\lambda)\right)(x)\right\|^2 = \int_{-\infty}^{\infty} |\xi(\lambda)|^2 \mu_x(d\lambda) \tag{E.6}$$

and

$$\left\langle\left(\int_{-\infty}^{\infty} \xi(\lambda)E(d\lambda)\right)(x),\ x\right\rangle = \int_{-\infty}^{\infty} \xi(\lambda)\mu_x(d\lambda). \tag{E.7}$$

By a standard argument of approximation, the definition is extended to a more general class of $\xi(\cdot)$. More precisely, for any Borel-measurable $\xi(\lambda)$ on $\mathbb{R}$, the linear operator

$$\left(\int_{-\infty}^{\infty} \xi(\lambda)\mathbb{E}(d\lambda)\right)(x) \qquad x \in \mathcal{D}_\xi \tag{E.8}$$

is defined on the domain $\mathcal{D}_\xi \subset H$ given by

$$\mathcal{D}_\xi = \left\{x \in H;\ \int_{-\infty}^{\infty} |\xi(\lambda)|^2 \mu_x(d\lambda) < \infty\right\}.$$

Further, (E.6) and (E.7) hold for every $x \in \mathcal{D}_\xi$. By the fact that μ_x is a finite measure and by Cauchy-Schwarz inequality, we get

$$\int_{-\infty}^{\infty} |\xi(\lambda)|\mu_x(d\lambda) < \infty$$

whenever $x \in \mathcal{D}_\xi$. Consequently, the right hand side of (E.7) is well defined.

A striking fact is that the linear operator in (E.8) is self-adjoint (Theorem 2, Section XI.5, [**175**]).

Correspondent to (E.1), the following theorem (Theorem 1, Section XI.6, [**175**]) claims that the linear operators in the form of (E.8) are the only kind of self-adjoint operators.

THEOREM E.2. *Given a self-adjoint operator A, there is a unique resolution of identity $\{E(\lambda);\ -\infty < \lambda < \infty\}$, such that*

$$A = \int_{-\infty}^{\infty} \lambda E(d\lambda), \tag{E.9}$$

$$\mathcal{D}(A) = \left\{ x \in H;\ \int_{-\infty}^{\infty} |\lambda|^2 \mu_x(d\lambda) < \infty \right\}. \tag{E.10}$$

The equality (E.9) is called the *spectral integral representation* of the self-adjoint operator A. The discussion in the rest of the section is based on the representation (E.9). The following theorem is used multiple times in this book.

THEOREM E.3. *Let the self-adjoint operator A be given in (E.9). For $x \in H$, the spectral measure μ_x is supported by the interval $[\lambda_0', \lambda_0]$, where*

$$\lambda_0 = \sup_{\substack{x\in\mathcal{D}(A)\\ ||x||=1}} \langle x, Ax\rangle, \quad \text{and} \quad \lambda_0' = \inf_{\substack{x\in\mathcal{D}(A)\\ ||x||=1}} \langle x, Ax\rangle$$

are defined in the extended real line.

Proof. Due to similarity we only show that the supporting set of $\mu_x(d\lambda)$ is bounded by λ_0 from above. To this purpose we may assume that $\lambda_0 < \infty$. We use the argument by contradiction. We show that if there were $x_0 \in H$ and $\lambda_1 > \lambda_0$ such that $\mu_{x_0}(\lambda_1, \infty) > 0$, there would be $x^* \in \mathcal{D}(A)$ such that $||x^*|| = 1$ and that $\langle x^*, Ax^*\rangle > \lambda_0$.

Indeed, one can find $\lambda_2 > \lambda_1$ such that

$$\mu_{x_0}(\lambda_1, \lambda_2) > 0.$$

Let $x_1 = E(\lambda_2)x_0 - \mathbb{E}(\lambda_1)x_0$. By the definition of the resolution of identity, we get

$$\begin{aligned} ||x_1||^2 &= ||E(\lambda_2)x_0||^2 + ||E(\lambda_1)x_0||^2 - 2\langle E(\lambda_1)x_0, E(\lambda_2)x_0\rangle \\ &= ||E(\lambda_2)x_0||^2 + ||E(\lambda_1)x_0||^2 - 2||E(\lambda_1)x_0||^2 \\ &= \langle x_0, E(\lambda_2)x_0\rangle - \langle x_0, E(\lambda_1)x_0\rangle = \mu_{x_0}(\lambda_1, \lambda_2] > 0. \end{aligned} \tag{E.11}$$

In particular, $x_1 \neq 0$.

In addition, it is easy to check that $\langle x_1, E(\lambda)x_1\rangle = ||x_1||^2$ when $\lambda \geq \lambda_2$, and $\langle x_1, E(\lambda)x_1\rangle = 0$ when $\lambda < \lambda_1$. Since $\mu_{x_1}(\mathbb{R}) = ||x_1||^2$, μ_{x_1} is supported by $(\lambda_1, \lambda_2]$. Thus,

$$\int_{-\infty}^{\infty} |\lambda|^2 \mu_{x_1}(d\lambda) = \int_{(\lambda_1, \lambda_2]} |\lambda|^2 \mu_{x_1}(d\lambda) < \infty.$$

Consequently, $x_1 \in \mathcal{D}(A)$.

For the same reason,

$$\langle x_1, Ax_1\rangle = \int_{-\infty}^{\infty} \lambda \mu_{x_1}(d\lambda) = \int_{(\lambda_1, \lambda_2]} \lambda \mu_{x_1}(d\lambda) \geq \lambda_1 \mu_{x_0}(\lambda_1, \lambda_2] = \lambda_1 ||x_1||_2^2$$

where the last step follows from (E.11).

Finally, letting $x^* = ||x_1||^{-1} x_1$ completes the proof. □

In the light casted by the representation in (E.4), given a self-adjoint operator A in the form of (E.9) *the function* $\xi(A)$ of A is defined as the self-adjoint operator

$$\xi(A) = \int_{-\infty}^{\infty} \xi(\lambda)E(d\lambda) \tag{E.12}$$

where $\xi(\cdot)$ is a Borel function on $\mathbb{R}$.

The following are some immediate observations. The domain of $\xi(A)$ is determined as

$$\mathcal{D}\big(\xi(A)\big) = \Big\{x \in H;\ \int_{-\infty}^{\infty} |\xi(\lambda)|^2 \mu_x(d\lambda) < \infty\Big\}. \tag{E.13}$$

By (E.6) and (E.7), for any $x \in \mathcal{D}\big(\xi(A)\big)$,

$$\big|\big|\xi(A)x\big|\big|^2 = \int_{-\infty}^{\infty} |\xi(\lambda)|^2 \mu_x(d\lambda), \tag{E.14}$$

$$\langle \xi(A)x,\ x\rangle = \int_{-\infty}^{\infty} \xi(\lambda)\mu_x(d\lambda). \tag{E.15}$$

By the definition given in (E.12), some standard properties enjoined by integrals, such as the linearity, hold naturally (see Theorem 3, Section XI.12, [**175**]). In addition to that, there are some truly remarkable properties in operational calculus, such as the one given in the following theorem (part (iv) of Theorem 3, Section XI.12, [**175**]).

THEOREM E.4. *Let A be a self-adjoint operator. For two Borel functions $\xi(\lambda)$ and $\eta(\lambda)$ on $\mathbb{R}$, and for $x \in \mathcal{D}\big(\xi(A)\big)$, we have that $\xi(A)x \in \mathcal{D}\big(\eta(A)\big)$ if and only if $x \in \mathcal{D}\big((\eta\cdot\xi)(A)\big)$. In this case,*

$$\big(\eta(A)\circ\xi(A)\big)(x) = (\eta\cdot\xi)(A)(x).$$

In some special cases, Theorem E.4 connects different approaches for creating the functions of self-adjoint operators. To show our point, we assume that the self-adjoint operator A is bounded. Given an integer $m \geq 2$, on the one hand, the power A^m can be defined by the spectral integral

$$\int_{-\infty}^{\infty} \lambda^m E(d\lambda).$$

On the other hand, A^m can be defined as the composition

$$A^m = \overbrace{A\circ\cdots\circ A}^{m}. \tag{E.16}$$

A natural question is whether these two approaches agree. Observe that the domain of a bounded operator is the whole space H, we have the following direct corollary from Theorem E.4.

COROLLARY E.5. *Given the bounded self-adjoint operator A in the form (E.9), we get*

$$A^m = \int_{-\infty}^{\infty} \lambda^m E(d\lambda) \qquad m = 1, 2, \cdots$$

where A^m is defined in (E.16).

In general, identifying the operator $\xi(A)$ given in (E.12) with the ones defined in other ways (such as defining $\xi(A)$ by Taylor expansion or by polynomial approximation) appears to be highly non-trivial. Such problems have to be treated on a case-by-case basis. By comparing (E.10) and (E.13), for example, it is not hard to imagine (see Section 4.1 for an example where $\xi(\lambda) = e^{t\lambda}$) that some function $\xi(A)$ of an unbounded self-adjoint operator A may have a domain $\mathcal{D}(\xi(A))$ genuinely larger than $\mathcal{D}(A)$. On the other hand, the domain of any polynomial of A of degree at least 1 is a subset to $\mathcal{D}(A)$. Consequently, approximation of by polynomials of A leads to a different operator in this case. A big advantage of defining $\xi(A)$ by (E.12) is that it largely reduces the investigation of the operator $\xi(A)$ to the ordinary integration problems through the relations (E.5), (E.13) and (E.14).

Bibliography

[1] de Acosta, A. (1985). Upper bounds for large deviations of dependent random vectors. *Z. Wahrsch. Verw. Gebiete* **69** 551–565.

[2] Asselah, A. (2008). Large deviation estimates for self-intersection local times for simple random walk in $\mathbb{Z}^3$. *Probab. Theor. Rel. Fields* **141** 19-45.

[3] Asselah, A. (2008). On large deviation and self-Intersection local times in dimension five or more. (preprint).

[4] Asselah, A. and Castell, F. (2007) Self-intersection local times for random walk, and random walk in random scenery in dimension $d \geq 5$. *Probab. Theor. Rel. Fields* **138** 1-32.

[5] Bass, R. F. and Chen, X. (2004). Self-intersection local time: critical exponent and laws of the iterated logarithm. *Ann. Probab.* **32** 3221-3247.

[6] Bass, R. F., Chen, X. and Rosen, J. (2005). Large deviations for renormalized self-intersection local times of stable processes. *Ann. Probab.* **33** 984-1013.

[7] Bass, R. F., Chen, X. and Rosen, J. (2006). Moderate deviations and laws of the iterated logarithm for the renormalized self-intersection local times of planar random walks. *Electron. J. Probab.* **11** 933-1030.

[8] Bass, R. F., Chen, X. and Rosen, J. (2009). Moderate deviations for the range of planar random walks. *Memoirs of AMS.* **198** No. 929.

[9] Bass, R. F., Chen, X. and Rosen, J. (2009). Large deviations for Riesz potentials of Additive Processes. *Annales de l'Institut Henri Poincaré* **45** 626-666.

[10] Bass, R. F. and Kumagai, T. (2002). Laws of the iterated logarithm for the range of random walks in two and three dimensions. *Ann. Probab.* **30** 1369-1396.

[11] van den Berg, M., Bolthausen, E. and den Hollander, F. (2001). Moderate deviations for the volume of Wiener sausage. *Ann. Math.* **153** 335-406.

[12] van den Berg, M., Bolthausen, E. and den Hollander, F. (2004). On the volume of the intersection of two Wiener sausages. *Ann. Math.* **159** 741-785.

[13] Bertoin, J. (1996). *Lévy Process.* Cambridge Univ. Press.

[14] Biskup, M. and König, W. (2001). Long-time tails in the parabolic Anderson model with bounded potential. *Ann. Probab.* **29** 636-682.

[15] Bolthausen, E. (1989). A central limit theorem for two-dimensional random walks in random sceneries. *Ann. Probab.* **17** 108-115.

[16] Bolthausen, E. (1990). On the volume of Wiener sausage. *Ann. Probab.* **18** 1576-1582.

[17] Bolthausen, E. (1999). Large deviations and interacting random walks. *Lecture Notes in Math.* **1781** 1-124.

[18] Breiman, L. (1968). *Probability.* Addison-Wesley Massachusetts.

[19] Brydges, D. C. and Slade, G. (1995). The diffusive phase of a model of self-interacting walks.

[20] Bucklew, J. A. (1990). *Large Deviations Techniques in Decision, Simulation, and Estimation.* Wiley, New York.

[21] Buffet, E. and Pulé, J. V. (1997) A model of continuous polymers with random charges. *J. Math. Phys.* **38** 5143–5152.

[22] Calais, J. Y. and Yor, M. (1987). Renormalization et convergence en loi pour certaines intégrales multiples associées au mouvement brownien dans $\mathbb{R}^d$. *Séminaire de Probabilités XXI. Lecture Notes in Math.* **1247** 375–403. Springer, Berlin.

[23] Carlen, E. A. and Loss, M. (1993). Sharp constant in Nash's inequality. *Internat. Math. Res. Notices* **1993** 213-215.

[24] Carmona, R. and Molchanov, S. (1994). *Parabolic Anderson Problem and Intermittency. Amer. Math. Soc.*, Providence, RI.

[25] Castell, F. Large deviations for intersection local times in critical dimension. *Ann. Probab.* (to appear).

[26] Chen, X. (1999). How often does a Harris recurrent Markov chain recur? *Ann. Probab.* **3** 1324-1346.

[27] Chen, X. (2004). Exponential asymptotics and law of the iterated logarithm for intersection local times of random walks. *Ann. Probab.* **32** 3248-3300.

[28] Chen, X. (2005). Moderate deviations and law of the iterated logarithm for intersections of the range of random walks. *Ann. Probab.* **33** 1014–1059.

[29] Chen, X. (2006). Self-intersection local times of additive processes: large deviation and law of the iterated logarithm. *Stoc. Proc. Appl.* **116** 1236-1253.

[30] Chen, X. (2006). Moderate and small deviations for the ranges of one-dimensional random walks. *J. Theor. Probab.* **19** 721-739.

[31] Chen, X. (2007). Large deviations and laws of the iterated logarithm for the local times of additive stable processes. *Ann. Probab.* **35** 602-648.

[32] Chen, X. (2007). Moderate deviations and laws of the iterated logarithm for the local times of additive Levy processes and additive random walks. *Ann. Probab.* **35** 954-1006.

[33] Chen, X. (2008). Limit laws for the energy of a charged polymer. *Annales de l'Institut Henri Poincaré* **44** 638-672.

[34] Chen, X. and Khoshnevisan, D. (2008). From charged polymers to random walk in random scenery. *Proceedings of the Third Erich L. Lehmann Symposium, IMS Lecture Notes-Monograph Series.* **57** 237-251.

[35] Chen, X. and Li, W. V. (2004). Large and moderate deviations for intersection local times. *Probab. Theor. Rel. Fields.* **128** 213-254.

[36] Chen, X., Li, W. V. and Rosen, J. (2005). Large deviations for local times of stable processes and stable random walks in 1 dimension. *Electron. J. Probab.* **10** 577-608.

[37] Chen, X. and Mörters, P. (2009). Upper tails for intersections of random walks in supercritical dimensions. *J. London Math. Soc.* **79** 186-210.

[38] Chen, X., and Rosen, J. (2005). Exponential asymptotics for intersection local times of stable processes and random walks. *Ann. Inst. H. Poincaré* **41** 901-928.

[39] Clifford, W. K. (1882). On Bessel's functions, *Mathematical Papers* 246-349.

[40] Cordero-Erausquin, D., Nazaret, B. and Villani, C. (2004). A mass-transportation approach to sharp Sobolev and Gagliardo-Nirenberg inequalities. *Adv. Math.* **182** 307-332.

[41] Cramér, H. (1938). Sur un nouveau théorème-limite de la théorie des probabilités. In *Actualités Scientifiques et Industrielles, number 736 in Colleque Consacré à la Théorie des Probabilités, pp. 5-23*, Hermann, Paris.

[42] Cranston, M., Mountford, T. S. and Shiga, T. (2005). Lyapunov exponent for the parabolic Anderson model with Lévy noise. *Probab. Theory Related Fields* **132** 321–355.

[43] Csáki, E., König, W. and Shi Z. (1999). An embedding for Kesten-Spitzer random walk in random scenery. *Stochastic Process. Appl.* **82** 283-292.

[44] Csörgö, M., Shi, Z. and Yor, M. (1999). Some asymptotic properties of the local time of the uniform empirical process. *Bernoulli* **5** 1035-1058.

[45] Del Pino, M. and Dolbeault, J. (2002). Best constants for Gagliardo-Nirenberg inequalities and applications to nonlinear diffusions. *J. Math. Pures Appl.* **81** 847-875.

[46] Del Pino, M. and Dolbeault, J. (2003). The optimal Euclidean L^p-Sobolev logarithmic inequality. *J. Funct. Anal.* **197** 151-161.

[47] Dembo, A. and Zeitouni, O. (1998). *Large Deviations Techniques and Applications. (2nd ed.)* Springer, New York.

[48] Dembo, A., Peres, Y., Rosen, J. and Zeitouni, O. (2001). Thick Points for Planar Brownian Motion and the Erdös-Taylor Conjecture on Random Walk. *Acta Math.* **186** 239-270.

[49] Dembo, A., Peres, Y., Rosen, J. and Zeitouni, O. (2004). Cover times for Brownian motion and random walks in two dimensions. *Ann. Math.* **160** 433-464.

[50] Dembo, A., Peres, Y. and Rosen, J. (2007). How large a disc is covered by a random walk in n steps? *Ann. Probab.* **35** 577-601.

[51] Derrida, B., Griffiths, R. B. and Higgs, R. G. (1992). A model of directed walks with random self-interactions. *Europhys. Lett.* **18** 361-366.

[52] Derrida, B. and Higgs, P. G. (1994) Low-temperature properties of directed walks with random self-interactions. *J. Phys. A* **27** 5485-5493.

[53] Deuschel, J-D. and Stroock, D. W. (1989). *Large Deviations* Academic Press, Boston.

[54] Donoghue, W. (1969). *Distributions and Fourier transforms.* Academic Press, New York.

[55] Donsker, M. D. and Varadhan, S. R. S. (1974). Asymptotic evaluation of certain Wiener integrals for large time. *Functional Integration and Its Applications: Proceedings of the International Conference at London.* pp. 15-33.

[56] Donsker, M. D. and Varadhan, S. R. S. (1975). Asymptotic evaluation of certain Markov process expectations for large time. I. *Comm. Pure Appl. Math.* **28** 1-47, 279-301.

[57] Donsker, M. D.and Varadhan, S. R. S. (1975). Asymptotics for the Wiener sausage. *Comm. Pure Appl. Math.* **28** 525–565

[58] Donsker, M. D. and Varadhan, S. R. S. (1976). Asymptotic evaluation of certain Markov process expectations for large time. III. *Comm. Pure Appl. Math.* **29** 389-461.

[59] Donsker, M. D. and Varadhan, S. R. S. (1979). On the number of distinct sites visited by a random walk. *Comm. Pure Appl. Math.* **32** 721-747.

[60] Donsker, M. D. and Varadhan, S. R. S. (1981). The polaron problem and large deviations. *New stochastic methods in physics. Phys. Rep.* **77** 235-237.

[61] Donsker, M. D.and Varadhan, S. R. S. (1983). Asymptotics for the polaron. *Comm. Pure Appl. Math.* **36** 505-528.

[62] Dudley, R. M. (1967). The sizes of compact subsets of Hilbert space and continuity of Gaussian processes. *J. Funct. Anal.* **1** 290-330.

[63] Duplantier, B. (1986). Exact critical exponents for two-dimensional dense polymers. *J. Phys. A* **19** L1009-L1014.

[64] Duplantier, B. (2003). Higher conformal multifractality. *J. Statist. Phys.* **110** 691–738.

[65] Dvoretzky, A. and Erdös, P. (1951). Some problems on random walk in space. *Proc. Secon Berkeley Symposium on Math. Statistics and Probability* 353-367. Berkeley, CA: University of California Press.

[66] Dvoretzky, A., Erdös, P. and Kakutani, S. (1950). Double points of paths of Brownian motions in n-space. *Acta Sci. Math. Szeged* **12**, Leopoldo, Fejér et Frederico Riesz LXX annos natis dedicatus, Pars B, 75-81.

[67] Dvoretzky, A., Erdös, P. and Kakutani, S. (1954). Multiple points of paths of Brownian motion in the plane. *Bull. Res. Council Israel.* **3**, 364-371

[68] Dynkin, E. B. (1981). Additive functionals of several time-reversible Markov processes. *J. Funct. Anal.* **42** 64-101.

[69] Dynkin, E. B. (1985). Random fields associated with multiple points of the Brownian motion. *J. Funct. Anal.* **62** 397-434

[70] Dynkin, E. B. (1988). Regularized self-intersection local times of planar Brownian motion. *Ann. Probab.* **16** 58–74.

[71] Edwards , R. E. (1979) *Fourier Series. A Modern Introduction* Vol. **1**, 2nd Edition, Springer, New York.

[72] Edwards , R. E. (1982) *Fourier Series. An Modern Introduction.* Vol. **2**, 2nd Edition, Springer, New York.

[73] Ellis, R. S. (1984). Large deviations for a general class of random vector *Ann. Probab.* **12** 1-12.

[74] Ellis, R. S. (1985). *Entropy, Large Deviations and Statistical Mechanics* Springer-Verlag, New York.

[75] Erdös, P. and Taylor, S. J. (1960). Some problems concerning the structure of random walk paths. *Acta Sci. Hung.* **11** 137-162.

[76] Feller, W. E. (1971). *An Introduction to Probability Theory and Its Applications*, Vol. **II**, 2nd edition. Wiley, New York.

[77] Feng, J. and Kurtz, T. G. (2006). *Large Deviations for Stochastic Processes* Mathematical Surveys and Monographs **131** Amer. Math. Soc. Providence, RI.

[78] Fernández, R., Fröhlich, J. and Sokal, A. D. (1992). *Random Walks, Critical Phenomena, and Triviality in quantum field theory.* Springer, New York.

[79] Fleischmann, K., Mörters, P. and Wachtel, V. (2008). Moderate deviations for random walk in random scenery. *Stoch. Proc. Appl.* **118** 1768-1802.

[80] Freidlin, M. I. and Wentzell, A. D. (1984). *Random Perturbations of Dynamical Systems.* Springer-Verlag, New York

[81] Fukushima, M. (1980). *Dirichlet Forms and Markov Processes.* North-Holland, Amsterdam.

[82] Gantert, N., König, W. and Shi, Z. (2007). Annealed deviations of random walk in random scenery. *Ann. Inst. H. Poincaré* **43** 47-76.

[83] Gantert, N., van der Hofstad, R. and König, W. (2006). Deviations of a random walk in a random scenery with stretched exponential tails. *Stochastic Process. Appl.* **116** 480–492.

[84] Gärtner, J. (1977). On large deviations from the invariant measures. *Theor. Probab. Appl.* **22** 24-39.

[85] Geman, D. and Horowitz, J. (1980). Occupation densities. *Ann. Probab.* **8** 1-67.

[86] Geman, D., Horowitz, J. and Rosen, J. (1984). A local time analysis of intersections of Brownian paths in the plane. *Ann. Probab.* **12** 86-107.

[87] Hamana, Y. (1997). The fluctuation result for the multiple point range of two-dimensional recurrent random walks. *Ann. Probab.* **25** 598-639.

[88] Hamana, Y. (1998). An almost sure invariance principle for the range of random walks. *Stochastic Proc. Appl.* **78** 131-143.

[89] Hamana, Y. (2001). Asymptotics of the moment generating function for the range of random walks. *J. Theor. Probab.* **14** 189-197.

[90] Hamana, Y. and Kesten, H. (2001). A large-deviation result for the range of random walk and for the Wiener sausage. *Probab. Theory Related Fields* **120** 183-208.

[91] Hamana, Y. and Kesten, H. (2002). Large deviations for the range of an integer valued random walk. *Ann. Inst. H. Poincaré Probab. Statist.* **38** 17-58.

[92] Hoel, P. G., Port, S. C. and Stone, C. J. (1972), *Introduction to Stochastic Processes.* Houghton Mifflin Co., Boston.

[93] van der Hofstad, R., den Hollander, F. and König, W. (2003). Large deviations for the one-dimensional Edwards model. *Ann. Probab.* **31** 2003–2039.

[94] van der Hofstad, R. and König, W. (2001). A survey of one-dimensional random polymers. *J. Statist. Phys.* **103** 915–944.

[95] van der Hoftad, R., König, W. and Mörters, P. (2006). The universality classes in the parabolic Anderson model. *Comm. Math. Phys.* **267** 307-353.

[96] den Hollander, F. (1996). Random polymers. *Stat. Nederl* **50** 136-145.

[97] den Hollander, F. (2000). *Large Deviations.* Amer. Math. Soc., Providence, RI.

[98] den Hollander, F. (2009) *Random Polymers.* Lecture Notes in Mathematics **1974** Springer, Heidelberg.

[99] Hu, Y. Z. and Nualart, D. (2005). Renormalized self-intersection local time for fractional Brownian motion. *Ann. Probab.* **33** 948-983.

[100] Ito, K. and McKean, H. P., jr (1974). *Diffusion Processes and Their Sample Paths.* Springer, New York.

[101] Jain, N. C. and Orey, S. (1968). On the range of random walk. *Israel J. Math.* **6** 373-380.

[102] Jain, N. C. and Pruitt, W. E. (1971). The range of transient random walk. *J. Anal. Math.* **24** 369-393.

[103] Jain, N. C. and Pruitt, W. E. (1972). The law of the iterated logarithm for the range of random walk. *Ann. Math. Statist.* **43** 1692-1697.

[104] Jain, N.C. and Pruitt, W. E. (1983). Asymptotic behavior of the local time of a recurrent random walk. *Ann. Probab.* **11**64-85.

[105] Kac, M. (1951). On some connections between probability theory and differential and integral equations. *Proc. Second Berkeley Symposium, L. M. Le Cam and J. Neyman editors.* University of California Press, Berkeley, pages 189-215.

[106] Kantor, Y. and Kardar, M. (1991). Polymers with self-interactions. *Europhys. Lett.* **14** 421-426.

[107] Kesten, H. and Spitzer, F. (1979). A limit theorem related to a new class of self-similar processes. *Z. Wahrsch. Verw. Gebiete* **50** 2-25.

[108] Khanin, K. M., Mazel, A. E., Shlosman, S. B. and Sinai, Ya. G. (1994). Loop condensation effect in the behavior of random walks. *Markov process and their applications (Freidlin, ed.), Progr. Probab.* **34**, Birkhäuser, Boston, pp. 167-184.

[109] Khoshnevisan, D. (2002). *Multiparameter Processes. An Introduction to Random fields.* Springer, New York.

[110] Khoshnevisan, D. (2003). Intersection of Brownian motions. *Exposition. Math.* **21** 97-114.

[111] Khoshnevisan, D. and Lewis, T. M. (1998). A law of the iterated logarithm for stable processes in random scenery. *Stochastic Proc. Appl.* **74** 89-121.

[112] Khoshnevisan, D., Xiao, Y. M. and Zhong, Y. (2003a). Local times of additive Lévy processes. *Stoch. Proc. Their Appl.* **104** 193-216.

[113] Khoshnevisan, D., Xiao, Y. M. and Zhong, Y. (2003b). Measuring the range of an additive Lévy process. *Ann. Probab.* **31** 1097-1141.

[114] König, W. and Mörters, P. (2002). Brownian intersection local times: Upper tail asymptotics and thick points. *Ann Probab.* **30** 1605-1656.

[115] König, W. and Mörters, P. (2006). Brownian intersection local times: exponential moments and law of large masses. *Trans. Amer. Math. Soc.* **358** 1223–1255.

[116] Lawler, G. F. (1991). *Intersections of Random Walks, Probability and Its applications.* Birkhäuser, Boston

[117] Lawler, G. F. (2005). *Conformally Invariant Processes in the Plane. Mathematical Surveys and Monographs,* **114**. American Mathematical Society, Providence, RI.

[118] Lawler, G. F., Schramm, O. and Werner, W. (2001). Values of Brownian intersection exponents. I. Half-plane exponents. *Acta Math.* **187** 237–273.

[119] Lawler, G. F., Schramm, O. and Werner, W. (2001). Values of Brownian intersection exponents. II. Plane exponents. *Acta Math.* **187** 275–308.

[120] Lawler, G. F., Schramm, O. and Werner, W. (2002). Values of Brownian intersection exponents. III. Two-sided exponents. *Ann. Inst. H. Poincaré Probab. Statist.* **38** 109–123.

[121] Ledoux, M. and Talagrand, M. (1991). *Probability in Banach spaces.* Springer, Berlin.

[122] Le Gall, J-F. (1985). Sur le temps local d'intersection du mouvement brownien plan et la méthode de renormalisation de Varadhan. *Séminaire de Probabilités, XIX. Lecture Notes in Math* **1123**, 314-331. Springer, Berlin.

[123] Le Gall, J-F. (1986). Sur la saucisse de Wiener et les points multiples du mouvement brownien. *Ann. Probab.* **14** 1219-1244.

[124] Le Gall, J-F. (1986). Propriétés d'intersection des marches aléatoires. I. Convergence vers le temps local d'intersection. *Comm. Math. Phys.* **104**, 471-507,

[125] Le Gall, J-F. (1986). Propriétés d'intersection des marches aléatoires. II. Étude des cas critiques. *Comm. Math. Phys.* **104**, 509-528.

[126] Le Gall, J-F. (1987). Temps locaux d'intersection et points multiples des processus de Lévy. *Séminaire de Probabilités, XXI. Lecture Notes in Math* **1247**, 341-375. Springer, New York.

[127] Le Gall, J-F. (1988). Fluctuation results for the Wiener sausage. *Ann. Probab.* **16** 991-1018.

[128] Le Gall, J-F. (1990). Wiener sausage and self-intersection local time. *J. Funct. Anal.* **88** 299-341.

[129] Le Gall, J-F. (1992). Some properties of planar Brownian motion. *École d'Été de Probabilités de Saint-Flour XX. 1990. Lecture Notes in Math* **1527** 111-235. Springer, Berlin.

[130] Le Gall, J-F. (1994). Exponential moments for the renormalized self-intersection local time of planar Brownian motion. *Séminaire de Probabilités, XXVIII. Lecture Notes in Math* **1583**, 172-180. Springer, Berlin.

[131] Le Gall, J-F. and Rosen, J. (1991). The range of stable random walks. *Ann. Probab.* **19** 650–705.

[132] Levine, H. A. (1980). An estimate for the best constant in a Sobolev inequality involving three integral norms. *Ann. Mat. Pura Appl.* **124** 181-125.

[133] Li, W. V. and Shao, Q. M. (2001). Gaussian processes: inequalities, small ball probabilities and applications. In *Stochastic processes: Theory and Methods. Handbook of Statistics.* (C. R. Rao and D. Shanbhag eds.) **19** 533-598. North-Holland, Amsterdam.

[134] Madras, N. and Slade, G. (1993). *The Self-avoiding Walk.* Birkhäuser, Boston.

[135] Mansmann, U. (1991). The free energy of the Dirac polaron, an explicit solution. *Stochastics & Stochastics Report.* **34** 93-125.

[136] Marcus, M. B. and Rosen, J. (1997). Laws of the iterated logarithm for intersections of random walks on Z^4. *Ann. Inst. H. Poincaré Probab. Statist.* **33** 37–63.

[137] Martinez, S. and Petritis, D. (1996). Thermodynamics of a Brownian bridge polymer model in a random environment. *J. Phys. A* **29** (1996), 1267–1279.

[138] Peres, Y. (2002). Brownian intersections, cover times and thick points via trees. *Proceedings of the International Congress of Mathematicians, Vol. III*, Beijing, 73-78,

[139] Pinsky, R. G. (1995). *Positive Harmonic Functions and Diffusion.* Cambridge University Press, Cambridge.

[140] Remillard, B. (2000). Large deviations estimates for occupation time integrals of Brownian motion. *Stochastic models (Ottwa, ON)* 375-398, *CMS Cof. Proc.* **26** Amer. Soc. Providence. RI.

[141] Revuz, D. and Yor, M. (1994). *Continuous Martingale and Brownian Motion*, 2nd Edition. Springer, Berlin.

[142] Révész, P. and Shi, Z. (2000). Strong approximation of spatial random walk in random scenery. *Stochastic Proc. Appl.* **88** 329-345.

[143] Robert, A. (1976). A short proof of the Fourier inversion formula. *Proc. Amer. Math. Soc.* **59** 287-288.

[144] Rogers, L. C. G. and Williams, D. (1987). *Diffusions, Markov processes, and Martingales, Vol.* **2**. *Ito Calculus*, John Wiley & Sons, Inc., New York.

[145] Rogers, L. C. G. and Williams, D. (1994). *Diffusions, Markov processes, and Martingales, Vol.* **1**. *Foundations*, John Wiley & Sons, Inc., New York.

[146] Rosen, J. (1983). A local time approach to the self-intersections of Brownian paths in space. *Comm. Math. Physics* **88** 327-338.

[147] Rosen, J. (1984). Self-intersections of random fields. *Ann. Probab.* **12** 108-119.

[148] Rosen, J. (1986). Tanaka formula and renormalization for intersections of planar Brownian motion. *Ann. Probab.* **14** 1245-1251.

[149] Rosen, J. (1986). A renormalized local time for multiple intersections of planar Brownian motion. *Séminaire de Probabilités, XX, 1984/85, Lecture Notes in Math.*, **1204** 515-531.

[150] Rosen, J. (1987). Joint continuity of the intersection local time of Markov processes. *Ann. Probab.* **12** 108-119.

[151] Rosen, J. (1990). Random walks and intersection local time. *Ann. Probab.* **18** 959-977.

[152] Rosen, J. (1997). Laws of the iterated logarithm for triple intersections of three-dimensional random walks. *Electron. J. Probab.* **2** 2, 1–32

[153] Rosen, J. and Yor, M. (1991). Tanaka formulae and renormalization for triple intersections of Brownian motion in the plane. *Ann. Probab.* **19** 142-159.

[154] Saloff-Coste, L. (2002). *Aspects of Sobolev-Type Inequalities.* London Math. Soc. LNS Vol. **289**, Cambridge University Press, Cambridge.

[155] Spitzer, F. (1964). *Principles of Random Walk*, Van Nostrand, Princeton, New Jersey.

[156] Stroock, D. W. (1984). *An Introduction to the theory of large deviations.* Springer, Berlin.

[157] Sznitman, A. S. (1990). Long time asymptotics for the shrinking Wiener sausage. *Comm. Pure & Appl. Math.* **43** 809-820.

[158] Sznitman, A. S. (1998). *Brownian Motion, Obstacles and Random media.* Springer Berlin

[159] Varadhan, S. R. S. (1984). *Large Deviations and Applications.* SIAM, Philadelphia.

[160] Varadhan, S. R. S. (1969). Appendix to "Euclidean quantum field theory", by Symmanzik, K. *Local Quantum theory* (R. Jost ed). Academic New York.

[161] Varadhan, S. R. S. (2008). Large deviations. *Ann. Probab.* **36** 397-419.

[162] Vanderzande, C. (1998) *Lattice models of Polymers.* Cambridge University Press, Cambridge.

[163] Watson, G. N. (1995) *A Treatise on the Theory of Bessel Functions*, Second Edition, Cambridge University Press. Cambridge.

[164] Weinstein, M. I. (1983). Nonlinear Schrödinger equations and sharp interpolation estimates. *Comm. Math. Phys.* **87** 567-576.

[165] Westwater, J. (1980). On Edwards' model for long polymer chains. *Comm. Math. Phys.* **72** 131-174.

[166] Westwater, J. (1981). On Edwards' model for polymer chains. II. The self-consistent potential. *Comm. Math. Phys.* **79** 53-73.

[167] Westwater, J. (1982). On Edwards' model for polymer chains. III. Borel summability. *Comm. Math. Phys.* **84** 459-470.

[168] Wolpert, R. (1978). Wiener path intersections and local time. *J. Funct. Anal.* **30** 329-340.

[169] Xiao, Y. M. (2006). Properties of local nondeterminism of Gaussian and stable random fields and their applications. *Ann. Fac. Sci. Toulouse Math. (6)* **15** 157–193.

[170] Yor, M. (1985). Compléments aux formules de Tanaka-Rosen. *Séminaire de Probabilités XIX. Lecture Notes in Math.* **1123** 332–349. Springer, Berlin.

[171] Yor, M. (1985). Renormalisation et convergence en loi pour les temps locaux d'intersection du mouvement brownien dans $\mathbb{R}^3$. *Séminaire de Probabilités XIX. Lecture Notes in Math.* **1123** 350–365. Springer, Berlin.

[172] Yor, M. (1986). Précisions sur l'existence et la continuité des temps locaux d'intersection du mouvement brownien dans $\mathbb{R}^d$ *Séminaire de Probabilités XX. Lecture Notes in Math.* **1204** 532-541. Springer, Berlin.

[173] Yor, M. (1986). Sur la représentation comme intégrales stochastiques des temps d'occupation du mouvement brownien dans $\mathbb{R}^d$. *Séminaire de Probabilités XX. Lecture Notes in Math.* **1204** 543-552. Springer, Berlin.

[174] Yor, M. Remarques sur l'intégrabilité du temps local d'intersection renormalisé. (unpublished).

[175] Yosida, K. (1966). *Functional Analysis.* Springer, Berlin.

[176] Ziemer, W. P. (1989). *Weakly Differentiable Functions.* Springer, New York.

List of General Notations

$(\Omega, \mathcal{A}, \mathbb{P})$ a complete probability space
$1_A(\cdot)$ indicator on A
Δ Laplacian operator
$\delta_x(\cdot)$ Daric function at x
$\emptyset$ empty set
$\lambda \cdot x$ inner product between $\lambda, x \in \mathbb{R}^d$
$\langle \cdot, \cdot \rangle$ inner product in Hilbert space
∇ gradient operator
$\mathbb{R}$, $\mathbb{R}^d$ real line, d-dimensional Euclidean space
$\mathbb{R}^+$ set of all non-negative numbers
Σ_m group of the permutations on $\{1, \cdots, m\}$
$\widehat{f}(\lambda)$ Fourier transform of $f(x)$
$\mathbb{Z}$, $\mathbb{Z}^d$ set of integers, d-dimensional lattice space
$\mathbb{Z}^+$ set of all non-negative integers
$C(T)$ space of real continuous functions on T
$C\{T, \mathbb{R}^d\}$ space of continuous functions on T taking values in $\mathbb{R}^d$
$l^p(\mathbb{Z}^d)$ space of all p-square summable functions on $\mathbb{Z}^d$
$W^{1,2}(\mathbb{R}^d)$ space of the functions f such that $f, \nabla f \in \mathcal{L}^2(\mathbb{R}^d)$
$\mathcal{F}_d$, $\mathcal{F}$ subspace of $W^{1,2}(\mathbb{R}^d)$ with $|f|_2 = 1$, $\mathcal{F} = \mathcal{F}_1$
$\mathcal{L}^p(\mathbb{R}^d)$ space of all p-square Lebesgue-integrable functions on $\mathbb{R}^d$
$\mathcal{L}^p(E, \mathcal{E}, \pi)$ space of all p-square integrable functions on $(E, \mathcal{E}, \pi)$
a.s. almost surely

Index

Titles in This Series

157 **Xia Chen,** Random walk intersections: Large deviations and related topics, 2010

156 **Jaime Angulo Pava,** Nonlinear dispersive equations: Existence and stability of solitary and periodic travelling wave solutions, 2009

155 **Yiannis N. Moschovakis,** Descriptive set theory, 2009

154 **Andreas Čap and Jan Slovák,** Parabolic geometries I: Background and general theory, 2009

153 **Habib Ammari, Hyeonbae Kang, and Hyundae Lee,** Layer potential techniques in spectral analysis, 2009

152 **János Pach and Micha Sharir,** Combinatorial geometry and its algorithmic applications: The Alcála lectures, 2009

151 **Ernst Binz and Sonja Pods,** The geometry of Heisenberg groups: With applications in signal theory, optics, quantization, and field quantization, 2008

150 **Bangming Deng, Jie Du, Brian Parshall, and Jianpan Wang,** Finite dimensional algebras and quantum groups, 2008

149 **Gerald B. Folland,** Quantum field theory: A tourist guide for mathematicians, 2008

148 **Patrick Dehornoy with Ivan Dynnikov, Dale Rolfsen, and Bert Wiest,** Ordering braids, 2008

147 **David J. Benson and Stephen D. Smith,** Classifying spaces of sporadic groups, 2008

146 **Murray Marshall,** Positive polynomials and sums of squares, 2008

145 **Tuna Altinel, Alexandre V. Borovik, and Gregory Cherlin,** Simple groups of finite Morley rank, 2008

144 **Bennett Chow, Sun-Chin Chu, David Glickenstein, Christine Guenther, James Isenberg, Tom Ivey, Dan Knopf, Peng Lu, Feng Luo, and Lei Ni,** The Ricci flow: Techniques and applications, Part II: Analytic aspects, 2008

143 **Alexander Molev,** Yangians and classical Lie algebras, 2007

142 **Joseph A. Wolf,** Harmonic analysis on commutative spaces, 2007

141 **Vladimir Maz′ya and Gunther Schmidt,** Approximate approximations, 2007

140 **Elisabetta Barletta, Sorin Dragomir, and Krishan L. Duggal,** Foliations in Cauchy-Riemann geometry, 2007

139 **Michael Tsfasman, Serge Vlăduţ, and Dmitry Nogin,** Algebraic geometric codes: Basic notions, 2007

138 **Kehe Zhu,** Operator theory in function spaces, 2007

137 **Mikhail G. Katz,** Systolic geometry and topology, 2007

136 **Jean-Michel Coron,** Control and nonlinearity, 2007

135 **Bennett Chow, Sun-Chin Chu, David Glickenstein, Christine Guenther, James Isenberg, Tom Ivey, Dan Knopf, Peng Lu, Feng Luo, and Lei Ni,** The Ricci flow: Techniques and applications, Part I: Geometric aspects, 2007

134 **Dana P. Williams,** Crossed products of C^*-algebras, 2007

133 **Andrew Knightly and Charles Li,** Traces of Hecke operators, 2006

132 **J. P. May and J. Sigurdsson,** Parametrized homotopy theory, 2006

131 **Jin Feng and Thomas G. Kurtz,** Large deviations for stochastic processes, 2006

130 **Qing Han and Jia-Xing Hong,** Isometric embedding of Riemannian manifolds in Euclidean spaces, 2006

129 **William M. Singer,** Steenrod squares in spectral sequences, 2006

128 **Athanassios S. Fokas, Alexander R. Its, Andrei A. Kapaev, and Victor Yu. Novokshenov,** Painlevé transcendents, 2006

127 **Nikolai Chernov and Roberto Markarian,** Chaotic billiards, 2006

126 **Sen-Zhong Huang,** Gradient inequalities, 2006

125 **Joseph A. Cima, Alec L. Matheson, and William T. Ross,** The Cauchy Transform, 2006

124 **Ido Efrat, Editor,** Valuations, orderings, and Milnor K-Theory, 2006